Lecture Notes in Physics

Bisher erschienen/Already published

Vol. 1: J. C. Erdmann. Wärmeleitung in Kristallen, theoretische Grundlagen und fortgeschrittene experimentelle Methoden. II, 283 Seiten. 1969.

Vol. 2: K. Hepp, Théorie de la renormalisation. III, 215 pages. 1969.

Vol. 3: A. Martin, Scattering Theory: Unitarity, Analyticity and Crossing. IV, 125 pages. 1969.

Vol. 4: G. Ludwig, Deutung des Begriffs „physikalische Theorie" und axiomatische Grundlegung der Hilbertraumstruktur der Quantenmechanik durch Hauptsätze des Messens. 1970. Vergriffen.

Vol. 5: Schaaf, The Reduction of the Product of Two Irreducible Unitary Representations of the Proper Orthochronous Quantummechanical Poincare Group. IV, 120 pages. 1970.

Vol. 6: Group Representations in Mathematics and Physics. Edited by V. Bargmann. V, 340 pages. 1970.

Vol. 7: R. Balescu, J. L. Lebowitz, I. Prigogine, P. Résibois, Z. W. Salsburg, Lectures in Statistical Physics. V, 181 pages. 1971.

Vol. 8: Proceedings of the Second International Conference on Numerical Methods in Fluid Dynamics. Edited by M. Holt. 1971. Out of print.

Vol. 9: D. W. Robinson, The Thermodynamic Pressure in Quantum Statistical Mechanics. V, 115 pages. 1971.

Vol. 10: J. M. Stewart, Non-Equilibrium-Relativistic Kinetic Theory. III, 113 pages. 1971.

Vol. 11: O. Steinmann, Pertubation Expansions in Axiomatic Field Theory. III, 126 pages. 1976.

Vol. 12: Statistical Models and Turbulence. Edited by C. Van Atta and M. Rosenblatt. Reprint of the First Edition. VIII, 492 pages. 1975.

Vol. 13: M. Ryan, Hamiltonian Cosmology. VII, 169 pages. 1972.

Vol. 14: Methods of Local and Global Differential Geometry in General Relativity. Edited by D. Farnsworth, J. Fink, J. Porter, and A. Thompson. V, 188 pages.

Vol. 15: M. Fierz, Vorlesungen zur Entwicklungsgeschichte der Mechanik. V, 97 Seiten. 1972.

Vol. 16: H.-O. Goergii, Phasenübergang 1. Art bei Gittergasmodellen. IX, 167 Seiten. 1972.

Vol. 17: Strong Interaction Physics. Edited by W. Rühl and A. Vancura. V, 405 pages. 1973.

Vol. 18: Proceedings of the Third International Conference on Numerical Methods in Fluid Mechanics, Vol. I. Edited by H. Cabannes and R. Temam. VII, 186 pages. 1973.

Vol. 19: Proceedings of the Third International Conference on Nemerical Methods in Fluid Mechanics, Vol. II. Edited by H. Cabannes and R. Temam. VII, 275 pages. 1973.

Vol. 20: Statistical Mechanics and Mathematical Problems. Edited by A. Lenard. VIII, 247 pages. 1973.

Vol. 21: Optimization and Stability Problems in Continuum Mechanics. Edited by P. K. C. Wang. V, 94 pages. 1973.

Vol. 22: Proceedings of the Europhysics Study Conference on Intermediate Processes in Nuclear Reactions. Edited by N. Cindro, P. Kulišic and Th. Mayer-Kuckuk. XIV, 329 pages. 1973.

Vol. 23: Nuclear Structure Physics. Proceedings 1973. Edited by U. Smilansky, I. Talmi, and H. A. Weidenmüller. XII, 296 pages. 1973.

Vol. 24: R. F. Snipes, Statistical Mechanical Theory of the Electrolytic Transport of Non-electrolytes. V, 210 pages. 1973.

Vol. 25: Constructive Quantum Field Theory. The 1973 "Ettore Majorana" International School of Mathematical Physics. Edited by G. Velo and A. Wightman. III, 331 pages. 1973.

Vol. 26: A. Hubert, Theorie der Domänenwände in geordneten Medien. XII, 377 Seiten. 1974.

Vol. 27: R. K. Zeytounian, Notes sur les Ecoulements Rotationnels de Fluides Parfaits. XIII, 407 pages. 1974.

Vol. 28: Lectures in Statistical Physics. Edited by W. C. Schieve and J. S. Turner. V. 342 pages. 1974.

Vol. 29: Foundations of Quantum Mechanics and Ordered Linear Spaces. Advanced Study Institute, Marburg 1973. Edited by A. Hartkämper and H. Neumann. VI, 355 pages. 1974.

Vol. 30: Polarization Nuclear Physics. Proceedings 1973. Edited by D. Fick. IX, 292 pages. 1974.

Vol. 31: Transport Phenomena. Sitges International Schools of Statistical Mechanics, June 1974. Edited by G. Kirczenow and J. Marro. XIV, 517 pages. 1974.

Lecture Notes in Physics

Edited by J. Ehlers, München, K. Hepp, Zürich,
R. Kippenhahn, München, H. A. Weidenmüller, Heidelberg,
and J. Zittartz, Köln
Managing Editor: W. Beiglböck, Heidelberg

61

Photonuclear Reactions I

International School on Electro- and
Photonuclear Reactions,
Erice, Italy 1976
Edited by S. Costa and C. Schaerf

Springer-Verlag Berlin Heidelberg GmbH 1977

Editors

Prof. Sergio Costa
Instituto di Fisica, Università di Torino
Corso M. d'Azeglio, 46
10125 Torino, Italia

Prof. Carlo Schaerf
Instituto di Fisica, Università di Roma
Piazzale delle Scienze, 5
00185 Roma, Italia

Library of Congress Cataloging in Publication Data

International School on Electro and Photonuclear Reactions, Erice, Italy, 1976.
   Photonuclear reactions.

   (Lecture notes in physics ; 61-62)
   1. Photonuclear reactions--Congresses. 2. Electro-
magnetic interactions--Congresses. I. Costa, Sergio.
II. Schaerf, Carlo, 1935-   III. Ettore Majorana
Centre for Scientific Culture. IV. Title. V. Series.
QC794.8.P4I55 1976   539.7'56   77-4456

   ISBN 978-3-540-08139-5   ISBN 978-3-540-37388-9 (eBook)
   DOI 10.1007/978-3-540-37388-9

PREFACE

Our understanding of the nature of fundamental interactions has been greatly helped by experiments involving electrons and photons. These tools had, and still have, indeed, a relevant impact on the investigation of atoms and molecules, as well as nuclear structure.

In the case of nuclear systems held together by forces not yet completely understood, the electromagnetic probes, whose interaction with the nucleons in the nucleus is basically well known, provided us with matrix elements giving direct information on the nuclear wave functions.

The study of the electromagnetic porperties of nuclear states has been crucial for the development of nuclear models and the good knowledge of the quantum numbers associated with specific multipoles has notably simplified the analysis of the fundamental types of nuclear motion.

A large number of laboratories are currently carrying out experiments using both electron and real photon beams and, in many of these, a noticeable effort is being made to improve the quality of the existing instrumentation, as well as to set up new facilities.

The purpose of the School on Electro- and Photonuclear Reactions is, therefore, to review and discuss the most significant achievements in the study of nuclear properties using electromagnetic probes with special emphasis on the most recent experimental and theoretical results obtained in this field.

Because of the vastness of the subject, the first course of the School was essentially devoted to photonuclear interactions from few MeV up to intermediate energy.

Ten series of lectures (collected in Volume I) and a number of seminars were given. Most of the seminars were presented in the form of status reports from important laboratories, and these reports are collected in Volume II.

In the lecture sessions, the classic topic of the giant dipole resonance was first reviewed and the phenomenology discussed together with the mechanisms leading to the excitation of the giant states. Collective and microscopic models were proposed in order to push the theoretical description closer to the experimental results. Isotopic spin effects and sum rules completed the study of the doorway states through which the G.D.R. is formed.

The competition between the decay channels from the G.D.R. was analysed and the existing knowledge of MI and E2 resonances, both isoscalar and isovector, was surveyed.

Nuclear elastic photon scattering was discussed in detail, including the contributions of Thomson, Rayleigh and Delbruck scattering. Suggestions for future experimentation, mainly with polarized photons, were also made.

A new way of calculating the transition matrix at intermediate energy was presented, underlining the relevance of gauge terms and describing the direct coupling of the photon to nucleon-nucleon correlations.

Exchange-current phenomena were also discussed in connection with the integrated photo-absorption cross section, which is shown to contain information on mesonic degrees of freedom in nuclei. The effects of explicitly introducing isobars on nuclear constitutents were explored, in particular, in the case of electromagnetic interactions in the two-nucleon system.

Finally the few-body systems and their interaction with real and virtual photons and with hadrons were discussed as a check of our present understanding of nuclear properties in terms of the basic n-n force.

During the course special theoretical topics were introduced by some participants; these are also summarized in Volume II.

The course could have not been so rewarding as it was without the enthusiastic collaboration of all the lecturers and participants and the entire staff of the Centro di Cultura Scientifica "Ettore Majorana". It is a pleasure to thank in particular Prof. A. Zichichi, Director of the Centre, Dr. S.A. Gabriele, Miss P. Savalli and Miss M. Zaini for their generous help.

The organization of the course was made possible by the financial contributions of the National Research Council (CNR) and the National Institute of Nuclear Physics (INFN). We wish to express our gratitude to Prof. E. Amaldi, President of the National Committee for the Physical Sciences of CNR and Prof. A. Gigli, President of INFN.

TABLE OF CONTENTS

LECTURES

BERGERE, R.   :   Features of the Giant E1 Resonance 1

RICCO, G.   :   Photonuclear Reactions above the Giant Dipole Resonance.
A Survey 223

HANNA, S.   :   Giant Multipole Resonances 275

HAYWARD, E.   :   Photon Scattering in the Energy Range 5-30 MeV 340

HEBACH, H.   :   Mechanisms of Photonuclear Reactions at Intermediate
Energies (40 - 140 MeV) 407

BOSCO, B.   :   Real and Virtual Photons 461

WEISE, W.   :   Sum Rules in Photonuclear Physics 484

LEONARDI, R.   :   Isospin Structure of the Dipole Giant Resonance 501

CIOFI DEGLI ATTI, C.   :   Electromagnetic and Hadronic Interactions with
the Few-Body Systems at Intermediate Energies ... 521

ARENHÖVEL, H.   :   Bayron Resonances in Nuclei 586

TABLE OF CONTENTS
(to Volume II)

INVITED SEMINARS

BERTOZZI, W.   :   Recent Developments at M.I.T. 1

CATILLON, Ph.   :   News from Saclay 47

DE VRIES, C.   :   Electron Scattering Work at Amsterdam - past, present
and future activities 62

DRECHSEL, D.   :   Electronuclear Sum Rules 92

LINDGREN, K.   :   A Review of Present Photonuclear Research at Lund and
future Accelerator Plans 132

MATONE, G.   :   A Monochromatic and Polarized Photon Beam for Photonuclear Reactions, The Ladon Project at Frascati 149

RICHTER, A.   :   Latest from Dalinac 165

SANZONE, M.   :   Preliminary Results on the Annihilation Photon Beam at
the Frascati Linac Laboratory 199

SOLODUKHOV, G.V.   :   Some Experimental Results on the Measurement of the
the total Photoabsorption Cross Sections 216

TORIZUKA, Y.   :   Electroexcitation of Giant Multipole Resonances 258

SUMMARIES OF CONTRIBUTED SEMINARS

BOHIGAS, O.   :   Description of Isoscalar Resonances. A Sum-Rule Approach .. 294

DELSANTO, P.P.   :   The Center of Mass Problems in Continuum 295

GIANNINI, M.   :   A Simple Model for Resonance Shifts 297

PROSPERI, D.   :   Nucleon Polarizabilities and Deep Inelastic Electron
Scattering 298

CHRISTILLIN, P. and ROSA-CLOT, M.   :   Exchange Effects in Photon
Scattering in Nuclei 301

List of Participants

ARENHOEVEL, H., Mainz

BERGERE, R., Bures sur Yvette

BERTOZZI, W., Cambridge (USA)

BOHIGAS, O., Orsay

BOSCO, B., Arcetri

CALOI, R., Roma

CARCHON, R., Gent

CATILLON, P., Gif sur Yvette

CHEW, S.H., Birmingham

CIOFI DEGLI ATTI, C., Roma

COSTA, S., Torino

DELSANTO, P.P., Cagliari

D'ERASMO, G., Bari

DEVOS, J., Gent

DE VRIES, C., Amsterdam

DRECHSEL, D., Mainz

EPPEL, D., Hamburg

FABRE DE LA RIPELLE, M., Orsay

FINDLAY, D.J.S., Glasgow

GIANNINI, M., Genova

GIUSTI, C., Pavia

GOERINGER, H., Mainz

HANNA, S.S., Stanford (USA)

HAYWARD, E., Washington, DC

HEBACH, H., Bochum

JOHNSSON, B., Lund

LAUTERBACH, C., Garching

LEONARDI, R., Bologna

LEPRETRE, A., Gif sur Yvette

LINDGREN, K., Lund

LIPPARINI, E., Trento

MATONE, G., Frascati

MATTHEWS, J.L., Cambridge (USA)

MECKING, B., Bonn

PANTALEO, A., Bari

PROSPERI, D., Frascati

RICCO, G., Genova

RICHTER, A., Darmstadt

ROSA-CLOT, M., Geneva

ROSS, C., Ottawa

SANZONE, M., Genova

SCHAERF, C., Roma

SOLODUKHOV, G.V., Moscow

STRANGIO, C., Roma

TERRANOVA, M.L., Roma

TORIZUKA, Y., Sendai

TRAINI, M., Trento

VAN CAMP, E., Gent

WEISE, W., Erlangen

WOLYNEC, E., Sao Paulo

FEATURES OF THE GIANT E I - RESONANCES

R. BERGERE
Département de Physique
Nucléaire
CEN/ SACLAY

INTRODUCTION

CHAPTER I : THE GIANT MULTIPOLE RESONANCES

I.A. What is a so called Giant Resonances ?

I.B. A brief classification of the Giant Resonances.

I.B.1 Isoscalar giant resoances.

I.B.2 Compression modes.

I.B.3 Polarization modes.

I.C. Excitation of a Giant Resonance mode.

I.C.1. Real photon induced reactions.

I.C.2. Radiative capture (p, γ), (α, γ).

I.C.3. Inelastic scattering of electrons.

I.C.4. Inelastic scattering of hadrons.

I.C.5. Muon capture.

I.C.6. Pion radiative capture.

I.C.7. Virtual excitations.

CHAPTER II : THE PHOTON BEAMS

I. Photons produced in nuclear excitations.

I.A. (p, γ) sources.

I.B. (n, γ) sources

I.B.1. Photons sources with discrete energies.

I.B.2. Compton scattering of neutron capture γ.

I.B.3. Nuclear resonance scattering.

I.C. Nuclear resonance scattering of Bremstrahlung γ.

.../...

II. Bremstrahlung photon beams.

  II.A. Use of the tip of Bremstrahlung spectra.

  II.B. Total absorption experiments.

  II.C. Tagged photon beams.

  II.D. Unfolding of Bremstrahlung yield data.

III. Monochromatic photon beams with variable energy.

CHAPTER III : THE COLLECTIVE MODELS OF THE EI
                GIANT RESONANCE

  III.A. The nucleon effective charges.

  III.B. The static collective model for spherical nuclei.
      III.B.1. The Goldhaber - Teller model.
      III.B.2. The Steinwedel - Jensen model.
      III.B.3. Comparison with experimental data.
      III.B.4. Refinements of the theoretical prediction .
          a) consideration of a realistic nuclear surface
          b) introduction of a variable K.

  III.C. The static collective model for permanently deformed nuclei.
      III.C.1. The Q_0 splitting.
      III.C.2. The Lorentz line fit.
      III.C.3. Comparison with experimental data.

  . III.D.   The dynamic collective model.
      III.D.1. The dynamic collective model for vibrationnal nuclei
          a) theoretical summary
          b) comparison with experimental data :
              - Sn isotopes
              - Te, Cd, Pd
              - Nd isotopes
              - general features of experimental data.

III.D.2. The improved dynamic collective model

    a) theoretical summary

    b) comparison with experimental data :

        - transition region around N = 89

        - transition region around A = 190

CHAPTER IV : THE MICROSCOPIC MODELS OF THE EI
GIANT RESONANCE

IV.A. The schematic model of Brown - Bolsterli.

IV.B. Effective 1 p - 1 h calculations of the G.D.R. for closed
shell nuclei

    1) ^{16}O

    2) ^{208}Pb.

IV.C. Effective 1 p - 1 h calculations of the GDR for non closed
shell nuclei.

IV.D. 1 p - 1 h self consistent models of the GDR.

IV.E. The 1 p - 1 h continuum models of the GDR.

IV.F. Consideration of np - nh states in the fine structure of the GDR

    IV.F.1. Coupling of the 1 p - 1 h dipole state to the first low
energy 2^{+} state.

    IV.F.2. The effect of the quasi bound states.

    IV.F.3. A 3 p - 3 h model for the GDR of ^{16}O.

    IV.F.4. The fine structure of the GDR as a consequence of the
"nuclear coexistence".

IV.G. A microscopic description of the damping width.

IV.H. The microscopic models of the sum rules

    IV.H.1. The energy weighted sum-rule σ_0

    IV.H.2. The Bremstrahlung weighted sum-rule σ_{-1}

    IV.H.3. The σ_{-2} sum-rule.

CHAPTER V : THE DECAY CHANNELS FROM THE EI GIANT STATES

V.A. Competition between the (γ, n) and $(\gamma, 2n)$ decay modes in heavy nuclei.

V.B. Competition between the (γ, n) and $(\gamma, \text{fission})$ channels in fissile nuclei.

V.C. The statistical competition between the (γ, n) and (γ, p) channels.

V.D. The branching ratios towards the various levels in residual nuclei.

INTRODUCTION

The main properties of the giant dipole E1 resonance (GDR) were so far mostly obtained through experimental channels using real photons. The curve of Fig. 1 summarizes the typical behaviour of real photons when absorbed by a nucleus.

1- Up to $E_\gamma \approx 9$ MeV ($\lambda \approx 40$fm) one mostly observes photoexcitations of individual bound or unbound states whose microscopic natures can be connected to shell-model effects.

2- From 10 to 25 MeV ($\lambda \approx 10$ fm) approximately, one observes the systematic excitation of the collective mode known as the giant dipole resonance even, surprisingly, for nuclei as light as ^{6}Li or ^{4}He. The only true exception is the deuteron case.

3- Above $E_\gamma = 30$ MeV one expects to "feel" phenomena, such as the short range correlations between nucleons, in connection with the shorter wavelength of the incident photons ($\lambda = 2$ fm at $E_\gamma \approx 100$ MeV).

4- Above the photopion threshold (≈ 140 MeV) one reaches the region of the nucleon resonances.

The GDR of point 2 above is now reasonably well known and its main characteristics can be summarized as follows.

1- Its average localization in energy E_D shows a smooth variation versus the mass number A (at least for medium and heavy nuclei) in fairly good agreement with the predictions of the various collective models. In chapter III, which deals with collective models, this point will be taken up in some detail.

2- Its width is strongly modulated by the effects governing the shapes of nuclei (spherical nuclei, vibrationnal nuclei, permanently deformed nuclei). This property will also be reviewed in chapter III.

3- The position in energy and the fine structure of the GDR are more closely connected to shell model predictions as nuclei get lighter. The various experimental integrated cross sections are also more easily understood by comparisons with microscopic models. These properties will be reviewed in chapter IV " The microscopic models of the GDR ".

4- Finally, many things can be learnt about the properties of the GDR by studying and comparing the various decay channels, which will be done in chapter V.

I must point out that, in this review, I will hardly mention the following important characteristics of the GDR since they will be exposed in detail by other participants in this conference :

- The isospin splitting of the GDR (and the isospin mixing problems) by R. Leonardi and S. Hanna.
- The detailed examination of the sum-rule problems by W. Weise.
- The (γ, γ) and (γ, γ') channels by E. Hayward.

Since most of the available experimental data concerning the GDR have been obtained with real photons, I thought it useful to devote one whole lecture (chapter II) to review the characteristics of the various sources of real photons, with an attempt to connect the experimental particularities of each photon source to the specific parameters of the GDR which it is best suited to reach.

Chapter I is meant as a somewhat extended general introduction. In particular, I hope to show that the well known E1 GDR is just one member of the family of giant resonances, and that the study of this E1 resonance by photoexcitation, although privileged, is just a particular means of study among many others.

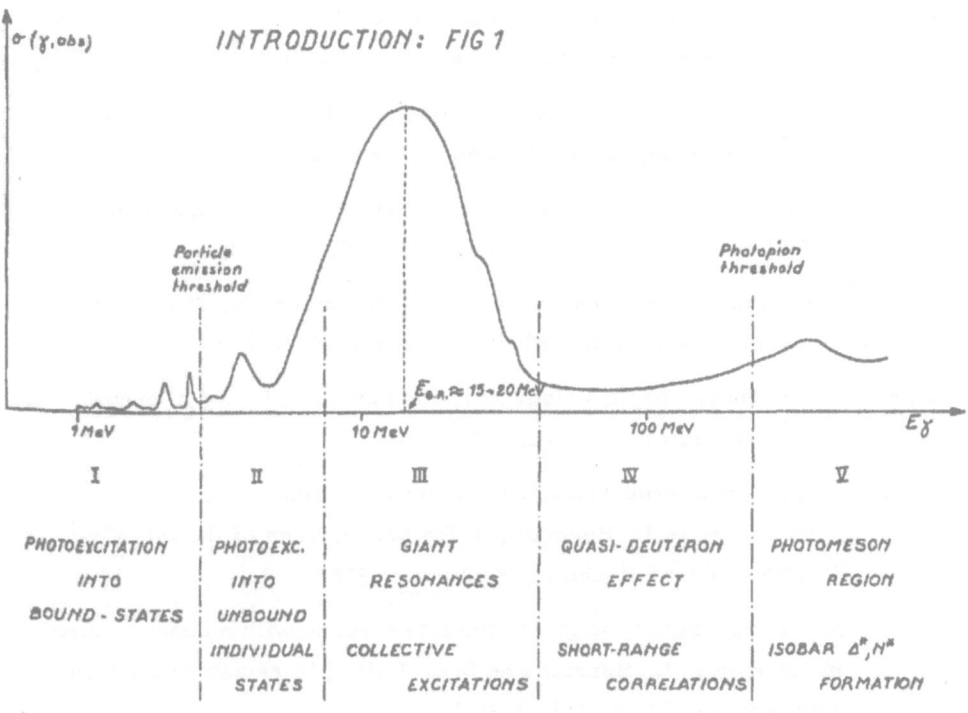

INTRODUCTION: FIG 1

CHAPTER I
THE GIANT MULTIPOLE RESONANCES

During the last 20 years several review papers have been published, which summarize the state of knowledge of photonuclear reactions in the region of the E1 giant dipole resonance. Even if a complete list of these review papers cannot be given, one can quote the following ones :

[I. 1] - Nuclear Photodisintegration by D. H. Wilkinson (Physica XXII, 1039, Amsterdam nuclear reaction conference 1956)

[I. 2] - The present status of photoneutron cross section measurements by B. M. Spicer (Supplements al Nuovo Cimento, vol. II, 3, 1964).

[I. 3] - Photonuclear reactions by E. Hayward (in Nuclear Structure and electromagnetic interactions, Oliver and Boyd 1964).

[I. 4] - Photonuclear reactions by M. Danos and E. G. Fuller (in annual review of nuclear science, vol. 15, 29, 1965).

[I. 5] - The giant dipole resonance, by B. M. Spicer (in advance in nuclear physics, vol. 2, 39, 1969, Plenum Press).

[I. 6] - Low energy photonuclear reactions by F. W. Firk (in annual review of nuclear science, vol. 20, 39, 1970).

[I. 7] - Systematic properties of the giant resonance by R. L. Bramblett, S. C. Fultz, B. L. Berman (in the proceedings of the Int. Conf. on Photonuclear Reaction, Asilomar 1973).

[I. 8] Measurements of the giant dipole resonance with monoenergetic photons by B. L. Berman and S. C. Fultz (in review of Modern Physics, vol. 47, 3, 713, 1975).

Of course, any new attempt to summarize the present knowledge concerning the E1 giant dipole resonance should first try to complement the above papers with new experimental properties and recently proposed theoretical descriptions of the E1 giant resonance. But when one considers the wealth of recent papers which deal with the collective oscillation modes of nuclei, in the energy region of excitation ranging from 10 MeV to 35 MeV approximately, one realises that most of these papers do not concern the E1 giant resonance. Actually, in the last five years a lot of experimental evidence, obtained through various reaction channels, showed that several other, highly collective oscillations, could be found in this energy range. Several " giant resonances " could thus be identified or are in the process of being studied. Let us therefore have a closer look at this family of " giant resonances " and try to find some way of " identifying " each of its members, and in particular the giant electric dipole resonance E1, by means of some specific characteristics.

The following features of this question will therefore be discussed in chapter I :

IA - What is a so-called " giant resonance " ?

I B - How can one classify the various giant resonances, and what is the place of the classical E1 GDR in such a classification ?

I C - Since numerous reaction channels have been recently used to excite the various giant resonances, what are the characteristics of each channel for a specific excitation of the E1 G.R. ?

I.A - WHAT IS A SO CALLED GIANT RESONANCE ?

First of all, what do we call an electric (or magnetic) giant resonance with a multipolarity λ ? A clear summary has been given by G.R. Satchler [I.9] who first put the energy factor into perspective by looking at a typical reaction channel summarized as

$$A + a \longrightarrow B + b$$

for which a typical cross section is given in Fig. 1a. If the energy of the incident particle, a, is about 100 MeV, the overal spectrum covers approximately 100 MeV and includes as E_b decreases (or E_B^* increases).

1- Some discrete, low lying, bound levels ($E_B^* \simeq$ a few MeV) which are studied in conventional nuclear spectroscopy.

2- The intermediate region (E_B^* $\simeq$ 10 to 30 MeV) where some gross structure, determined by average nuclear properties. can appear . These fairly wide structures are localized at excitation energies for which particle emission can happen. Therefore, as G. R. Satchler again pointed it out, [I. 9] the underlying microstructure cannot generally show up since the individual states overlap and one can just " see " the envelope of these individual states as a large resonance. This is typical of the doorway state description which is very well suited to the gross structure seen in this energy region, such as the isobaric analog resonances or the giant resonances.

3- Finally at a higher excitation energy, the use of a high energy projectile leads to a kind of statistical equilibrium which can result in an evaporation peak in the energy spectrum of the particle b $\left(\dfrac{dN_b}{dE_b}\right)''$ which is generally the information directly provided by the experiment.

Once such a gross structure has been localized in energy, how is it characterized as a giant resonance with the multipolarity $E\lambda$ or $M\lambda$? This gross structure (as well as the individual states $| n \rangle$ which build it up) has a spin and parity $J_n^{\pi_n}$. Then a transition probability T_λ exists for emission of a photon of energy E_n $= \hbar \omega$ ($k = \dfrac{\omega}{c}$), angular momentum $\lambda\mu$ and of electric or magnetic type, and connecting the ground state $|J_i^{\pi_i} M_i\rangle$ of the nucleus B to this particular excited states $|J_m^{\pi_m}, M_n\rangle$ at energy E_n (Fig. 1b). One knows that one can write

$$T_{in}(\lambda) = \frac{8\pi(\lambda+1)}{\lambda[(2\lambda+1)!!]^2} \frac{k^{2\lambda+1}}{\hbar} B(\lambda)$$

with $$B(\lambda)_{J_i \to J_m} = \frac{1}{2J_i+1} \sum_{M_i, M_m} |\langle n | \mathcal{O}_{\lambda\mu} | i \rangle|^2$$

and $$|J_m - J_i| < \lambda < J_m + J_i$$
$$\mu + M_i = M_n$$

$O_{\lambda\mu}$ = electric ($Q_{\lambda\mu}$) or magnetic ($M_{\lambda\mu}$) multipole
operators.

This allows one to characterize the electromagnetic transition as

λ- pole electric if $\quad \pi^{i} = \pi^{m}_{\cdot}(-1)^{\lambda}$

λ- pole magnetic if $\quad \pi^{i} = \pi^{m}_{\cdot}(-1)^{\lambda+1}$

One can thus characterize the giant resonance, which is built up
with individual states of the above nature, as an electric λ - pole
or magnetic λ - pole giant resonance.

In the above expression the word " giant " refers to the collectivity of the observed gross state. Two criteria are currently
used to ascertain   this collectivity (Fig. 1c).

1-   The transition rates for the excitation (or deexcitation) of such
a state must be much larger than some " single particle " transition rate which would represent the effect of a single-nucleon
jump between two shell-orbits. For example, one knows that for
electromagnetic transitions these single particle transition rates
are customarily introduced as the Weisskopf units

$$T_w(E_\lambda) = \frac{2(\lambda+1)}{\lambda\left[(2\lambda+1)!!\right]^2}\left(\frac{3}{\lambda+3}\right)^2 \frac{e^2}{\hbar c}\left(\frac{\omega R}{c}\right)^{2\lambda} \omega \quad sec^{-1}$$

$$T_w(M_\lambda) = \frac{2(\lambda+1)}{\lambda\left[(2\lambda+1)!!\right]^2}\left(\frac{3}{\lambda+3}\right)^2 \frac{e^2}{\hbar c}\left(\frac{\hbar}{M_c R}\right)^2\left(\frac{\omega R}{c}\right)^{2\lambda} \omega \quad sec^{-1}$$

where R is the nuclear radius.

One also knows that for inelastic scattering interactions ,
one considers the S=T=0 single particle operator $r^\lambda Y_\lambda^\mu(\theta, \varphi)$
which gives the isoscalar part $Q_{o\lambda\mu} = \sum\limits_{i=1}^{A} r_i^\lambda Y_\lambda^\mu(\theta_i, \varphi_i)$
for the electric E λ radiation, whereas the isovector part is
given by $Q_{1\lambda\mu} = \sum\limits_{i=1}^{A} \tau_i^3 r_i^\lambda Y_\lambda^\mu(\theta_i, \varphi_i)$ where
τ_i^3 is the 3- component of the isospin of nucleon i whose
spherical coordinates are r_i, θ_i, φ_i . Once again, to ascertain
the collectivity of the resonance observed in an inelastic scattering

interaction one verifies that its transition rate is much larger than the single particle transition rate corresponding to the excitation of a $J_i = 0$ nucleus to a state $J_F = \lambda$ by the operator $r_i^\lambda Y_\lambda^\mu (\theta_i, \varphi_i)$ for which the reduced transition probability is : [I. 9]

$$B^{SP}(\lambda, S=T=0) = \frac{2\lambda+1}{4\pi} \left(\frac{3R^\lambda}{\lambda+3} \right)^2$$

2- The other useful criterion, implies a comparison with some sum rule, the most useful of which is the so-called energy weighted sum rule (EWSR) (because it is the most model independent sum-rule) , written by Lane [I. 10] in the case of the above E λ operators, as

$$S_{EW}^{T\lambda} = \sum_m [E_m - E_o] |\langle n | Q_{T\lambda o} | o \rangle|^2$$
$$= \frac{1}{2} \langle o | [Q_{T\lambda o}, [H, Q_{T\lambda o}]] | o \rangle$$
$$\text{with } T=0 \text{ or } 1$$

For the observed gross resonance , the strength of the observed transition must be large enough to exhaust a large fraction of the above sum rule, for which several detailed expressions can be found in the litterature (the expressions used in the case of the E1 giant resonance will be briefly recalled later).

I.B - A BRIEF CLASSIFICATION OF THE GIANT RESONANCES -

Since these giant resonances are, by definition, highly collective modes, it comes as no surprise that a simple hydrodynamical description of the collective motion of the nucleons provides a clarifying classification of the giant resonances. The summary which is given below, follows the line of explanation proposed by Bohr and Mottelson [I. 11] rather closely.

I.B.1 - Isoscalar giant resonances (T = 0)

These collective modes are called isoscalar because they are build up by considering in-phase motions of all the nucleons, protons or neutrons. Following the description of the liquid drop model one thus considers a _surface oscillation_ characterized by the quantum multipole numbers λ and μ

$$R = R_o \left[1 + \sum_\mu \alpha_{\lambda\mu}(t) \, Y^*_{\lambda\mu}(\theta,\varphi) \right]$$

as characterizing the collective isoscalar vibration of the spherical nucleus. One knows that the $\lambda = 0$ and $\lambda = 1$ isoscalar mode respectively describe a compression mode and a simple displacement of the center of gravity of the nucleus. Therefore genuine isoscalar giant resonances correspond to $\lambda \geqslant 2$. The isoscalar quadrupole mode has now been extensively studied, theoretically and experimentally $[I.9]$ and is known to lie around the energy $E_2 \approx 60 \, A^{-1/3}$ MeV at least for medium and heavy nuclei.

I.B.2 - Compression modes.

Bohr and Mottelson $[I.11]$ consider also " compression modes " in which small deviations of nuclear density from its equilibrium value ρ_o are allowed. For these " sound waves " a sound velocity V_c can be defined and is related to a compressibility coefficient b_c ($\approx$ 15 MeV)

$$V_c = \sqrt{\frac{b_c}{M}}$$

The hydrodynamical equation is then a standard wave equation

$$\nabla^2 \delta\rho(\vec{r},t) - \frac{1}{V_c^2} \frac{\partial^2 \delta\rho(\vec{r},t)}{\partial t^2} = 0$$

whose spherical solution

$$\delta\rho = \rho_o \, \alpha_{m\lambda\mu}(t) \cdot j_\lambda(k_{m\lambda} r) \, Y^*_{\lambda\mu}(\theta,\varphi)$$

must be fitted to the limit condition $\delta\rho = 0$ at the nuclear surface

$r = R_o = 1.2\,A^{-1/3}$ fm to give the energies of the compression eigen-modes

$$\hbar\,\omega_{n\lambda} = \hbar\,k_{n\lambda}\,v_c = \begin{cases} 65\,A^{-\frac{1}{3}}\text{ MeV} & \text{for } \lambda = 0 \quad n = 1 \\[2ex] 94\,A^{-\frac{1}{3}}\text{ MeV} & \text{for } \lambda = 1 \quad n = 1 \\[2ex] 120\,A^{-\frac{1}{3}}\text{ MeV} & \text{for } \lambda = 2 \quad n = 1 \end{cases}$$

I. B. 3 - Polarisation modes .

Always following the same hydrodynamical approach Goldhaber-Teller [I. 12] described the giant dipole resonance as a polarisation mode of the isovector type (T = 1) in which the density variation has an isovector character due to the out of phase motion of the neutrons and protons within the volume of the nucleus

$$\delta\rho(\hbar) = \rho_p(\hbar) - \rho_m(\hbar)$$
$$\rho_o = \rho_p(\hbar) + \rho_m(\hbar)$$

But one can generalize the G. T. model of the giant multipole resonances by introducing more than 2 fluids. Following the approximate SU (4) symmetry one can thus consider 4 interpenetrating fluids :

neutrons with spin up

neutrons with spin down

protons with spin up

protons with spin down

Foldy and Walecka showed that one expects the following collective excitations for $J = 0^+$, T = 0 nuclei   [I. 13]

$$\Delta S = 1 \qquad \Delta T = 0 \qquad \Delta S_3 = \pm 1, 0 \qquad \text{3 spin resonances}$$

$$\Delta S = 0 \qquad \Delta T = 1 \qquad \Delta T_3 = \pm 1, 0 \qquad \text{3 isospin resonances}$$

$$\Delta S = 1 \qquad \Delta T = 1 \quad \begin{cases} \Delta T_3 = \pm 1, 0 \\ \Delta S_3 = \pm 1, 0 \end{cases} \qquad \text{9 spin-isospin resonances}$$

Thus each giant resonance with multipolarity λ^π in $J^\pi = 0^+$, T=0 nuclei can be shown to form a 15 dimensional (SU-4 vector) supermultiplet. The 15 corresponding energies would be degenerate if nuclear forces were spin and charge independent. Uberall [I. 14] described these problems, and gave an example of classification of these polarisation modes in the case of the collective vibrations of a $J^\pi = 0^+$ nucleus (table I and Fig. 2)

Table I

Mode	ΔT	λ	ΔS	J_f	number of resonances
isospin	1	λ^π	0	λ^π	3
spin-isospin	1	λ^π	1	$(\lambda-1)^\pi$ λ^π $(\lambda+1)^\pi$	9
spin	0	λ^π	1	$(\lambda-1)^\pi$ λ^π $(\lambda+1)^\pi$	3

(total 15)

The well known giant electric dipole (isovector) resonance is the $\Delta S = 0$, $\Delta T = 1$, $\Delta T_3 = 0$ mode in the case of a multipolarity $\lambda^\pi = 1^-$.

NB - Although the above classification makes a clear distinction between isoscalar and isovector modes, reality is much less clear cut, especially for $N \rangle Z$ nuclei. Bohr and Mottelson pointed out that a coupling between λ-pole isoscalar moment and λ-pole isovector moment is then to be expected according to

$$\mathcal{M}_\lambda(\Delta T = 1) = \frac{N-Z}{A} \mathcal{M}_\lambda(\Delta T = 0)$$

This very coupling between isoscalar and isovector oscillations has also
been studied through a more microscopic approach by Goulard and Fallieros
$\begin{bmatrix} I. 15 \end{bmatrix}$. Starting from sum rule considerations they computed the
isoscalar. λ -pole transition densities which are associated to any
isovector λ -pole excitation in a N $>$ Z nucleus.

There is another way to look at such a phenomenon. The simple
hydrodynamical classification connects mostly isoscalar modes to surface
vibrations with an induced isoscalar potential $\delta V_{\lambda} \neq 0$ only at the
surface of the nucleus - on the other hand isovector oscillations are
associated with volume oscillations of the nucleus with $\delta V_{\lambda} (T=1) \neq 0$
in the whole nucleus. A recent typical R. P. A calculation with reasonable
Skyrme interaction by G. Bertsch $\begin{bmatrix} I. 16 \end{bmatrix}$ showed that the transition
densities associated to each giant resonance are only to some extent
surface peaked or volume spread (Fig. 3).

I. C- EXCITATION OF A "GIANT RESONANCE " MODE -

To summarize the above classification it is clear that the excitation of any giant resonance can be considered as a self-consistent
process summarized in Fig. 4. Let us consider a specific reaction channel
which brings in λ units of angular momentum. Most of the time the
operator acting in this reaction channel is a one-body operator promoting
a particle out of its ground state orbit and thus leading to the excitation
of a simple (1p-1h) or (2p-1h) state.

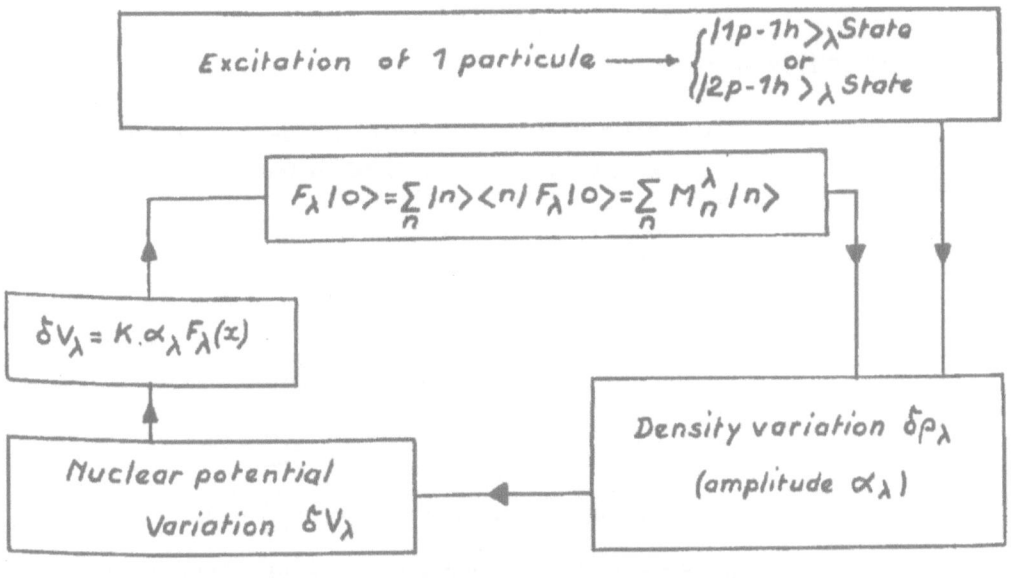

$$FIG. 4$$

This corresponds to a variation $\delta\rho_\lambda$ in the nucleonic density which generates a corresponding variation δV_λ in the average nuclear potential

$$\delta V_\lambda = K . F_\lambda (x) \alpha_\lambda$$

$F_\lambda (x)$ is the associated field operator which can, in turn, act on the individual particle in the nucleus ground state and, as pointed out by A. Bohr and B. K. Mottelson [I. 11], organize the collective motion of the particles and build up the giant resonance collective state $\sum_m M_m^\lambda |m\rangle$ It is thus quite understandable that many reaction channels, inducing some initially simple $|1p\text{-}1h\rangle_\lambda$ or $|2p\text{-}1h\rangle_\lambda$ states, will finally induce several of the various giant resonances in a more or less specific way. As an example, the consideration in the dipole case ($\lambda = 1$) of some specific operators inducing the initial particle state immediately shows that the dipole giant resonances $\Delta T = 1$, $\Delta S = 0$ (cf Table I) can be excited with several final isospin component (see for example Bressani [I. 17]).

1- The dipole operator $\frac{1}{2} \sum_{i=1}^{A} \tau_i^{(3)} x_i$

induces a particle-hole pair $|p^{-1}p\rangle$ or $|n^{-1}n\rangle$ in which the particle and hole isospin are coupled to

$$t_{ph} = 1 \quad , \quad t_{ph}^{(3)} = 0$$

2 - The 1-body operator $\frac{1}{2} \sum_{i=1}^{A} \tau_i^{\pm} x_i$

induces particle-hole states such as

$$|n^{-1}p\rangle \quad \text{coupled to } t_{ph} = 1 \quad t_{ph}^{(3)} = -1$$

$$|p^{-1}n\rangle \quad \text{coupled to } t_{ph} = 1 \quad t_{ph}^{(3)} = 1$$

thus creating states analog to the classical dipole state

Fig. 5

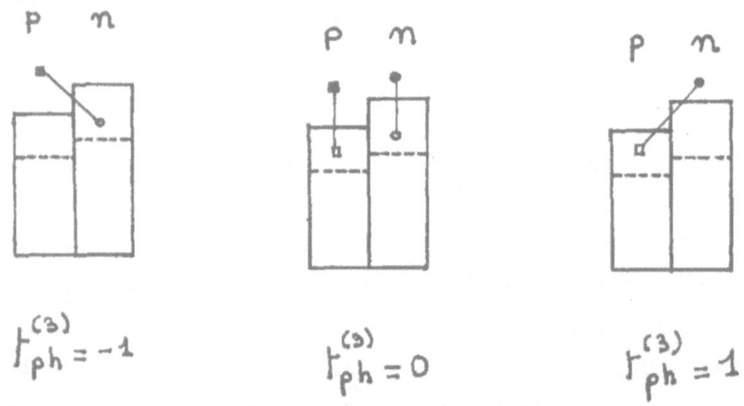

$$t_{ph}^{(3)} = -1 \qquad t_{ph}^{(3)} = 0 \qquad t_{ph}^{(3)} = 1$$

One thus sees that the giant dipole resonance can have two isospin states $T = T_0$ and $T = T_0 + 1$ in nuclei A_Z^N whose ground state has an isospin $T_0 \neq 0$ and that dipole giant resonances with isospin ($T = T_0 + 1$) or ($T = T_0-1$, T_0, $T_0 + 1$) can be excited as analog dipole states in the nuclei

A^{N+1}_{Z-1} and A^{N-1}_{Z+1} respectively (Fig. 6).

It is well known that many reaction channels have been used to excite the various giant resonances (G. R.) namely.

I. C. 1 - real photon induced reaction.

I. C. 2 - radiative capture (p, γ_0) (α, γ_0) ...

I. C. 3 - inelastic scattering of electrons.

I. C. 4 - inelastic scattering of hadrons.

I. C. 5 - μ - capture.

I. C. 6 - radiative capture of π^- mesons.

I. C. 7 - virtual excitations of G. R. in nuclei.

Let us try to see very briefly how each of the above reaction channels is more or less suited for a specific excitation of the giant E 1 isovector resonance.

<u>I. C. 1</u> - The huge majority of the data concerning the giant dipole E1 resonance have been obtained through the processes 1) and 2) in which <u>real photons</u> are involved. Actually since a photon is a zero mass particle, when it brings an energy $\hbar \omega$ into a nucleus it only brings a small momentum $\frac{\hbar \omega}{c}$. It will therefore excite mostly the low multipolarity giant resonances E1, M1 and to some extent E2. Sum rule considerations show that,for photon absorption cross sections σ_T^λ, one has

$$\int \sigma_{T=0}^{\lambda=2} (\gamma, \text{tot.}) \frac{dE}{E^2} \approx 4 \; 10^{-4} \; Z^2 \; A^{-1/3} \quad \text{Mev}^{-1}. \text{mb}.$$

$$\int \sigma_{T=1}^{\lambda=2} (\gamma, \text{tot}) \frac{dE}{E^2} \approx 4 \; 10^{-4} \; NZ \; A^{-1/3} \quad \text{Mev}^{-1}. \text{mb}.$$

$$\int \sigma_{T=1}^{\lambda=1} (\gamma, \text{tot}) \frac{dE}{E^2} \approx 2.25 \; A^2 \; A^{-1/3} \quad \text{Mev}^{-1}. \mu\text{b}.$$

Hence for A $\simeq$ 40 , the E1 cross section for γ absorption is at least 20 times larger than the E2 absorption cross section. If we except some possible excitation of M1 giant states at an energy roughly equal to half the energy of the E1 giant resonance, one thus sees that real photons are best fitted for a specific excitation of the E1 G.R.

Just as an example fig. 7 shows how the (γ, absorption) channel, a partial channel such as (γ, n)+(γ, pn) and the partial channel (γ, po), obtained through the reverse reaction (p, γ_0), complement each other.

I. C. 2 - The (p, γ_0) channel brings about the high resolution obtainable with incident beams of charged particle and is well suited to angular distribution measurements which allows to sort out a small E2 component. On the other hand the coulomb barrier generally limits its use to light nuclei. Of course the (γ, po) cross section of a A_Z nucleus is thus obtainable only if the $(A-1)_{Z-1}$ nucleus exist. ^{39}K (γ, po) is therefore not obtainable through this reverse reaction. On the contrary the absorption of real photons can be studied for all existing nuclei thus leading if some difficulties are correctly overcome (see chapter II), to a direct comparison between the integrated measured cross section $\int \sigma \, (\gamma, abs) \, dE$ and the corresponding predictions for the energy weighted sum rule

$$\sum_m (E_m - E_o) \, B \, (E_1, |0\rangle \longrightarrow |m\rangle)$$

Unfortunately, as it will also be seen in chapter II, the energy resolution provided by the available real photons beams is usually fairly poor.

I. C. 3 - The situation is entirely different when one turns to the excitations of giant resonances by virtual photons , which happen in (e, e') experiments. Whereas a real photon with an energy $\hbar\omega$ = 20 MeV only transfers a momentum $q_o = \frac{1}{\hbar} \left(\frac{\hbar\omega}{c} \right)$ = 0.1 fm^{-1} to a nucleus, a relativistic electron with initial energy E_0 = kc = 300 MeV will transfer the same energy $\hbar\omega$ = 20 MeV = E_0 - E = (k_0-k) c to the nucleus with a transferred momentum q

$$q \sim \left[\frac{1}{\hbar^2} \, (\, k_o^2 + k^2 - 2 k \, k_o \, \cos\theta) \right]^{1/2} F \, \text{fm}^{-1}$$

variable from 0.1 fm^{-1} (scattering angle $\theta = 0°$) to 3. fm^{-1}($\theta = 180°$).
Hence the excitation of λ pole collective modes with $\lambda > 1$ will be much
easier with virtual photons than with real ones. Moreover, if one classically writes for the (e, e') process

$$\frac{d\sigma}{d\Lambda} = 4\pi \, \sigma_{Mott.} \left\{ |F_L(q)|^2 + \left(\frac{1}{2} + tg^2 \frac{\theta}{2} \right) |F_T(q)|^2 \right\}$$

one knows that the collective monopole mode can be excited through the
longitudinal form factor $F_L(q)$ whereas, when θ is getting close
to 180°, the magnetic transitions are preferentially excited through the
transverse form factor F_T. A clear illustration of the excitation of
different giant resonances, when the transferred momentum varies, is
provided by the Darmstadt experiment ^{140}Ce (e, e') [I. 18] (Fig. 8).
When θ increases one thus observes the successive showing up of the
isovector E1 GR at $E_X \simeq 15.5$ MeV, then of the isoscalar E2 GR
at $E_X \simeq 12$ MeV and finally of a giant M1 resonance around $E_X \simeq 8.7$ MeV.
Therefore, one can conclude than the E1 GR will be generally strongly
excited on (e, e') reactions simultaneously with other GR and the sorting
out of the various GR require very careful procedures which will be
described during this meeting by Y. Torizuka. Another classical result
was obtained by (e, e') excitation of the GDR at Darmstadt and shows, in the
case of the ^{26}Mg nucleus, the isospin splitting of the GDR into a T = 1
and a T = 2 component (Fig. 9a) which was found later in a (γ, n) experiment with real photons Fig. 9b [I. 19, I. 20] .

I. C. 4 - It is by using <u>inelastic scattering of hadrons</u> that most of the recent
data concerning the giant resonances were produced. There, the high
momentum which is brought in by the hadron projectile allows the excitation
of collective modes with multipolarities $\lambda \gg 1$. Moreover the interaction
potential " projectile-nucleus " $U = U_o + \underset{A}{t.T} U_1 + U_c$ indicates that,
through U_o, the isoscalar oscillations will be mostly excited. The isovector
modes will be more weakly excited through the U_c terms and also the $\underset{A}{t.T} U_1$ term,

provided that the isospin t of the projectile is not zero. As an example, Fig. 10 shows how, at Oakridge, a simultaneous excitation of the giant dipole E1 mode and of the isoscalar E2 modes was observed by inelastic scattering of 65 MeV protons on ^{58}Ni whereas the E1 isovector mode was practically no longer excited by deuteron scattering [I. 21] .

I. C. 5 - If one writes [I. 17, I. 22] the matrix elements for a μ^- capture by a nucleus (in a 1s orbit) one sees that the operators acting on a nucleus with isospin T_0, will excite, the isospin and spin-isospin modes with isospin ($T_0 + 1, T_3 = T_0 + 1$). Therefore the reaction

$$\mu^- + {}^A_Z X \longrightarrow \nu_\mu + {}^A_{Z-1} X'^*$$

will excite in the residual nucleus X'^* , the analog states of the giant isospin and spin-isospin resonances which may be excited in the nucleus $^A_Z X$. Unfortunately, since spectroscopy of neutrinos ν_μ is impossible, one must then turn either to secondary γ spectroscopy or to difficult fast neutron spectroscopy. As an example of the first method, which was suggested by Raphael [I. 23] , Kaplan [I. 24] identified in ^{16}N* the isospin dipole state analog to the classical dipole states in ^{16}O by showing the identity of the γ_i and γ_j spectra in the two reactions

$$^{16}O + \mu^- \longrightarrow \nu_\mu + {}^{16}N^*_{T=1} \longrightarrow {}^{15}N + n + \gamma_i$$

$$^{16}O + \gamma \longrightarrow {}^{16}O^*_{T=1} \longrightarrow {}^{15}N + p + \gamma_j$$

It must also be pointed out that μ^- capture on N $>$ Z nuclei should provide a clean identification of the $T_>$ isospin component of the giant dipole state [I. 25, I. 26] .

I.C.6 - The <u>radiative pion capture</u> process

$$\pi^- + {}^A_Z X \longrightarrow \gamma + {}^A_{Z-1} X'^*$$

depends on Gamov-Teller type transitions only and should therefore excite only spin-isospin resonances without exciting the isospin resonances [I. 17] (namely the spin-isospin dipole modes L = 1, S= 1 with $J^\pi = 0^-, 1^-, 2^-$ if the target nucleus is a $J^\pi = 0^+$ nucleus). However a typical example of the data obtained at Berkeley [I. 27] is given in fig. 11 corresponding to the reaction ${}^{12}C + \pi^- \longrightarrow \gamma + {}^*B^{12}$. The spectroscopy of the high energy γ shows several peaks at 124.7 MeV ($J^\pi = 1^+$), 120.3 MeV (2^-) and 116.9 MeV (1^-) which are the analogs ($T_3 = 1$) of the collective states ($T_3 = 0$) in ${}^{12}C$ at 15.1, 19.9 and 23.6 MeV respectively, with the latter being very likely the giant dipole state

I.C.7 - Turning to the <u>virtual excitation</u> of the giant resonances in nuclear reactions it is well known that such a process occurs in the " semi-direct " capture (n, γ) through the dipole states [I. 28, I. 29] . Recently it was even claimed that by using the GDR parameters of the final nucleus rather than those of the target nucleus, the neutron capture in ${}^{40}Ca$ could be better explained by implying just the virtual excitation of the $T_<$ part of the GDR in ${}^{41}Ca$ [I. 30] .

But the most systematic study of the virtual excitations of giant resonances in nuclear reactions was carried out at Julich by Von Geramb and his collaborators. In addition to the direct inelastic scattering to a giant resonance state (fig. 12 a),V. Geramb considers also the two step inelastic scattering (p, p') towards a bound low lying state, schematised in fig. 12 b,where a giant resonance fast decaying intermediate state is virtually excited. The transferred momentum J and the isospin T (at transferred energy Q) characterizing this GR state can possibly be obtained from the angular distribution σ (θ) or the analyzing power A (θ) of the outgoing protons [I. 31] . Particularly the enhancement of σ (θ) for backward angles can be clearly traced back to the virtual excitation of λ - pole giant resonances. In the corresponding transition matrix element one parametrizes the coupling constant γ_λ (Q) as

$$\gamma_\lambda(\overline{Q}) = \frac{-\beta_\lambda^2}{\overline{Q} - \hbar\omega_\lambda + \frac{1}{2}\Gamma_\lambda}$$

where $\overline{Q}$ is the energy transferred via the resonance with multipolarity λ, energy $\hbar\omega_\lambda$ and width Γ_λ (associated with a deformation parameter β_λ). In V. Geramb's formalism these parameters are left free as adjustable quantities to fit the various experimental data σ (θ), A (θ), etc... From the ^{16}O (p, p') ^{16}O * (2^-; 8.88 MeV) data he thus extracted the shapes of the λ = 1, 2, 3, 4 virtually excited giant resonances (fig. 13). One sees that a strong giant dipole resonance (λ = 1) is thus strongly virtually excited around the very energy (E_x = 22 MeV) found in the study with real photons. Similarly Von Geramb extracted from various ^{12}C(p, p') experiments a shape for the virtually excited E1 GR in good agreement with the direct ^{11}B (p, γ) measurement (fig. 14) [I. 32] .

I. C. 8 - Finally, some more exotic ways to excite the giant dipole resonance are also possible. As an example, N-A. Dadajan recently computed the possible excitation of the G. D. R. by neutral weak currents in neutrino scattering on ^{12}C and ^{16}O [I. 33] .

- REFERENCES OF CHAPTER I -

[I.1]   Nuclear photodisintegration by D.H. Wilkinson
         (Physica XXII, 1039, Amsterdam nuclear reaction
         conference 1956).

[I.2]   The present status of photoneutron cross section measure-
         ments by B.M. SPICER.
         (Supplements al Nuovo Cimento, vol. II, 3, 1964)

[I.3]   Photonuclear reactions by E. Hayward
         (in Nuclear Structure and electromagnetic interactions,
         Oliver and Boyd 1964).

[I.4]   Photonuclear reactions by M. Danos and E.G. Fuller
         (in annual review of nuclear science, vol. 15, 29, 1965).

[I.5]   The giant dipole resonance by B.M. Spicer
         (in advance in nuclear physics, vol. 2, 39, 1969, Plenum
         Press).

[I.6]   Low energy photonuclear reactions by F.W. Firk
         (in annual review of nuclear science, vol. 20, 39, 1970).

[I.7]   Systematic properties of the giant resonance by
         R.L. Bramblett, S.C. Fultz, B.L. Berman
         (in the proceedings of the Int. Conf. on photonuclear reaction,
         Asilomar 1973).

[I.8]   Measurements of the giant dipole resonance with monoener-
         getic photons by B.L. Berman and S.C. Fultz
         (in review of Modern Physics, vol. 47, 3, 713, 1975).

[I.9]   G.R. Satchler, Comments on Nu. Part. Phys., 5, 145, 1972
         G.R. Satchler, Physics Reports, 14 , 97 , 1974.

[I.10]   A.M. Lane, Nuclear Theory, W.A. Benjamin 1964.

[I.11]   A. Bohr, B. Mottelson in " Neutron-capture gamma ray
         spectroscopy " (Studvisk 1969) IAEA, Vienna 1969.
         A. Bohr, B. Mottelson " Nuclear Structure " tome 2
         (à paraître).

[I. 12]   M. Goldaber and E. Teller, Phys. Rev. 74 , 1046, 1948.

[I. 13]   L. Foldy and J. D. Walecka, Nuovo Cim. 34 , 1026, 1964.

[I. 14]   H. Uberall, Electron scattering from complex nuclei,
           Academic Press, 1971.

[I. 15]   B. Goulard, S. Fallieros, Can. Journ. of Physics 45 (1967)
           3221.

[I. 16]   O. Bertsch, S. F. Tsai, Phys. Lett. 50 B (1974) 319.

[I. 17]   T. Bressani, Rivista del Nuovo Cimento 1 , 2 (1971) 268.

[I. 18]   R. Pitthan, Th. Walcher, Phys. Lett. 36 B (1971) 563
           Z. Naturforsch 27 A (1972) 1683
           I. K. D. A. Report 7 3/2 .

[I. 19]   O. Titze, A. Goldmann, E. Spamer, Phys. Lett. 31B (565)
           1970.

[I. 20]   S. C. Fultz, R. A. Alvarez, B. L. Berman, M. A. Kelly,
           D. R. Lasher and T. W. Phillips, Phys. Rev. C4 -149 (1971).

[I. 21]   C. C. Chang, F. E. Bertrand and D. C. Kocher
           ORNL, Technical Report Number 75. 043 , december 1974.

[I. 22]   A. Fuji and H. Primakoff, Nuovo Cimento 12 , 327, 1959.

[I. 23]   R. Raphael et al, Phys. Lett. 24B , 15, 1967.

[I. 24]   S. N. Kaplan et al, Phys. Rev. Lett. 22 , 795, 1969.

[I. 25]   B. Goulard et al, Phys. Rev. Lett. 27 , 1238, 1971.

[I. 26]   O. Nalcioglu et al, Nucl. Phys. A 218 , 495, 1974.

[I. 27]   J. A. Bistirlich et al, Phys. Rev. Lett. 25 , 689, 1970.
           H. W. Baer and K. M. Crowe, Proceed. Int. Conf. Photonuclear
           Reactions, Asilomar 1973.
           H. W. Baer et al, Phys. Rev. C10 , 1140, 1974.

[I. 28]   G. Brown, Nucl. Phys. 57 , 339, 1964.

[I. 29]   G. Longo et al, Nucl. Phys. A 199 , 530, 1973
           and II Nuovo Cimento 20 A , 373, 1974.

[I.30]   L. Nilsson and J. Eriksson, Phys. Lett. 49 B , 165, 1974.

[I.31]   V. Geramb et al - Nucl. Phys. A 199 , 545, 1973

          - Villars Meeting on Nucl. Phys. 1974

          - Int. Conf. on Nucl. Struct., Amsterdam 1974

          - Phys. Lett. 52 B , 138, 1974.

[I.32]   V. Geramb and K. Amos - Phys. Rev. C12 , 1697, 1975.

[I.33]   N. A. Dadajan - JINR, E2 - 8936, 1975.

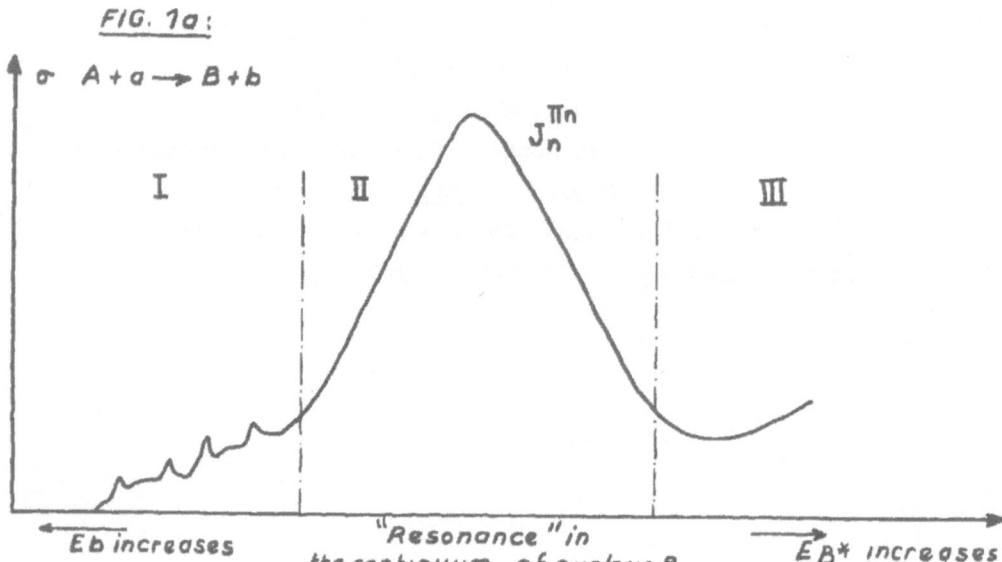

FIG. 1a :

σ   A + a ⟶ B + b

$J_n^{\pi n}$

I   II   III

E b increases

"Resonance" in
the continuum of nucleus B

E_{B^*} increases

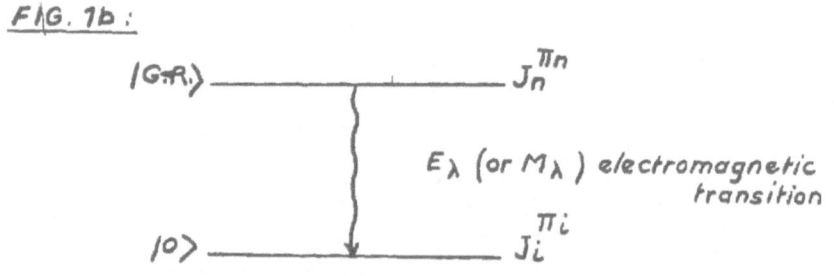

FIG. 1b :

$|G.R.\rangle$ _____ $J_n^{\pi n}$

E_λ (or M_λ) electromagnetic
                                   transition

$|0\rangle$ _____ $J_i^{\pi i}$

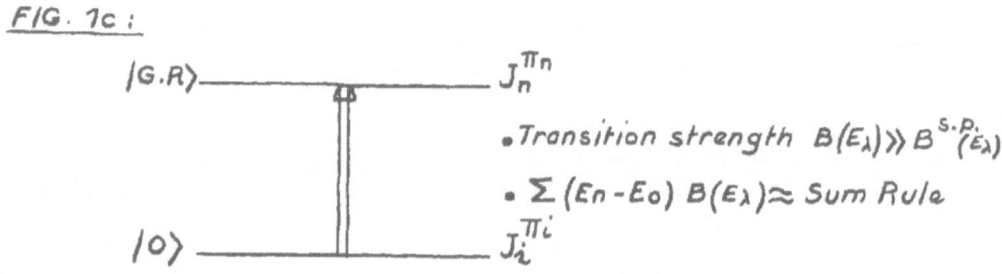

FIG. 1c :

$|G.R\rangle$ _____ $J_n^{\pi n}$

• Transition strength $B(E_\lambda) \gg B^{S.P.}(E_\lambda)$

• $\sum (E_n - E_0) B(E_\lambda) \approx$ Sum Rule

$|0\rangle$ _____ $J_i^{\pi i}$

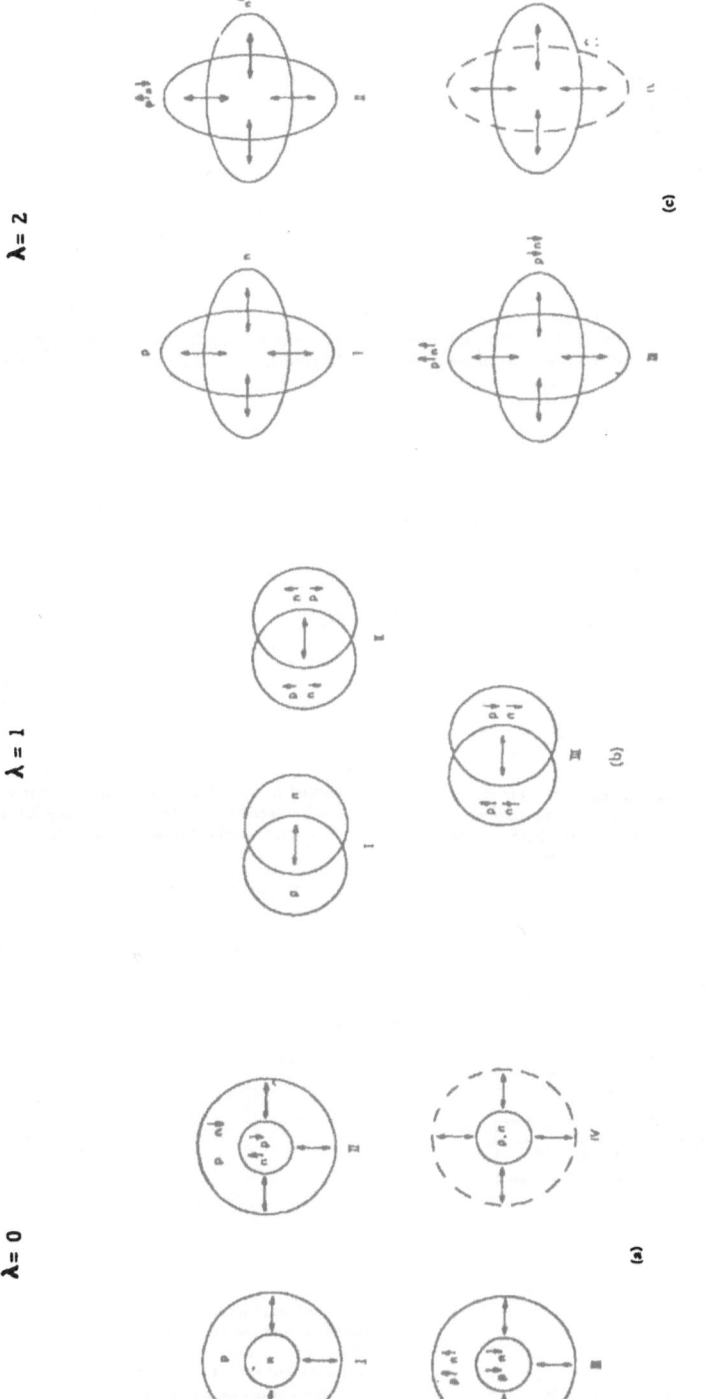

Fig.2 - From H. UBERALL'S electron scattering from complex nuclei (Academic Press 1971):collective multipole vibrations of the nucleus (generalized Goldhaber-Teller model) : I)isospin mode II) spin-isospin mode III) spin mode IV) in phase vibration.

FIG. 3 : Reproduced from ref. I.16

ISOVECTOR MODE

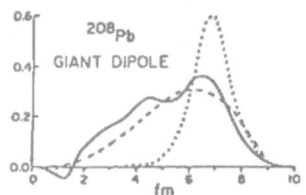

$T = 1$ dipole transition densities $(\times r^2)$ to the lower state at 12.2 MeV with SkI interaction. Velocity-dependent terms in the particle-hole interaction are neglected. The dashed curve corresponds to the SJ model, with $q = 0.5$ fm^{-1}. The dotted curves are obtained from the GT model, . Solid curves represent the RPA result,

ISOSCALAR MODES

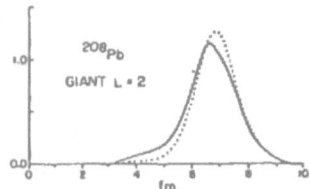

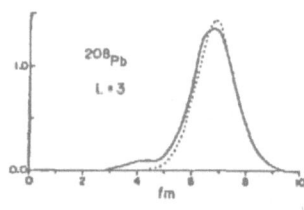

$T = 0$ quadrupole transition density $(\times r^2)$ in ^{208}Pb with the SkI interaction to the giant quadrupole state predicted at 11.0 MeV. This state as given by the RPA calculation exhausts 70% of the energy-weighted isoscalar sum rule.

.. $T = 0$ octupole transition density $(\times r^2)$ in ^{208}Pb from ground to the collective state predicted at 2.7 MeV with the SkI interaction. The RPA calculation gives this state 17% of the energy-weighted isoscalar sum rule.

COMPRESSION MODE

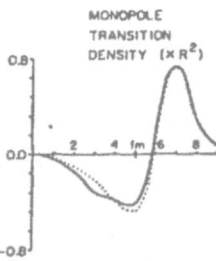

. $T = 0$ monopole transition density $(\times r^2)$ in ^{208}Pb, between ground and the 20 MeV giant monopole state. The solid curve is the calculation with the SkI interaction; the dotted curve is the prediction of the Tassie model, normalized for a visual fit.

FIG. 6 : reproduced from ref. I.17.

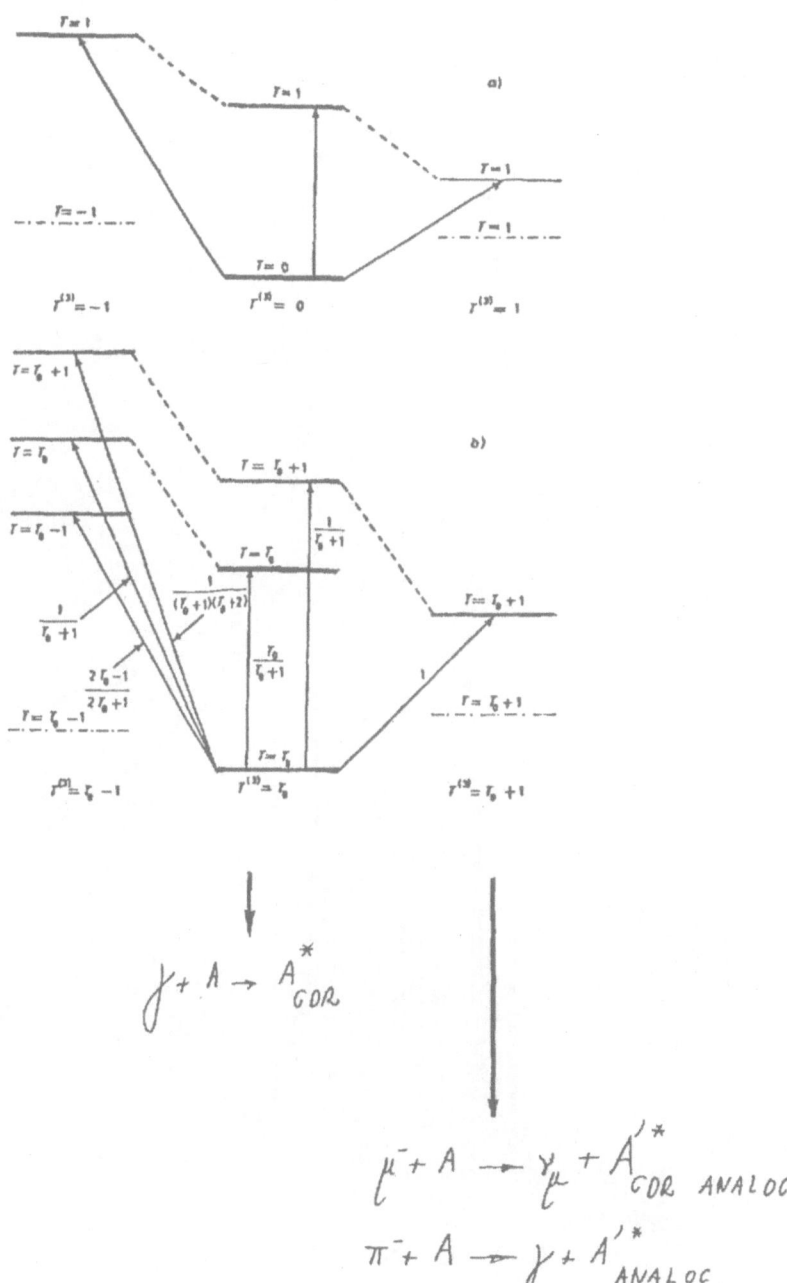

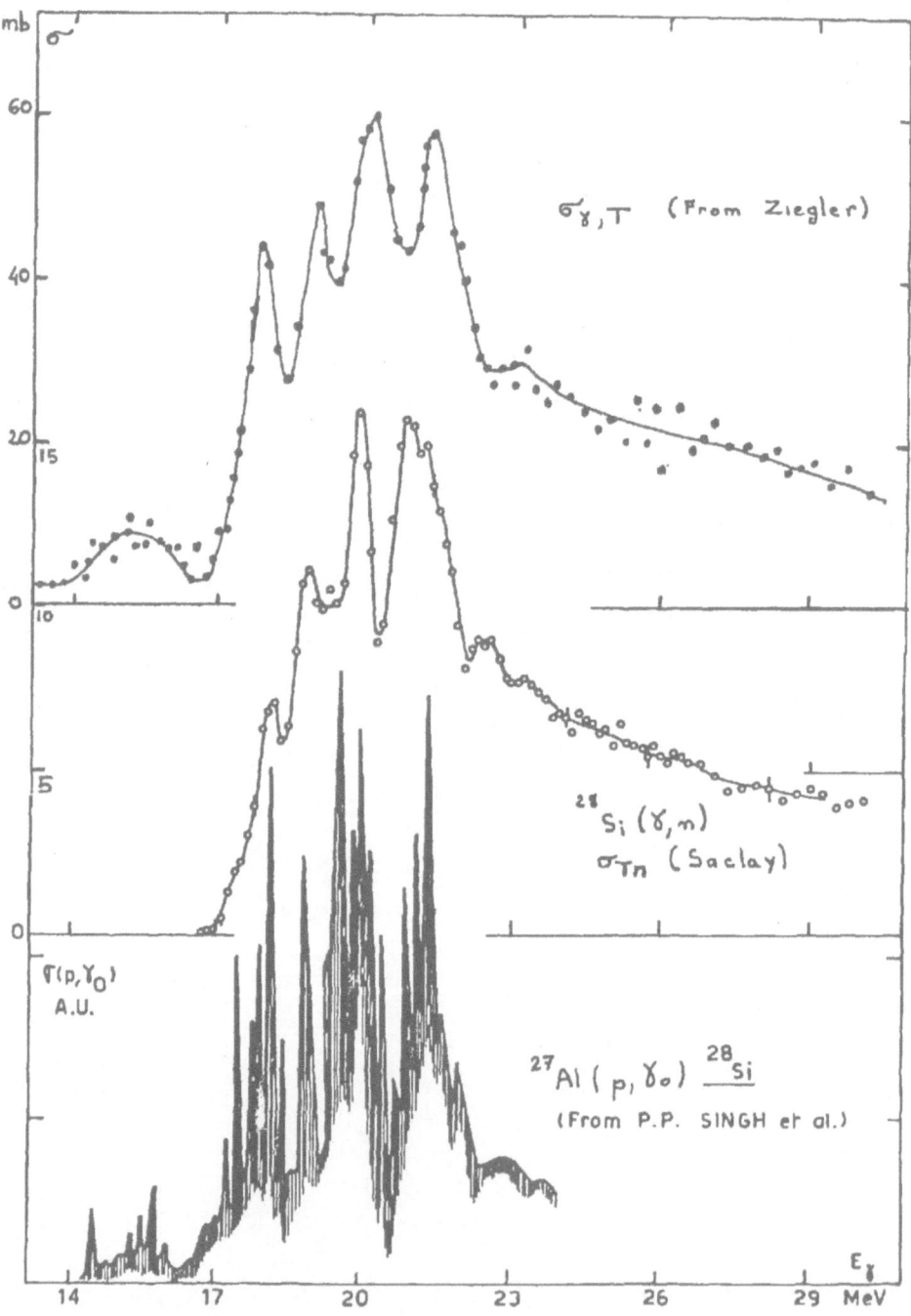

FIG 7.

· FIG 8   ref [I.18]

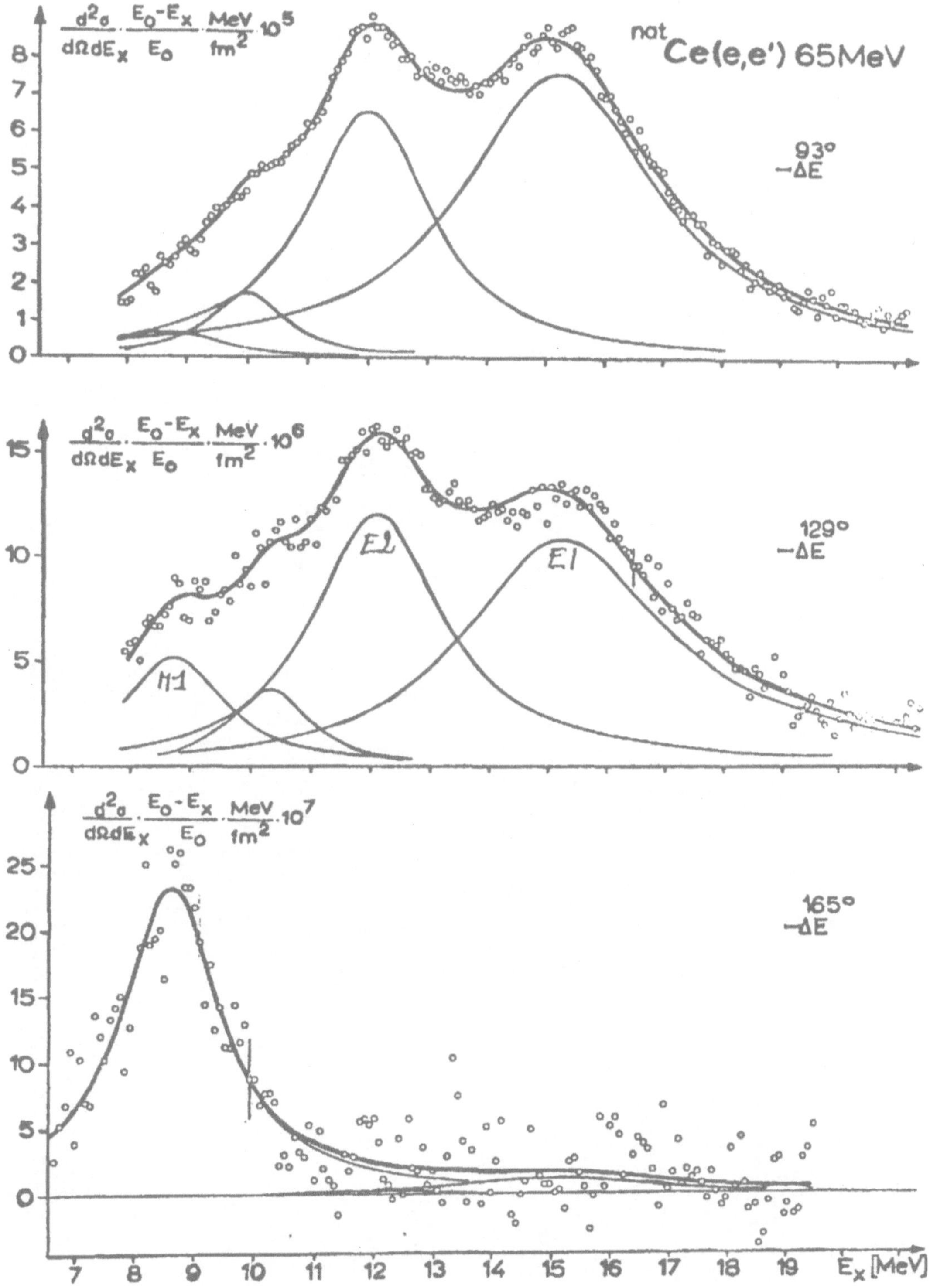

$$\frac{d^2\sigma}{d\Omega dE_x} \cdot \frac{E_0-E_x}{E_0} \cdot \frac{MeV}{fm^2} \cdot 10^5$$

$^{nat}Ce(e,e')$ 65MeV

93°
—ΔE

$$\frac{d^2\sigma}{d\Omega dE_x} \cdot \frac{E_0-E_x}{E_0} \cdot \frac{MeV}{fm^2} \cdot 10^6$$

E2

E1

M1

129°
—ΔE

$$\frac{d^2\sigma}{d\Omega dE_x} \cdot \frac{E_0-E_x}{E_0} \cdot \frac{MeV}{fm^2} \cdot 10^7$$

165°
—ΔE

7 8 9 10 11 12 13 14 15 16 17 18 19 E_x [MeV]

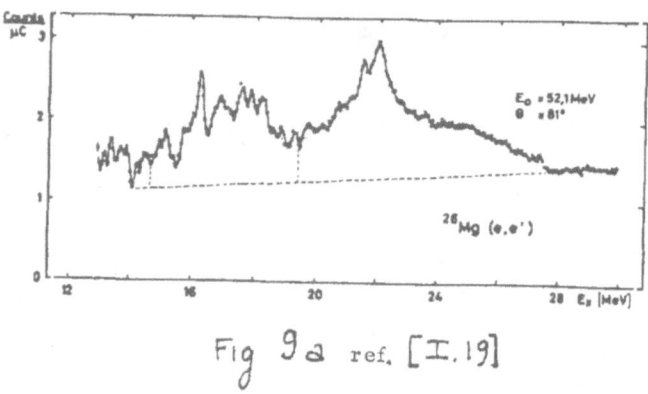

Fig 9a ref. [I.19]

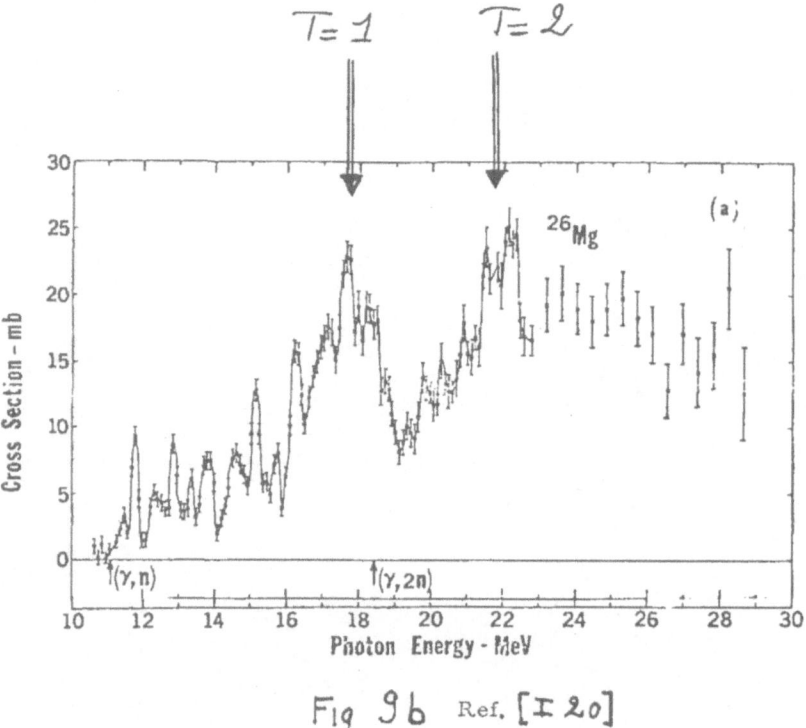

Fig 9b Ref. [I 20]

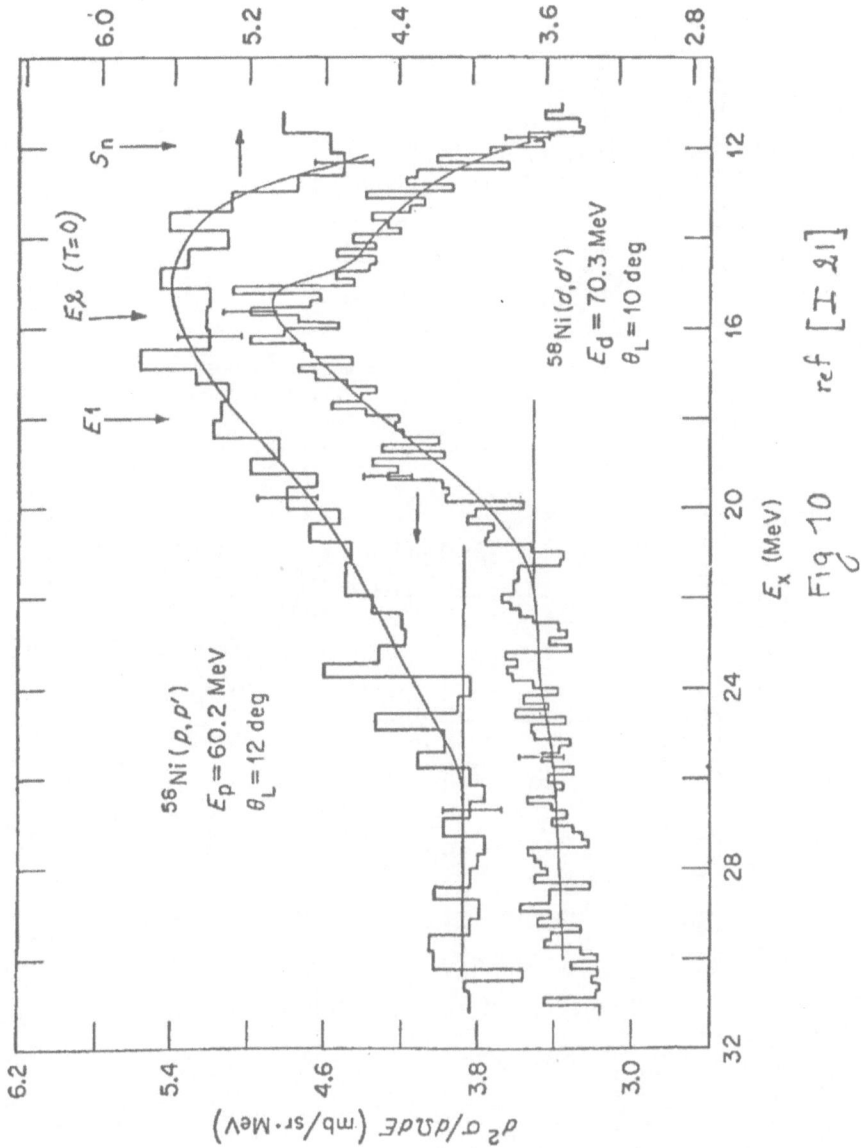

Fig 10   ref [I 2,1]

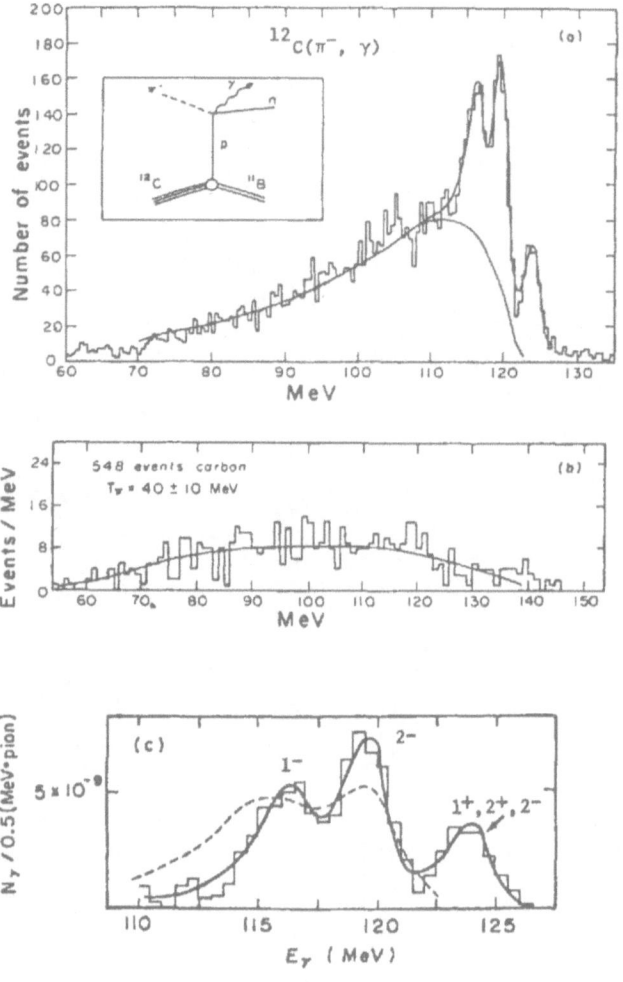

XBL 733-307

Fig 11   ref [I. 27]

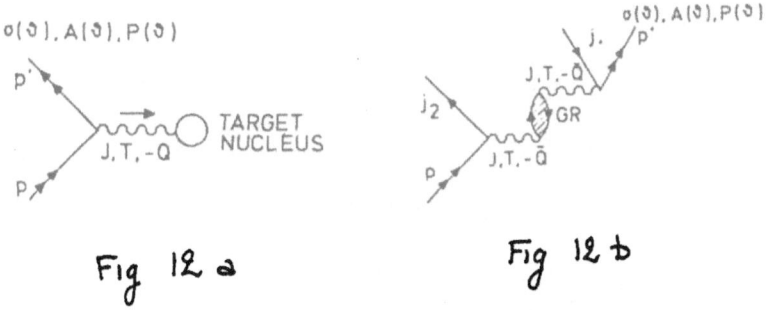

Fig 12 a   Fig 12 b

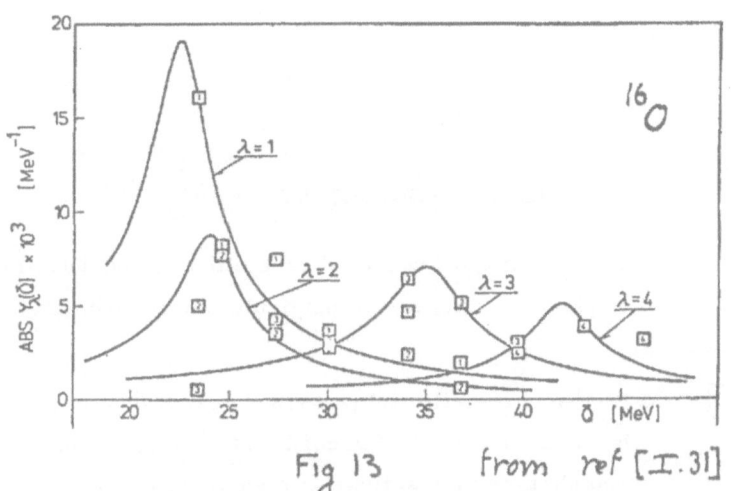

Fig 13   from ref [I.31]

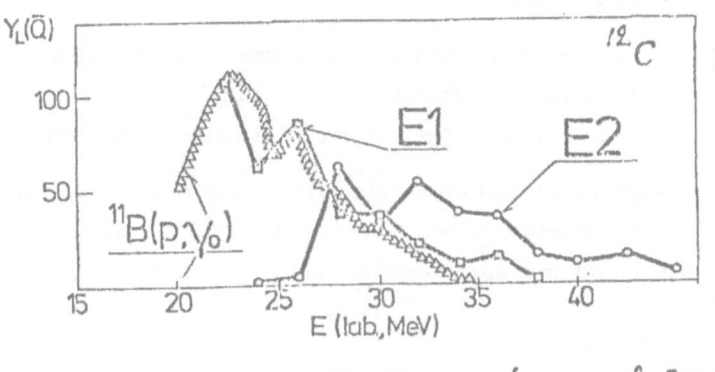

Fig 14   from ref [I.32]

CHAPTER II

THE PHOTON BEAMS

Several review papers, of which two are cited below, have been specially devoted to photon beams and in particular monochromatic photon beams.

[II. 1]   " Monochromatic and polarized beams of γ quanta by G. Diambrini in Proceed. Int. Symp. on electrons and photons interactions at high energy, Stanford September 1967.

[II. 2]   Survey of monochromatic photon capabilities, by C. Schuhl, in Proceed. Int. Conf. on photonuclear reactions and applications, Asilomar March 1973.

In this chapter, special emphasis will be paid to some relevant photon beam characteristics, which make a particular photon probe more suitable than others for measurements of some specific G. D. R. parameters.

I - PHOTONS PRODUCED IN NUCLEAR EXCITATION -

I. A - <u>Proton capture monochromatic photon sources</u> -

The $^{7}_{3}$ Li (p, γ) $^{8}_{4}$ Be radiative proton capture reaction is an important source of " monochromatic " photons of energy E_{γ} = 17. 6 MeV. This
reaction produces hard photons of energy E_{γ} = Q + E_{p} which is a radiative transition directly to the ground state of ^{8}Be. It also produces photons
of E_{γ} = (Q - E_{ex}) + E_{p} to an excited state of ^{8}Be. In these expressions
Q = 17. 2 MeV, E_{ex} = 2. 95 MeV and E_{p} is the incident proton energy in
center of mass coordinates so that the actual proton energy (E'_{p}) in
laboratory coord. is E_{p} ($\dfrac{M_{p} + M_{Li}}{M_{Li}}$) where M_{p} and M_{Li} are the proton
and Li masses respectively. This process shows a strong resonance at
E_{p} = 0. 37 MeV (i. e. E'_{p} = 0. 44 MeV) whose measured width is approximately 12 keV and which generates photons of E_{γ} = 17. 6 MeV contaminated
with a broad band of photons centred around E_{γ} = 14. 8 MeV (fig. 1).
Estimates indicate $\left[\text{II. 3}\right]$ that, even with a relatively thick ($\sim$ 500 keV)
Lithium target, one can only hope to obtain approximately 10^{-10} monochromatic photons of about 20 MeV per proton per unit solid angle

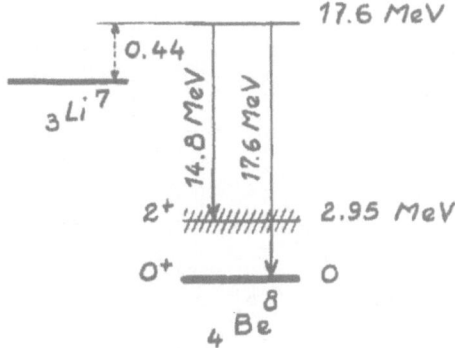

FIG. 1

Moreover, as E_p increases the obtainable photon intensity drops sharply so that at $E_p \simeq$ 900 keV, only 5% of the intensity at E_p = 370 keV is still available. The relative intensity between the 17.6 MeV and 14.8 MeV photons also changes roughly by a factor of four for the same change in E_p. Low thresholds for competing interactions, such as E_p = 1.63 MeV for the (p, n) channel, further limit the proton energy range and hence the corresponding photon range of this particular photon source. Of course since the photon energy of 17.5 MeV is practically invariable, the photonuclear data that one can get with such a beam are useful only if some systematics can be achieved for several targets or for several output channels. As an example let us quote the photofission studies of ^{238}U studied with a flux of 10^6 γ/cm^2. sec (photons of 17.5 MeV) by J. L. Meason and P. K. Kuroda [II.4] . By radiochemical techniques they thus got the mass yield curve for 17.5 MeV photofission of ^{238}U (fig. 1 bis)

FIG 1 bis   (ref II.4)

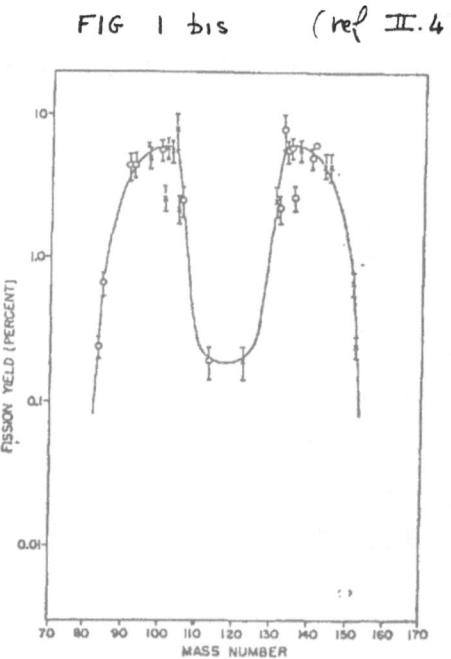

The mass-yield curve for 17.5-MeV photofission of U²³⁸

The $^{13}_{6}$C (p, γ) $^{14}_{7}$N and ^{27}Al (p, γ) ^{28}Si reactions have also been used as sources of monochromatic photons of 9.18 MeV and 12.33 MeV in experiments on the resonant absorption of photons [II.5, II.6] .

Another source of high energy monochromatic photons of the same type is the reaction $^{3}_{1}$H (p, γ) $^{4}_{2}$He. Since no bound excited states of ^{4}He seem to exist however, both the photon yield and the monochromatic photon energy are smoothly rising functions [II.7, II.8, II.9] of the proton energy E_p . The photon yield of emitted photons at a resonance emission angle $\theta' = 90°$ shows a broad maximum in the $E'_p = 5$ MeV region and drops off slightly at higher energies. But the small differential cross section of $\approx 10^{-29}$ cm^2 per steradian at $\theta' = 90°$ and $E'_p = 5$ MeV shows clearly that intensities will be rather small ($\simeq 10^3$ " useful " monochromatic γ rays per second for a proton current $\sim 15\mu$A). In particular, the fact that the threshold of the competing reaction $^{3}_{1}$H (p, n) $^{3}_{2}$He is to be found at only $E'_p = 1.019$ MeV will further limit the effective energy range of the emitted monochromatic photons. Nevertheless, precise experiment techniques using measurements of the induced radioactivity allowed Lochstet and Stephens to measure the ^{12}C(γ, n) ^{11}C giant resonance cross sections [II.10] using such a $^{3}_{1}$H (p, γ) $^{4}_{2}$He source in the range $21 \leqslant E_\gamma \leqslant 26.7$ Mev with a resolution of about 0.1 MeV at $E_\gamma = 22$ MeV and 0.2 MeV at $E_\gamma = 26$ MeV. Similarly, Del Bianco et al [II.11] measured the ^{50}Cr (γ, n) ^{49}Cr cross sections (fig. 2a) over the 20.43 $\leqslant E_\gamma \leqslant 22.22$ MeV energy range with a photon energy resolution of 110 keV over the whole range (fig. 2b).

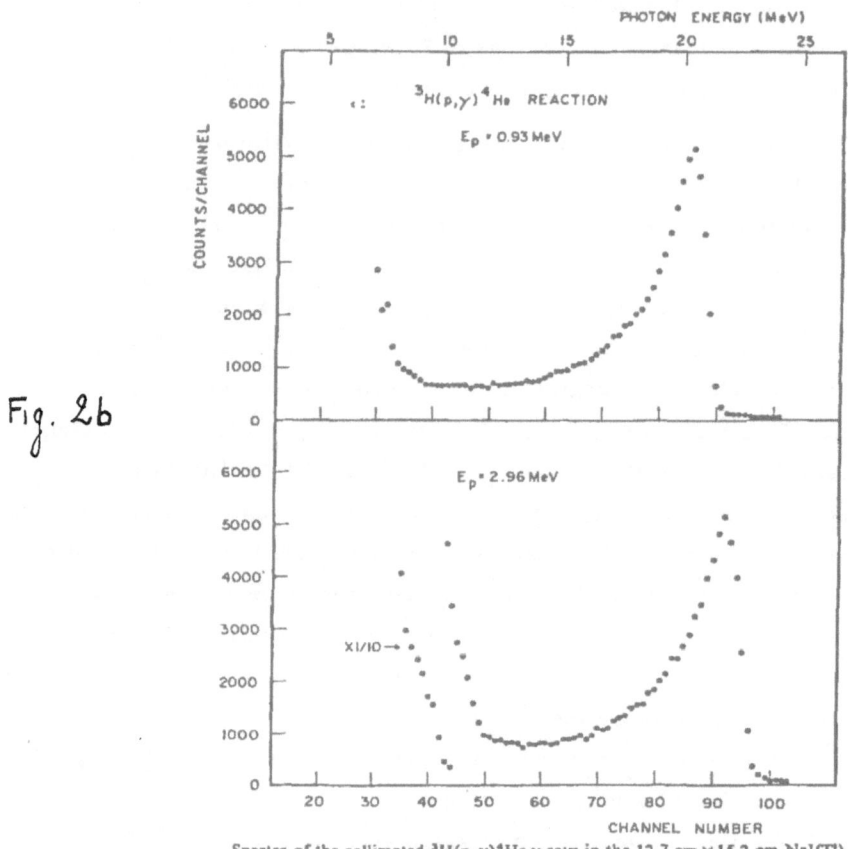

Fig. 2b

Spectra of the collimated $^3H(p,\gamma)^4He$ γ-rays in the 12.7 cm × 15.2 cm NaI(Tl) crystal.

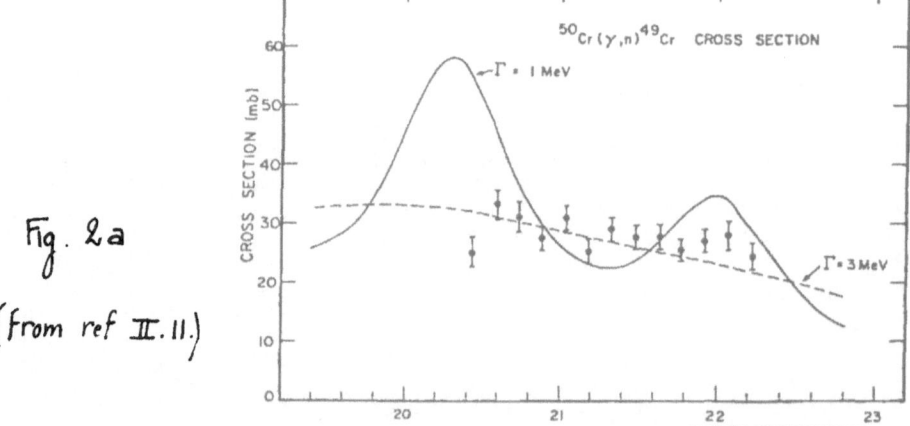

Fig. 2a

(from ref II.11.)

The $^{50}Cr(\gamma, n)^{49}Cr$ cross section. The points are the results of this experiment.

I. B - Neutron capture monochromatic photon sources -

I. B. 1 - Photon sources with discrete energies.

The kinematics of radiative neutron capture are essentially the same as those previously discussed for radiative proton capture and both recoil and Doppler effects must be taken into account in the precise evaluation of the observed electromagnetic transitions. The widths of neutron capture photon rays is, in general, only a few ev and is determined largely by the Doppler width $\Delta = E_0 \left[\frac{2 k T}{M c^2} \right]^{1/2}$.
If one considers the 7.64 MeV line from Fe (n, γ), one can obtain approximately 10^8 photons cm^{-2} sec^{-1} at the target for a thermal neutron flux of roughly 2×10^{13} neutrons cm^{-2} sec^{-1} near the source.

However since these γ energies are generally limited below 10 MeV these sources were used only for some (γ, n) and (γ, γ) experiments at the very threshold of the GDR by D. J. Donahue et al [II. 12] . To day they are also used for studying Delbruck and nuclear Raman scattering. As an interesting example, let us just quote the threshold photoneutron and photofission studies by O. Y. Mafra [II. 13] et al . Using the discrete γ lines shown in Table I.

TABLE I

Targets employed, principal γ-ray energies flux incident on the samples

Element	Energy (MeV)	ϕ(y/cm² sec)
^{33}S	5.43	$(2.6 \pm 0.2) \times 10^4$
^{89}Y	6.07	$(7.3 \pm 0.7) \times 10^3$
^{40}Ca	6.42	$(6.8 \pm 0.7) \times 10^3$
^{48}Ti	6.73	$(7.7 \pm 0.6) \times 10^4$
^{9}Be	6.83	$(8.5 \pm 1.1) \times 10^3$
^{55}Mn	7.23	$(3.5 \pm 0.4) \times 10^4$
^{207}Pb	7.38	$(2.8 \pm 0.3) \times 10^3$
^{56}Fe	7.64	$(2.8 \pm 0.3) \times 10^4$
^{27}Al	7.72	$(1.4 \pm 0.1) \times 10^4$
^{64}Zn	7.88	$(1.1 \pm 0.1) \times 10^4$
^{63}Cu	7.91	$(2.8 \pm 0.3) \times 10^4$
^{55}Ni	9.00	$(1.5 \pm 0.1) \times 10^4$

they obtained the σ (γ, n) and σ (γ, f) points of Fig. III a and Fig. III b.

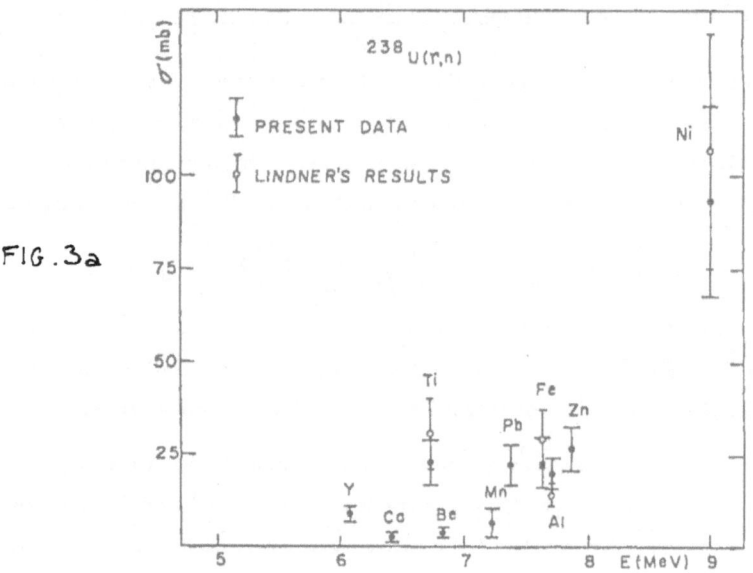

FIG. 3a

Photoneutron cross sections of ^{238}U compared with Lindner's results. Element symbols indicate the sources of neutron capture γ-rays whose energies are listed in Table 1.

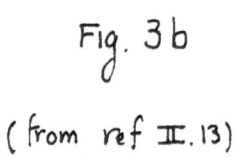

Fig. 3 b

(from ref II. 13)

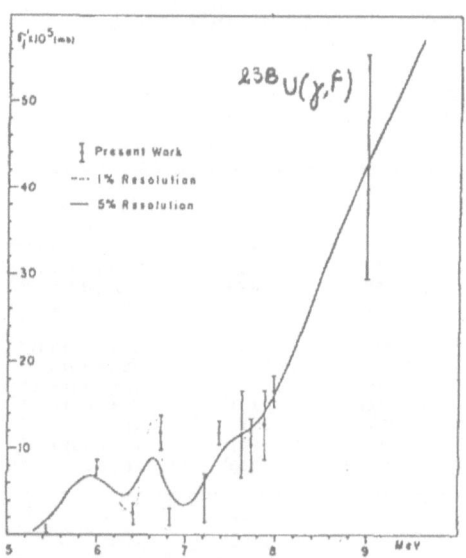

. Photofission cross sections folded with 1 % and 5 % resolution functions.

It should be mentionned that R. Moreh et al $\left[\text{II. 14}\right]$ where able to produce a 11.4 MeV monoenergetic photon beam from the ^{59}Ni (n, γ) thermal neutron capture, where the ^{59}Ni target itself was created as a by product of the original ^{58}Ni (n, γ) reaction from a prolonged ($\approx$ 6 years) neutron irradiation of natural nickel in the Israel Research Reactor .

I. B. 2 - Compton scattering of neutron capture γ-rays .

When a photon of energy E_0 is Compton scattered through an angle α, its energy is given by

$$E = E_0 \left[1 + \left(E_0 / mc^2 \right)\left(1 - \cos \theta \right) \right]^{-1}$$

which is the Compton scattering equation. This equation is used to relate the energy spread in the scattered beam ΔE , to the geometric angular spread $\Delta \theta$.

$$\Delta E = -(E^2/mc^2) \sin \theta \, \Delta \theta$$

where E_0 = energy of incident radiation.

   E = energy of scattered radiation at α.

   θ = Compton scattering angle.

   mc^2 = rest mass energy of electron.

One can use a neutron capture γ ray source associated with a curved scatterer as shown in Fig. 4a and 4b $\left[\text{II. 15}\right]$

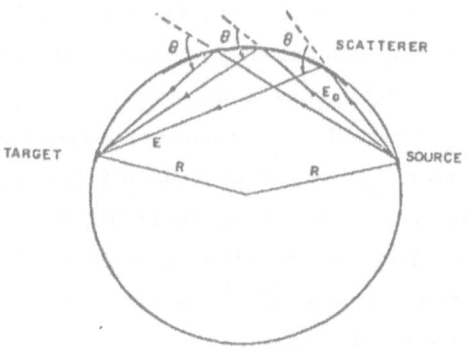

Focusing principle of a Compton scattering facility.

FIG 4a

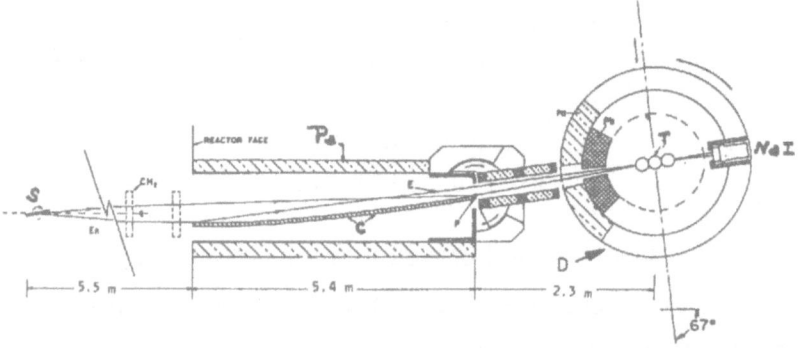

The variable energy γ-ray monochromator consists of a nickel source S in a water-cooled jacket and a curved aluminium scatterer C enclosed by Pb and boron-paraffin (Pa) shielding. Target T, fission ion chambers and NaI radiation detector are enclosed in shielding drum D of lead and boron paraffin. The shielding drum, target and detector move as a unit at an angle of 67° to the direction of the unscattered radiation from the source. The drum rotates about pivot P so that the target face remains normal to the direction of the incident beam. Boron loaded paraffin plates CH₂ inside the reactor shielding filter out slow and epithermal neutrons.

FIG 4b .   (ref II.16)

In a recent paper $\left[\text{II. 16}\right]$. Knowles then describes measurements of photofission cross sections of ^{232}Th, ^{238}U and ^{235}U between 5 and 8.3 MeV once again at the very threshold of the GDR of fissile nuclei. In this set up, a slab of nickel, S, located in the thermal column of the NRU reactor, Chalk River, provides a spectrum of sharply defined photons predominantly from the ^{58}Ni (n, γ) ^{59}Ni reaction. Radiation of variable energy $E = E_0\left[1 + \left(E_0/_{mc^2}\right)\left(1 - \cos\theta\right)\right]^{-1}$ is incident on the target from all parts of the curved aluminium plate scatterer if the point source, point target and line scatterer are located on the circumference of a circle. The energy of the radiation, incident on the target, is changed by changing the target position relative to the source and scatterer. The overall $(\Delta E/_E)$ value is determined by the widths of the source, target and scatterer. For example, between 5 and 8 MeV one obtains about ($\Delta E/_E$) = (F. W. H. M.) = 3% for a target width of 1.2 cm but this increases to $\approx$ 4 and 5% respectively for 1.7 cm and 3.8 cm wide targets. Energies E and intensities I of the more intense groups of ^{59}Ni neutron capture photon-rays [II. 16, II. 17, II. 18] emitted by the source S relative to the 9 MeV photons are listed in Table II.

Energies E_{nm} and relative intensities α_{nm} of source γ-rays and of Compton-scattered radiation

Source γ-rays[a])		Radiation scattered at angle θ_m (rad)					
		$\theta_1 = 0.100$		$\theta_2 = 0.167$		$\theta_3 = 0.250$	
E_n (MeV)	α_{n0}	E_{n1} (MeV)	α_{n1}	E_{n2} (MeV)	α_{n2}	E_{n3} (MeV)	α_{n3}
9.00	1.00	8.26	1.00	7.25	1.00	5.81	1.00
8.50	0.50	7.87	0.50	6.94	0.50	5.62	0.51
8.11[b])	0.17	7.52	0.16	6.67	0.17	5.43	0.17
7.81	0.26	7.25	0.26	6.45	0.26	5.29	0.27
7.54	0.17	6.99	0.17	6.25	0.17	5.15	0.18
6.84	0.33	6.41	0.32	5.78	0.33	4.83	0.35

The number of photons incident on the target for a thermal neutron flux of 3×10^{12} neutrons cm^{-2} sec^{-1} at the source S in this particular facility varies from 10^6 photons MeV^{-1} cm^{-2} sec^{-1} for E = 3 MeV to about 10^4 photons MeV^{-1} cm^{-2} sec^{-1} for 8 MeV photons.

Fig. 5a and 5b show how such a beam with a continuously variable energy improves the photonuclear data with respect to the case where one uses only discrete γ lines $\left[\text{II, 16, II. 19, II. 20}\right]$.

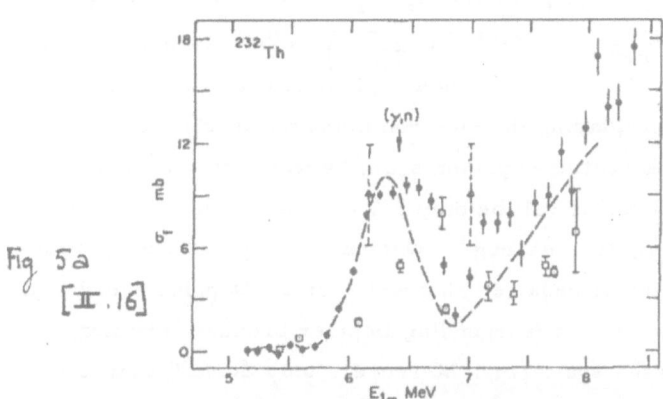

Fig 5a
[II.16]

Absolute photofission cross-section measurements σ_f of ^{232}Th as a function of γ-radiation energy E_{1m}. The error bars, estimated from the statistical count, indicate relative precision only. The photoneutron separation energy is indicated by (γ, n). The broken line is the corresponding photofission cross-section spectrum of Rabotnov ; the points $\square$ and $\triangle$ are measurements of Manfredini and Huizenga respectively.

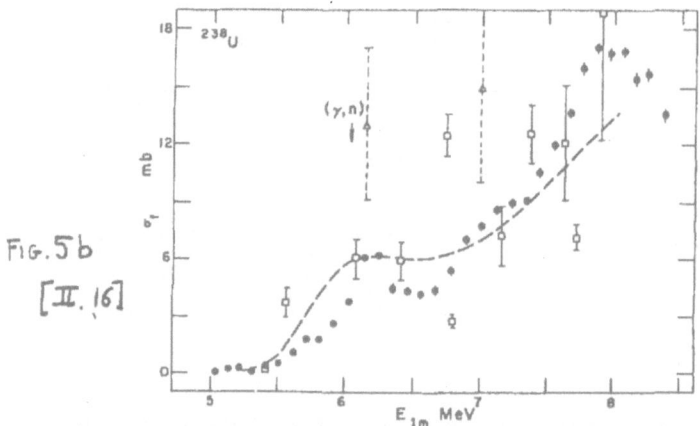

FIG. 5 b
[II. 16]

Absolute photofission cross-section measurements σ_f of ^{238}U as a function of E_{1m}. The broken line represents measurements of Rabotnov , the points $\square$ and $\triangle$ are measurements of Manfredini and Huizenga respectively.

I. B. 3 - <u>Nuclear resonance scattering of neutron capture γ-rays.</u>

A tremendous improvement in energy resolution has recently been achieved by R. Moreh [II. 21] . The basic equation for the energy E of a scattered photon is $E = E_o \left[1 + \left(E_o / Mc^2 \right) \left(1 - \cos \theta \right) \right]^{-1}$ and it follows that $\Delta E = \left(- \frac{E_o^2}{Mc^2} \right) \sin \theta \Delta \theta \approx \left(- \frac{E_o^2}{Mc^2} \right) \sin \theta \Delta \theta$ which is identical to the previously discussed Compton equations with M, the mass of the recoiling nucleus, replacing the electron mass m. It follows that the total energy variation of the scattered photon beam between $\theta = 0°$ and $180°$ for the 7.28 MeV photon line will be only 547 eV thus giving an average energy variation of $\left(\frac{\Delta E}{\Delta \theta} \right) \approx 3$ eV deg^{-1}. One can thus get an energy resolution $\frac{\Delta E}{E} \leqslant 10^{-6}$. But it requires an overlap of the Doppler broadened incident photon line and the corresponding Doppler broadened nuclear level which performs the scattering. Moreover, only those " source - scatterer " combinations which yield high scattering intensities will have a certain usefulness in practice (Fe-Pb for example at 7.28 MeV). Such a γ beam is of course useless for a study of the collective GDR and has been mostly used for the study of the spacing of bound levels in stable isotopes around 7 MeV.

I. C - <u>Nuclear resonance scattering of Bremsstrahlung photons</u> -

The possibility of creating monochromatic photons by means of resonance scattering of a continuous Bremsstrahlung spectrum [II. 22] was suggested by Schiff. He also pointed out that this type of experiment would meet with very serious background problems. The first successful experiments were performed [II. 23, II. 24] by E. Hayward and E. G. Fuller with the 15. 1 MeV excited state of ^{12}C. These earliest experiments however, were meant to demonstrate the feasibility of observing resonance fluorescence from isolated bound states by means of Bremsstrahlung photons rather than produce usable monochromatic photon beams as such. A few typical examples of such resonance scattering experiments [II. 25,] [II. 26, II. 27] can be found in the literature.

However, the idea of producing a " beam " of monochromatic 15. 1 MeV plane polarized photons, by scattering a continuous Bremsstrahlung spectrum through 90° from a carbon target, as a means of investigating the giant resonances of heavy nuclei, was proposed [II. 28] in 1967 by H. Arenhövel and E. Hayward. A similar scattering level at 11. 4 MeV in ^{28}Si was also suggested. In subsequent papers [II. 29, II. 30] , E. Hayward then reported data concerning the observation of such nuclear scattering effects by means of just such a monochromatic plane polarized incident photon beam.

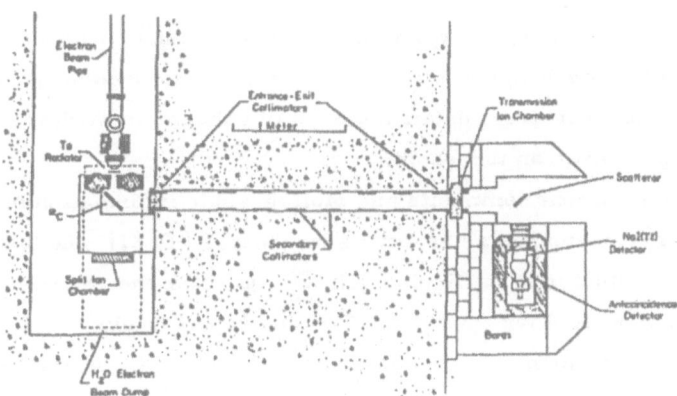

FIG. 6 .

The physical arrangement of these experiments is shown in fig. 6. This particular set-up produces approximatively 10^4 monochromatic, plane-polarized (95%) 15.1 MeV photons sec $^{-1}$ at the second scattering target for roughly 50 μA of 20 MeV electrons incident on the tantalum radiator The counting rates in the NaI detector arrangement, uncorrected for backgrounds, never exceeds a few counts per hour. E. Hayward [II.29] used such a γ beam to study the predictions of the dynamic model which imply a coupling of the GDR to the quadrupole oscillations of the nuclear surface.

II - BREMSSTRAHLUNG PHOTON BEAMS -

From 1950 to 1960 the most widely used photon beams were bremsstrahlung photon beams, which easily cover the GDR energy region. However one knows that the large drawback of such beams lies in their continuous nature. But, methods for extracting or identifying narrow or sufficiently narrow energy bands from such primary continuous photon spectra, exist. Some of them shall be briefly discussed below, in connection with the particular photonuclear experiments for which they are particularly suited.

II.A - Use of the high energy tip of pulsed bremsstrahlung photon spectra.

In recent years, the quasi-monochromatic high-energy tip of an appropriately chosen Bremsstrahlung spectrum has become a popular means of probing in great detail the band of unbound individual states just above the photoneutron threshold.

For pertinent comments and bibliographic information on the status of low-energy photonuclear reactions in general [II.31] and threshold photoneutron studies in particular [II.32] we refer the reader to the review articles of F.W.K. Firk and H.E. Jackson respectively.

The basic principles of the threshold technique, in its simplest form are shown schematically in fig. 7. The major attraction of this type

of experiment is the possibility of studying the photoabsorption process
near threshold with the extremely high resolution characteristic of neutron
time of flight measurements [II. 31, II. 32]. The target of interest is
bombarded by an intense pulsed Bremsstrahlung beam of very precisely
defined maximum energy.

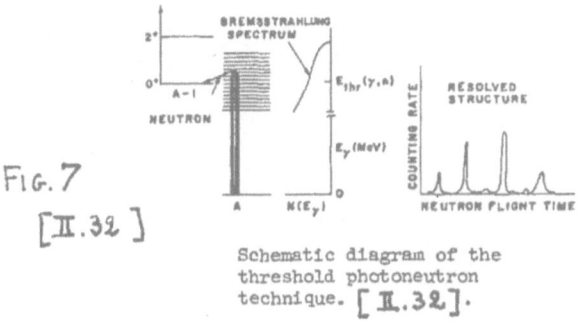

FIG. 7
[II. 32]

Schematic diagram of the
threshold photoneutron
technique. [II. 32].

If this energy is less than the sum of the neutron binding energy and the
energy of the first excited state of the daughter nucleus, then photoexcitation
of the levels just above the (γ, n) threshold, allows one to identify unambiguously all resonances seen in the neutron energy spectrum, as representing levels in the target nucleus having an excitation energy equal to the
neutron binding energy plus the neutron energy. One can thus study the
individual levels which are close to the photoneutron threshold and whose
widths are smaller than the average spacing. For each observed resonance,
the resonance yield is proportionnal to

$$\frac{\Gamma_{\gamma_0} \Gamma_n}{\Gamma}$$

with Γ_{γ_0} = ground state radiation width (proportionnal to the area below
the resonance if $\frac{\Gamma_n}{\Gamma} \simeq 1$);one can thus obtain , at the very threshold
of the GDR (Fig. 8a, 8 b) [II. 32], the photon strength function and by angular
distributions the spin and parities of each resonance.

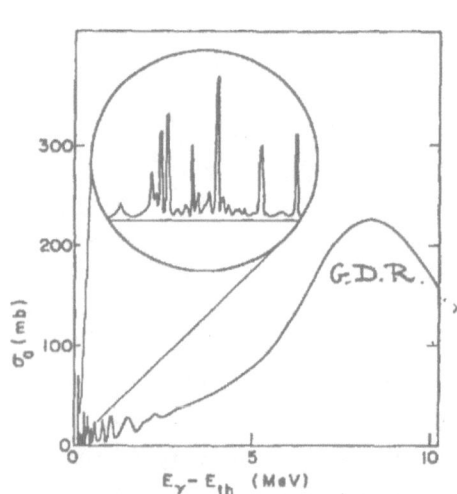

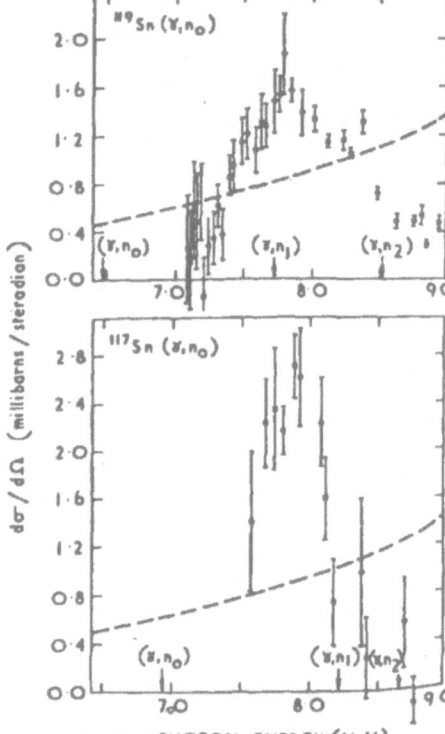

Schematic plot of the photon-absorption cross section in the region above the neutron threshold for a nucleus with A ≈ 60.

Cross section for $^{119}Sn(\gamma,n_0)$ and $^{117}Sn(\gamma,n_0)$. The dashed lines represent the extrapolation of Lorentz fits.

FIG. 8a

FIG. 8b.

(from ref II.32)

Assuming that the absorption cross-section is determined by the tail of the GDR, Axel [II. 33] found for the photon strength function a law:

$$\frac{\overline{\Gamma_{\gamma_o}}}{D} \simeq 6.1 \ 10^{-15} \ (E_{\gamma_{Mev}})^5 \ A^{8/3}$$

Bowman et al [II. 34] found 5 M 1 resonance in ^{208}Pb (with $\Sigma \ \Gamma_{\gamma_o}$ = 50 eV = 5 Weisskoph units). Just to show that this method is still very fruitful at the threshold of the GDR let us just quote that H. Jackson in recent polarization measurements, confirmed only 2 of these resonances as M1. [II. 61]

II. b - Total photoabsorption experiments with bremsstrahlung beams -

In such experiments one tries to measure as a function of E_γ the total photoabsorption

$$\sigma_{Tot} = \left[\sigma_{nucl} + \sigma_{elect} \right]$$

A later evaluation or computation of σ_{elect} should allow to get σ_{nucl} (E_γ). It turns out that such experiments can be envisaged because the final photon detector can identify the energy of the photons to some extent. The experiments are carried out with and without the studied target in the photon beam (fig. 9)

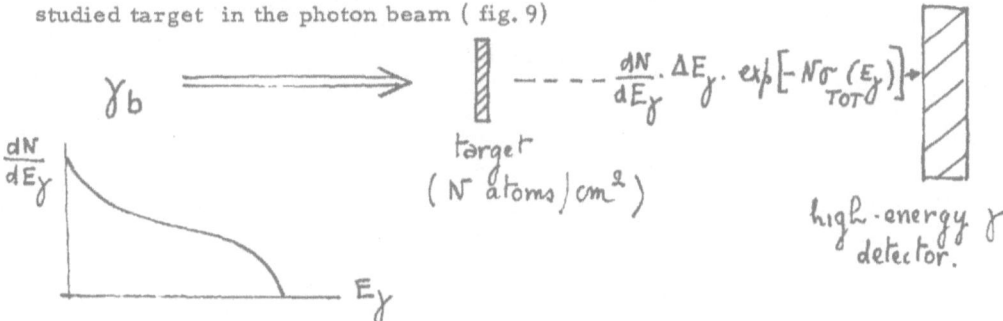

Such experiments can be classified according to the type of photon spectrometer used.

II. B. 1 - Magnetic pair spectrometer.

High resolution experiments ($\Delta E_\gamma \approx$ 120 keV at 10 MeV and 220 keV at 20 MeV) have been carried out at Moscow by L. E. Lazareva and her collaborators [II. 35] . Unfortunately the detection efficiency of such a device is fairly low ($\approx 10^{-6}$). Fig. 10 shows an example of the σ_{nucl} (Total) obtained .

Fig 10

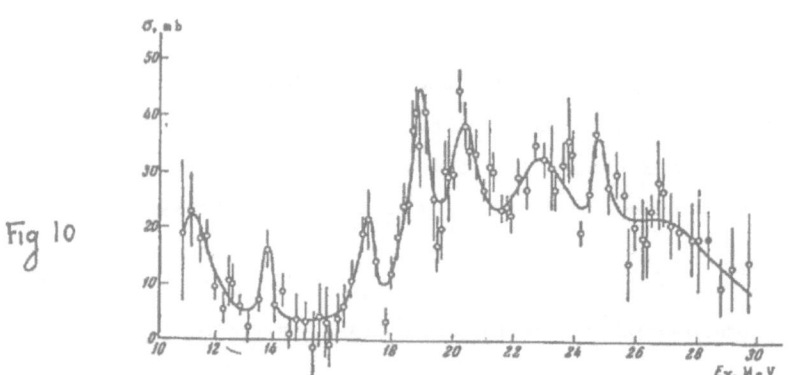

Nuclear absorption cross section of magnesium.

II. B. 2 - Compton spectrometer.

A single Compton spectrometer device was used at Ljubljana [II. 36, II. 37] with a better efficiency (10^{-4}) but a poorer resolution (500 keV at 20 MeV). Fig. 11 gives an example of the data obtained on ^{28}Si .

Fig. 11

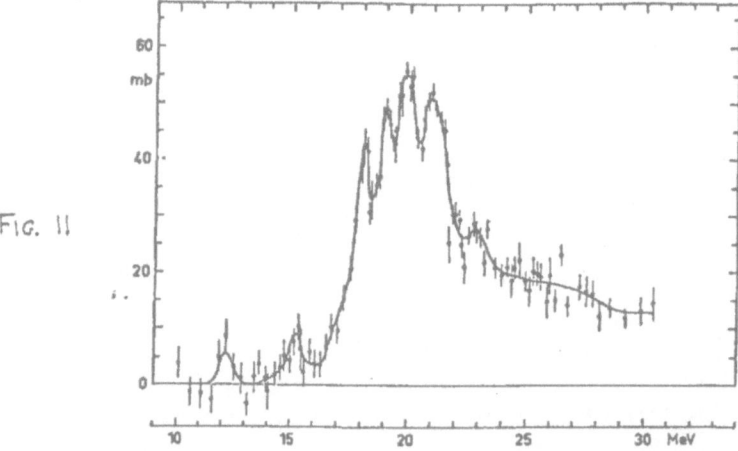

. Nuclear absorption cross section for silicon. The estimated error in the zero line is ± 1.5 mb.

II. B. 3 - <u>Spectrometer using INa crystals</u>.

With a single crystal total absorption spectrometer one can
obtain an efficiency $\geqslant$ 95% but with an energy resolution $\approx$ 10%
which is too large. To overcome this difficulty J. M. Wyckoff [II. 38]
used a two crystal scintillation pair spectrometers device which, through
coincidence requirements, finally provide a detection efficiency of 4%
with an energy resolution of 2 % . Fig. 12a, 12b show some of the raw
data and the σ_{nucl} which can be extracted from them as obtained
with this specific device at the National Bureau of Standard.

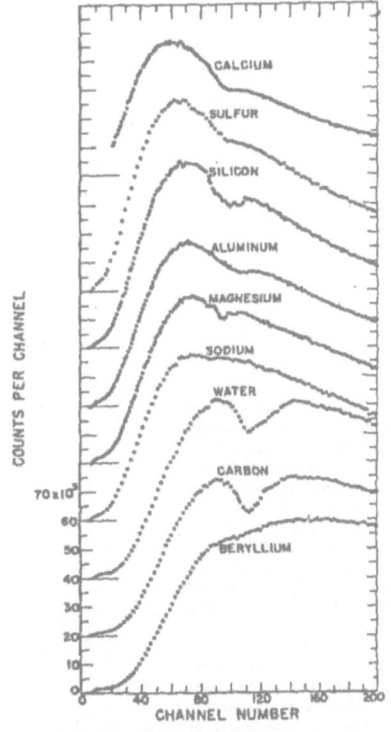

Pulse-height distributions for nine elements.

Successive elements have been displaced by 20 000 counts per
channel in order to separate these data.

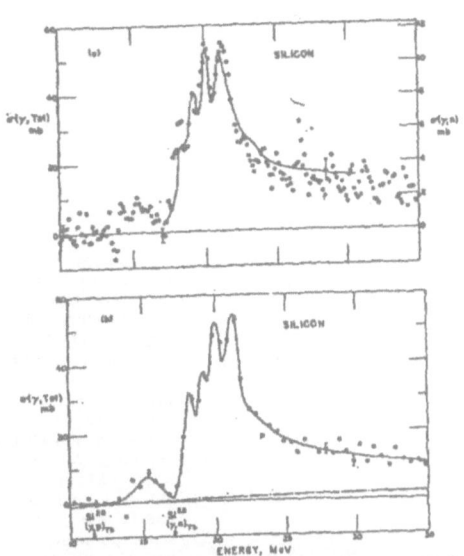

Silicon total photonuclear cross section. The solid line
on (a) represents the $\sigma(\gamma,n)$ data from the Livermore group shifted
up in energy by 250 keV using the right-hand ordinate scale.

FIG. 12 a [from ref II 38]   FIG. 12 b

II. B. 4 - <u>The Mainz double magnetic Compton spectrometer</u> .

A double magnetic Compton spectrometer arrangement, actually in operation at the Mainz Linac and described in detail in reference $\left[\text{II. 39}\right]$, is shown in fig. 13.

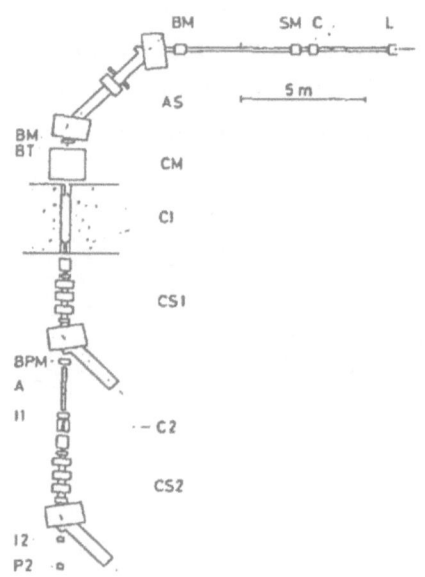

Experimental set-up for bremsstrahlung- and absorption measurements. The primary electron beam from the linear accelerator L is analysed and deflected by 90° in an achromatic beam handling system AS. The bremsstrahlung beam from the target BT (3 × 3 mm² Pt, 0.035 radiation lengths) enters the experimental hall through a tapered lead collimator Cl (4 mm Ø at the entrance into a 3 m thick concrete wall, 6.4 × 6.4 mm² at the exit). Two identical spectrometers CS1 and CS2 are placed before and behind the removable absorber A. Two crossed differential ion chambers which are used as a beam position monitor BPM in x- and y-direction are connected to two steering magnets SM at the end of the linac in a feedback loop in order to automatically stabilize the beam position. Two ion chambers I1 and I2 monitor the fraction of the beam passing through the second collimator C2. The chamber P2 measures the total radiant flux. C is an electron beam collimator, BM are beam current monitors, and CM is a cleaning magnet. The optimum length of absorber material is 2.6 attenuation lengths (e.g. 120 to 150 cm water).

FIG. 13 : reproduced from ref. II. 39.

This sophisticated version of a Bremsstrahlung spectrometer has been used for absolute total photonuclear absorption cross section measurements [II. 40] in the 10 to 300 MeV photon energy range. It can also be used for absolute photon flux measurements with an energy resolution of 1% and a statistical accuracy of roughly 1.5%. When used in relative photonflux and associated relative photon absorption cross-sections, accuracies of the order of 0.1% can be obtained. The experimental parameters of this particular spectrometer were chosen in such a way that ($\Delta E/_E$) $\approx$ 1% for an optimal photon detection efficiency $\mathcal{E}$ over as wide an energy range as possible . ($\mathcal{E}$ $\approx$ a few 10^{-5}).

In a typical experimental run, the first spectrometer serves as a beam monitor for normalizing the measurements. The number of photons hitting the converter foil is of the order of ($\emptyset$ (k)) $\approx 10^9$ photons of energy (k) per MeV, per sec. The efficiency η_i (k_0), for a white photon spectrum, is about one count per detector for 10^5 photons per MeV at k = 40 MeV. With these characteristics, one hour of running time is sufficient to obtain 11 energy points simultaneously with an accuracy of 0.5%. One should note that the limiting factor for these counting rates is not given by the available number of photons $\left[\emptyset (k)\right]$ but by the dead time losses of the counting system itself. Finally the high accuracy required by these transmission experiments made the use of a second Compton spectrometer as a normalizing instrument inevitable.

Fig. 14 shows how accurate σ_{nucl} (γ, total) data can be obtained with such a device.

FIG 14 (from ref II. 40)

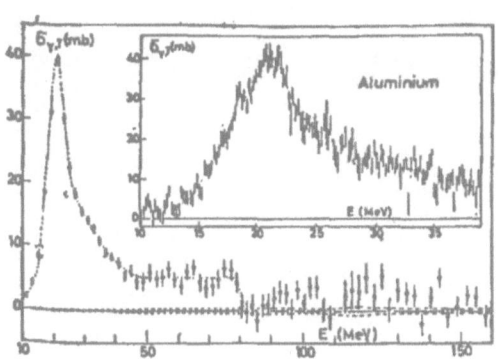

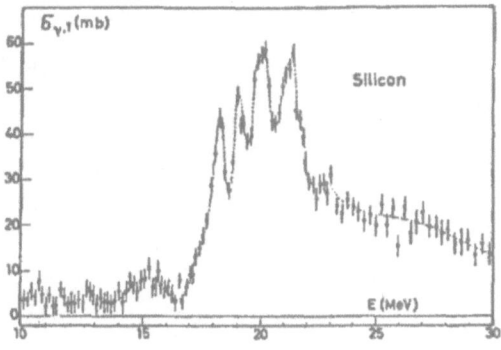

$$\sigma_{TOT} = \sigma_{NUC} + \sigma_{ELEC}$$

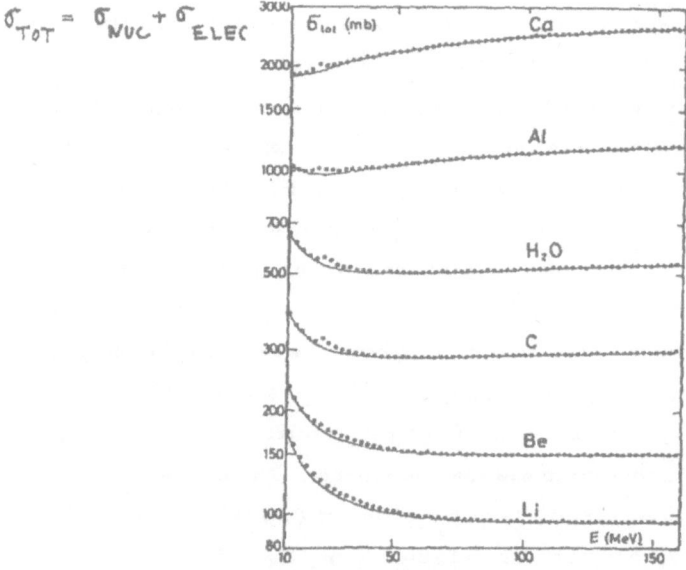

Measured total attenuation cross sections σ_{tot}. The error bars are smaller than the symbols. The solid lines give the calculated non-nuclear cross sections.

FIG. 15 : reproduced from ref. II.40

$$\sigma_{NUC}$$

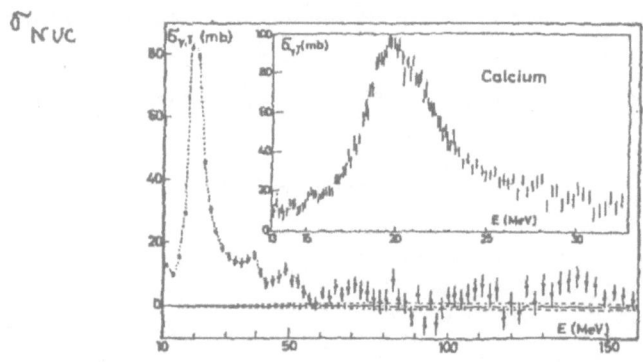

But once again one must emphasize that σ_{elect} is much larger than σ_{nucl} (cf. fig. 15) . For example for ^{28}Si $\frac{\sigma_{elect}}{\sigma_{nucl}} \approx \frac{1\ 100}{12} \approx$ 90 at 30 MeV and 20 at 20 MeV. One thus sees that especially on both sides of the GDR and when A increases, the final accuracy is provided more by the theoretical knowledge of σ_{elec} than by the experimental statistics.

II. C - " Tagged photon beams " .

Such a system is summarized in fig. 16 given by J. S. O'Connell et al $\begin{bmatrix} II.41 \end{bmatrix}$. A monochromatic beam of electrons with energy E_0 crosses a thin bremsstrahlung converter. When a bremsstrahlung photon with energy E_γ is radiated by one electron, the electron energy E_n becomes $E_0 - E_\gamma$ which is measured in the β magnetic spectrometer. The nuclear product which is then measured in coincidence with this ($E_0 - E_\gamma$) electron is then attributed to a photoreaction induced by a photon with energy E_γ. With such a device the counting rate is limited only by the chance coincidences and not by the intensity of the electron beam. This means that one needs a d. c beam associated to many electron counters such as M. W. P. C. With a superconducting d. c linac and an array of 100 electron counters one could hope to identify up to 10^8 γ/sec with an efficiency of 100% and a resolution of a few tens keV. Such a device is being built at Illinois University $\begin{bmatrix} II.42 \end{bmatrix}$. So far nuclear data have been obtained for the Giant Resonance at Illinois betatron by studying ^{197}Au (γ, γ') fig. 17 $\begin{bmatrix} II.41 \end{bmatrix}$ and at Illinois superconducting linac by studying ^{238}U (γ, f) below the giant resonance fig. 18 $\begin{bmatrix} II.43 \end{bmatrix}$.

In these 2 experiments 2.10^4 tagged monochromatic γ-rays struck the target per second.

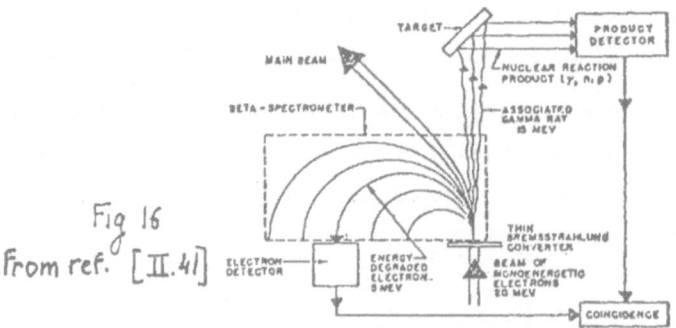

Fig 16
From ref. [II.41]

Schematic of Bremsstrahlung Monochromator.

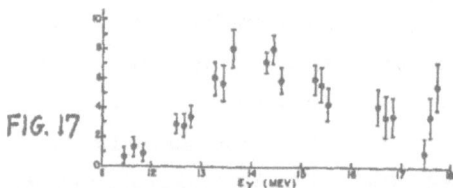

FIG. 17

Differential Scattering Cross Section of Au[197] at 135°.
The cross section includes both elastic scattering and high energy
inelastic scattering. The ordinate is the differential cross section in
units of 10^{-28} cm²/sr; the abscissa is the gamma ray energy in
Mev. The energy resolution varied from point to point but was
about 150 kev±30 kev. [II.41]

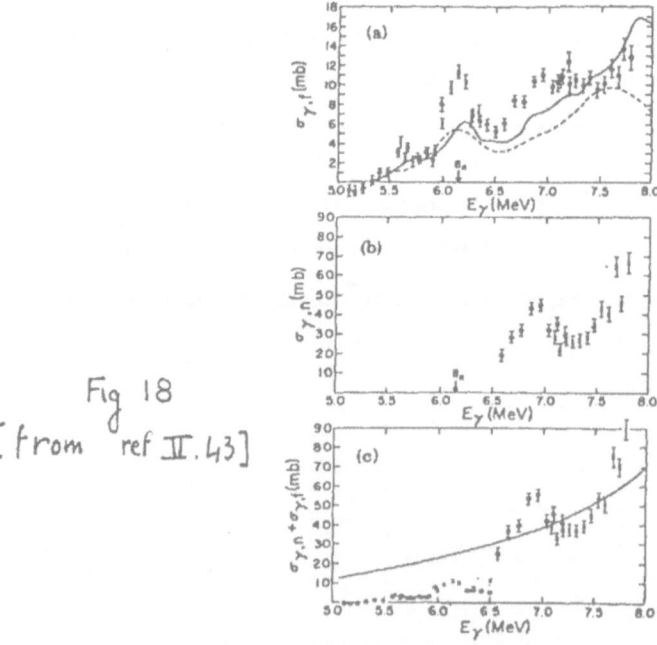

Fig 18
[from ref II.43]

Photon-induced cross sections for [238]U

II. D - <u>Unfolding of yield data obtained with bremsstrahlung spectra</u> .

From 1950 to approximately 1964 most of the photonuclear cross sections in the region of the GDR were obtained by using as an incident photon beam, a bremsstrahlung photon beam whose end energies E_i is increased step by step to produce a so-called yield curve Y_{exp} (E_i). To summarize the methods allowing to extract from a set of experimental yield points Y_{exp} (E_i) the photonuclear cross section σ(E_i) which is looked for, I will reproduce the approach used by E. Bramanis et al in their review paper " The analysis of photonuclear yield curves " [II. 44] .

In a typical bremsstrahlung experiment one measures the reaction yield $Y_{exp}(E_i)$ per unit response of a suitable beam monitor, where Ei is the energy of the electrons incident on the bremsstrahlung target. The true yield function Y_T (E) and the reaction cross section σ(k) are related by

$$Y_T(E) = A \int_0^E N_{exp}(E, k) . \sigma(k) . dk$$

where A is a constant containing target and efficiency factors and N_{exp} (E, k) gives the number of photons with energies k to k+ dk incident on the target per unit monitor response. It is related to the spectrum N (E, k) produced by an electron striking the bremsstrahlung target by

$$N_{exp}(E, k) = \frac{N(E, k) . f_s(k)}{M(E)}$$

where f_s (k) corrects for absorbing materials between target and sample and M(E), the monitor response function, normalizes the spectrum to unit monitor response. As it is customary, we define a " reduced " yield function Y (E) such that

$$Y(E) = \int_0^E N(E, k) . \sigma(k) . dk$$

with $\quad Y(E) = Y_T(E) . \dfrac{M(E)}{A}$

and $\quad\quad \sigma(k) = \mathcal{N}(k) . f_i(k)$.

Clearly one must solve the integral equation in order to get the useful photonuclear information σ (k) .

A first sophisticated method has been proposed by Penfold- Leiss [II. 45] , which leads to a matrix equation in which the various bremsstrahlung spectra are represented by a square matrix with elements $N_{ij}(E_i k_j)$. Most of the difficulties lie in an inversion of this matrix

$$ N^{-1} Y(E_i) \approx \sigma(k_i) $$

The straight matrix approximation is likely to produce excessive scatter in the cross section solution due to its failure to treat the (Y_i) as statistical information and to the effect of the approximate form used for N (E, k). The Cook Least Structure Method of reducing the measured yield function to the cross section, is an attempt to overcome some of these shortcomings and to produce a conservative (" least structure ") estimate of σ .

Actually these methods produced a lot of experimental data [II. 46, II. 47]. For example by means of such methods E. Hayward and E. G. Fuller [II. 47] could study the splitting of the GDR of permanently deformed nuclei (fig. 19) .

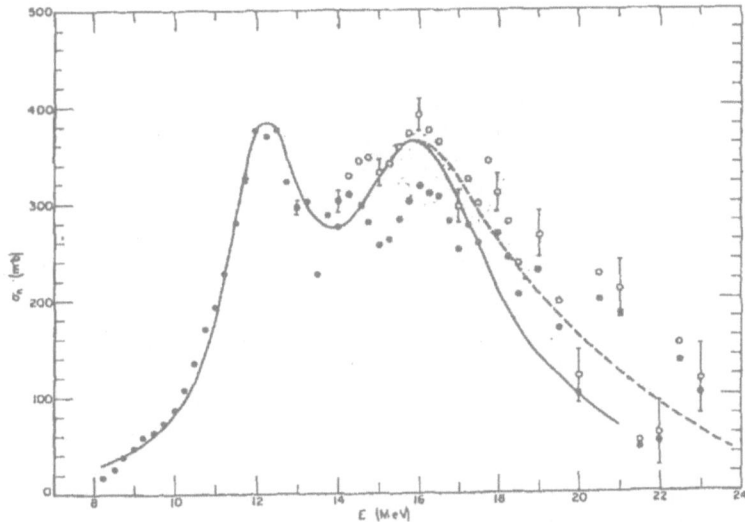

The sum of the neutron producing cross sections for holmium. The errors indicated again represent only statistical uncertainties. The two sets of points above 13.5 MeV correspond to the multiplicity correction factors A and B   The edges of the shaded region correspond to smooth curves drawn through these two sets of points. The solid curve is the sum of two Lorentz lines having the parameters $\sigma_a^o = 318$ mb, $E_a = 12.2$ MeV, $\Gamma_a = 2.33$ MeV, $\sigma_b^o = 328$ mb, $E_b = 16$ MeV, and $\Gamma_b = 4.5$ MeV. The dashed curve drawn above 16 MeV is a smooth curve drawn through the experimental points.

FIG. 19 : reproduced from ref. II.47

However one sees on such an example that another model is also necessary to get σ_k when several partial decay channels are open. For such heavy nuclei, the measured neutron yield leads, after unfolding, only to the knowledge of the quantity

$$\sigma(\gamma, n) + 2. \sigma(\gamma, 2n)$$

and a more or less statistical model is needed to connect $\sigma(\gamma, 2n)$ to $\sigma(\gamma, n)$ and thus get the interesting absorption cross section $\sigma(\gamma, n) + \sigma(\gamma, 2n)$.

The following data, provided by H.H. Thies and Spicer, show clearly the difficulties of the method [II.48] . Fig.20 a shows the rough neutron yield data for ^{159}Tb on which no structure at all can be seen. Fig. 20b shows the result of an unfolding procedure , $\sigma(\gamma, n) +$ 2 $\sigma(\gamma\ 2n)$. Fig. 20c shows the absorption cross section after correction for the neutron multiplicity.

FIG. 20 a

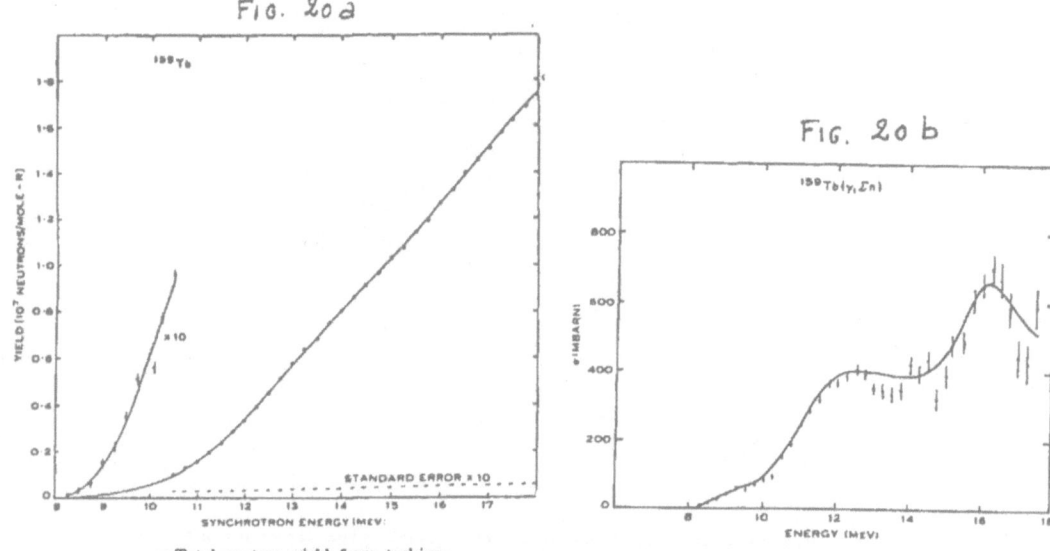

—Total neutron yield from terbium.

FIG. 20 b

—Total neutron production cross section for ^{159}Tb.

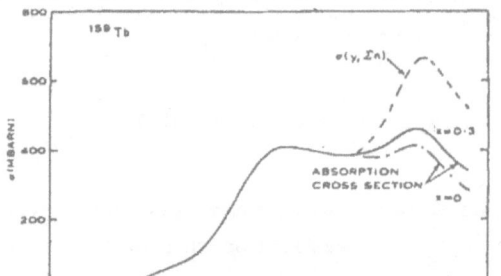

—Cross section for photon absorption in ^{159}Tb, using statistical theory of nuclear reactions to correct for neutron multiplicity. The factor x is a measure of the probability that a direct photoeffect will occur.

FIG. 20 c.

(from ref II 48)

But one clearly realizes that the main problem arises when one tries to extract too much information from a bremsstrahlung yield curve. One now knows that the usable information included in such a yield curve is intrinsically limited by the statistical errors which are amplified in a non random way by the chosen mathematical procedure [II. 45, II. 49] . A striking example arose when the structure found in the GDR of ^{141}Pr by bremsstrahlung techniques could not be confirmed by any measurements with monochromatic γ -rays (Fig. 21) [II. 50] .

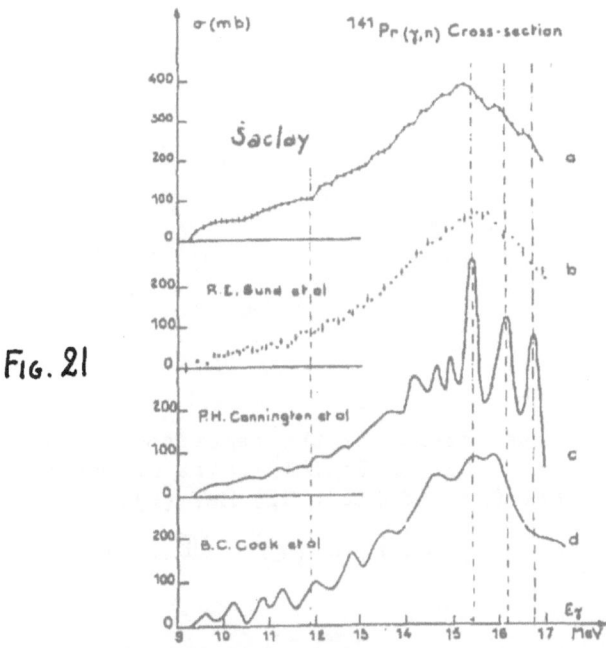

FIG. 21

Fig. 4. The σ(γ, n) cross sections for ^{141}Pr as observed by several authors. Curves a) and b) are obtained with monochromatic photons whereas c) and d) are obtained with bremsstrahlung

Systematic studies such as the one by Goriachev [II. 51, II. 52] summarized (fig. 22-23)showed later that when one tries to decrease the analyzing energy bins too much, it is the unfolding procedure itself which can generate a spurious structure . However since such a spurious generation is somehow connected to the size of the matrix to be inverted it is clear that, when one is studying a cross section close enough to its threshold, bremsstrahlung beams are indeed capable of providing good resolution (fig. 24) [II. 53] .

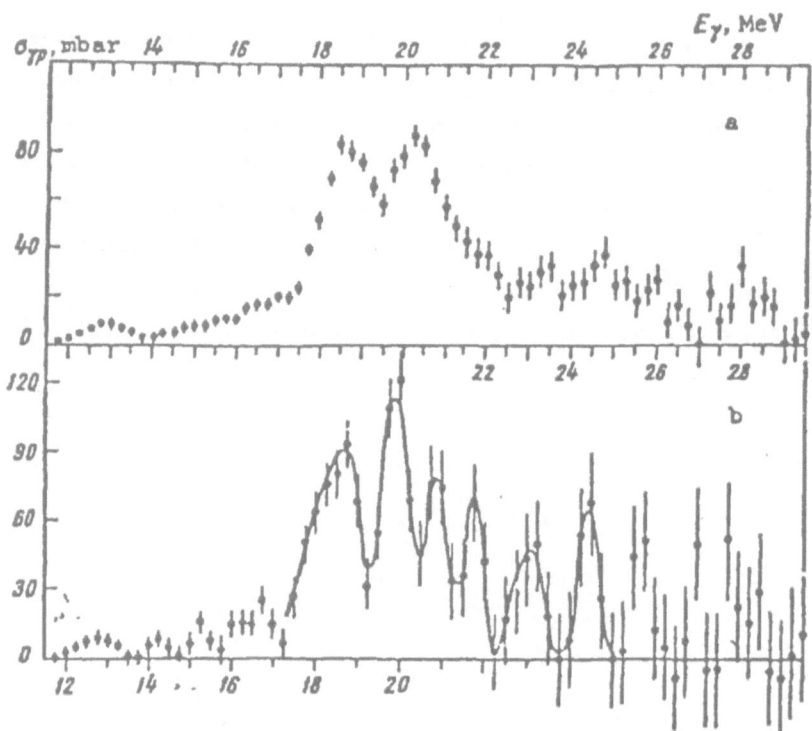

FIG. 22 : Cross section of the reaction Ca⁴⁰(γ, p) ($E_p \geq 2$ MeV) calculated by the Penfold-Leiss method with steps ΔE_γ = 1.0 MeV (a) and ΔE_γ = 0.5 MeV (b).

(from ref II.51)

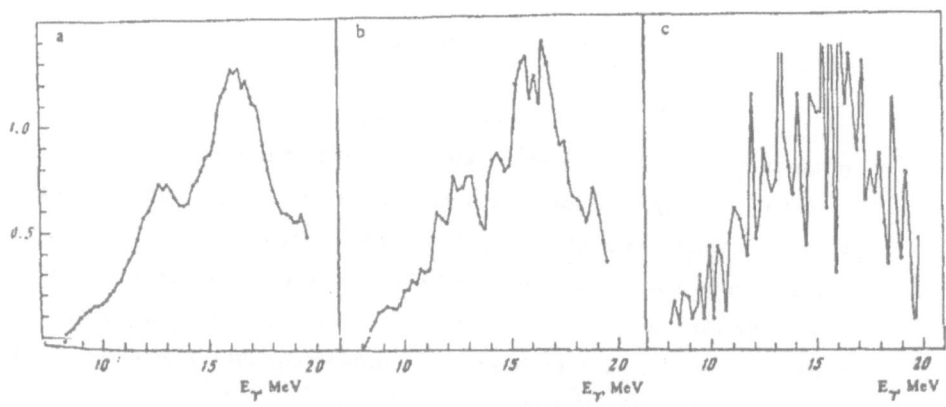

FIG. 23 : Calculations of the cross section for Sm¹⁵² by different methods: a, b, c—the Penfold-Leiss method with 1-, 0.5-, and 0.2-MeV spacings;

[Ref II.51]

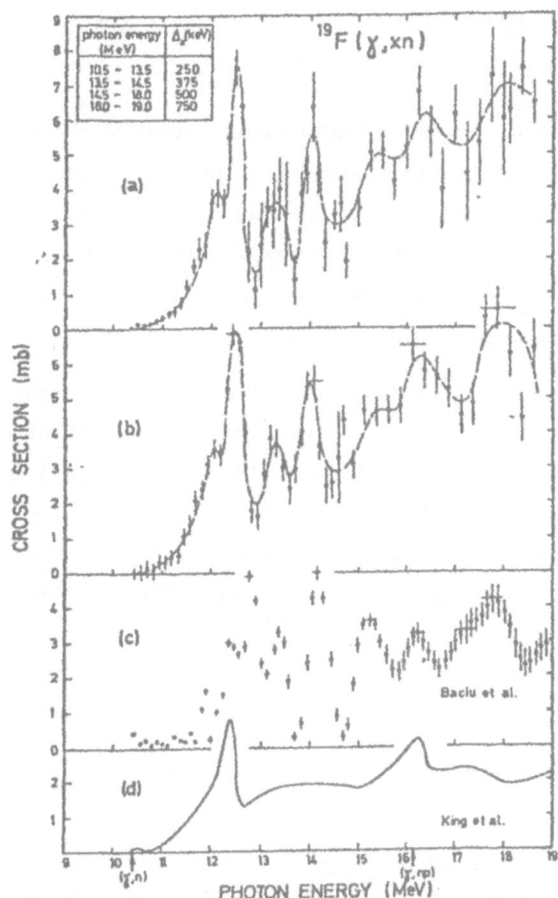

FIG. 24: Total photoneutron cross section $\sigma[(\gamma, n) + (\gamma, np)]$ for ^{19}F: (a) shows our results using the unfolding procedure, while (b) gives the results from a Cook analysis of the same data (the lines serve merely to guide the eye); (c) and (d) represent the other existing photoneutron cross-section data.

(from ref II.53)

III - USE OF " MONOCHROMATIC PHOTON BEAMS " WITH VARIABLE
ENERGY .

With the exception of monochromatic photons which are
produced by back compton scattering of laser photons by high energy
electrons (a technique used so far only in high energy physics [II. 1]
but only planned at Frascati for further studies in the GDR region [II, 54])
practically all the photonuclear data produced during the last 14 years
with monochromatic photons used in-flight annihilation of positrons.

As soon as the availability of high power electron accelerators
made feasible some positron beams with a reasonable intensity C. Tzara
at Saclay [II. 55]   and S. C. Fultz at Livermore [II. 56] began to build
such experimental facilities.

Fig. 25 show   the Saclay installation which was used from
1966 to 1974 for a systematic study of the σ(γ , xn) cross-sections
[II. 57, II. 58] .

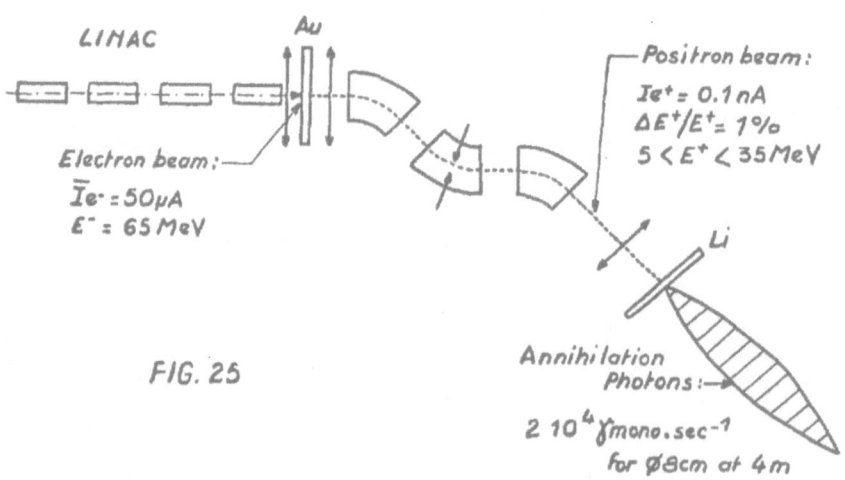

FIG. 25

The principle of the production of annihilation photons by the predominant processes

$$e^+(\beta c \neq 0) + e^-(\beta c = 0) \longrightarrow \gamma_1 + \gamma_2$$

is summarized by the equations below (fig. 26).

FIG. 26

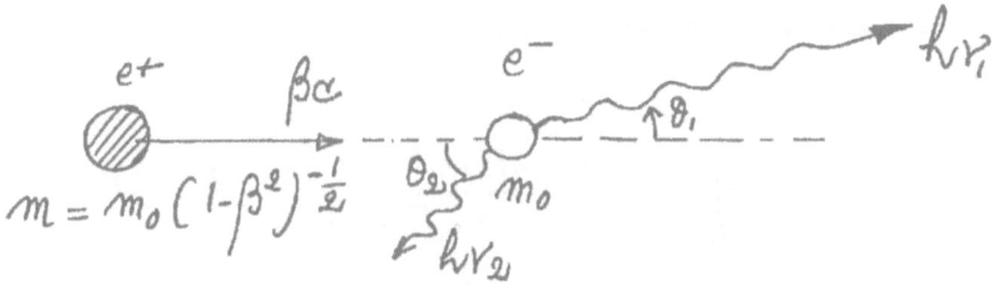

$$e^+ \qquad \beta c \qquad e^-$$

$$m = m_0(1-\beta^2)^{-\frac{1}{2}}$$

$$h\gamma_1 + h\gamma_2 = m_0 c^2 \left(\frac{1}{\sqrt{1-\beta^2}} + 1\right)$$

$$\frac{h\gamma_1}{c} \cos\theta_1 - \frac{h\gamma_2}{c} \cos\theta_2 = \frac{m_0 \beta c}{\sqrt{1-\beta^2}}$$

$$\frac{h\gamma_1}{c} \sin\theta_1 - \frac{h\gamma_2}{c} \sin\theta_2 = 0$$

$$E^+ = \frac{m_0 c^2}{\sqrt{1-\beta^2}}$$

$$\beta'_{CM} = \sqrt{\frac{E^+ - m_0 c^2}{E^+ + m_0 c^2}}$$

Hence: $\qquad h\gamma_1 = \dfrac{m_0 c^2}{1 - \beta' \cos\theta_1}$

Hence for the forward emitted photons ($h\gamma_1$, θ_1 in lab. coordinates)
a maximum energy obtained when $\theta_1 = 0$

$$\left(\hbar\gamma_1\right)_{MAX} (\theta_1 = 0) = E^+ + \frac{m_o c^2}{2}.$$

Unfortunately when a positron with total energy E^+_o cross an infinitely thin
annihilation target made of molecules m_i (A_i, Z_i) they produce annihilation
photons and bremsstrahlung photons according to the probability laws (if
$\Delta x = 1g/cm^2$)

$$-\left(\frac{dE_o}{dx}\right)_{Brems.} = E_o \frac{\mathcal{N} r_o^2}{137} \left[4 \log\left(183 Z^{-1/3}\right) + 2/9\right] \frac{\sum m_i z_i^2}{\sum m_i A_i}$$

$$\left(\frac{d N_{\gamma annih.}}{d\lambda}\right)_{\theta=0} = \mathcal{N} \pi r_o^2 \frac{\mu}{E_o^+}\left(\log \frac{2 E_o^+}{\mu} - 1\right)\frac{\sum m_i Z_i}{\sum m_i A_i}$$

Hence the photon spectrum produced at $\theta = 0$ by e^+ of energy E^+ resembles
the typical spectrum of fig. 27, where the width ΔE^+ of the annihilation
line depends on

- the energy spread of the " monochromatic positron beam "

- the collision losses ΔE_c in the target annihilation

$$-\left(\frac{dE_o}{dx}\right) = 2 N Z \pi \mu r_o^2 \, Log\left(\frac{E^3}{2I^2 z\mu}\right)$$

($I =$ excitation energy $\simeq$ 11 eV)

- the angular divergence of the incident e^+ beam and of the output
  γ -beam.

FIG 27

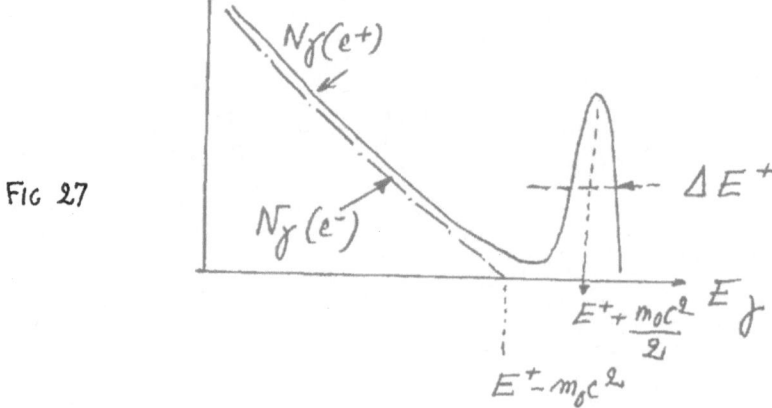

As it can be seen from formula the ratio of monochromatic γ to bremsstrahlung γ varies as $\frac{Z}{Z^2} = \frac{1}{Z}$ (cf. fig. 28a and b showing two typical γ spectra produced when e$^+$ cross two annihilation targets of Li H and Cu of the same radiation length.

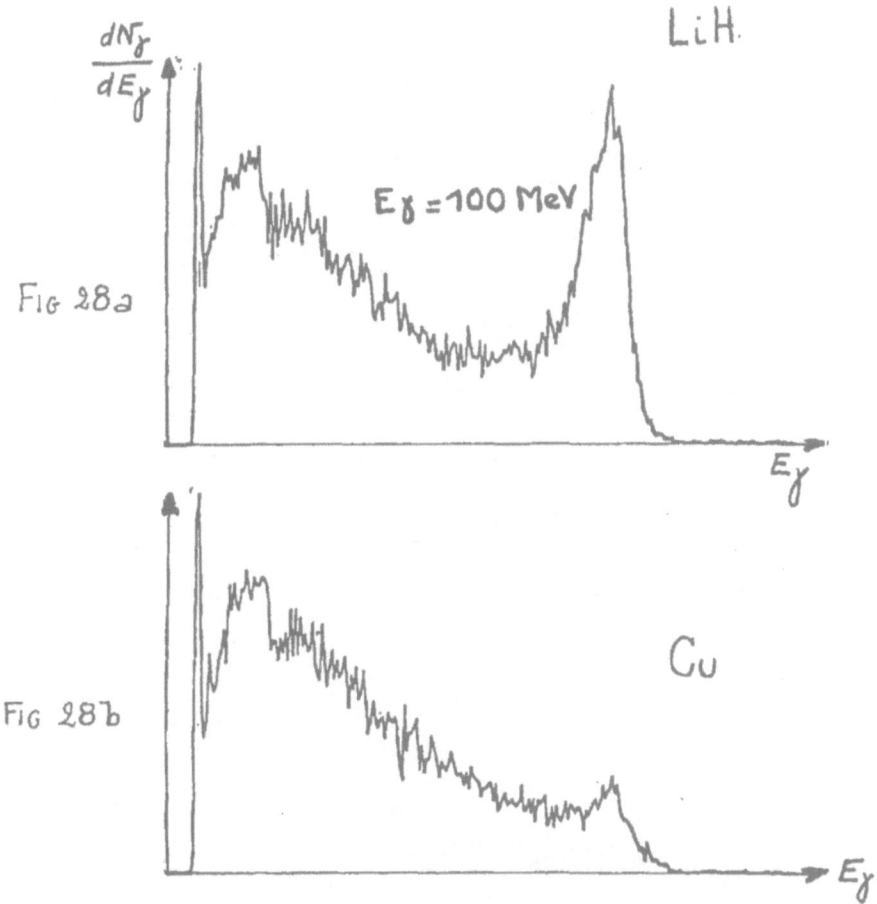

$\frac{dN_\gamma}{dE_\gamma}$ LiH

$E_\gamma = 100 \text{ MeV}$

Fig 28a

E_γ

Fig 28b   Cu

E_γ

To the first Born approximation the same number of electrons with the same kinetic energy crossing the annihilation target will produce only the bremsstrahlung part of the above spectrum.

Therefore a difference of the e$^+$ produced γ spectrum and the e produced γ spectrum will provide a monochromatic γ line of variable energy (FIG 29 c)

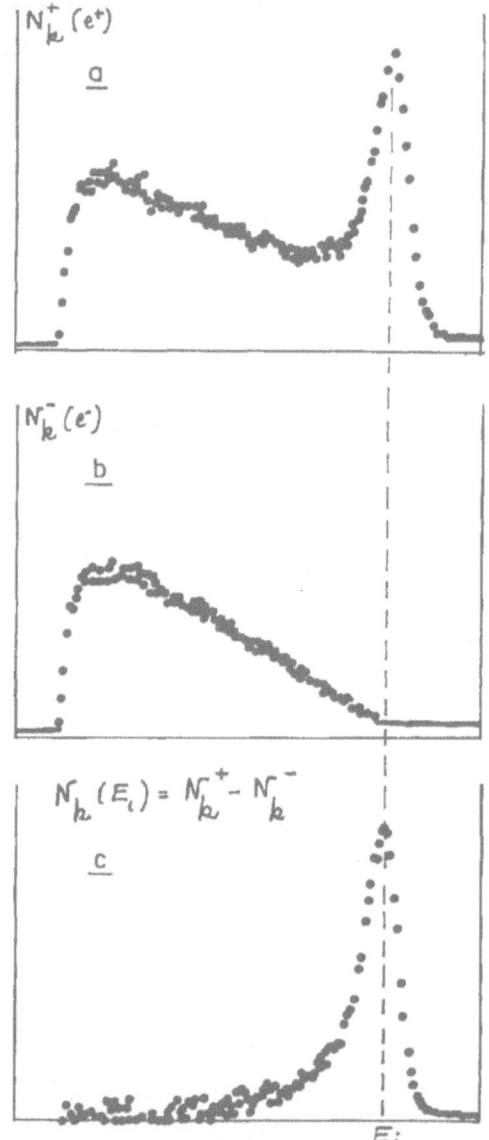

FIG. 29.

Therefore, refering to Fig. 30, one sees that any photonuclear measurement will be a two-step measurement.

1- measurements of the photonuclear products $X_P^+(E_i)$ given by a γ-spectrum produced by e^+ of energy E_L

2- measurements of the photonuclear products $X_N^-(E_i)$ given by a γ-spectrum produced by e^- of the same energy E_L.

Hence $X_p^+(E_i)$ - $X_N^-(E_i)$ divided by the number of " monochromatic "
photons, i. e. $N(E_i) = \sum_k N_k^+(e^+) - \sum_k N_k^-(e^-)$ is directly proportionnal
to the looked for cross section without any unfolding treatment but this
simple difference.(FIG. 30 c)

FIG. 30

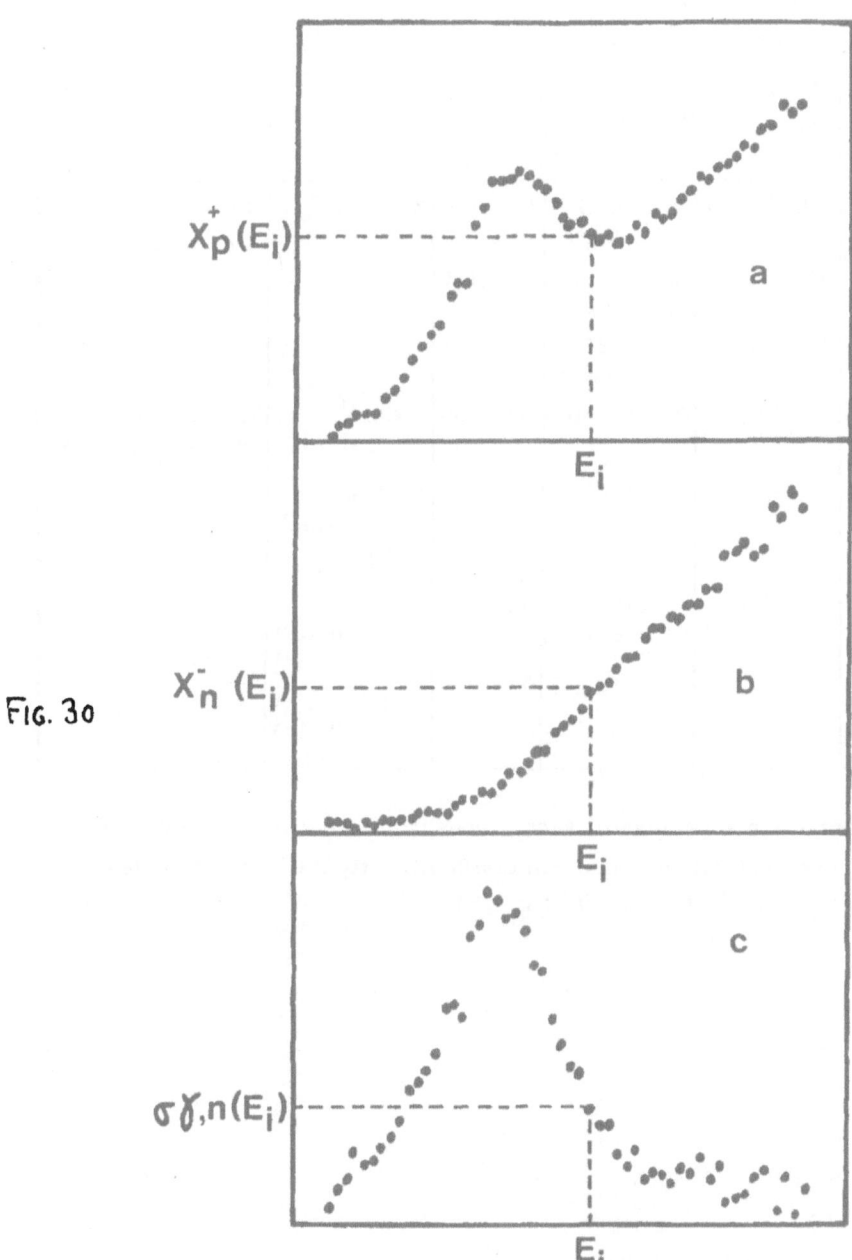

One sees on fig. 30 that fig. 30 a is the " yield curve " associated with this technique but now it shows directly the structure of the giant resonance.

Some typical figures of the monochromatic photon beams which have been or are still in used is given in table III.

TABLE III

	e^+ average current	annihilation target	energy range (MeV)	Nb of γ mono/sec	energy resolution
Saclay I	10^{-10} A	Li (3mm)	$6 \longrightarrow 40$	2 to 4 10^4	150 kev at 10 MeV 400 kev at 30 MeV
Saclay II	10^{-8} A	Li H 2mm Cu 30 μ	$20 \longrightarrow 120$		
Livermore I	10^{-10}A	Be (0. 25 0. 75 mm)	$6 \longrightarrow 35$	$\sim 10^4$	
Livermore II	10^{-9} to 10^{-8}	Be (0.13mm)	$5 \longrightarrow 100$	$< 10^6$	75 kev at 10 MeV 160 kev at 35 MeV
Mainz	$5. 10^{-10}$	(Be) (Al)		2.10^4 at 10 MeV 10^6 at 30 MeV	
Giessen	10^{-10} at 12 MeV 10^{-9} at 40 MeV	Be(0. 25, 0. 5 mm)	$12 \longrightarrow 40$	$5. 10^3$ at 10 MeV 10^5 at 40 MeV	

Another main advantage of such monochromatic γ-rays is that, if one uses a convenient detector, one can divide directly the counted nuclear product $\qquad Y(E) = Y^+(E, e^+) - Y^-(E, e^-)$

into the partial components. For example the use at Saclay of a high
efficiency neutron counter associated with a separate counting of the
(γ, n) (γ, 2n) (γ, 3n) (γ, 4n) events allowed to measure the corresponding
partial cross sections directly and simultaneously. Fig. 31 a and b show
that these simultaneous measurements are absolutely necessary if one
wants to get the total absorption cross section, for medium and heavy
nuclei at least, in a reliable manner. Indeed careful simultaneous
measurements of σ(γ, n) and σ (γ, 2n) for ^{150}Sm and ^{152}Sm are necessary to show that the splitting of the total photoabsorption curves shows up
only in the deformed ^{152}Sm nucleus(fig. 32) [II. 59]

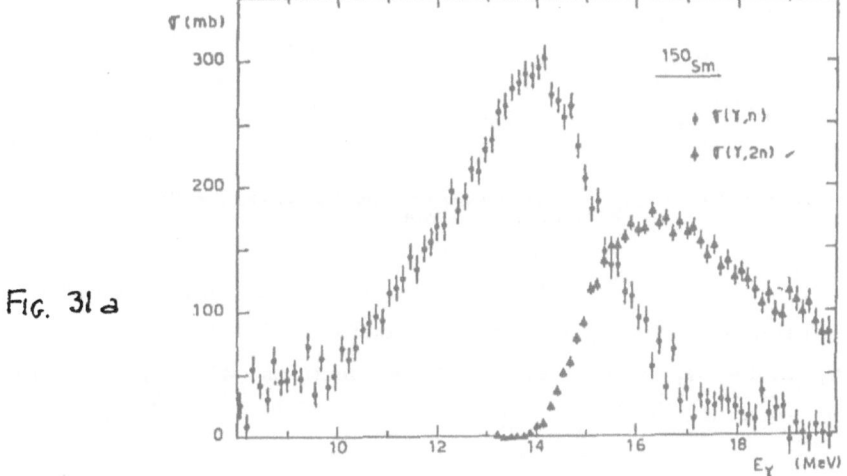

Fig. 31 a

Partial photoneutron cross sections $[\sigma(\gamma, n) + \sigma(\gamma, np)]$ and $\sigma(\gamma, 2n)$ of ^{150}Sm

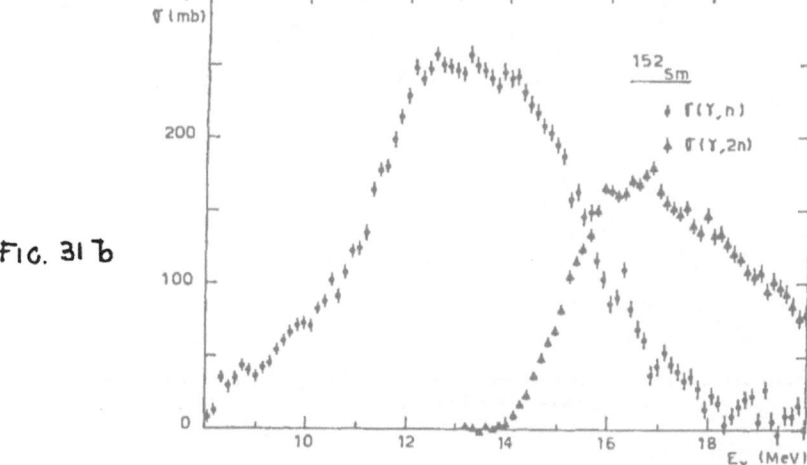

Fig. 31 b

Partial photoneutron cross sections $[\sigma(\gamma, n) + \sigma(\gamma, np)]$ and $\sigma(\gamma, 2n)$ of ^{152}Sm

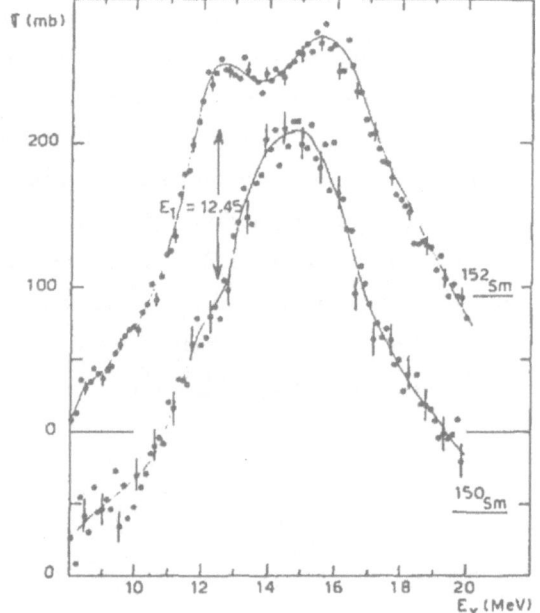

Comparison of the total photoneutron cross sections $\sigma_T(E)$ of ^{150}Sm with $\sigma_T(E)$ of ^{152}Sm.
The solid lines are only to guide the eye.

Fɪɢ. 32

. Similarly the deformation splitting of the total photoabsorption curve of ^{238}U is demonstrated only by adding the separately measured cross-sections $\sigma(\gamma, n)$, $\sigma(\gamma, 2n)$ and $\sigma(\gamma, \text{fission})$ (fig. 33) [II. 60].

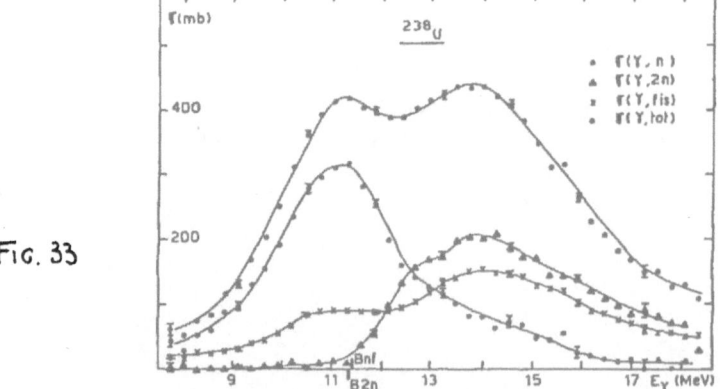

Fɪɢ. 33

Partial and total photonuclear cross sections $\sigma(\gamma, n)$, $\sigma(\gamma, 2n)$, $\sigma(\gamma, F)$ and $\sigma_{tot} = \sigma(\gamma, n) + \sigma(\gamma, 2n) + \sigma(\gamma, F)$ of $^{238}_{92}$U.

Some improvements are now being tested for the production of monochromatic γ-rays by in flight annihilation of positrons and are summarized below (fig. 34).

1° - Use of the same e⁺ beam with two annihilation targets (one with low Z, the other with high Z) to produce a monochromatic photon line. If one can overcome some normalization problems the difference (Fig 34 c) of the 2 γ-spectra gives directly a monochromatic photon line.

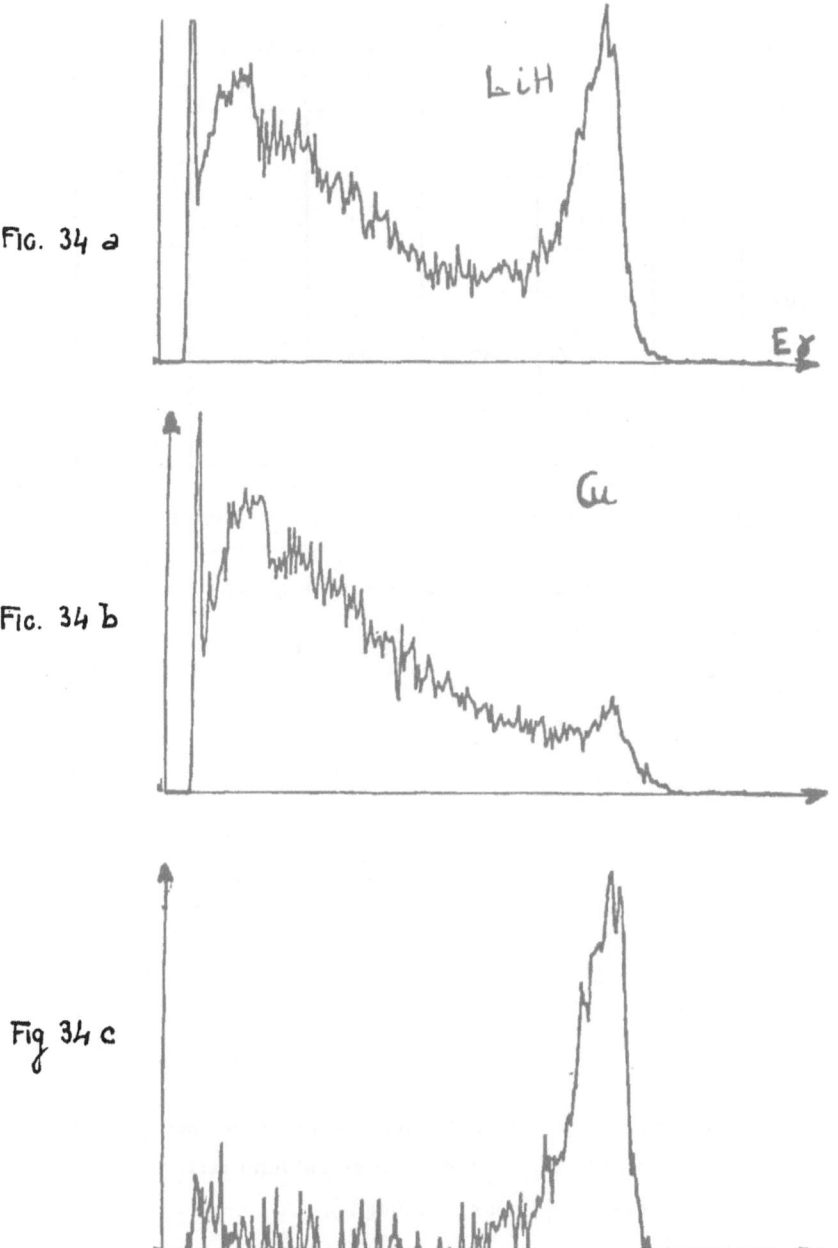

FIG. 34 a

FIG. 34 b

Fig 34 c

2° - Choice of an angle $\theta \neq 0°$ at which the useful photons are produced with respect to the direction of the incident e⁺.

When θ increases the number of bremsstrahlung γ produced in dΩ around θ decreases more rapidly than the number of annihilation photons, although the energy of this annihilation line is shifted downwards while the energy resolution is slightly spoiled (table IV).

TABLE IV

$E(e^+)$ = 50 MeV

θ	0°	1°	2°	3°	4°	5°
S	1301	534	146	52	21	10
B	304	64	7.4	1.5	0.5	0.2
$\frac{S}{B}$ (normalized)	1	2	4.6	8	9.8	11.7
E_γ (MeV)	50	49.6	47.7	44.6	40.9	36.9
ΔE_γ (MeV)	0.68	1	22	2.9	3.3	3.4

S = Nb of γ annihilation for 10^8 e⁺ incident on LiH 3 mm *(in dΩ = 4.10^{-5}st.Rad).*
B = Nb of γ bremsstrahlung in 500 kev at 20 MeV for 10^8 e⁺ incident on LiH 3 mm

To decrease the bremsstrahlung tail should be a great advantage to study the higher part of the GDR (fig.35).

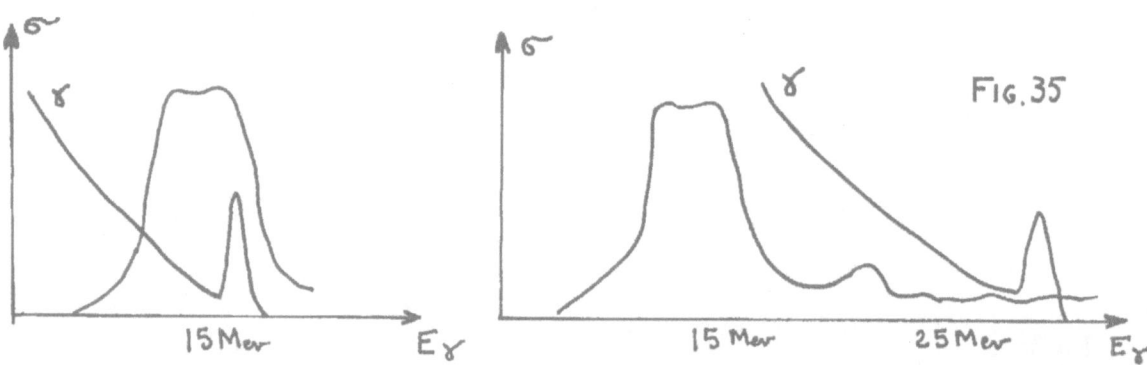

FIG. 35

3° - Finally a tagging of the second annihilation photon, would also provide a mean to get rid of most of the bremsstrahlung tail (if the duty cycle of the accelerator is large enough).

REFERENCES II -

[II. 1]   G. Diambrini - Palazzi
           Proceedings 1967 - International Symposium on electrons
           and photons interactions at high energy - Stanford,   Sept. 67

[II. 2]   C. Schull
           Survey of monochromatic photon capabilities, technologies
           and uses in proceed. Int. Conf. on photonuclear reactions
           and application - Asilomar, March 1973.

[II. 3]   O. V. Bogdankevich and F. A. Nikolaev
           Methods in bremsstrahlung research - Academic Press,
           New-York, London.

[II. 4]   J. L. Meason and P. K. Kuroda
           Physical Review, 142, 691, (1966).

[II. 5]   S. S. Hanna and L. Meyer - Schützmeister
           Phys. Rev. 108, 1644, (1957).

[II. 6]   L. Meyer-Shützmeister and S. S. Hanna
           Bull. Amer. Phys. Soc., 3, 188, (1958).

[II. 7]   M. M. Wolff and W. E. Stephens
           Phys. Rev. 112, 890, (1958).

[II. 8]   D. S. Gemmell and G. A. Jones
           Nucl. Phys. 33, 102, (1962).

[II. 9]   J. E. Perry and S. V. Bame
           Phys. Rev. 99, 1368, (1955).

[II. 10]   W. A. Lochstet and W. E. Stephens
           Phys. Rev. 141, 1002, (1966).

[II. 11]   W. Del Bianco, S. Kundu and P. Boucher
           Nuclear Physics- A 209, 181-188, (1973).

[II. 12]   R. E. Welsh and D. J. Donahue- Phys. Rev. 121, 880, (1961).
           L. Green and D. J. Donahue - Phys. Rev. 134, B701, (1964).
           R. R. Hurst and D. J. Donahue - Nucl. Phys. A91, 365, (1967).

[II. 13]   O. Y. Mafra, S. Kuniyoski and J. Goldemberg
           Nucl. Phys. A 186, 110-126, (1972).

[II. 14] R. Moreh and T. Bar-Noy
Nucl. Instr. Meth., 105 , 557,(1972).

[II. 15] R. A. Anderl, J. E. Hall, R. C. Morrison, R. G. Strùss,
M. V. Yester and D. J. Zaffarano
Nucl. Instr. and Meth. 102 , 101-108 (1972).
J. Fagot, R. Lucas, H. Nifenecker and M. Schneeberger
Nucl. Instr. and Meth. 95 , 421-427 (1971).

[II. 16] A. M. Khan and J. W. Knowles
Nucl. Phys. A 179 , 333 (1972).

[II. 17] L. V. Groshew and al
" Atlas   of γ-ray spectra from radiative capture of
thermal neutrons " - Pergamon Press, London.

[II. 18] " Neutron capture gamma-ray spectroscopy "
Studsvik, Aug. 1969 - IAEA, Vienna, Proc. Int. Symp.

[II. 19] A. Manfredini, L. Fiore, Cramorino, H. G. de Carvalho,
W. Wölfli
Nucl. Phys. A 127 , 687-692, (1969).

[II. 20] J. R. Huizenga, K. M. Clarke, J. E. Gindler and R. Vandenbosch
Nucl. Physics 34 , 439-456, (1962).

[II. 21] R. Moreh, I. Jacob and R. Mourad
Nucl. Instr. and Meth. 127-193 (1975).

[II. 22] L. Schiff
Phys. Rev. 70, 761 (1946).

[II. 23] E. G. Fuller and E. Hayward
Phys. Rev. 101, 692 (1956).

[II. 24] E. Hayward and E. G. Fuller
Phys. Rev. 106, 991 (1957) .

[II. 25] O. Beckmann and R. Sandstrom
Nucl. Phys. 5 , 595 (1958).

[II. 26] F. Seward, H. Koch, R. Shafer and S. Fultz
Bull. Am. Phys. Soc. 5 , 68 (1960).

[II. 27] E. C. Booth
Nucl. Phys. 19 , 426 (1960).

[II. 28]   H. Arenhövel and E. Hayward
Phys. Rev. <u>165</u> , 1170 (1968).

[II. 29]   E. Hayward, W. C. Barber and J. Sazama
Phys. Rev. <u>C8</u> , 1065 (1973).

[II. 30]   E. Hayward, W. C. Barber and J. J. Mac Carthy
Phys. Rev. <u>C10</u>, 2652 (1974).

[II. 31]   F. W. K. Firk
Ann. Rev. Nucl. Sci. <u>20</u> , 39 (1970).

[II. 32]   H. E. Jackson
Proc. Asilomar Conference, March 1973.

[II. 33]   P. Axel
Phys. Rev. <u>126</u>, 671 (1962).

[II. 34]   C. D. Bowman, R. J. Baglan, B. L. Berman and
T. W. Phillips
Phys. Rev. Letters <u>25</u>, 1302 (1970).

[II. 35]   B. S. Dolbilkin, V. A. Zapevalov, V. I. Korin, L. E. Lazareva
and F. A. Nikolaev.
Bull. Acad. Sci. USSR Phys. <u>30</u> , 354 (1966).

[II. 36]   N. Bezic, A. Brinsek, G. Kernel and J. Snajder
Nucl. Instr. and Meth. <u>75</u> , 190-196 (1969).

[II. 37]   N. Bezic, D. Jamnik, G. Kernel, J. Krajnik and J. Snajder
Nucl. Phys. <u>A 117</u>, 124-128 (1968).

[II. 38]   J. M. Wyckoff, B. Ziegler, H. W. Koch and R. Uhlig
Physical Review <u>137</u> , 576 B (1965).

[II. 39]   J. Ahrens, H. Borchert, A. Zieger and B. Ziegler
Nucl. Inst. and Meth. <u>108</u> , 517 (1973).

[II. 40]   J. Ahrens, H. Borchert, K. M. Czock, H. B. Eppler, M. Gimm,
H. Gundrum, M. Kroning, P. Richn, G. Sita Ram, A. Zieger
and B. Ziegler
Nucl. Phys. <u>A 251</u>, 479 (1975).

[II. 41]   J. S. O'Connell, P. A. Tipler and P. Axel
Physical Review <u>126</u>, 228 (1962).

[II. 42]   An electron accelerator with a 100% duty factor - University
of Illinois at Urbana-Champaian - April 1975.

[II.43] P.A. Dickey and P. Axel
Physical Review Letters 36 , 501 (1975).

[II.44] E. Bramanis, T.K. Deague, R.S. Hicks, R.J. Hughes,
E.G. Muirhead, R.H. Sambell and R.J.J. Stewart
N.I.M. 100, 59-71 (1972).

[II.45] A.S. Penfold and J.E. Leiss
Physical Review 114, 1332 (1959).

[II.46] D.W. Anderson, B.C. Cook and T.J. Englert
Nucl. Phys. A 127 ,474 -480 (1969).
R. Carchon, J. Devos, R. Van de Vyrer, C. Van Deynse
and H. Ferdinande
Nucl. Phys. A 223,416- 422 (1974).

[II.47] E.G. Fuller and E. Hayward (1962) - Nuclear Reaction II
(Ed. P.M. Endt and P.B. Smith) North Holland Publishing
Company, Amsterdam chap. 3.

[II.48] H.H. Thies and B.M. Spicer
Australian J. Phys. 13, 505 (1960).

[II.49] R.E. Sund, V.V. Verbinski, H. Weber and L.A. Kull
Phys. Rev. C2 , 1129 (1970).

[II.50] P.H. Cannington, R.J.J. Stewart, B.M. Spicer and
M.G. Huber
Nuclear Physics, A 109, 385-392 (1968).
H. Beil, R. Bergère, P. Carlos, A Leprêtre, A. Veyssière
and A. Parlag
Nuclear Physics A 172 , 426-436 (1971).

[II.51] A.M. Goryachev, G.N. Zalesnyi, S.F. Semenko and
B.A. Tulupov
Sov. J. Nucl. Phys. vol. 17, n°3, p.236 (1973)

[II.52] B.N. Goryachev, B.S. Ishklanov, I.M. Kapitonov,
I.M. Piskarev, V.G. Shevchenko and O.P. Shevchenko
JETP Pis'ma 5, n°7, 225-227 (1967).

[II.53] R.E. Van de Vyrer, H. Ferdinande, G. Knuyt, R. Carchon
and J. Devos
Nucl. Phys. A 198, 144-152 (1972).

$\left[\text{II. 54}\right]$ Proceedings of the working group on the study of photonuclear reactions with monochromatic and polarized gamma ray Frascati (1973).

$\left[\text{II. 55}\right]$ J. Miller, C. Schuhl et C. Tzara
Nucl. Phys. 32 ,236 , (1962)

$\left[\text{II. 56}\right]$ S. C. Fultz, R. L. Bramblett, J. T. Caldwell and N. A. Kerr
Phys. Rev. 127 , 1273 (1962).

$\left[\text{II. 57}\right]$ H. Beil, R. Bergère and A. Veyssière
Nucl. Instr. 67 , 293 (1969).

$\left[\text{II. 58}\right]$ G. Audit, N. de Botton, G. Tamas, H. Beil, R. Bergère and A. Veyssière
Nucl. Instr. 79 , 203 (1970).

$\left[\text{II. 59}\right]$ P. Carlos, H. Beil, R. Bergère, A. Leprêtre, A. de Miniac and A. Veyssière
Nuclear Physics A225 , 171-188 (1974).

$\left[\text{II. 60}\right]$ A. Veyssière, H. Beil, R. Bergère, P. Carlos and A. Leprêtre
Nuclear Physics A 199, 45-64 (1973).

II. 61 R. J. Holt and H. E. Jackson
Phys. Rev. Lett, 36, 5 (1976), 244.

CHAPTER III

THE COLLECTIVE MODELS OF THE E I GIANT RESONANCE

Since, from the very definition of a giant resonance, specific and coherent motions of all nucleons in a nucleus must be considered, it should come as no surprise that collective models, even of the simplest hydrodynamical nature, provide good representations of some experimental features of the GDR. In this chapter, some systematic properties gathered from experimental data of the GDR will be compared with various predictions of the collective models, namely.:

. The average energy of the GDR in the paragraph III.B entitled "The static collective model for spherical nuclei."

. The splitting of the GDR in the paragraph III.C entitled "The static collective model for permanently deformed nuclei."

. The broadening of the GDR in the paragraph III.D entitled "The dynamic collective model."

III.A The nucleon effective charges.

Classically, when one studies the GDR, the nucleus is considered as a system of non-relativistic point nucleons with charge e_i, mass M_i, momentum p_i and spin σ_i. One thus does not consider the interaction of the incident photon with the exchange charged mesons and the total hamiltonian between the nucleus and an incident photon field with vector potential $\vec{A} = A_0 \vec{\varepsilon} e^{-i\vec{k}.\vec{z}}$ can then be expressed as

$$H_T = \sum_i \left\{ \frac{1}{2M_i} \left(\vec{p}_i - e_i \frac{\vec{A}}{c} \right)^2 + \mu_i \vec{\sigma}_i \left(\vec{\nabla} \times \vec{A} \right) \right\} + \sum_{ij} V_{ij} + H_{RAD}$$

Therefore, the interaction hamiltonian for the case of an electric interaction is written as

$$H' = \sum_{\iota} \left[-\frac{e_{\iota}}{M_{\iota}c} \vec{p_{\iota}}.\vec{A} + \frac{e_{\iota}^{2}}{2M_{\iota}c^{2}} \vec{A}^{2} \right]$$

One knows that in the above expression the first term H_1 describes the emission or the absorption of a photon while the second one describes photon scattering.

Moreover, when one proton is submitted to the electric field of an incident photon, its induced motion corresponds to a recoil of the other nucleons in the center of mass frame. The effect is normally described in terms of "effective charges" which can be summarized as follows (cf. E. Hayward |III.1|) :

$$\vec{A} = A_{0}\vec{\epsilon}\, e^{-i\vec{k}.\vec{z}} \simeq A_{0}\vec{\epsilon}.$$

$$\vec{p_{\iota}} = M_{\iota}\vec{v_{\iota}}$$

$$H_{1} = -\sum_{\iota=1}^{Z} \frac{A_{0}e_{\iota}}{c}\,\vec{\epsilon}.\vec{v_{\iota}}$$

$$\vec{V} = \vec{V}_{C.M} = \sum_{n=1}^{A} \frac{\vec{V_{n}}}{A} = \text{velocity of CM.}$$

$$Ze\vec{V} = \text{total nuclear current}$$

$$E_{CM} = -\frac{A_{0}}{c}\,Ze\,(\vec{\epsilon}.\vec{V}).$$

$$E_{CM} = -\frac{A_{0}Ze}{cA}\left[\sum_{\iota=1}^{Z}(\vec{\epsilon}.\vec{v_{\iota}}) + \sum_{j=1}^{N}(\vec{\epsilon}.\vec{v_{j}}) \right]$$

hence $\quad H_{1} = \frac{-eA_{0}}{c}\left[\sum_{\iota=1}^{Z}(\vec{\epsilon}.\vec{v_{\iota}}) - \frac{Z}{A}\sum_{\iota=1}^{Z}\vec{\epsilon}.\vec{v_{\iota}} - \frac{Z}{A}\sum_{j=1}^{N}\vec{\epsilon}.\vec{v_{j}} + Z\vec{\epsilon}.\vec{V} \right]$

$$H_{1} = \frac{-eA_{0}}{c}\left[\underbrace{\frac{N}{A}\sum_{\iota=1}^{Z}(\vec{\epsilon}.\vec{v_{\iota}})}_{(1)} - \underbrace{\frac{Z}{A}\sum_{j=1}^{N}(\vec{\epsilon}.\vec{v_{j}})}_{(2)} + \underbrace{Z\,(\vec{\epsilon}.\vec{V})}_{(3)} \right]$$

One can thus associate the above terms :

(1) with electric dipole transitions of protons with effective charge $\frac{Ne}{A}$

(2) with electric dipole transitions of neutrons with effective charge $-\frac{Ze}{A}$

(3) with Thomson scattering by the whole nucleus.

Therefore, due to their opposite effective charges, the protons and the neutrons will move out of phase in an electric dipole collective mode. One thus recognizes the fundamental hydrodynamic description of the E1 mode as a polarization (isovector) mode which was actually the first useful description given more than 20 years ago in the so-called static collective model.

III.B The static collective model for spherical nuclei.

As early as 1948 Goldhaber-Teller, followed in 1950 by Steinwedel-Jensen, proposed a model of the GDR in which all the protons move out of phase with respect to all the neutrons. This model was considerably refined year after year, especially by M. Danos and by the German School led by W. Greiner. In what follows I will use mostly the book "Nuclear theory" |III.2| by W. Greiner and J. Eisenberg, a very clarifying paper published by M. Huber |III.3| under the title "Collective model of the giant resonance." and a critical studie of the GDR energy by R. Bach and C. Werntz |III.4|.

III.B.1 The Goldhaber-Teller model. (Fig. 1)

In the Goldhaber-Teller model |III.5|, the proton sphere moves as a whole with respect to the neutron sphere and this displacement is characterized by the vector $\vec{d}(r)$. Hence, for the dipole vibration, the transition charge density $\rho'_{p} (\rho'_{n})$, corresponding to the local excess density of protons (neutrons) with respect to the ground state spherical densities $\rho_{o}(r)$, is given by

$$\rho_h'(z,t) = Z\left[\rho_0\left(\vec{z} - \frac{N}{A}\vec{d}\right) - \rho_0(\vec{z})\right] \simeq -\frac{NZ}{A}\vec{d}(t)\cdot\vec{\nabla}\rho_0(\vec{z})$$

$$\rho_n'(z,t) = N\left[\rho_0\left(\vec{z} + \frac{Z}{A}\vec{d}\right) - \rho_0(\vec{z})\right] \simeq +\frac{NZ}{A}\vec{d}(t)\cdot\vec{\nabla}\rho_0(\vec{z})$$

If one then considers the equations of continuity

$$\frac{\partial\rho_h'}{\partial t} \simeq -Z\,\nabla\cdot(\vec{v}_p\rho_0)$$

$$\frac{\partial\rho_n'}{\partial t} \simeq -N\,\nabla\cdot(\vec{v}_m\cdot\rho_0).$$

one sees that they are satisfied by the relative velocities

$$\vec{v}_p = \frac{N}{A}\frac{d}{dt}(\vec{d}) \qquad\qquad \vec{v}_m = -\frac{Z}{A}\frac{d}{dt}(\vec{d})$$

which keep track of the effective charges computed above.

One thus finds a charge transition density ρ_h' (ρ_n') and its associated potential both peaked at the nuclear surface as $\vec{\nabla}\rho_0(z)$ a fact which is not quite compatible with the statement made in chapter I that " isovector modes are mostly volume modes ". In an harmonic vibrator description, the restoring force will thus be proportional to the nuclear surface $A^{2/3}$ whereas the mass parameter is of course proportional to A. Hence, the average frequency of the giant dipole mode will be :

$$\omega_D \simeq \sqrt{\frac{A^{2/3}}{A}} = A^{-\frac{1}{6}}$$

III.B.2 The Steinwedel-Jensen model.

In the Steinwedel-Jensen |II.6| (S.J.) model, the proton and neutron fluids are two compressible interpenetrating fluids moving within the rigid surface of the initial nucleus (Fig. 2). Then the incident photon field "sets up" two different evolutions for the proton density ρ_h and the neutron density ρ_m inside the nucleus, whereas the total mass distribution of the nucleons is not perturbed by the vibrations of the proton and the neutron fluids.

Fig: 1   G.T. Model

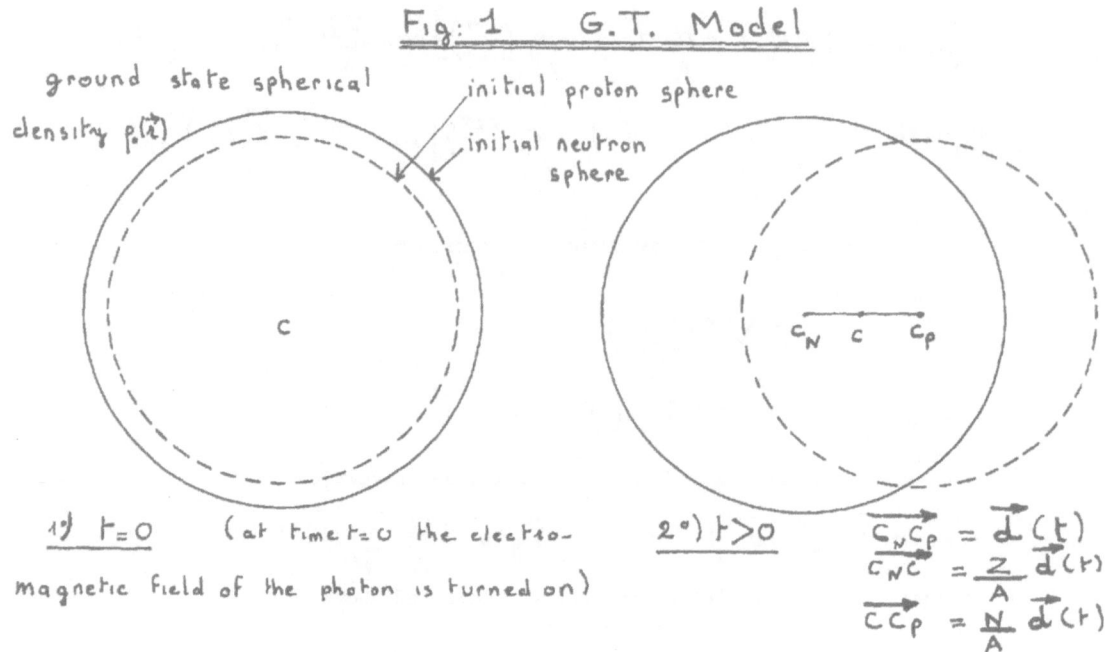

ground state spherical density $\rho_o(\vec{x})$

initial proton sphere

initial neutron sphere

C

c_N c c_P

1) $t=0$ (at time $t=0$ the electro-magnetic field of the photon is turned on)

2°) $t>0$

$$\overrightarrow{c_N c_P} = \vec{d}(t)$$
$$\overrightarrow{c_N c} = \frac{Z}{A}\vec{d}(t)$$
$$\overrightarrow{c c_P} = \frac{N}{A}\vec{d}(t)$$

Fig: 2   S. J. Model

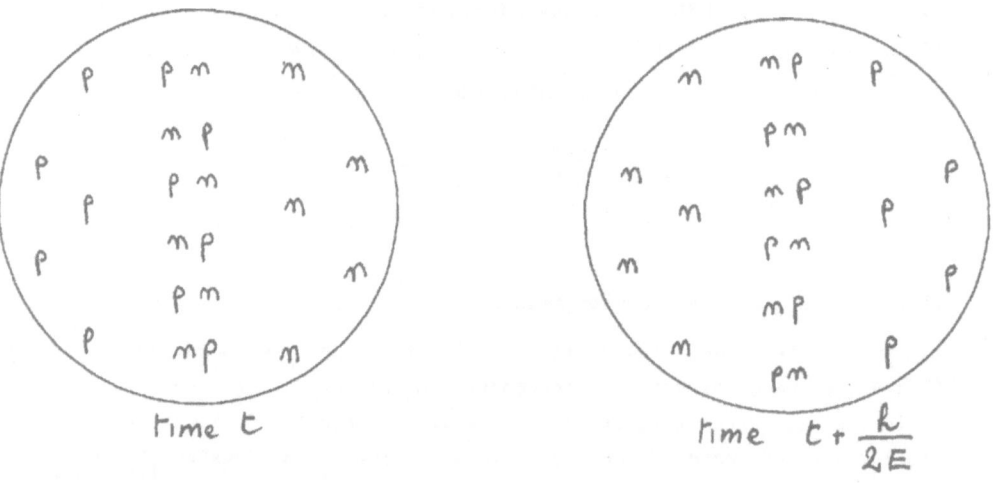

p   p m   m
m p
p   p m   m
p   m
m p
p   m
p m
p   m p   m

time t

m   m p   p
p m
m   p
m p   p
m   p m
m   m p
p m
m   p
p m   p

time $t + \frac{k}{2E}$

One can write the relations :

$$\rho_{h}(\vec{r},t) = \frac{Z}{A}\rho_0 \left[1 + \frac{A}{Z}\eta(\vec{r},t) \right]$$

$$\rho_{m}(\vec{r},t) = \frac{N}{A}\rho_0 \left[1 - \frac{A}{N}\eta(\vec{r},t) \right]$$

$$\rho_{h}(\vec{r},t) + \rho_{m}(\vec{r},t) = \rho_0(\vec{r}) = \left(\frac{4}{3}\pi R_0^3 \right)^{-1}.A.\ f(r)$$

where ρ_0 is the normal mass density of the nucleus A, whose radius is R_0 and $f(r)$ describes a radial dependence of the mass (or charge) distribution, such as the one obtained from elastic electron scattering experiments.

The heart of the theory is that there is a restoring force, proportional to the symmetry term B_S in the Bethe-Weizsäcker mass formula :

$$B_S = K\frac{(N-Z)^2}{A}$$

which tends to restore the normal value of ρ_h and ρ_n and therefore to cancel the deviations $\pm\,\eta(\vec{r},t)$

Once again, for an harmonic vibrator one can write :

$$\eta(\vec{r},t) = e^{i\omega t}\zeta(r)$$

where $\zeta(r)$ obeys a classical Helmholtz equation :

$$\nabla^2\zeta(r) + k^2\zeta(r) = 0$$

According to the prediction of the hydrodynamical model, the following connection exists between the energy E_D of the dipole vibration, k the wave number of the vibration, and the velocity u with which the vibration propagates through the nucleus :

$$u = \left[\frac{8K}{M}.\frac{NZ}{A^2} \right]^{\frac{1}{2}}$$

$$E_D = \hbar\omega_D = \hbar k u\ .$$

The solution of the wave equation can be expanded in spherical waves with multipolarity numbers $\lambda\mu$:

$$\zeta_{\lambda\mu}(z) = j_\lambda(k_\lambda z) \, Y_{\lambda\mu}(\hat{z})$$

The eigenmodes of the above vibration are given by the boundary condition implying a rigid surface of the nucleus. Hence the radial flux of matter through the nuclear surface must be zero :

$$\left[\frac{\partial}{\partial z} \zeta(z,t) \right]_{z=R_0} = 0$$

The possible wave numbers k for a dipole mode $\lambda = 1$ are then such that :

$$k_1 = \frac{2.08}{R_0}$$ (fundamental mode carrying 86 % of the transition strength for the dipole vibration)

$$\left. \begin{array}{l} k_2 = \dfrac{5.95}{R_0} \\[2mm] k_3 = \dfrac{9.25}{R_0} \end{array} \right\}$$ overtones carrying 6 % and 2 % of the dipole transition strength respectively

Therefore the energy of the fundamental dipole mode is predicted by the S.J. model as :

$$E_D \simeq \hbar \left[\frac{8K}{M} \cdot \frac{NZ}{A^2} \right]^{\frac{1}{2}} \frac{2.08}{R_0} \simeq \frac{constant}{R_0}$$

which clearly points to a $A^{-\frac{1}{3}}$ law for E_D. If one assumes K = 23 MeV and $R_0 = 1,2 \, A^{1/3}$ fm then :

$$E_D(S.J) \simeq 70 \, A^{-\frac{1}{3}} \; MeV.$$

III.B.3 Comparison with experimental data.

It must first be pointed out that several definitions of the average experimental energy E_D of the GDR exist. In medium and heavy nuclei where the total absorption cross section $\sigma(\gamma, tot)$ can be fairly well represented by one Lorentz line (see III.C.2) :

$$\sigma(E) - \sigma_0 \frac{E^2 \Gamma^2}{(E^2 - E_0^2)^2 + E^2 \Gamma^2}$$

one can take the central energy E_O of the Lorentz line. If several Lorentz lines are necessary to fit the experimental $\sigma(\gamma,tot)$ curve, one can take the weighted average $\overline{E_O}$ value. The following figures, gathering the data obtained at Livermore and Saclay, will use the above definitions. However other definitions can be used such as those provided by the ratio of different sum rules :

$$E_1 = \frac{\int^{E_M} \sigma(E)\,dE}{\int^{E_M} \sigma(E)\,\frac{dE}{E}}$$

$$E_2 = \frac{\int^{E_M} \sigma(E)\,\frac{dE}{E}}{\int^{E_M} \sigma(E)\,\frac{dE}{E^2}}$$

As an example, Table I shows the E_0, E_1 and E_2 values for some nuclei studied at Saclay, up to $E_M = 30$ MeV.

Table I

	^{56}Fe	^{82}Se	^{89}Y	^{116}Sn	^{140}Ce	^{175}Lu	^{208}Pb
E_0 (MeV)	18.3	16.2	16.70 ± 0.1	15.6	15	14.4	13.4
E_1 (MeV)	19.2	16.9	17.7 ± 0.2	16.5	15.6	14.6	14
E_2 (MeV)	18.6	16	17.5 ± 0.2	15.8	15	.14.1	13.1

One sees immediately from this presentation of the average dipole energies E_0 versus A that the experimental data fit neither the $A^{-\frac{1}{3}}$ (S.J.) nor the $A^{-\frac{1}{6}}$ law (G.T.) very well. (Fig. 3a, b, c).

III.B.4 <u>Refinements of the theoretical predictions.</u>

Several attempts were made to reconcile the theoretical predictions and the experimental data. Two such attempts are particularly interesting.

III.B.4a <u>Consideration of a realistic nuclear surface.</u>

In the above (G.T.) and (S.J.) models, the surface of the nucleus

Fig. 3c

[ref III.23]

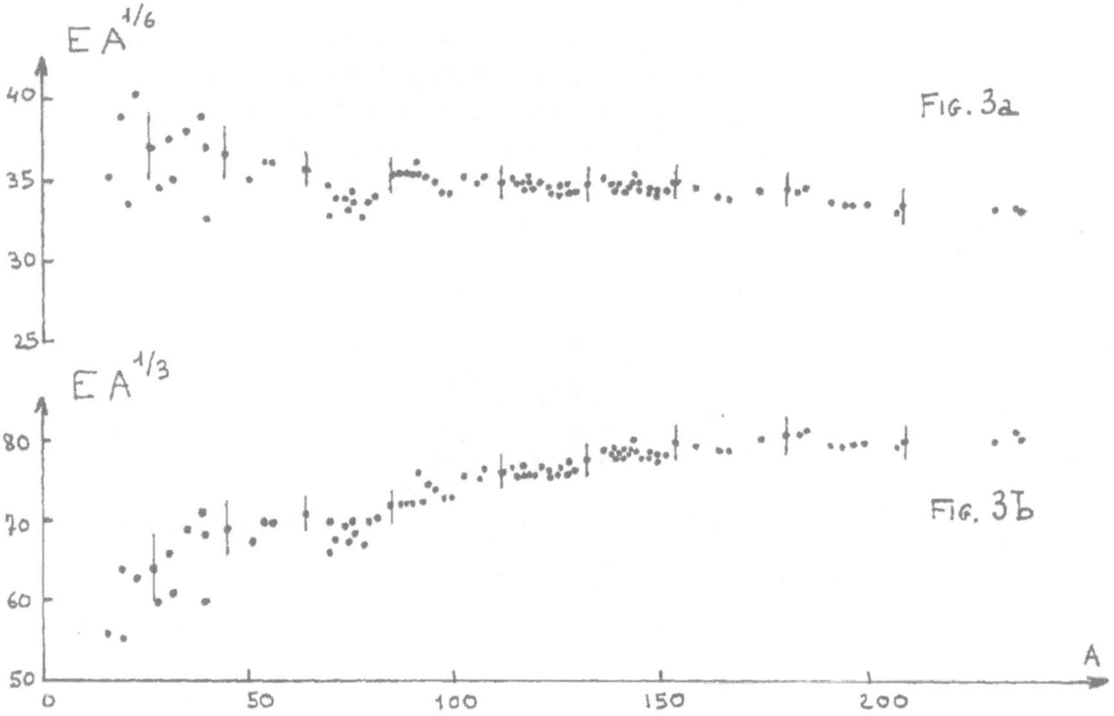

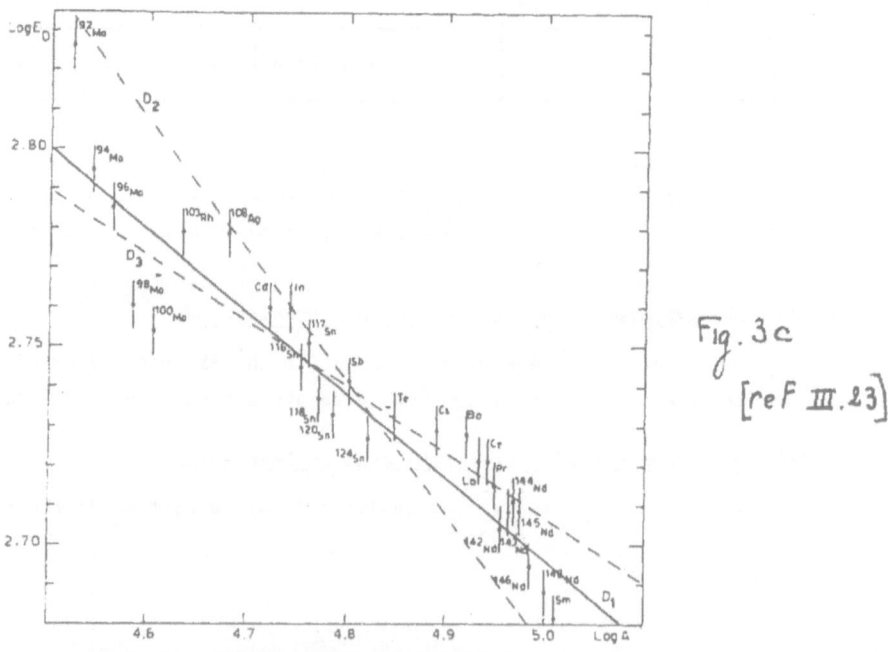

Experimental results of the average energy E_D of the GDR versus the mass number A with the best fit $E_D = 41.8\, A^{-1/4.8}$ (line D_1). Two fits with $A^{-1/3}$ and $A^{-1/6}$ laws are also shown as lines D_2 and D_3 respectively.

plays an important role in the definition of the energy E_D of the dipole
mode. But how well is the surface of a nucleus defined ? We all know that
the surface thickness t, as obtained by (e,e) experiments, cannot be neglected especially in the light nuclei (Fig. 4). Actually the observation
that for a light nucleus, such as ^{12}C, the inside region with a constant
density practically does not exist makes dubious the application of the
above "collective models" to nuclei with only a small number of nucleons.

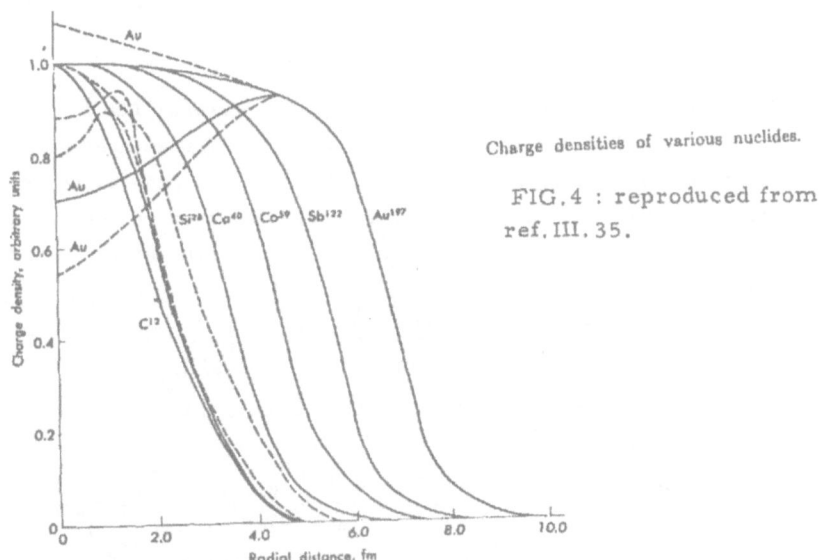

Charge densities of various nuclides.

FIG. 4 : reproduced from
ref. III. 35.

Starting from the Lagrangian introduced by Wilets for the ground state
density of the nucleus :

$$\mathcal{L}_s = -\left[E(\rho_N, \rho_m) + \frac{\xi}{8M} \left(\frac{(\nabla \rho_m)^2}{\rho_m} + \frac{(\nabla \rho_N)^2}{\rho_N} \right) \right]$$

R. Back and C. Werntz |III.4| expanded this Lagrangian to second order in
neutron and proton densities and the gradient of these densities to obtain
a realistic restoring force for fluctuations around the ground state densities. The above parameter ξ was fixed so that the correct surface thickness
t is reproduced.

Back and Werntz then used as trial solutions the transition charge densities of the (G. T.) model, which were considered as more realistic than the
(S. J.) ones since ρ'_N (S. J.) switches abruptly from the value
to the value 0 as   crosses the value R_0 of the nucleus radius. They thus
found again :

$$E_0 \simeq \left[\frac{4NZ}{A^2} \right] A^{-\frac{1}{6}}$$

Since for actual nuclei the factor $\frac{4NZ}{A^2}$ decreases from the value 1 (for self-conjugate light nuclei) to the value 0.96 for heavy nuclei such as ^{208}Pb, one sees that the above formula explains (although not completely) the decrease of $E_D A^{\frac{1}{6}}$ experimentally observed when A increases. (Fig. 3a)

III.B.4b Introduction of a variable K.

Starting from the opposite point of view, B. Berman |III.7| and S. Fultz adopted   the (S.J.) point of view in which the fundamental parameter to be considered in expressing the energy of the dipole mode is the nuclear symmetry energy K. They used an improved expression for E_D which was given in 1958 by M. Danos |III.8| and takes into account the width Γ (FMHW) of the GDR :

$$E_D \simeq \frac{\hbar k}{A} \left\{ \frac{8KNZ}{M} \left[1 - \left(\frac{\Gamma}{2E_D} \right)^2 \right] \right\}^{\frac{1}{2}}$$

With kR = 2.08 for spherical nuclei with a sharp nuclear radius R = 1.2 $A^{1/3}$ fm one finds :

$$K_{Mev} = 9.93 . 10^{-4} \left(\frac{A^{8/3}}{NZ} \right) . \frac{E_D^2}{1 - \left(\frac{\Gamma}{2E_D} \right)^2}$$

Hence, B. Berman considered that an "experimental" K value could be obtained from the experimental E_D and Γ values of the GDR of the nucleus (A,Z). Fig. 5 shows these experimental K values, whose increasing values with A exactly corresponds to the discrepancy between experimental E_D and a law 78 $A^{-1/3}$ MeV. But, once again, one is brought back to the fundamental role of the surface of the nucleus. B. Berman fitted these experimental K by an analytical form :

$$K_{Mev} = K_0 \left(1 - c A^{-\frac{1}{3}} \right) \simeq 42.3 - 86.6.A^{-\frac{1}{3}}$$

thus modifying the "volume" symmetry energy term K_0 by a surface symmetry energy $c K_0 A^{-\frac{1}{3}}$, in agreement with the mass formula introduced by Myers and Swiatecki |III.9|. One thus predicts E_D for every nucleus rather precisely (± 100 keV), and presents the nuclear symmetry energy K as the fundamental parameter in the giant dipole resonance exactly as the nuclear compressibility is presented as the fundamental parameter in the monopole giant resonance.

Fig : 6

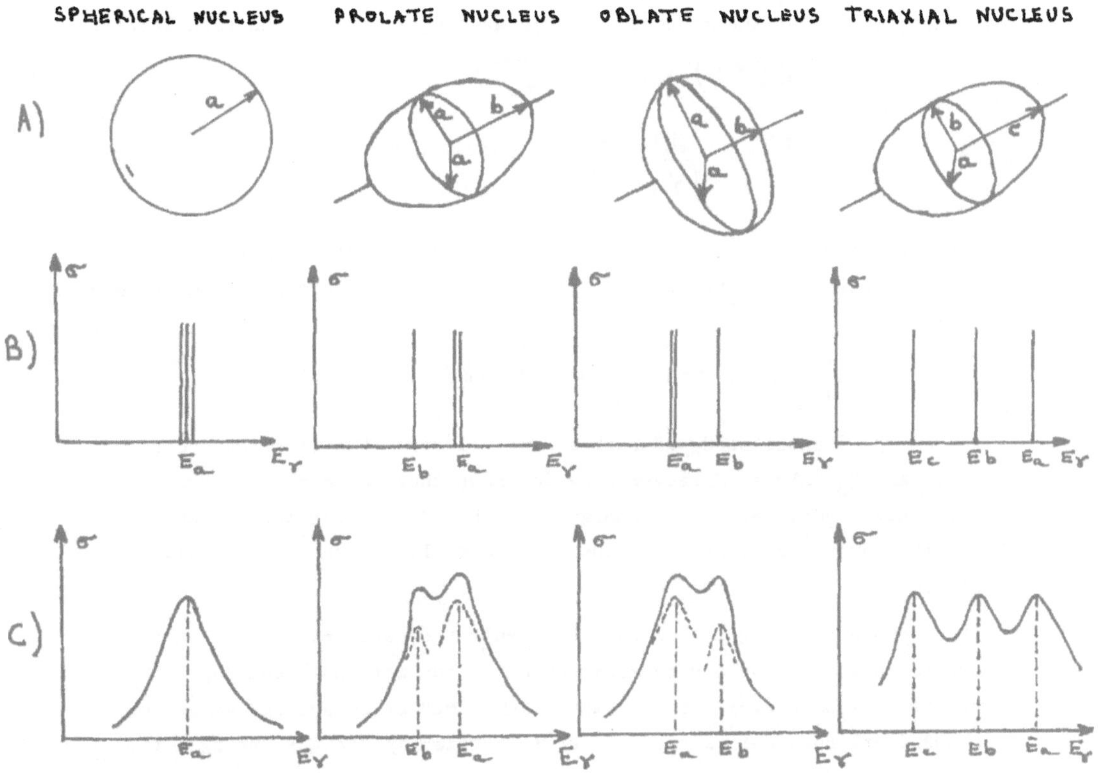

SPHERICAL NUCLEUS   PROLATE NUCLEUS   OBLATE NUCLEUS   TRIAXIAL NUCLEUS

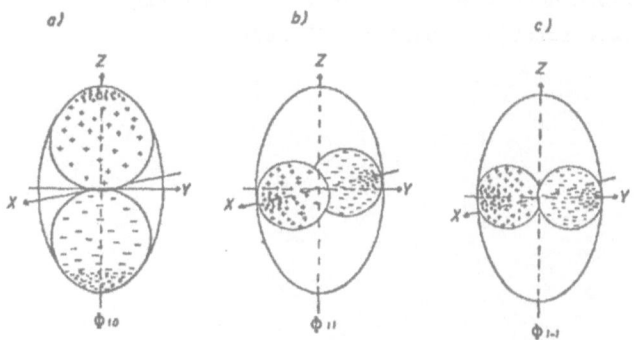

Pictorial representation of the normal modes of giant dipole oscillations in a
deformed nucleus   The protons are indicated by + + +, the neutrons
                                    by – – –.

FIG 7   (from ref III. 2)

It is well known in nuclear spectroscopy that β_0 is connected to the intrinsic quadrupole moment Q_0 of the nucleus, which is normally reached by a measurement of the reduced electric quadrupole transition rate B(E2) :

$$B(E2) = \frac{5}{16\pi} e^2 Q_0^2$$

$$Q_0 = \frac{3Z}{\sqrt{5\pi}} R_0^2 \beta_0 (1 + 0.36\beta_0).$$

Danos and Okamoto |III.8, III.10| showed that a splitting ($E_a \neq E_b$) could be predicted for the GDR of axially symmetric deformed nucleus (a ≠ b) with :

$$\frac{E_b}{E_a} = 0.911 \frac{a}{b} + 0.089$$

The experimental study of such a GDR splitting thus provides access to the β_0 and Q_0 values. Moreover, Fig. 6c shows that for a prolate nucleus the strength under the higher energy peak should be twice the strength below the lower energy peak . The reverse would be true for an oblate nucleus.

One can easily understand why the highly inelastic process of exciting a dipole mode provides a good measurement of the intrinsic quadrupole Q_0 even for J = 0$^+$ nuclei. If one takes $\Gamma \simeq$ 6 MeV as an average value for the GDR width this deformation effect is studied in the GDR during a time :

$$\Delta t \simeq \frac{\hbar}{\Gamma} \simeq 10^{-22} \text{ sec}$$

much shorter than the characteristic time $\Delta t'$ corresponding to one rotation of a permanently deformed nucleus ($\Delta t' \sim 70 \Delta t$ if the energy of the first 2$^+$ of the ground state rotational band $\simeq$ 100 keV).

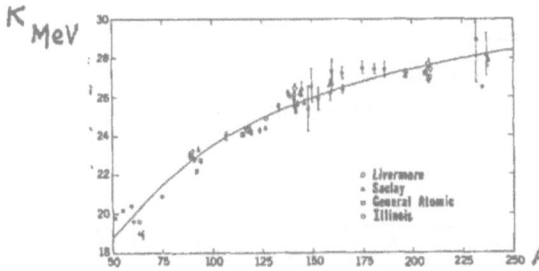

FIG. 5 : reproduced
from ref. III.7.

The nuclear symmetry energy derived from the Lorentz
parameters, fitted with the function $K = K_0(1 - cA^{-1/3})$ from which
$K_0 = 42.3$ MeV and $cK_0 = 86.6$ MeV (see text). The values without
error bars were not used in the fitting procedure.

III.C The static collective model for permanently deformed nuclei.

III.C.1 The "Q_0 splitting".

The three dipole modes characterized by $\mu = -1, 0, 1$ in the term
$Y_{1\mu}(\hat{r})$ are degenerate in a spherical nucleus. But since in the (S.J.)
model, the energy of the dipole mode depends on $1/R$, it is easily
understandable that 3 different dipole modes will characterize the dipole
oscillations along the various nuclear axis. The expected dipole modes
are summarized in Fig. 6 and 7.

Assuming a nuclear shape :

$$R(\theta,\varphi) = R_0\left[1 + \sum_{\mu} \alpha_{2\mu} Y_{2\mu}(\theta,\varphi)\right]$$

one knows that for an axially symmetric deformed nucleus :

$$\beta_0 = \alpha_{20}$$
$$R = R_0\left[1 + \beta_0 Y_{20}(\theta,\varphi)\right]$$
$$a = R(\theta = 0) = R_0\left(1 + 0.63\,\beta_0\right)$$
$$b = R\left(\theta = \frac{\pi}{2}\right) = R_0\left(1 - 0.32\,\beta_0\right)$$

a = length of the symmetry axis

b = length of an axis perpendicular to the symmetry axis

III.C.2 The Lorentz line fit.

It is useful to point out that most experimenters get the ratio E_b/E_a (and hence the ratio $d = a/b$) by fitting their experimental photoabsorption curve with a sum of two Lorentz lines :

$$\sigma(E) = \sigma_a \frac{(E\Gamma_a)^2}{(E^2 - E_a^2)^2 + (E\Gamma_a)^2} + \sigma_b \frac{(E\Gamma_b)^2}{(E^2 - E_b^2)^2 + (E\Gamma_b)^2}$$

It is traditional to fit one photoabsorption giant dipole resonance not by a Breit-Wigner line (Fig. 8) :

$$\sigma_{BW}(E) = \sigma_o \frac{E}{E_o} \frac{(\Gamma/2)^2}{(E-E_o)^2 + (\Gamma/2)^2}$$

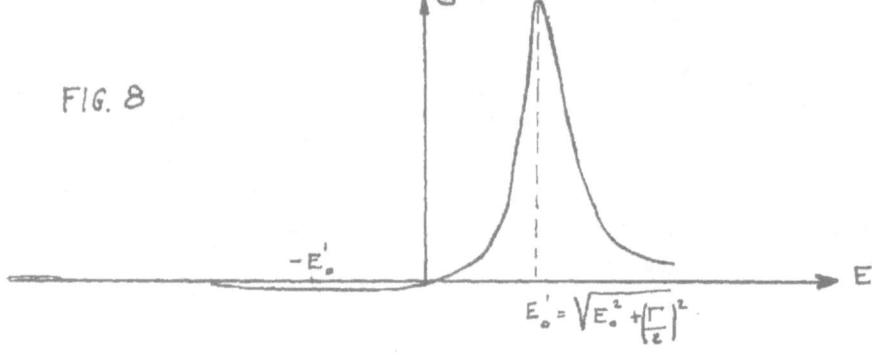

FIG. 8

$$E_o' = \sqrt{E_o^2 + \left(\frac{\Gamma}{2}\right)^2}$$

but by a Lorentz line σ_L :

$$\sigma_L(E) = \sigma_o \frac{E^2 \Gamma^2}{(E^2 - E_o^2)^2 + E^2 \Gamma^2}$$

Actually, one shows that a Lorentz line fit assures the time reversal conservation / III. 11 / for the case of a photon with a zero rest mass ($E_{tot} = h\nu$) since σ_L is the sum of two BW lines centered at E_o and $-E_o$ /III. 12/ :

$$\sigma_0 \; \frac{E}{E_0} \; \frac{(\frac{\Gamma}{2})^2}{(E-E_0)^2+(\frac{\Gamma}{2})^2} \;\; + \; \sigma_0 \; \frac{E}{(-E_0)} \; \frac{(\frac{\Gamma}{2})^2}{(E+E_0)^2+(\frac{\Gamma}{2})^2} \;\; =$$

$$= \; \sigma_0 \; \frac{E^2\Gamma^2}{\left\{ E^2 - \left[E_0^2 + (\frac{\Gamma}{2})^2 \right] \right\}^2 \; + \; E^2\Gamma^2}$$

$$= \; \sigma_L$$

$= \sigma_{Lorentz}$ with a maximum at $E_0' = \sqrt{E_0^2 + (\frac{\Gamma}{2})^2}$, exactly at the place of
the maximum for the BW line with $E_0 > 0$. (Fig 9)

FIG. 9

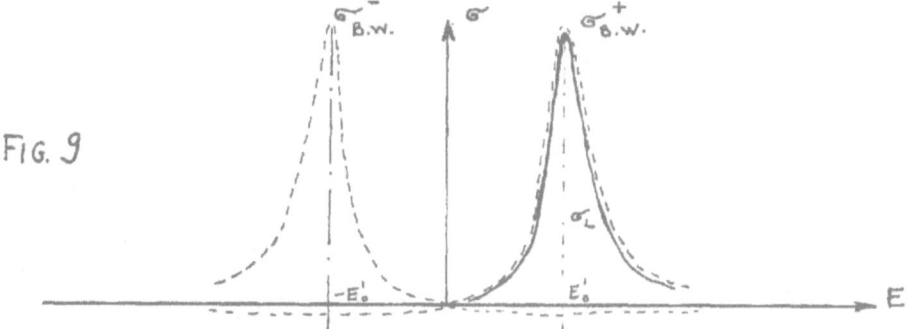

For $E > 0$, $\sigma_L(E, E_0') \simeq \sigma_{BW}(E, E_0)$ and experimental data can be fitted,
to within the same accuracy by either Lorentz or BW lines.

III.C.3 Comparison with experimental data.

The predictions of such a deformation splitting of the GDR have been
brilliantly confirmed by all experimental data. Usually, the experimenters have
to make sure that they do measure the total absorption cross section. For
medium and heavy nuclei this is generally achieved by measuring the cross
sections for the open neutron (fission) channels since the coulomb barrier
severely hinders the emission of charged particle (Fig. 10-11) |III.13|.

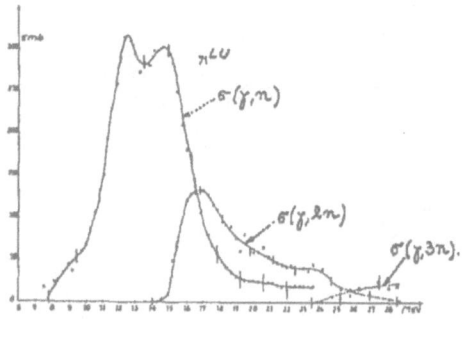

FIG. 10

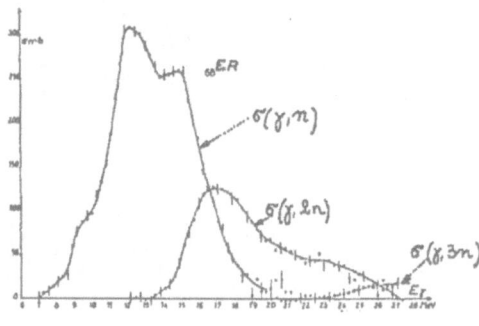

FIG. 11

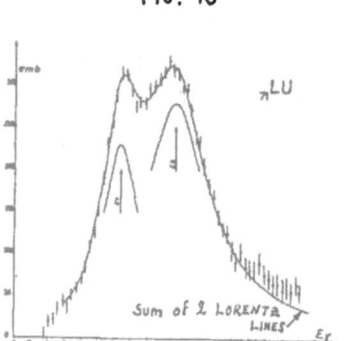

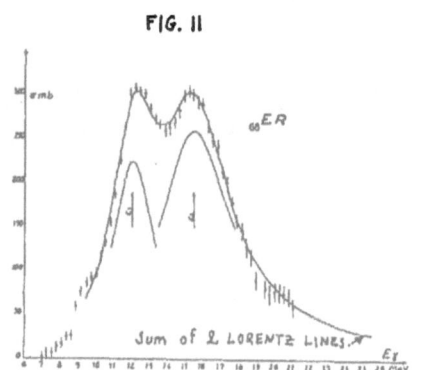

FIG. 10   FIG. 11

Then the "total" absorption cross section is fitted by two Lorentz lines by a chi-squared process. The central values E_a and E_b then provides the ratio $a/b = d$. Hence one computes :

$$Q_0 = \frac{2}{5} R_0^2 Z A^{\frac{2}{3}} \cdot \frac{d^2 - 1}{d^{2/3}}$$

with

$$R_0 \simeq 1.2 \, A^{\frac{1}{3}} \, fm.$$

Table II shows the remarkable agreement between :

- the Q'_0 values obtained by B (E2) measurements in nuclear spectroscopy.
- the Q''_0 values obtained through the rotational model evaluation by using experimental spectroscopic quadrupole moments Q_s from hyperfine atomic spectra.
- the Q_0 values obtained by the above described method.

$$Q_s = Q_0'' \frac{J(2J-1)}{(J+1)(2J+3)}$$

Table II

	Sb	^{127}I	^{143}Rh	^{150}Nd	^{152}Sm	^{154}Sm	Er
Q_0	−1.8±0.4	−2.3±0.4	1.7±0.2	6 ±1	5.9±0.4	6.6 ±0.4	6.96±0.4
Q_0'				5.1	5.9	6.65	7.6
Q_0''	−1.5	−2.2	2.2				

	Lu	W	Re	^{232}Th	^{237}Np	^{238}U
Q_0	6.93±0.3	6 ±0.5	6 ±0.5	10.2 ±1	11.3±1	11 ±1
Q_0'	7.2	6.2	5.9	9.66	10.9	11.3

Finally one must quote   one of the most beautiful experiment in GDR physics. At Livermore, Kelly et al. measured |III.14| :

$$\sigma(\gamma, abs) \simeq \sigma(\gamma, n) + \sigma(\gamma, pn) + \sigma(\gamma, 2n)$$

for a target made of ^{165}Ho nuclei aligned either parallel or perpendicular to the direction of the incident photon beam. If one calls K_i and K_n the projection on the nuclear symmetry axis of the spin I_i of the ground state and of the spin I_n of the excited state $|n\rangle$, induced by the dipole operator Q_1, then the GDR states associated with the nuclear symmetry axis (parallel modes) imply $\Delta K = K_i - K_n = 0$ while $\Delta K = \pm 1$ for states associated with the perpendicular modes. $(\sigma_{\parallel}$ and $\sigma_{\perp}$ cross. sections respectively$)$

$$\sigma_{\parallel} = \delta_{\Delta K, 0} \frac{4\pi \hbar}{3} \sum_m |\langle m|Q_1|i\rangle|^2 \frac{E E_m \Gamma_m}{(E^2 - E_m^2)^2 + E^2 \Gamma_m^2}$$

$$\sigma_{\perp} = \delta_{\Delta K, \pm 1} \frac{4\pi \hbar}{3} \sum_m |\langle m|Q_1|i\rangle|^2 \frac{E E_m \Gamma_m}{(E^2 - E_m^2)^2 + E^2 \Gamma_m^2}$$

Aligning   the targets will shift   the transition strengths towards these two groups of states

Kelly showed then that for ^{165}Ho ($I_i = K_i = 7/2$) one must observe the photoabsorption cross section :

$$\sigma_a (E, \theta, T) = \sigma_{||} + \sigma_{\perp} + 0.408 \left(\sigma_{\perp} - 2 \sigma_{||} \right) f(T) \, P_2 (\cos \theta)$$

where   f(T) expresses the degree of nuclear orientation

          θ = angle between the incident photon beam and the target quantization axis

One thus sees that :

for $\theta = 0°$ $\sigma_a = \sigma_{||} + \sigma_{\perp} + \left(0.408 \, \sigma_{\perp} - 0.816 \, \sigma_{||} \right) f(T)$

for $\theta = 90°$ $\sigma_a = \sigma_{||} + \sigma_{\perp} + \left(0.408 \, \sigma_{||} - 0.204 \, \sigma_{\perp} \right) f(T)$

Hence at $\theta = 0°$ the dipole modes $\sigma_{\perp}$, which are located at the higher energy E_b, will be enhanced while the dipole modes $\sigma_{||}$ associated with the lower energy E_a, will be decreased (and vice versa at $\theta = 90°$). This is exactly what was shown at Livermore. (Fig. 12)

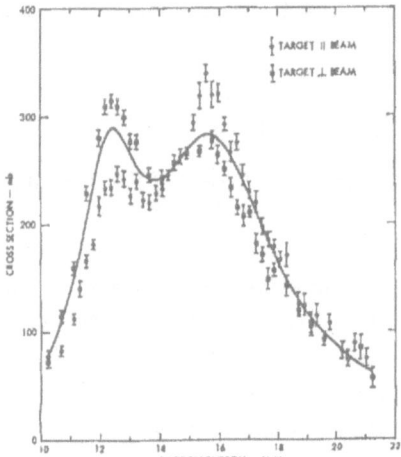

FIG. 12 : reproduced from ref. III.14.

The cross sections measured with the holmium target polarized parallel to the photon beam and perpendicular to the photon beam. The solid line is the two-component Lorentz curve fit to the unpolarized data.

III.D   <u>The dynamic collective model.</u>

Very soon after the first experimental confirmations of the above "static" collective models, associated with a rigid time-independent radius, it was realised by Le Tourneux |III.15| and Danos and

Greiner |III.16-17| that one should also take into account the nuclear
surface vibrations. Several theories were successively proposed to couple
the density fluctuations, characterizing the giant resonance motion, and
the nuclear surface vibrations. A very complete description of these
theories is given in W. Greiner and J. Eisenberg's Nuclear Theory (vol. 1,
Chapter II) |III.2|.

III.D.1 The dynamic collective model for vibrational nuclei.

III.D.1a Theoretical summary.

First one considered harmonic quadrupole surface vibrations of energy
E_2 and r.m.s. vibrational amplitude β_2 :

$$R = R_0 \left[1 + \sum_\mu \alpha_{2\mu}(t) \, Y^*_{2\mu}(\theta,\varphi) \right]$$

for which the hamiltonian H_Q can be written as a function of the coordinates $\alpha_{\lambda\mu}$ and their conjugate $\pi_{\lambda\mu}$ momenta :

$$H_2 = \frac{\sqrt{5}}{2} \left\{ \frac{1}{B_2} \left[\pi^{[2]} \times \pi^{[2]} \right]^{[0]} + C_2 \left[\alpha^{[2]} \times \alpha^{[2]} \right]^{[0]} \right\}$$

$$E_2 = \hbar \sqrt{\frac{C_2}{B_2}}$$

$$\beta_2^2 = \sum_\mu \langle 0 | \alpha_{2\mu}^2 | 0 \rangle = \frac{\hbar}{2\sqrt{B_2 C_2}}$$

$$B(E2, 2^+ \rightarrow 0^+) = \frac{9}{32\pi^2} Z^2 e^2 R_0^4 \frac{\hbar}{\sqrt{B_2 C_2}}$$

Similarly, one can write for the dipole vibration :

$$H_D = -\frac{\sqrt{3}}{2} \left\{ \frac{1}{B_1} \left[\pi^{[1]} \times \pi^{[1]} \right]^{[0]} + C_1 \left[\alpha^{[1]} \times \alpha^{[1]} \right]^{[0]} \right\}$$

with

$$E_D = \hbar \sqrt{\frac{C_1}{B_1}}$$

$$B(E1) = \frac{9}{32\pi^2} Z^2 e^2 R_0^2 \hbar \frac{1}{\sqrt{B_1 C_1}}$$

The next step was then to consider the coupling of these two vibrational modes which arises through the 1/R dependence of the dipole mode energy E_D when the boundary condition R slowly vibrates. Then :

$$E_D \simeq \frac{1}{R[\alpha_{2\mu}(H)]}$$

Since $w_Q = 1/10$ to $1/20$ of w_D one will use an adiabatic approximation and express such a coupling by introducing an interaction hamiltonian H_{DQ} in the total hamiltonian :

$$H = H_D + H_Q + H_{DQ}$$

which was then diagonalized on the basis of the hamiltonian $H_0 = H_D + H_Q$ (whose eigenfunctions are characterized by the number $N_{\jmath}$ of surface phonons) It was then shown that the non diagonal matrix elements, due to the coupling term H_{DQ}, impart a spreading of the dipole strength into several intermediate collective states. An exhaustive study of just these intermediate collective states as a function of β and E_{2+}, but considering simple harmonic vibrations only, has been undertaken by Huber |.III.17|, Arenhövel |III.18| and

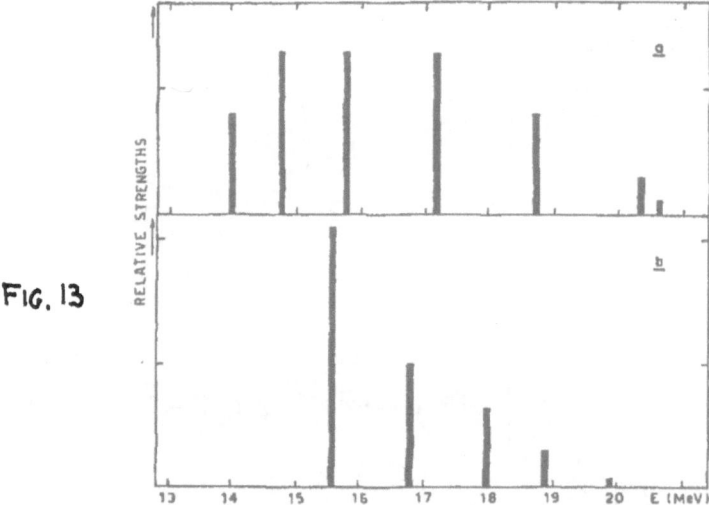

FIG. 13

Spreading of the dipole strength as predicted by Huber *et al.* within the framework of the dynamic collective model for two typical nuclei, (a) nucleus with $\beta = 0.25$ and $E_{2+} = 0.6$ MeV; (b) nucleus with $\beta = 0.50$ and $E_{2+} = 0.9$ MeV.

Urbas et al. |III.19|. Their results indicate that the spreading ought to increase whenever β increases and E_{2+} decreases. Fig. 13 reproduces Huber's computations which show these phenomena.

III.D.1b Comparison with experimental data.

A systematic attempt to check these theoretical prescriptions has been undertaken at Saclay and the corresponding results will be summarized below.

However,it is necessary to first point out that some attempts to identify those intermediate collective states,which were carried out |III.20| by unfolding bremsstrahlung yield data (Fig. 14) have not been confirmed by later experiments with monochromatic γ-rays |III.21|. (Fig. 15)

FIG. 14 :
reproduced from
ref.III.20.

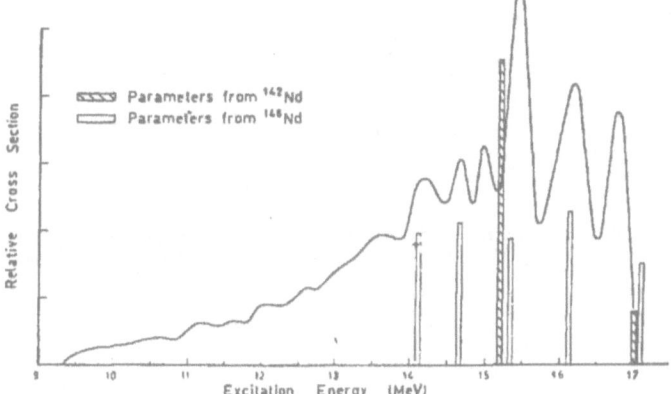

Comparison of the $^{141}Pr(\gamma, n)^{140}Pr$ cross section with the theoretical predictions based on the use of the low-energy properties of the ^{142}Nd and ^{146}Nd nuclei.

Therefore the only "indirect way" to compare the experimental data with the DCM predictions lies in a comparison with the experimental width of the GDR, Γ , which can be written as a convolution of 3 separate widths :

1) $\Gamma^{\uparrow}$ = direct decay width of the dipole state which is generally neglegible for medium and heavy nuclei ($\dfrac{\Gamma^{\uparrow}}{\Gamma} < 0.15$)

2) $\Gamma^{\downarrow}$ = the damping width of a dipole line at E_D which corresponds mainly to the dilution of the 1p-1h doorway states into more complicated 2p-2h states of density $\rho(E_D)$ at the dipole energy E_D :

$$\Gamma^{\downarrow} \simeq 2\pi \; \overline{< 2p \cdot 2h \,|\, V \,|\, 1p \cdot 1h >}_{J^{\pi}=1^-} \cdot \; \rho^{(E_D)}$$

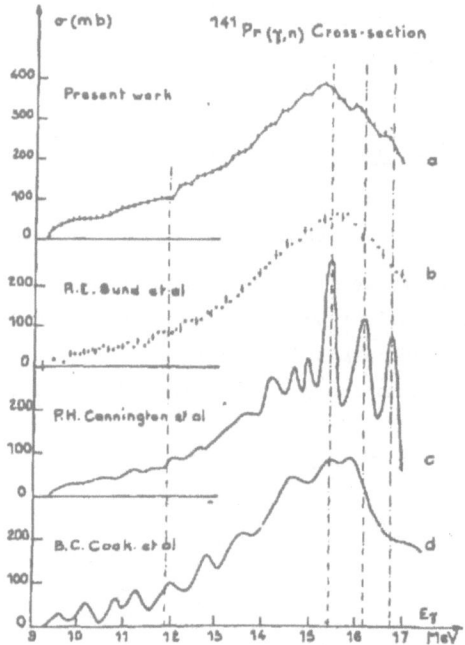

The $\sigma(\gamma, n)$ cross sections for ^{141}Pr as observed by several authors. Curves a) and b) are obtained with monochromatic photons whereas c) and d) are obtained with bremsstrahlung photons.

FIG. 15

It is known that $\Gamma^{\downarrow}$ can be, as a first approximation, represented as a slowly varying function of E_D fitted by the law $|III.22|$:

$$\Gamma^{\downarrow}_{MeV} \simeq (0.026 \pm 0.005) E^{1.91 \pm 0.1}_{MeV}$$

3) $\Delta\Gamma$ = a term which characterizes the splitting of the dipole strength predicted by the above version of the DCM.

Now, if one studies neighbour nuclei, and in particular neighbour isotopes, E_D will hardly vary and therefore the observed experimental width will be (nearly) directly dependent on $\Delta\Gamma$ only.

Such a comparison has been attempted in the region $100 < A < 135$ $|III.23|$ where some nuclei were found to have low energy "vibrational spectra" although there exists no nucleus which is truly a good harmonic spherical vibrator. Several series of isotopes will be discussed below.

1) The Sn isotopes.

For these nuclei |III.23|, the parameters β and E_{2^+} are practically constant, as can be seen in table III, and β is rather small. Hence it does not come as a surprise that Arenhövel's calculations predicted a small but essentially constant, spreading of the GDR for all tin isotopes as can be seen in Fig. 16.

Table III

Low energy parameters β and E_{2^+} of doubly even Sn isotopes.

	^{116}Sn	^{118}Sn	^{120}Sn	^{122}Sn	^{124}Sn
β	0.113	0.116	0.112	0.118	0.108
E_{2^+} (MeV)	1.293	1.230	1.175	1.140	1.131

FIG. 16

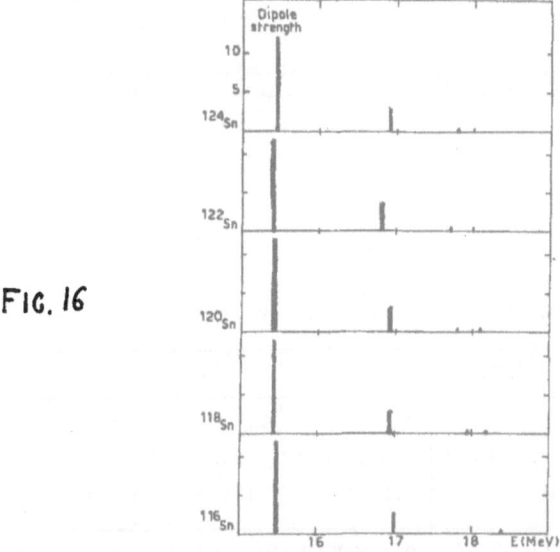

Spreading of the dipole strength for the tin isotopes as predicted by Arenhövel

The Saclay experimental data are summarized in Fig. 17 and Table IV gives the corresponding single Lorentz line parameters.

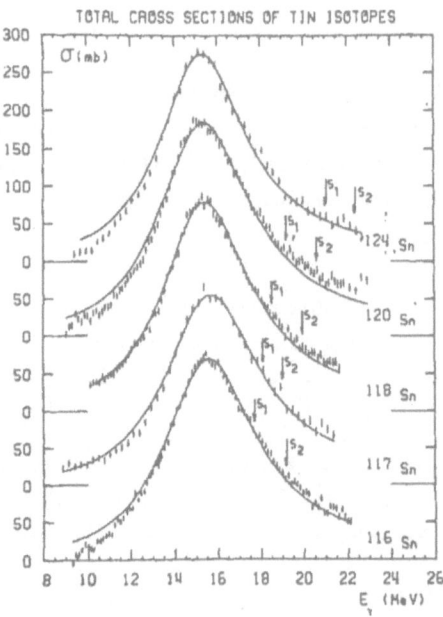

Best single Lorentz line fit to $\sigma_{exp}(\gamma, tot) = \sigma(\gamma, n) + \sigma(\gamma, pn) + \sigma(\gamma, 2n)$ for the tin isotopes. Arrows S_1 and S_2 refer to isobaric analogue states as observed by Sugawara using (e, e'p) reactions.

FIG. 17

Table IV

Lorentz line parameters for a single Lorentz line fit.

	^{116}Sn	^{117}Sn	^{118}Sn	^{120}Sn	^{124}Sn
σ_0 (mb)	270	255	278	284	275
Γ_0 (MeV)	5.21±0.1	5.30±0.1	4.99±0.1	5.25±0.1	4.96±0.1
E_0 (MeV)	15.57	15.67	15.44	15.38	15.29
Values of E_0 are given to within ±0.1 MeV. Values of σ_0 are given to within ±5 mb.					

One therefore found $\Gamma_{exp} = 5.15 \pm 0.15$ MeV for all Sn isotopes. The constant value can be taken to be a confirmation of the identical spreading predicted by the DCM for all the Sn isotopes.

2) The Te, Cd and Pd nuclei |III.23|.

A study of the GDR of vibrational nuclei close to the Sn isotopes, and for which β and E_{2+} are different from the values observed for these Sn isotopes, seems also very interesting. Meaningful results, showing a possible correlation with the predictions of the DCM, can be expected for Te, Cd and Pd nuclei, even with natural targets. For these nuclei, the values of the parameters β and E_{2+} for the main isotopes change very little with respect to the mean values $\bar{\beta}_0$ and $\bar{E}_{2+}$ presented in table V.

Table V

Mean values $\bar{\beta}_0$ and $\bar{E}_{2+}$ of the low energy parameters β and E_{2+} for Te, Cd and Pd nuclei

	Te	Cd	Pd
$\bar{\beta}_0$	0.145±0.018	0.190±0.010	0.230±0.020
$\bar{E}_{2+}$	0.750±0.090	0.580±0.070	0.460±0.080
$E_1 = 78A^{-\frac{1}{3}}$ (MeV)	15.4	16.0	16.3
$\beta'_0 = \bar{\beta}_0(\frac{1}{18} E_1)$	0.125	0.168	0.208

The β'_0 and E_1 notation is explained in the text.

Since Huber's calculations on the spreading were performed for E_1 = 18 MeV and since (β/E_1) is the parameter which contributes most to the spreading, due to the H_{DQ} coupling, one is obliged to use $\beta'_0 = \bar{\beta}_0 \frac{E_1}{18}$ so as to find theoretical cases as close as possible to the required Te, Cd and Pd nuclei. We thus show in Fig.18, a selection of possible theoretical cases illustrating the spreading of the GDR, each representing as closely as possible, one of the real nuclei Te, Cd and Pd. In particular, Fig. 18a shows the result, obtained with β'_0 = 0.150 and E_{2+} = 0.9 MeV, a case which closely resembles tellurium where :

$$\frac{\beta'_0}{E_{2+}} = \frac{0.125}{0.750} = \frac{1}{6} = \frac{0.15}{0.90}$$

In Fig. 18b the result shown is for β'_0 = 0.150 and E_{2+} = 0.6 MeV which resembles the cadmium case but with a somewhat smaller theoretically predicted spreading.

In Fig. 18c we used β'_0 = 0.2 and E_{2+} = 0.6 MeV, a case which is very similar to palladium but again with a somewhat smaller predicted spreading.

If one now compares the prognostications of Fig. 16 with those of Fig. 18, one observes that about 90 % of the dipole strength is concentrated in the following energy intervals ΔE :

$\Delta E = 1.5$ MeV for any of the Sn isotopes ;

$\Delta E = 2.2$ MeV for Te ;

$\Delta E = 2.7$ MeV for CD ;

$\Delta E = 4$ MeV for Pd.

Let us now assume that the damping widths, associated with each of the dipole states which characterize the intermediate collective strengths displayed in Figs. 16 and 18 are constant. It then is to be expected that the experimentally observed GDR widths Γ will show the following relations : Γ(any Sn isotopes) = constant = Γ(Sn) ; $\Gamma(^{Nat}Te) = \Gamma$(Sn) + 0.7 MeV ; $\Gamma(^{Nat}Cd) = \Gamma$(Sn) + 1.2 MeV ; $\Gamma(^{Nat}Pd) = \Gamma$(Sn) + 2.5 MeV.

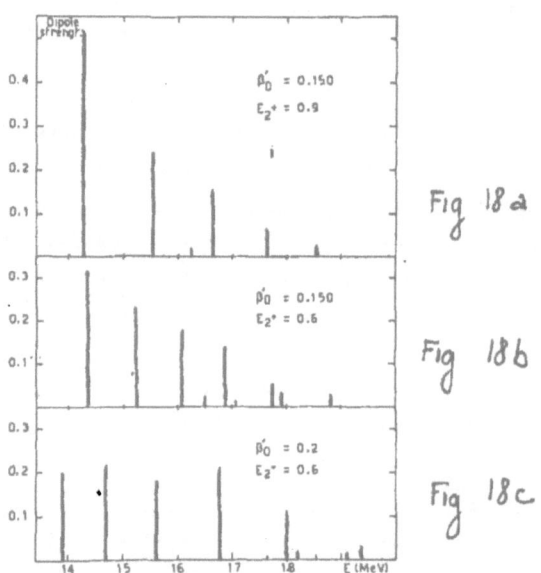

Theoretical spreading of the dipole strength as computed by Huber   for three typical vibrational nuclei.

FIG. 18

When one normalizes $\dfrac{\Gamma_{exp}}{\Gamma_{(DCM)}}$ for Sn to 1, Fig. 19 shows that the theoretical broadening predicted by the DCM for Te, Cd and Pd nuclei, is fairly well confirmed.

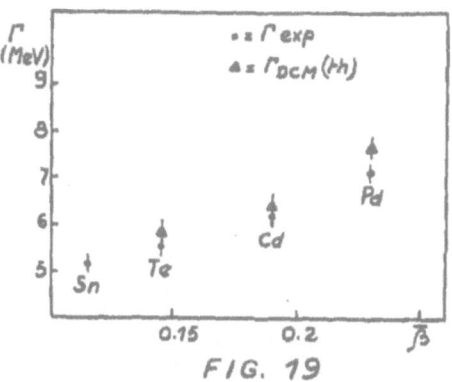

FIG. 19

3) The Mo isotopes |III.24|.

Here again, the total photoneutron cross sections show a variation of the FWHM value Γ (Fig. 20) which varies linearly with the two characteristic parameters of the theory β_2 and $\beta_2/E(2^+)$. (Fig. 21)

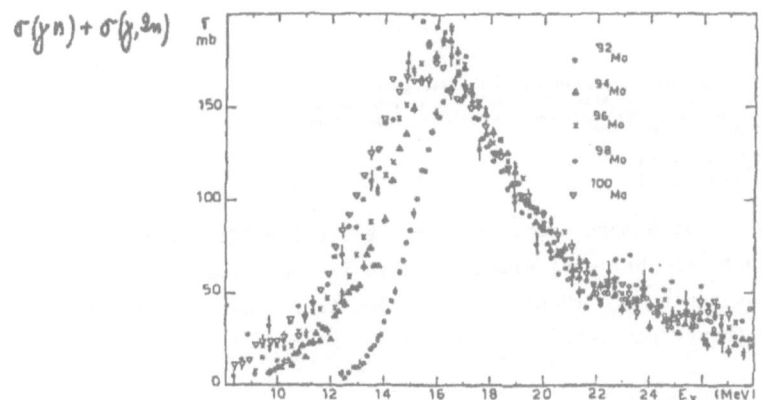

Experimental total photoneutron cross sections for molybdenum isotopes.

FIG. 20

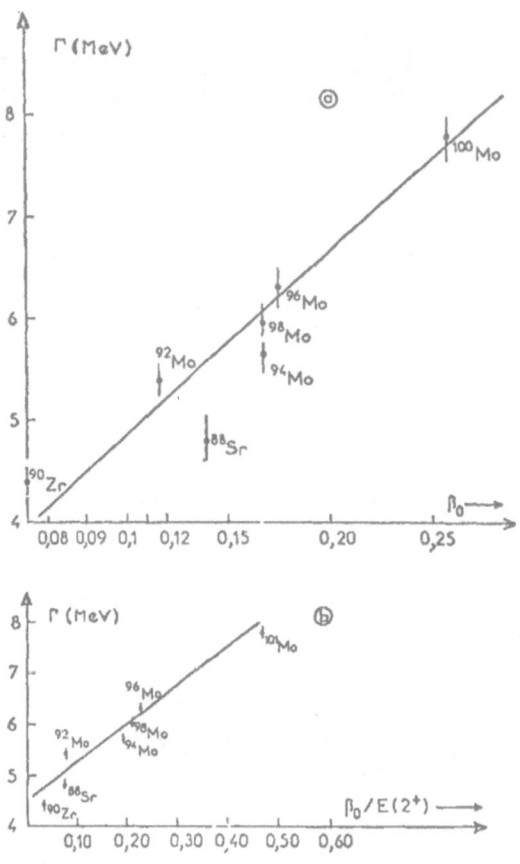

Fig. 21 : The experimental FWHM value $\Gamma = \Gamma$ of the giant dipole resonance as a function of the deformation parameter β (a) and as a function of $\beta/_{E_2+}$ for different nuclei around $A = 90$ (reproduced from ref. III.24).

4) The Nd isotopes |III.25|.

This case will be treated more completely in the section III.D.2 devoted to the improved DCM but, if they were simply treated in comparison with the above simple DCM, once again Γ_{exp} (GDR) will be shown to follow closely the variation of the low energy parameters β_2 and $1/E_{2+}$. (Fig 22a , 22b).

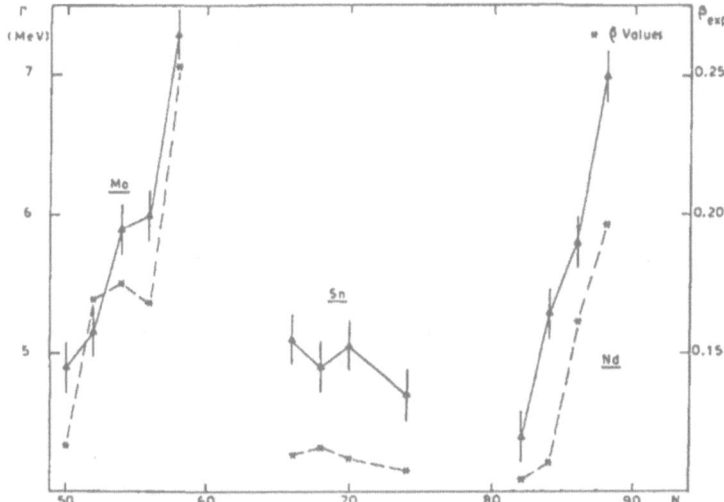

FIG. 22 a

Experimental Γ-values ($\blacktriangle$) compared to β-values ($\bullet$) obtained from the literature
both as a function of neutron number N.

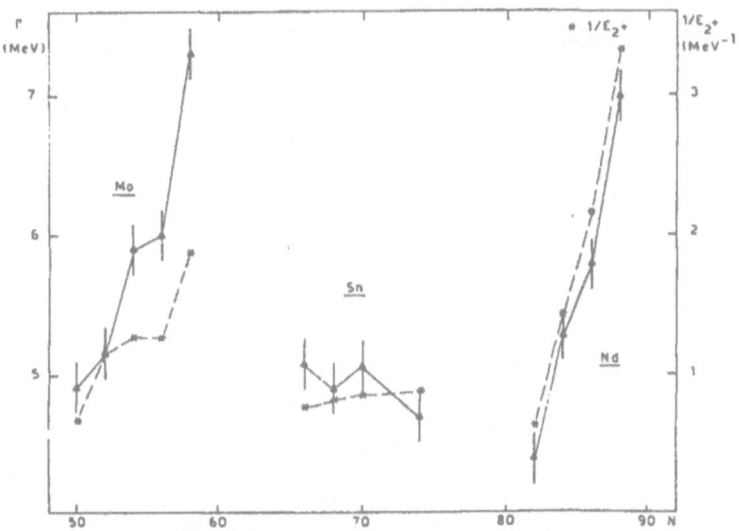

FIG. 22 b.

Experimental Γ-values ($\blacktriangle$) compared to $1/E_{2^+}$ ($\bullet$) values obtained from the literature
both as a function of neutron number N.

5) General summary of the experimental data.

For a series of nuclei studied at Saclay, Fig. 23 a shows the modulation of Γ (GDR) versus A. The modulation of the deformation parameter β_2 against A is shown to be very similar in Fig. 23 b.

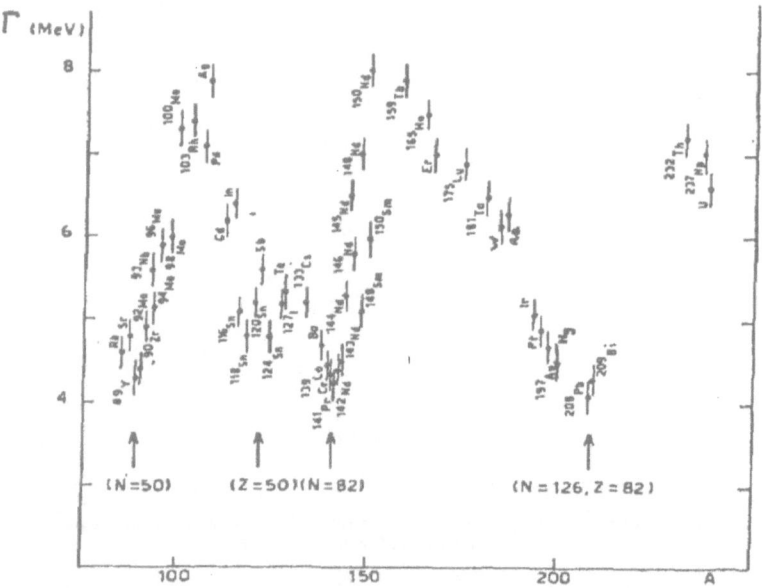

Fic 23a:Experimental FWHM values Γ of the σ/E curves for $90 \leqslant A \leqslant 238$ as obtained at Saclay from the experimental total photoneutron cross sections σ.

FIG. 23 b : (communicated by G. Solodukhov).

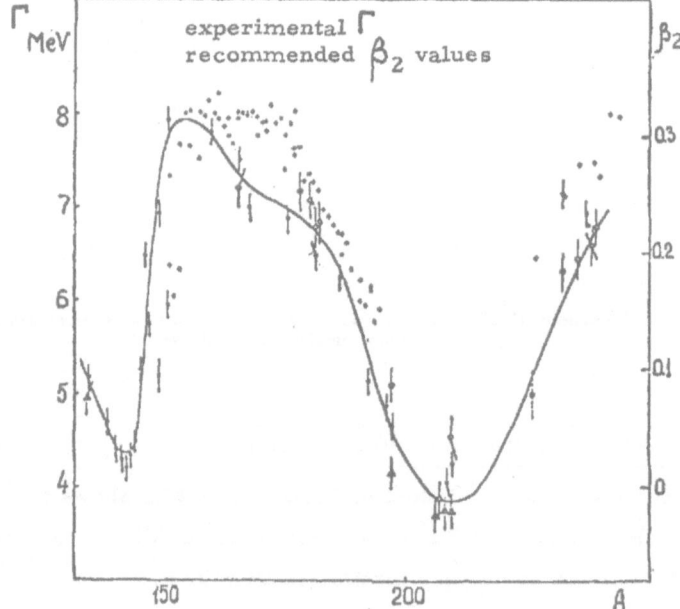

<u>NB</u> : Although such a parameter is not directly used in the DCM, one can notice that the ratio $E(4^+)/E(2^+)$, which is the fundamental parameter governing the "softness" of nuclei in the phase transition theory of Buck-Goldhaber and Mariscotti |III.26|, shows a remarkable correlation with the experimental width of the GDR. (Fig. 24)

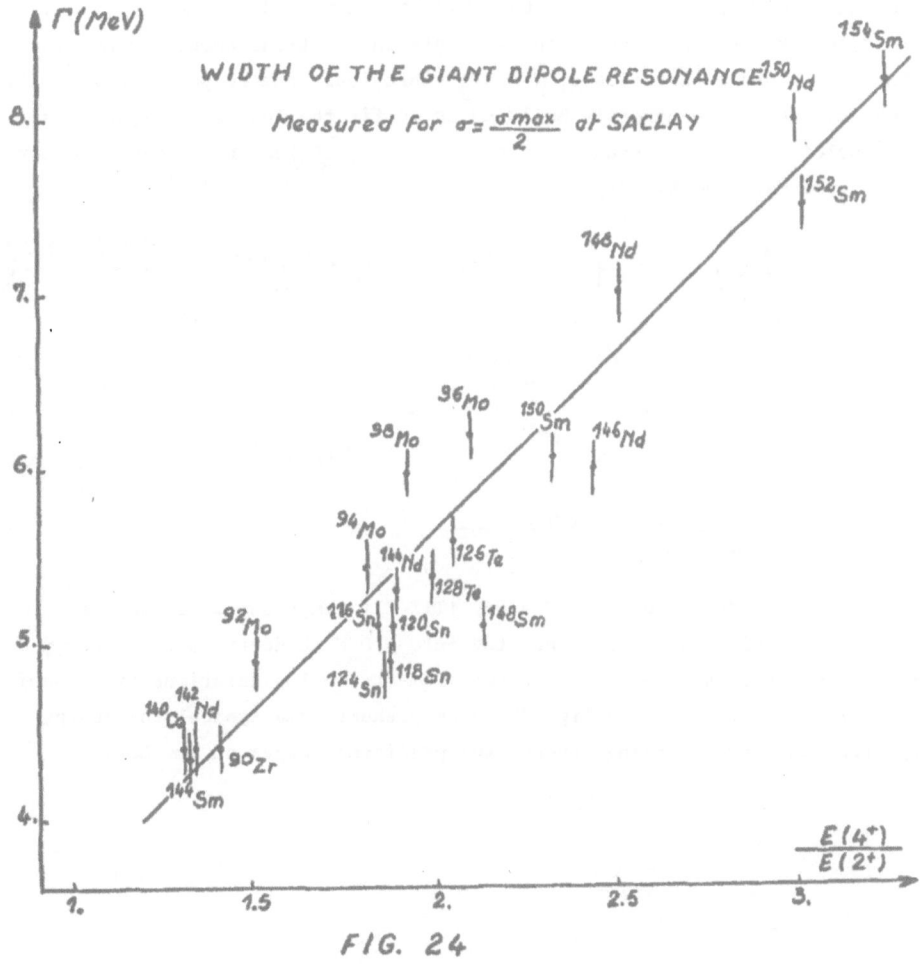

FIG. 24

III.D.2 The improved dynamic collective model.

III.D.2a Theoretical summary.

The previous simple DCM was applicable only to strongly deformed rotators or to spherical harmonic vibrators. They could take into account neither anharmonicities nor rotation-vibration-interactions. Therefore they could not predict any feature of the GDR in a transitional region such as the one around N ≃ 89 or the one around A ≃ 190.

Around 1970, the Frankfurt group of theoretical physics improved the DCM by using potential energy surfaces (PES) / III. 27 / to describe the low energy properties of (vibrational, rotational or transitional) nuclei.
It is well known that such surfaces supply informations about the nuclear shapes, the characteristics of β and γ vibrations, the shape isomerism, ...
Such surfaces are ordinarily built so as to fit the low energy spectra of the nuclei and are represented as a function $V(\beta,\gamma)$ of the intrinsic deformation parameters β , γ. (Fig. 25)

$$R(\theta,\varphi) = R_0 \left\{ 1 + \beta \cos\gamma \; Y_{20}(\theta,\varphi) + \frac{1}{\sqrt{2}} \beta \sin\gamma \left[Y_{22}(\theta\varphi) + Y_{2,-2}(\theta,\varphi) \right] \right\}$$

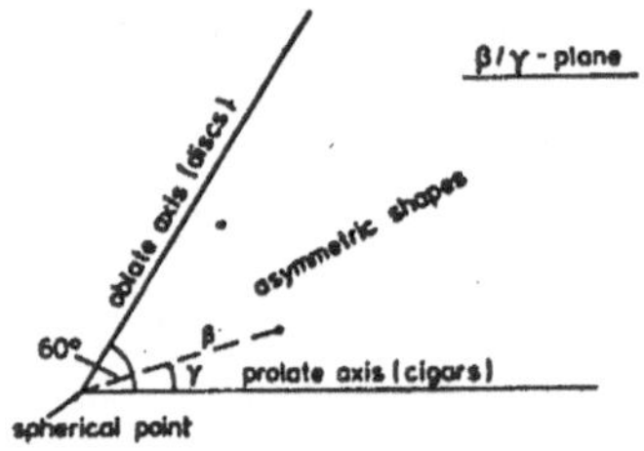

Fig. 25

From such PES, Gneuss and Greiner |III.27| construct a collective hamiltonian which may be incorporated into a DCM to describe high energy excitations such as the dipole states. Some typical predictions by Rezwani |III.27| are summarized in Fig. 26 which gathers some typical low energy spectra, the corresponding PES and the predicted shapes of the GDR.

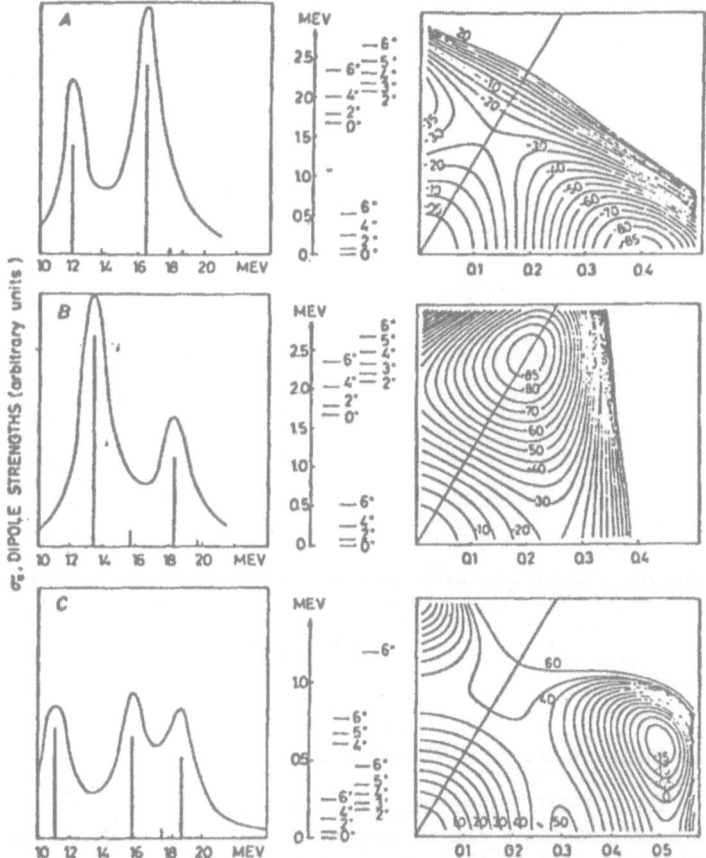

Fig. 26: Dipole strengths, γ-absorption cross sections, low-energy spectra, and potential energy surfaces for (a) a prolate, (b) an oblate, and (c) an asymmetric deformed nucleus reproduced from ref. III. 27.

Fig. 27 shows the spreading and the splitting of the dipole strength in some more realistic case :

A) soft spherical nucleus which exhibits a larger splitting than for a pure harmonic vibrator (dashed lines) ;

B) intermediate case between A and C ;

C) strongly deformed prolate nucleus where the higher energy modes are split because the low lying γ-band makes the nucleus dynamically triaxial ;

D) strongly deformed nucleus without appreciable excitation of vibrational satellites.

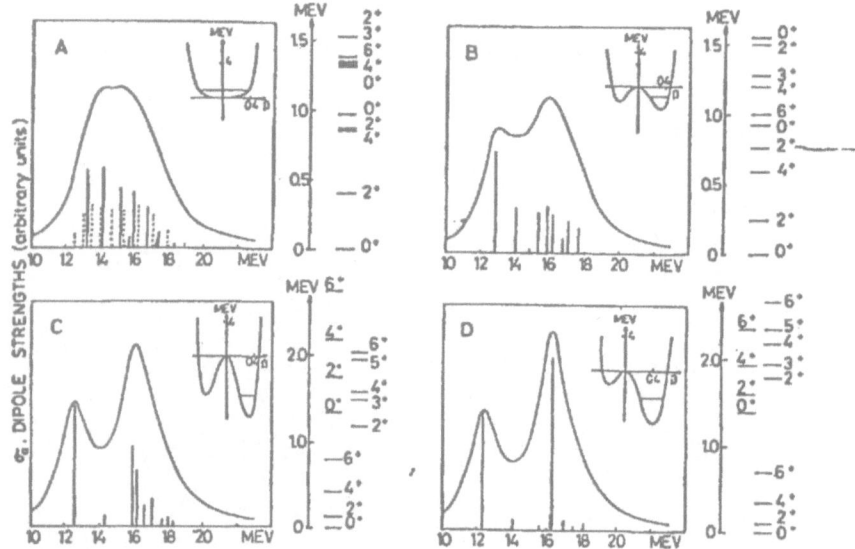

Dipole strengths, γ-absorption cross sections, low-energy spectra, and cuts through the potential energy surface along γ = 0 and γ = ½π for the transition study.

FIG. 27   [III. 27]

III.D.2b Comparison with experimental data.

III.D.2b.1 – The transition region N ≃ 89. Around N ≃ 89, a well known transition
region exists which links a rather spherical ground state shape for N ≪ 88
nuclei to a permanently deformed prolate shape for N ≥ 90 nuclei. The
softness of the nuclei increases from the spherical N = 82 nuclei up to
N = 88 nuclei. (Fig. 28)

Systematic studies of $\sigma(\gamma, n) + \sigma(\gamma, pn) + \sigma(\gamma, 3n) \simeq \sigma(\gamma, tot)$ were under-
taken at Saclay for Nd |III.25| and Sm |III.28| isotopes. The experimental
results are reproduced by Figs. 29 and 30.

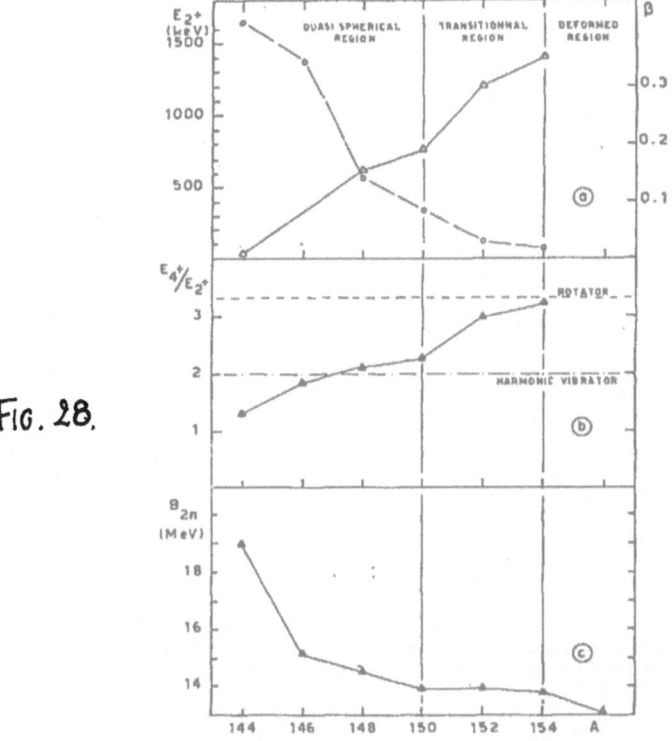

Fig. 28.

Variation of different parameters for doubly even samarium isotopes. (a) Deformation parameter β and $E(2^+)$ the energy of the first (2^+) excited level. (b) The ratio $[E(4^+)/E(2^+)]$ where $E(4^+)$ is the energy of the first (4^+) excited level. (c) The B_{2n} thresholds corresponding to the $(\gamma, 2n)$ exit channels.

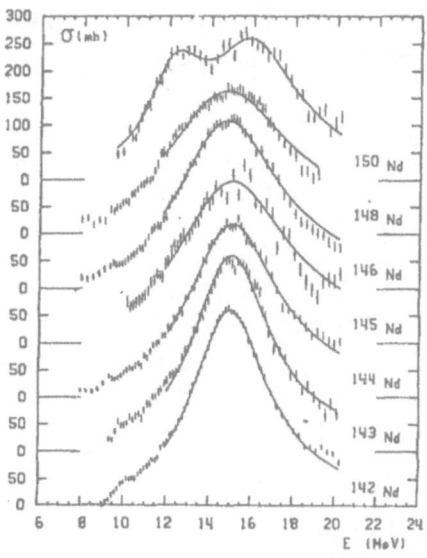

FIG. 29 :   Lorentz-line fits of the $\sigma_T(E)$ curves of neodymium isotopes.

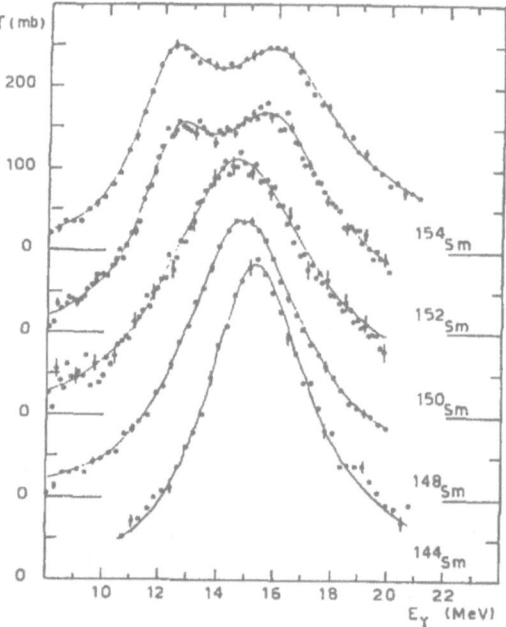

Fɪɢ. 30: Total photoneutron cross sections $\sigma_T(E)$ of the doubly even samarium isotopes. Best single Lorentz line fits are shown for ^{144}Sm, ^{148}Sm and ^{150}Sm. For ^{152}Sm and ^{154}Sm the best two Lorentz line fits are presented.

Fig. 31 shows that the IDCM / III. 29 / (predictions agree with) these experimental data and in particular with the continuous broadening of the GDR as N increases. The small discrepancy is certainly due to the theoretical consideration of too small a damping width $\Gamma^{\downarrow}$, which is not actually a part of the IDCM prediction.

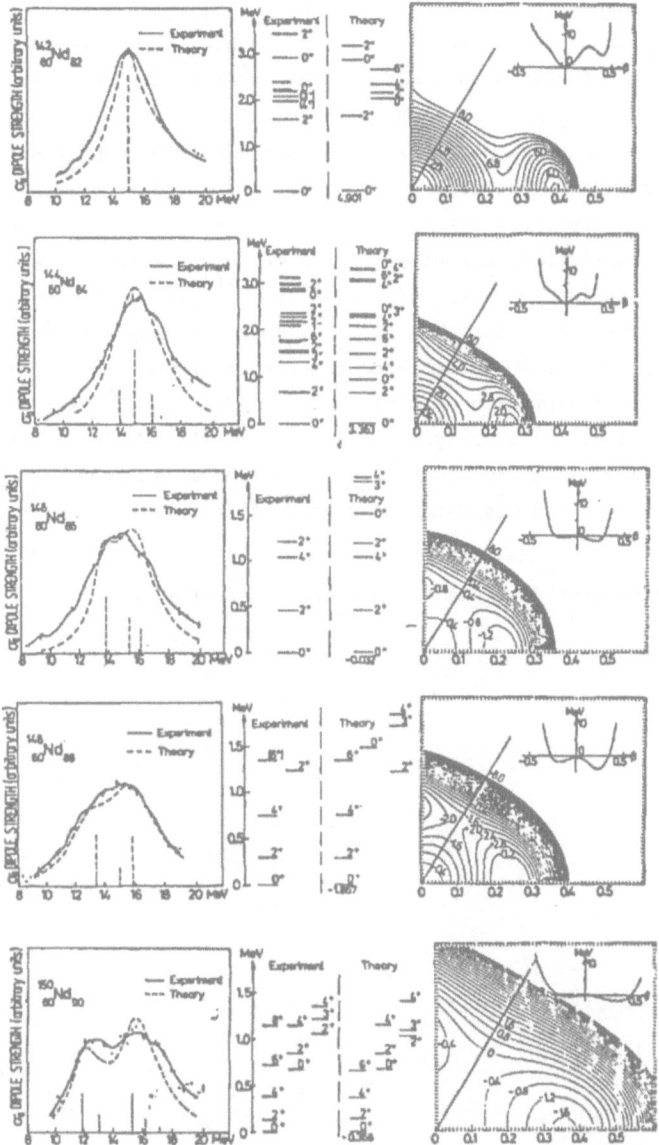

Fig. (31) : The 5 figures on the left compare the theoritical predictions of the IDCM (dotted lines)with the Saclay experimental determinations of the giant resonances of Nd isotopes (solid lines).
The 5 figures on the right show the potential energy surfaces (fitting the low-energy excited levels in Nd isotopes) which allowed the IDCM predictions of the shapes of the giant resonances. (Reproduced from ref. III.29).

III.D.2b.2 <u>The transitional region around A = 190</u> |III.30|. Potential
energy surfaces for nuclei in the A = 190 region are now sufficiently well
known for them to be used in an evaluation of the strength and energy dis-
tribution of the dipole states of the osmium |III.31| isotopes 184, 186, 188,
190 and 192. These theoretical results show a gradual GDR evolution from
(^{184}Os, ^{186}Os) towards a triaxial, gamma vibration unstable nucleus (^{192}Os).
It thus seems that any interpretation of the GDR form of natural osmium
will be rather complicated. One notes also a certain similarity between
the low energy rotational spectra of the (^{184}W - ^{184}Os) and (^{186}W - ^{186}Os)
nuclei, a behaviour also observable from the corresponding calculated P.E.S.
results. Further, a certain resemblance between the asymmetric rotational
spectra of the gamma vibration unstable nuclei ^{192}Os and (^{194}Pt - ^{196}Pt)
is again confirmed by the corresponding P.E.S. results obtained by Kumar
and Baranger |III.32|. It thus appears as if a systematic study of the GDR
in this transition region might not be devoid of interest even with natural
targets.

Saclay experimental data are again presented on Fig. 32.

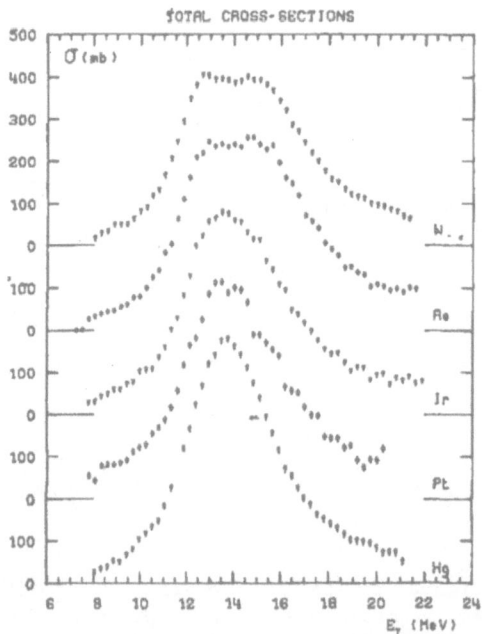

Fig. 32 — Sections efficaces totales $\sigma_T(E)$ des noyaux W, Re,
Ir, Pt, Hg.

As A increases in Fig. 30, one notes that the splitting of the GDR becomes less pronounced and Γ = 5.2, 5 and 4.4 MeV for Ir, Pt and Hg respectively. The GDR of these nuclei also shows a certain amount of a--symmetry with respect to the energy position E_0 of the maximum of the GDR. Such a asymmetry was not observed in the GDR of spherical nuclei |III.25,28| with a single (N = 82) or doubly |III.33| closed (^{208}Pb) shell. This additional strength of the GDR towards energies greater than E_0 could be interpreted within the framework of the DCM where for instance |III.31| several states at E $>E_0$ characterize the γ-vibration instability of ^{192}Os as shown in Fig. 33.

Fig. 33

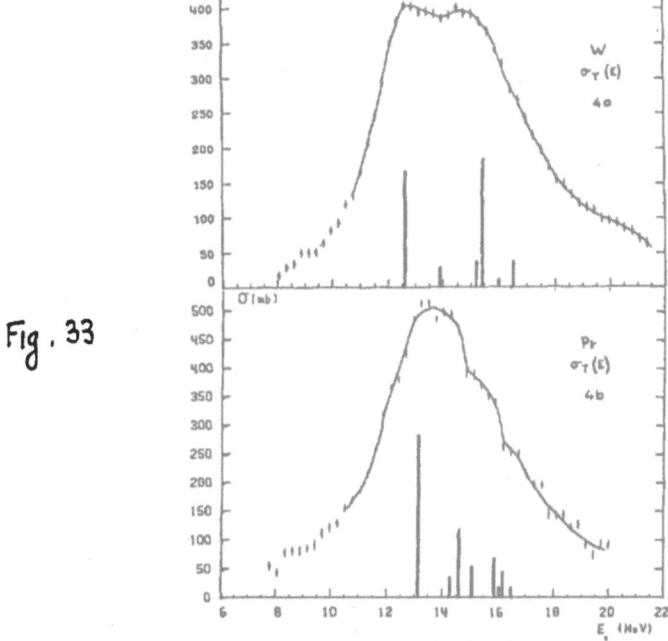

— Comparaison des sections efficaces totales expérimentales
$\sigma_T(E)$ des noyaux W et Pt avec les prédictions du modèle collectif
dynamique   positionnées de façon à faire coïncider les énergies
moyennes observées et calculées de la RGD.

Finally it should be remembered that the DCM predicts the strength and energy position of the principal dipole states but says nothing about their respective damping widths $\Gamma^{\downarrow}$. A comparison between recent experimental results |III.22| and a theoretical survey |III.34| led us to adopt a value of $\Gamma^{\downarrow}$ 3 MeV$< \Gamma^{\downarrow} <$ 4 MeV. It seems therefore quite improbable that photon absorption experiments will be capable to discern and identify the individual dipole states predicted by the DCM for just such, gamma vibration unstable, oblate or triaxial nuclei.

III.D.3 Conclusion.

Even if, as pointed out above, the effects of the damping width $\Gamma^{\downarrow}$ somewhat blur the direct connection of the experimental data with the predicted splitting of the dipole strength, it remains that the various collective models predict remarkably well (for medium and heavy nuclei at least) the fundamental interplay.

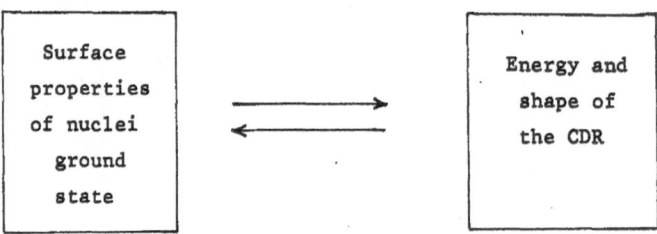

BIBLIOGRAPHY

|III.1| Photonuclear reactions, by E. HAYWARD in "Nuclear Structure and Electromagnetic Interactions." (Oliver and Boyd, 1964).

|III.2| J.M. EISENBERG and W. GREINER, Nuclear theory, vol. 1 (North Holland, 1970).

|III.3| M.G. HUBER, Am. J. Phys. 35, 685 (1967).

|III.4| R. BACH and C. WERNTZ, Phys. Rev. 173, 958 (1968).

|III.5| M. GOLDHABER and E. TELLER, Phys. Rev. 74, 1046 (1948).

|III.6| A. STEINWEDEL and J. JENSEN, Z. Naturforsch. 5, 413 (1950).

|III.7| B.L. BERMAN and S.C. FULTZ, Rev. Mod. Phys. 47, 713 (1975).

|III.8| M. DANOS, Nucl. Phys. 5, 23 (1958).

|III.9| W.D. MYERS and W.J. SWIATECKI, Nucl. Phys. 81, 1 (1966), and Ann. Phys. (N.Y.) 55, 395 (1969).

|III.10| K. OKAMOTO, Phys. Rev. 110, 143 (1958).

|III.11| M. DANOS and W. GREINER, Phys. Rev. 138, B876 (1965).

|III.12| A. BOHR and B.R. MOTTELSON, Nuclear structure (Benjamin).

|III.13| R. BERGERE, H. BEIL, P. CARLOS and A. VEYSSIERE, Nucl. Phys. A133, 417 (1969).

|III.14| M.A. KELLY, B.L. BERMAN, R.L. BROMBLETT and S.C. FULTZ, Phys. Rev. 179, 1194 (1969).

|III.15| J. LE TOURNEUX, Phys. Letters 13, 325 (1964).

|III.16| M. DANOS and W. GREINER, Phys. Letters 8, 113 (1964).

|III.17| M.G. HUBER, M. DANOS, H.J. WEBER and W. GREINER, Phys. Rev. 155, 1073 (1967).

|III.18| H. ARENHOVEL, M. DANOS and W. GREINER, Phys. Rev. 157, 1109 (1967).

|III.19| T. URBAS and W. GREINER, Z. Physik 196, 44 (1966).

|III.20| P.H. CANNINGTON, R.J.J. STEWARD, B.M. SPICER and M.G. HUBER, Nucl. Phys. A109, 385 (1968).

|III.21| H. BEIL, R. BERGERE, P. CARLOS, A. LEPRETRE, A. VEYSSIERE and A. PARLAG, Nucl. Phys. A172, 426 (1971).

|III.22| P. CARLOS, R. BERGERE, H. BEIL, A. LEPRETRE and A. VEYSSIERE, Nucl. Phys. A219, 61 (1974).

|III.23| A. LEPRETRE, H. BEIL, R. BERGERE, P. CARLOS, A. DE MINIAC, A. VEYSSIERE and K. KERNBACH, Nucl. Phys. A219, 39 (1974).

|III.24| H. BEIL, R. BERGERE, P. CARLOS, A. LEPRETRE, A. DE MINIAC and
A. VEYSSIERE, Nucl. Phys. A227, 427 (1974).

|III.25| P. CARLOS, H. BEIL, R. BERGERE, A. LEPRETRE and A. VEYSSIERE,
Nucl. Phys. A172, 437 (1971).

|III.26| M.A.J. MARISCOTTI, G.S. GOLDHABER and B. BUCK, Phys. Rev. 178,
1864 (1969).

G.S. GOLDHABER and A.S. GOLDHABER, Phys. Rev. Letters 24, 1349
(1970).

|III.27| V. REZWANI, G. GNEUSS and H. ARENHOVEL, Phys. Rev. Letters 25,
1667 (1970).

G. GNEUSS and W. GREINER, Nucl. Phys. A171, 449 (1971).

V. REZWANI, G. GNEUSS and H. ARENHOVEL, Nucl. Phys. A180, 254
(1972).

|III.28| P. CARLOS, H. BEIL, R. BERGERE, A. LEPRETRE, A. DE MINIAC and
A. VEYSSIERE, Nucl. Phys. A225, 171 (1974).

|III.29| V. REZWANI, M. BRAND, R. SEDLMAYR, L.V. BERNUS, U. SCHNEIDER and
W. GREINER, in "Nuclear Structure Studies Using Electron Scattering
and Photoreaction.", Sendai Conference 1972.

|III.30| A. VEYSSIERE, H. BEIL, R. BERGERE, P. CARLOS, A. LEPRETRE and
A. DE MINIAC, J. Phys. 36, 267 (1975).

|III.31| R. SEDLMAYR, M. SEDLMAYR and W. GREINER, Nucl. Phys. A232, 465
(1974).

|III.32| K. KUMAR and H. BARANGER, Nucl. Phys. A122, 273 (1968).

|III.33| A. VEYSSIERE, H. BEIL, R. BERGERE, P. CARLOS and A. LEPRETRE,
Nucl. Phys. A159, 561 (1970).

|III.34| E.D.M. SHELIA, K. ROOS and W. GREINER, Nucl. Phys. A212, 157 (1973).

|III.35| M.A. PRESTON, Physics of the Nucleus Addison, Wesley 1965.

CHAPTER IV

THE MICROSCOPIC MODELS OF THE GIANT E 1 RESONANCE

Since the electromagnetic interaction, which is responsible for the photonuclear excitation of the GDR, acts through a one-body operator, it first creates a 1 particle - 1 hole state in the target nucleus. Therefore such a state should be particularly amenable to a microscopic description.

Actually, in the simplest harmonic oscillator shell model, the $|1p-1h\rangle$ states (coupled to 1^-) induced in a (G.S.) = 0^+ nucleus, correspond to the jump of 1 particle to the next upper shell and all these $|1p-1h\rangle_{1^-}$ states are gathered in energy around $41 \, A^{-1/3}$ MeV. However one knows that the dipole strength is actually concentrated around an energy E_D twice as large.

IV.A The schematic model of Brown-Bolsterli.

Brown-Bolsterli $|IV.1|$ in their first schematic model, in 1959, showed that the residual particle-hole interaction builds up a collective state which is a coherent sum of the above $|1p-1h\rangle$ component states, which is pushed upwards in energy and carries almost all the dipole strength. A summary of this theory is given below, following G. Brown's book "Unified theory of nuclear models." $|IV.2|$.

The presence of a particle hole interaction implies that the particle-hole states $|mi\rangle$, $|nj\rangle$... will all get mixed. Hence the eigenstate $|\psi\rangle$ with energy E, associated to the hamiltonian H, will be given by the classical secular equation :

$$H|\psi\rangle = E|\psi\rangle$$

$$|\psi\rangle = \sum_{mi} c_{mi} |mi\rangle$$

$$\sum_{nj} \langle mi | H | nj \rangle c_{nj} = E c_{mi}$$

with non diagonal matrix elements such as :

$$\langle mj \,|\, V \,|\, ni \rangle \quad\quad \simeq$$

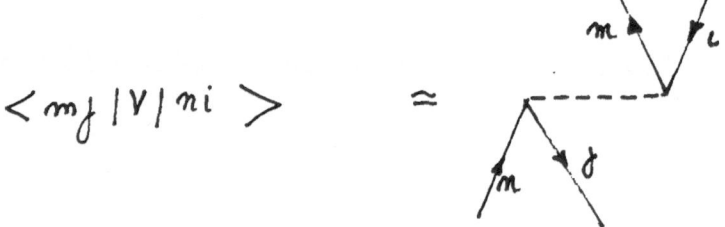

Brown and Bolsterli showed that in the simple case of the absorption of a γ with multipolarity J, for which one knows the interaction hamiltonian δH, and therefore the absorption amplitude D_{mi}^{J} :

$$\delta H = Ce \sum_{mi} \langle m \,|\, r^{J} Y_{JO} \,|\, i \rangle \, a_{m}^{+} a_{i}$$

$$D_{mi}^{J} = \langle mi \,|\, \delta H \,|\, 0 \rangle$$

G. Brown shows that: $\langle mj \,|\, V \,|\, ni \rangle \simeq \lambda \, D_{mi} \cdot D_{nj}$

One can therefore write, successively, with classical notations :

$$H = H_0 + V \qquad (H_0 = \text{single particle operator})$$

$$H_0 |mi\rangle = (\varepsilon_m - \varepsilon_i) |mi\rangle$$

$$(H_0 + V) |\psi\rangle = E |\psi\rangle$$

$$\sum_{nj} (H_0 + V) c_{nj} |nj\rangle = E \sum_{mi} c_{mi} |mi\rangle$$

$$(\varepsilon_m - \varepsilon_i) c_{mi} + \sum_{nj} \langle mi | V | nj \rangle c_{nj} = E c_{mi}$$

$$(E - \varepsilon_m + \varepsilon_i) c_{mi} = \sum_{nj} \langle mi | V | nj \rangle c_{nj}$$

$$= \lambda D_{mi} \sum_{nj} c_{nj} D_{nj}$$

$$\text{with} \quad \sum_{nj} c_{nj} D_{nj} = \text{Constant}$$

Hence :

$$c_{mi} = \frac{\lambda D_{mi}}{E - \varepsilon_m + \varepsilon_i} \sum_{nj} c_{nj} D_{nj}$$

$$\sum_{mi} c_{mi} D_{mi} = \sum_{mi} \frac{\lambda D_{mi}^2}{E - \varepsilon_m + \varepsilon_i} \sum_{nj} c_{nj} D_{nj}$$

$$1 = \sum_{mi} \frac{\lambda D_{mi}^2}{E - \varepsilon_m + \varepsilon_i}$$

This equation is classically solved graphically (Fig. 1) by plotting along Ox the unperturbed p-h energy $\varepsilon_{mi} = \varepsilon_m - \varepsilon_i$. If $\lambda > 0$, (repulsive interaction) a solution E, larger than all the components ε_{mi}, shows up. This is the dipole GDR.

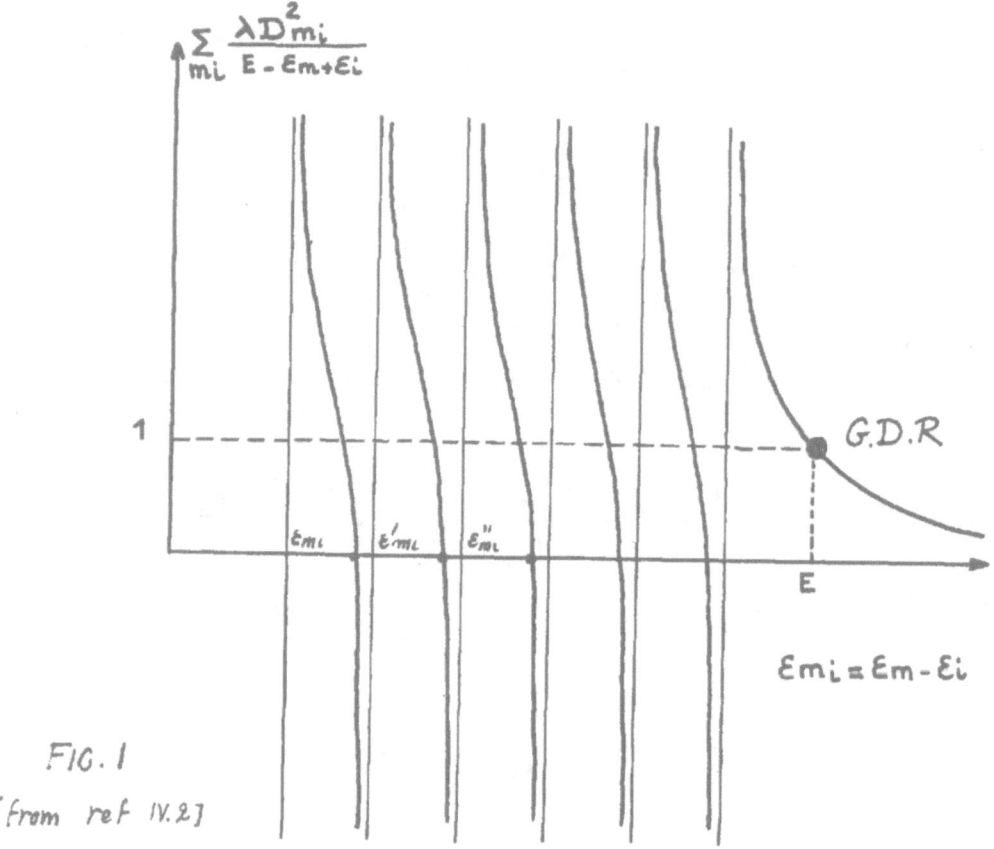

FIG. 1
(from ref IV.2)

If, as a further simplification, one assumes all $\varepsilon_{mi} = \varepsilon$ then :

$$E = \varepsilon + \sum_{mi} \lambda D^{2}_{mi}$$

i.e. E is pushed up above the energy of the individual $|$1p-1h$>$ component states by the sum of all the diagonal elements λD^{2}_{mi} of the particle-hole potential.

As for the multipole amplitude of this state at the energy E it is :

$$\mathcal{D} = \sum_{mi} c_{mi} D_{mi}$$

which is shown to be such that the transition probability :

$$\mathcal{D}^{2} = \sum_{mi} D^{2}_{mi}$$

captures the strengths of all the transition probabilities towards the unperturbed states $|$mi$>$. The gathering of all the dipole strengths on the state at energy E shows the collectivity of this " Giant Dipole " state .

IV.B Effective 1 particle – 1 hole calculations of the GDR for closed shell nuclei.

Following this idea of the schematic model, several specific particle-hole calculations were proposed after 1960. Among the first, one must quote Gillet and Vinh-Mau's results for closed shell nuclei ^{12}C and ^{16}O in 1964 |IV.3| and ^{208}Pb in 1966 |IV.4|. But here they used experimental single particle energies instead of the computed Hartree-Fock ones. These single particle and hole energies $(\varepsilon, \hat{\varepsilon})$ were obtained from the study of neighbour nuclei. The unperturbed (i.e. with zero particle-hole interaction) energy of a particle-hole pair is then $(\varepsilon - \hat{\varepsilon})$. In case of ^{208}Pb, the number of single states involved in all 1 $\hbar\omega$ transitions is 25 (table I) and large matrices will have to be diagonalised.

TABLE 1 reproduced from ref. IV.4

Experimental single-particle and hole energies

		Protons					Neutrons		
particles		ε	^{209}Bi	α	particles		ε	^{209}Pb	α
01	$1h_{\frac{9}{2}}$	3.77	0.00	0.432	13	$2g_{\frac{9}{2}}$	3.94	0.	0.403
02	$2f_{\frac{7}{2}}$	2.87	0.90	0.399	14	$1i_{\frac{11}{2}}$	3.15	0.79	0.434
03	$1i_{\frac{11}{2}}$	2.16	1.61	0.423	15	$1j_{\frac{15}{2}}$	2.53	1.41	0.427
04	$3p_{\frac{3}{2}}$	0.95	2.82	0.399	16	$3d_{\frac{5}{2}}$	2.36	1.58	0.395
05	$2f_{\frac{5}{2}}$	0.47		0.411	17	$4s_{\frac{1}{2}}$	1.91	2.03	0.391
06	$3p_{\frac{1}{2}}$	-0.53		unbound	18	$2g_{\frac{7}{2}}$	1.45	2.49	0.412
					19	$3d_{\frac{3}{2}}$	1.42	2.52	0.399
holes		$\hat{\varepsilon}$	^{207}Tl	α	holes		$\hat{\varepsilon}$	^{207}Pb	α
07	$3s_{\frac{1}{2}}$	8.03	0.00	0.387	20	$3p_{\frac{1}{2}}$	7.38	0.	0.396
08	$2d_{\frac{3}{2}}$	8.38	0.35	0.396	21	$2f_{\frac{5}{2}}$	7.95	0.57	0.405
09	$1h_{\frac{11}{2}}$	9.37	1.34	0.412	22	$3p_{\frac{3}{2}}$	8.27	0.89	0.394
10	$2d_{\frac{5}{2}}$	9.70	1.67	0.385	23	$1i_{\frac{13}{2}}$	9.01	1.63	0.418
11	$1g_{\frac{7}{2}}$	11.43	3.4	0.416	24	$2f_{\frac{7}{2}}$	9.72	2.34	0.394
12	$1g_{\frac{9}{2}}$	15.43		0.397	25	$1h_{\frac{9}{2}}$	10.85	3.47	0.425

The second feature of the approach by Gillet, Vinh-Mau and Sanderson was the choice of the effective particle-hole interaction. They chose a central force of gaussian shape with a range parameter $\mu = 1.67$ fm.

$$V(z) = V_0 \exp\left(-\frac{z^2}{\mu^2}\right) \cdot \left[W + M P^r + B P^\sigma + H P^r P^\sigma \right]$$

The parameters were obtained by a χ^{2} fit of the excited levels in ^{12}C, ^{16}O and ^{40}Ca by RPA calculations.

The results showed two different sets of data.

1) for ^{16}O (and other light nuclei).

The energies of the main dipole states are fairly well predicted at 25.3 and 22.5 MeV in ^{16}O (Fig. 2) but the transition probability B(E1) failed to be reproduced and was theoretically found, in similar computations, as twice that of experiment. Another interesting feature is that the resultant wave function of each of the two main dipole states is only a moderate mixture of the available configuration (table 2).

TABLE 2 reproduced from ref. 3.

^{16}O $\quad I^- T = 1$			$\frac{1p_{\frac{1}{2}}}{2s}$	$\frac{1p_{\frac{3}{2}}^{-1}}{1d_{\frac{5}{2}}}$	$\frac{1p_{\frac{1}{2}}^{-1}}{1d_{\frac{5}{2}}}$	$\frac{1p_{\frac{1}{2}}^{-1}}{1d_{\frac{3}{2}}}$	$\frac{1p_{\frac{1}{2}}^{-1}}{2s}$	$\frac{2s}{1p_{\frac{1}{2}}}$	$\frac{1d_{\frac{3}{2}}}{1p_{\frac{3}{2}}}$	$\frac{1d_{\frac{5}{2}}}{1p_{\frac{1}{2}}}$	$\frac{1d_{\frac{3}{2}}}{1p_{\frac{1}{2}}}$	$\frac{2s}{1p_{\frac{1}{2}}}$
E	μ	$\%$	18.52	22.73	17.65	16.59	12.38					
25.4	1.14	0.26	-0.131	0.943	-0.145	0.270	-0.006					
25.2	0.69	0.20	-0.116	0.963	-0.085	0.236	-0.013	0.003	0.009	0.057	-0.012	-0.017
22.7	2.98	0.68	0.180	0.259	0.880	-0.345	-0.088					
22.2	2.56	0.73	0.198	0.184	0.898	-0.354	-0.083	-0.022	0.037	-0.064	0.068	0.010
19.6	0.09	0.02	0.949	0.121	-0.266	-0.105	0.047					
19.6	0.08	0.02	0.951	0.115	-0.267	-0.097	0.047	-0.003	-0.010	-0.005	-0.006	0.012
18.1	0.05	0.01	0.221	-0.170	0.354	0.893	-0.018					
18.1	0.05	0.01	0.208	-0.164	0.340	0.903	-0.017	-0.001	0.012	0.026	-0.009	-0.007
13.6	0.14	0.03	-0.026	0.020	0.096	-0.008	0.995					
13.5	0.12	0.04	-0.026	0.019	0.090	-0.006	0.996	0.012	-0.008	0.008	-0.009	0.001

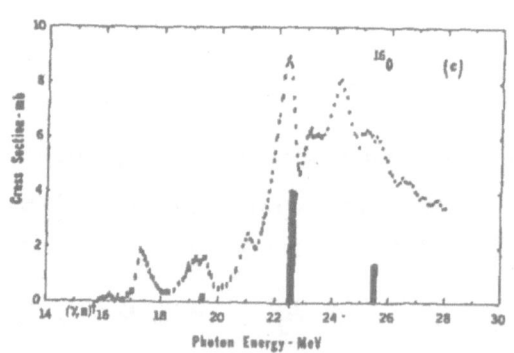

FIG. 2

Experimental data from |IV.36|
Theoretical lines from |IV.3|

2) <u>for</u> ^{208}Pb.

The GDR was shown |IV.4| to be built up mostly with |1p-1h> configurations whose energies lie between 5 and 9 MeV (Fig. 3). But here, the average evergy E_D is roughly found 2 MeV below the experimental value shown in Fig. 4 |IV.5|. On the other hand the computed transition rates showed a total integrated cross section in good agreement with experiments.

FIG. 3

(from ref IV. 4)

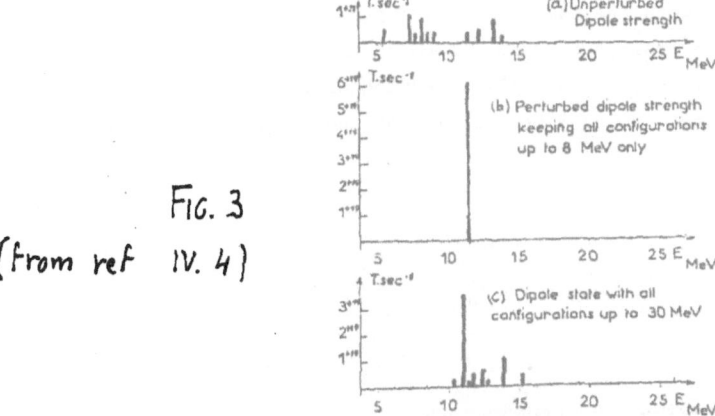

Transition rates (in sec⁻¹) of the dipole states as a function of the number of configurations; a) corresponds to pure shell-model states ($V_0 = 0$), b) and c) are obtained with $V_0 = -40$ MeV, COP, $\mu = 1.68$ fm, $\alpha = 0.43$ fm⁻¹.

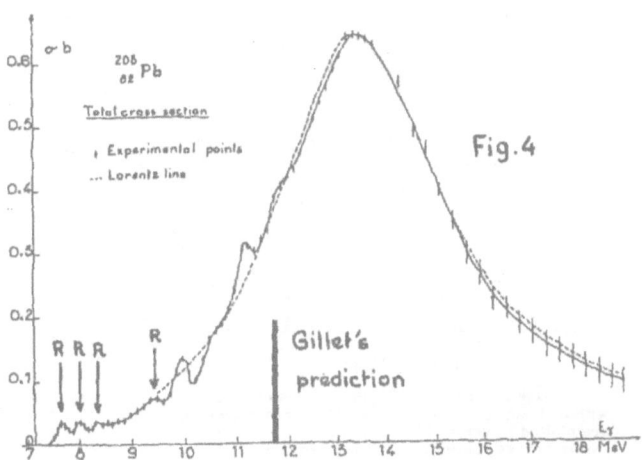

Total photonuclear cross section $\sigma_{\gamma,\tau}(E)$ of ^{208}Pb and best Lorentz line fit

FIG. 4 : (from ref IV. 5)

The small resonances, R, shown in fig. 4 between 7 and 9 MeV, can be connected to the simple 1p - 1h states which have not been completely denuded of their original strength by the build-up of the coherent dipole state at 13.5 MeV, as predicted by the schematic model.

Instead of using such a gaussian force, with strength and exchange mixture determined from fits to particle hole states in ^{16}O and ^{40}Ca, Kuo et al. |IV.6| in 1970 used the G-matrix elements of the Hamada-Johnston potential as effective interactions with the same configurations and unperturbed energies as Gillet. But once again, even when including the process of core polarization (Fig. 5c), referred to as self-screening of the exchange term, the dipole strength was still found 3 MeV too low on the average. (Fig. 6)

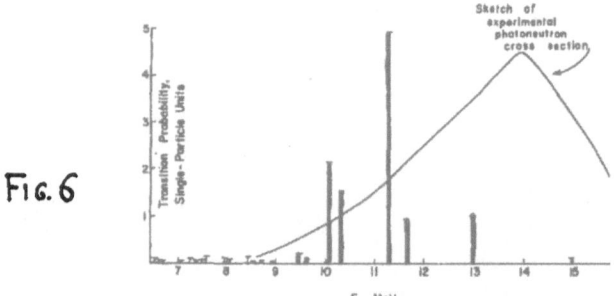

FIG. 5

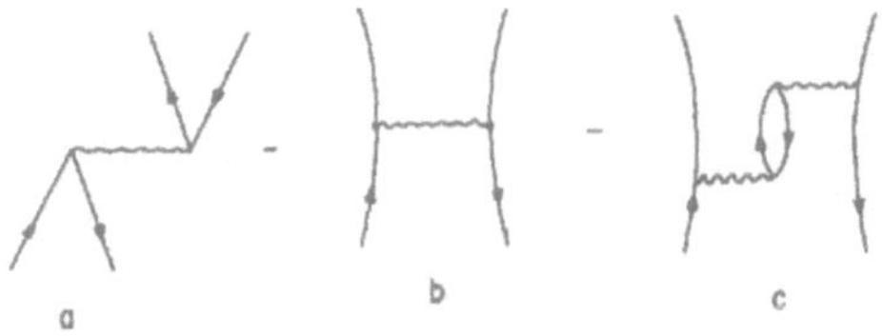

a   b   c

Processes included in our calculation of the dipole states. The wavy line represents the *G*-matrix element from the Hamada-Johnston potential. Fig. shows the process we refer to as "self-screening of the exchange term".

FIG. 6

Dipole strength in ^{208}Pb. Values calculated with inclusion of self-screening of the exchange term, are plotted as vertical bars. The experimental photoneutron cross section is sketched, for comparison.

FIG. 5 and FIG. 6 are both extracted from ref. IV.6

To overcome this difficulty associated with such a low theoretical dipole energy, several exotic considerations were proposed, some of which are summarized below, mostly to show which '' input nuclear parameters '' were proposed to be changed so as to fit the experimental data. Thus a somewhat '' ad hoc '' argument was proposed by Kuo, namely that the formation of α clusters on the surface of ^{208}Pb would restrict the volume left free for the oscillations of the protons against the neutrons (since p and n are strongly bound in α particle). Thus the effective volume available for the dipole oscillation is smaller than the normal spherical volume of ^{208}Pb, which results in an increase of E_D.

Kuo also suggested that it might be wrong to put protons and neutrons in the same "$\hbar\omega$" oscillator well because that gives an unrealistic r.m.s. neutron radius $R_N \simeq 7$ % larger than the proton radius R_p (in discrepancy with pion scattering data which showed $R_N/R_p = 0.980 \pm 0.015$). Thus, computing separately the neutron and proton central well radius parameters r_o^p, and r_o^n, (for the used Saxon-Woods potential well), Perez |IV.7| found two main dipole lines very close to the experimental value $E_D = 13.5$ MeV. (fig. 7)

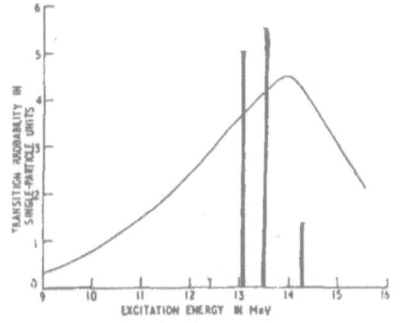

FIG. 7
(from ref IV.7)

The theoretical dipole strength in the region of the giant dipole resonance in ^{208}Pb. The experimental photoneutron cross-section is sketched for comparison.

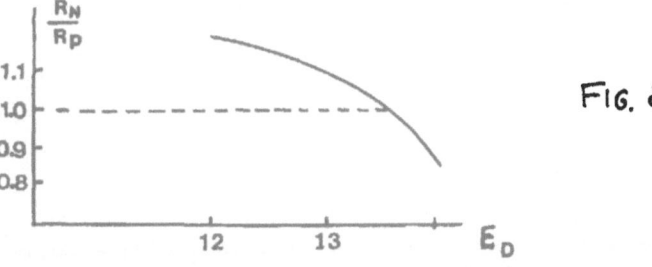

FIG. 8

Perez also found that the center of gravity of the dipole strength E_D depends strongly on the adopted ratio R_N/R_p. (Fig. 8)

As a conclusion, one can say that the Gillet type 1p-1h calculations of the GDR were successful to a large extent for closed shell nuclei but could not predict simultaneously all the properties of the GDR. One can think that this is due to an intrinsic lack of "self consistency". Actually, one cannot be sure that the interaction responsible for the single particle and single hole experimental energies is the same as the explicit effective interaction used in these calculations. In particular two such different interactions are considered in the apparently successful approach by Perez. However, the new self consistent method should allow to overcome this shortcoming (see paragraph IV.D).

IV.C <u>Effective 1p-1h calculations of the GDR for non-closed shell nuclei.</u>

For s-d shell nuclei between ^{16}O and ^{40}Ca, Bassichis has shown that the GDR states of 2s-1d shell nuclei can be split into a low energy group, where the main 1p-1h components correspond to a transition between the 2s-1d and the 2p-1f shells, and a high energy group, where the main 1p-1h components correspond to a transition between the 1p and 2s-1d shells |IV.8|.

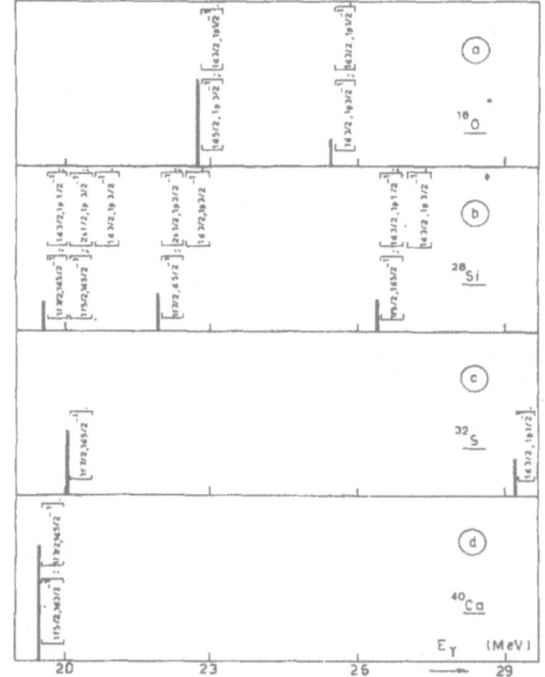

Fig. 9

Predicted dipole states for the $T = 0$ nuclei ^{16}O, ^{28}Si, ^{32}S and ^{40}Ca together with the principal particle-hole configurations

Gillet $|IV.3|$ had shown that for the GDR of ^{16}O, the two main dipole states, built on holes in the 1p shell, are to be found at 22.7 MeV for the $(1d_{\frac{5}{2}} 1p_{\frac{1}{2}}^{-1}, 1d_{\frac{5}{2}} 1p_{\frac{3}{2}}^{-1})$ configurations and at 25.4 MeV for the $(1d_{\frac{3}{2}} 1p_{\frac{1}{2}}^{-1}, 1d_{\frac{3}{2}} 1p_{\frac{3}{2}}^{-1})$ configurations, as represented in Fig. 9a. Also Farris et al. $|IV.9|$ and Blomqvist et al. $|IV.10|$ were able to show that, for ^{28}Si and ^{32}S on the one hand and for ^{40}Ca on the other hand, the respective dipole states built on holes in the 2s-1d shell were mainly located between 19 and 22 MeV and in particular, as can be seen from Figs. 9b, c and d :

Fig. 9b -
  $- (1f_{\frac{5}{2}} 1d_{\frac{3}{2}}^{-1}, 1d_{\frac{5}{2}} 1p_{\frac{3}{2}}^{-1}, 1f_{\frac{5}{2}} 1d_{\frac{5}{2}}^{-1}, 2s_{\frac{1}{2}} 1p_{\frac{3}{2}}^{-1}, 1d_{\frac{3}{2}} 1p_{\frac{3}{2}}^{-1})$ and $(1f_{\frac{5}{2}} 1d_{\frac{5}{2}}^{-1}, 2s_{\frac{1}{2}} 1p_{\frac{3}{2}}^{-1} 1d_{\frac{3}{2}} 1p_{\frac{3}{2}}^{-1})$ configurations in ^{28}Si at 19.5 MeV and 21.8 MeV respectively;

Fig. 9c - $\left(1f_{\frac{7}{2}} 1d_{\frac{5}{2}}^{-1}\right)$ configuration in ^{32}S at 20.07 MeV ;

Fig. 9d - $\left(1f_{\frac{5}{2}} 1d_{\frac{3}{2}}^{-1}, 1f_{\frac{7}{2}} 1d_{\frac{5}{2}}^{-1}\right)$ configurations in ^{40}Ca at 19.5 MeV.

These theoretical predictions were confirmed by some recent measurements achieved at Saclay.

Fig. 10 represents the partial $\sigma(\gamma,n) + \sigma(\gamma,p n) + \sigma(\gamma,2n)$ cross sections measured at Saclay $|IV.11|$ for which it has been checked that the average position in energy confirms that of the $\sigma(\gamma, tot)$ curves obtained by Ziegler. One immediately sees a clear shell effect :

- for 1p shell nuclei $(^{12}C, ^{16}O)$: $E_D \simeq 22$ to 25 MeV ;
- for 2s-1d shell nuclei : $E_D \simeq cte \simeq 19$-21 MeV ;
- for 2p-1f shell nuclei : $E_D \simeq 18$ MeV.

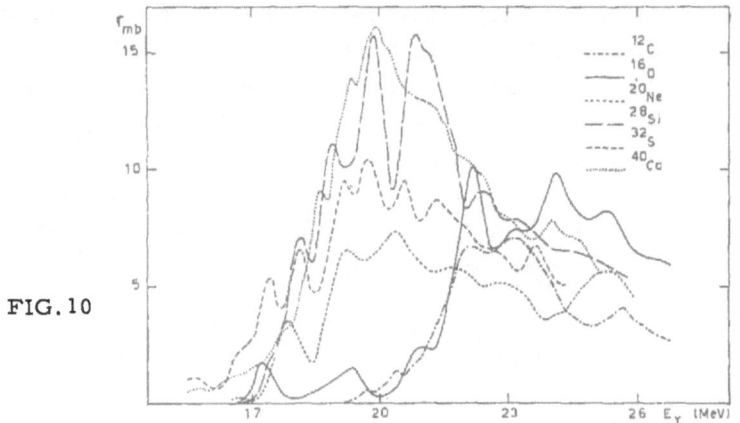

FIG. 10

Experimental σ_{Tn} curves showing the relative positions of the main dipole strength concentrations of ^{12}C, ^{16}O, ^{20}Ne, ^{28}Si, ^{32}S and ^{40}Ca.

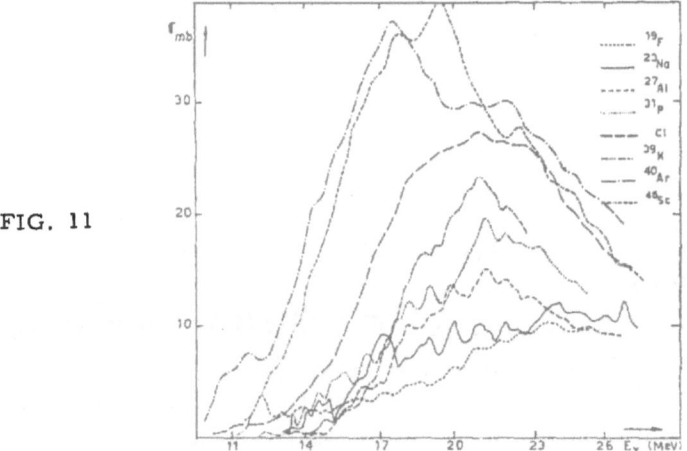

FIG. 11

Experimental σ_{Tn} curves showing the relative positions of the main dipole strength concentrations of ^{19}F, ^{23}Na, ^{27}Al, ^{31}P, Cl, ^{39}K, ^{40}Ar, and ^{45}Sc.

FIG. 10 and FIG. 11 are both extracted from ref. [IV. 11].

As the nucleons progressively fill up the s-d shell, the dipole strength concentration in the corresponding 18-22 MeV region ought to increase, a behaviour confirmed by our σ_{T_n} results for ^{28}Si, ^{32}S and in particular ^{40}Ca. As can be seen from Fig. 10, the ^{20}Ne result seems to be a transition case, where the main dipole strength is evenly divided between the 18-22 MeV region, corresponding to holes in the s-d shell and the 22-26 MeV region, corresponding to holes in the 1p shell.

In Fig. 11, we present the same type of experimental σ_{T_n} results for the T $\neq$ 0 nuclei ^{19}F, ^{23}Na, ^{27}Al, ^{31}P, Cl, ^{39}K, ^{40}Ar and ^{45}Sc. Here again it seems as if ^{19}F could be considered a transition case with the greatest dipole strength concentrated in the E = 22-27 MeV region corresponding to holes in the 1p shell.

The dipole strength of ^{23}Na, which has seven nucleons in the s-d shell, shows a rather uniform strength distribution in the E $\simeq$ 17-27 MeV region.

As to the ^{27}Al, ^{31}P, Cl and ^{39}K nuclei, their respective dipole strength concentrations show a certain tendency to peak at 21 MeV which might be due to an increase in the number of $p \cdot l$ states with holes in the s-d shell.

More recently, (1974) S.S.M. Wong et al. $\left| IV.12 \right|$ achieved open shell R.P.A. calculations for the GDR of s-d shell self-conjugate nuclei. Their results are given in Fig. 12 for different interactions (the discrete lines have been replaced by a gaussian to simulate particle emission broadening). The results are plotted as $\sigma(\gamma, abs)$.

One immediately sees that they predict a rather slowly decreasing value for the dipole average energy E_D when A increases whereas the above experimental data showed a much steeper variation when one passes from 1p shell nuclei to 2s-1d shell T = 0 nuclei, for all of which $E_D \neq$ constant = 21 $\pm$ 1 MeV. Moreover, once again, although the dipole strengths are considered up to 26 MeV at most, the theoretical B(E1) and the corresponding integrated cross sections exceed by 40 % the classical T.R.K. sum rule 60 NZ/A MeV.mb.

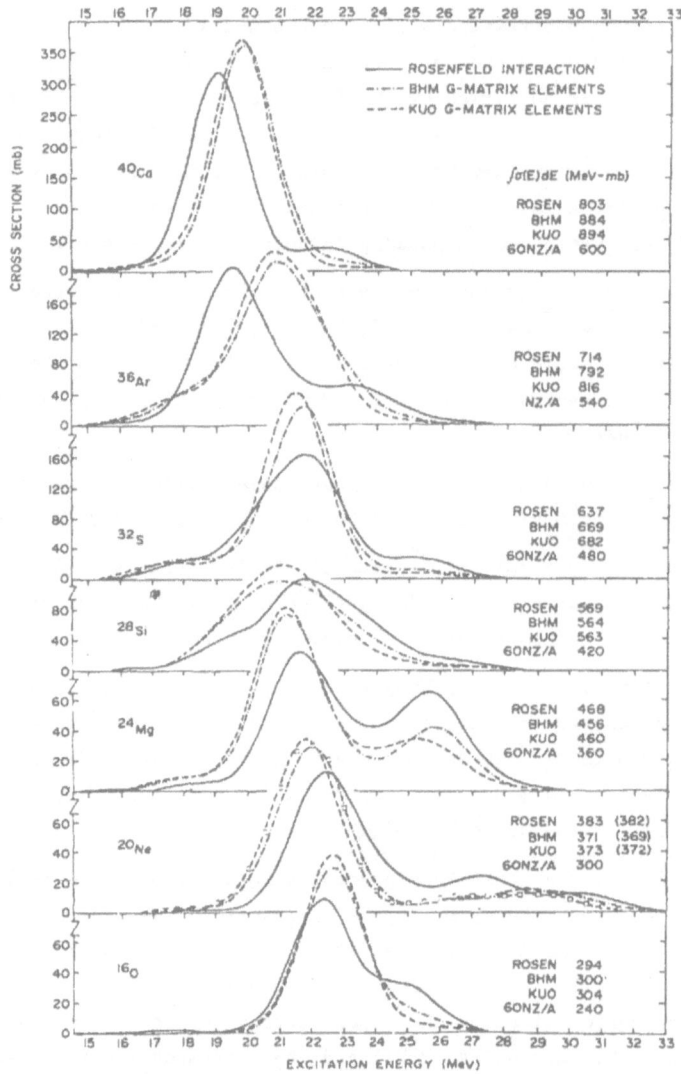

Gamma absorption cross sections to the giant dipole resonances of sd-shell nuclei calculated in the OSRPA. An oblate projected Hartree-Fock (PHF) ground state was used for the ^{28}Si calculations. Results for a PHF ground state and Kuo G-matrix elements are plotted for ^{20}Ne as squares.

FIG. 12 : reproduced from ref.[IV.12]

IV.D One particle-One hole self-consistent models of the GDR.

The recent use of self-consistent methods to compute the properties of a high energy collective mode such as the GDR may be considered as a

consequence of   Bohr-Mottelson's unified model. When the one-body
dipole operator creates a 1p-1h state, by removing a nucleon out of its
initial orbital, it also creates a density fluctuation and therefore an
immediate change in the average nuclear field through which all the nucleons may be involved in the excitation. Such self consistent processes
have been studied by Bertsch and Tsai |IV.13|. The summary given below
follows some recent work by J.P. Blaizot |IV.14| .which computes the cou-
pling between the density and the average field, in these excitations.
One starts from the effective interaction (such as a Skyrme
interaction)which was used to describe the ground state properties of
nuclei in the Hartree-Fock approximation. The 1p-1h spectrum is then deduced and J.P. Blaizot then computes the collective excitation of the
average nuclear field with the only parameters included in this effective
interaction. Preliminary data are presented in Fig. 13 and, for example, show
a concentration of the dipole strength around the experimental value
E_D = 14 MeV. in ^{208}Pb. Moreover the associated dipole strength is also
simultaneously and correctly predicted.

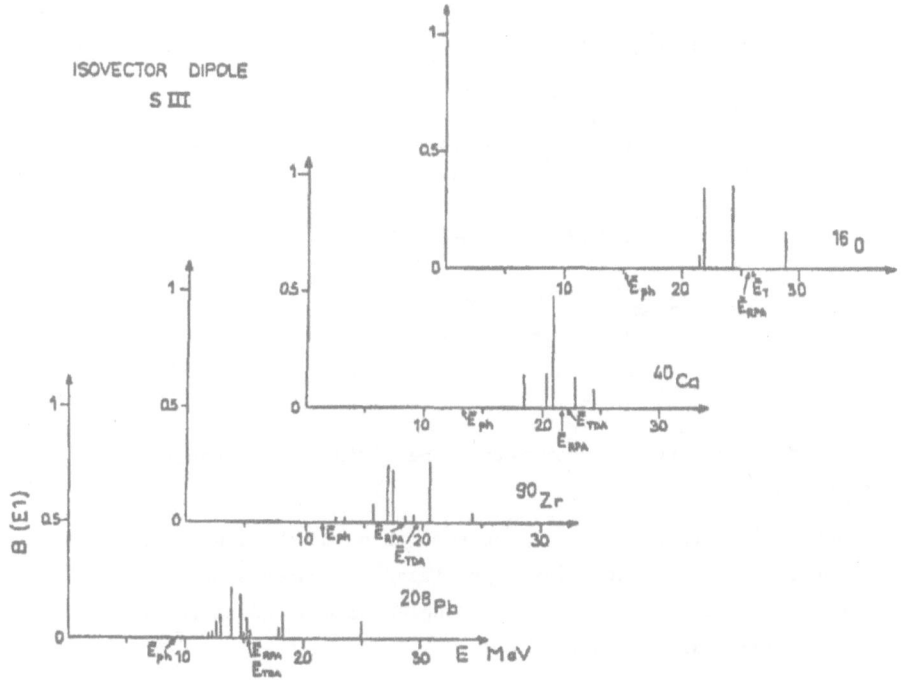

FIG. 13 : from J.P. BLAIZOT (IV.14)

IV.E <u>The 1 particle – 1 hole continuum models of the GDR.</u>

As soon as 1965 it appeared that conventional shell model calculations, using the eigenstates of the harmonic oscillator, were much too primitive since they consider only discrete states. With more realistic potential, e.g. Woods–Saxon potential, it was possible to obtain independent particle wave functions some of which correspond to bound–states and others to a continuum spectrum. Several methods were then developed to take into account the continuous character of the nuclear wave function at the energy of the GDR, for example :

– The coupled channel method with various mathematical treatments by Buck and Hill, Melkanoff–Raynal [IV. 15] , Saruis and Marangoni [IV. 16] , J. Birkholz [IV. 17].

– The eigenchannel method which requires a search for the eigenphases of the S-matrix, explained in detail in the review paper by R. F. Barrett [IV.18] .

As an example, let us summarize the coupled channel approach by A.M. Sarius and Marangoni for a system of A nucleons :

$$H = H_0 + V$$
$$H_0 = \text{independent particle hamiltonian} = \sum_{i=1}^{A} h_0(i)$$
$$h_0(i) = t(i) + V_0(i)$$
$$V = \text{residual interaction}$$
$$= \sum_{ij} V(ij) - \sum_{i=1}^{A} V_0(i)$$

In the 1p–1h continuum approximation one divides the eigenstates of H_0 into two groups :

$|\alpha\rangle$ bound eigenstates with all nucleons in bound shell model orbitals ;

$|\varepsilon c\rangle$ configurations (scattering eigenstates) in which a nucleon in a continuum state at energy ε is coupled to a bound state of the (A–1) residual nucleus.

Then the eigenstate $\psi(E)$ of the A nuclear system at energy E :

$$H|\psi(E)> = E|\psi(E)>$$

is expanded as

$$|\psi(E)> = \sum_\alpha a_\alpha |\alpha> + \sum_c \int d\varepsilon \, a_c(\varepsilon) |\varepsilon,c>$$

If the energy of the residual nucleus is E_c :

$$a_c(\varepsilon) = A_c \, \delta \, (\varepsilon - E + E_c) + b_c(\varepsilon)$$

Bloch [IV. 37] showed that the amplitudes a_α and $a_c(\varepsilon)$ can be computed from a system of coupled integral equations. One can therefore get the wave function $\psi(E)$ with a nucleon wave in channel c only and the integrated cross section for a photon dipole absorption (from the ground state into the channel c around energy E) can be written as

$$\frac{4\pi^2 e^2}{\hbar c} E \, | < \psi_c^+(E) \, | D | 0> |^2$$

$$D = \sum_{i=1}^{A} \frac{\tau_i^3}{2} \vec{3}_i$$

Table III shows the unperturbed 1p-1h configurations, computed by Saruis and Marangoni, within a real Saxon-Woods potential well with spin-orbit and coulomb terms.

TABLE 3

Choice of parameters for the single-particle potential and the residual interaction for ⁴⁰Ca

⁴⁰Ca	Protons				Neutrons			
nlj	ε	⁴¹Sc	$V_{ij}{}^{0}$	$V_{ij}{}^{s.o.}$	nlj	ε	⁴¹Ca	$V_{ij}{}^{0}$	$V_{ij}{}^{s.o.}$
1f$_{\frac{7}{2}}$	−1.63	0.0	52.88	7.0	1f$_{\frac{7}{2}}$	−8.36	0.0	52.45	7.0
p$_{\frac{3}{2}}$	[0.78]	[2.4]	55.9	6.8	2p$_{\frac{3}{2}}$	−6.30	2.06	56.51	6.8
p$_{\frac{1}{2}}$	2.47	4.1	55.9	6.8	2p$_{\frac{1}{2}}$	−4.23	4.13	56.49	6.8
f$_{\frac{5}{2}}$	4.77	6.4	53.6	7.0	1f$_{\frac{5}{2}}$	−1.96	6.4	53.32	7.0

	³⁹K					³⁹Ca			
1d$_{\frac{3}{2}}$	8.33	0.6	54.28	9.5	1d$_{\frac{3}{2}}$	15.6	0.0	53.76	9.5
2s$_{\frac{1}{2}}$	10.33	2.0	55.98		2s$_{\frac{1}{2}}$	18.2	2.6	56.19	
1d$_{\frac{5}{2}}$	14.73	6.4	54.77	9.5	1d$_{\frac{5}{2}}$	21.6	6.0	53.88	9.5

$r_0 = 1.25\,\text{fm}, \quad d = 0.53\,\text{fm}, \quad U_0 = 850\,\text{MeV}\cdot\text{fm}^3, \quad p = 0.46; \; W(\text{MeV}) = 0.06E_\gamma(\text{MeV}) - 0.5.$

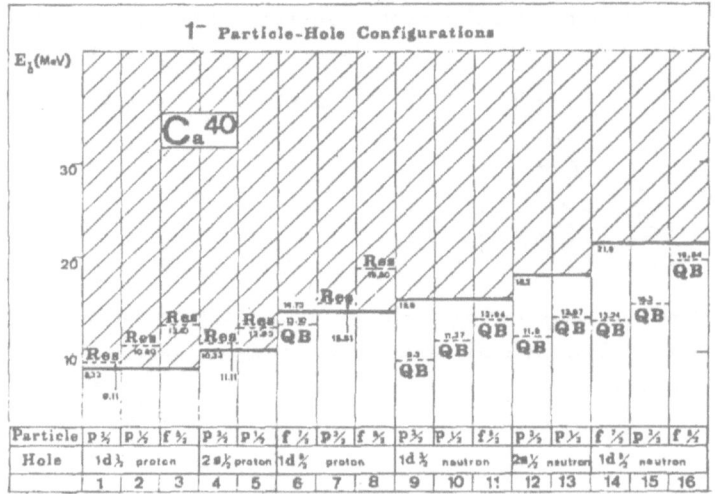

Energies of the 1⁻ 1p-1h unperturbed configurations of ⁴⁰Ca computed from the parameter values given in table 3. The full lines represent the experimental threshold energies. The dashed lines correspond either to quasi bound (Q.B.) states or to resonant (Res.) states. The correction for the Coulomb proton-proton residual interaction is taken into account.

(The above tables are reproduced from Nuclear Physics A. 132, 1969, 649.)

Then the (γ, p) and (γ, n) cross sections were calculated using a zero range residual force and are shown in Fig. 14.

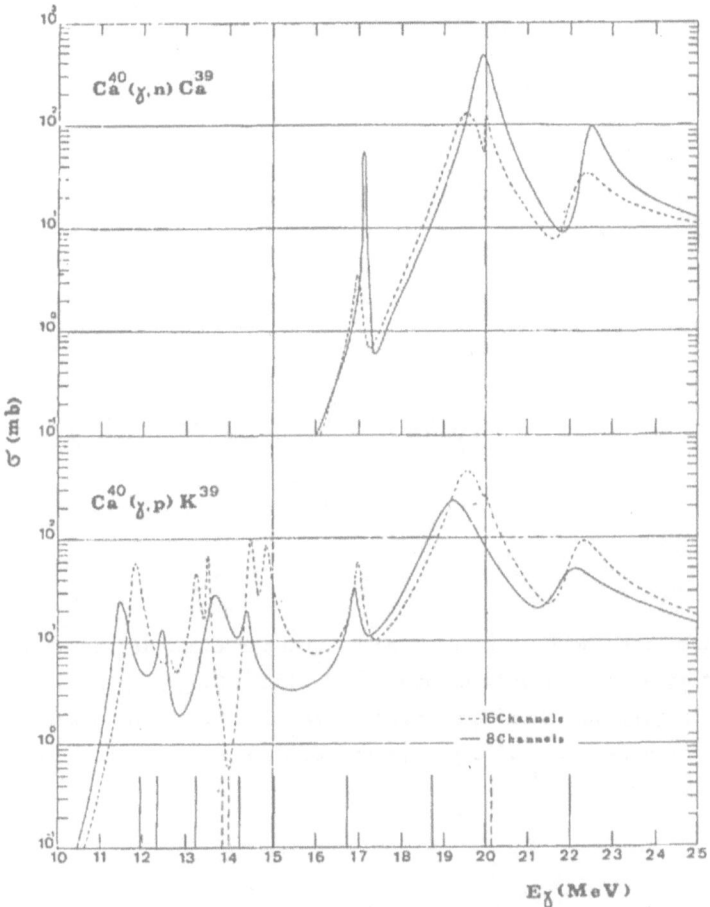

Total $^{40}Ca(\gamma, p)^{39}K$ and $^{40}Ca(\gamma, n)^{39}Ca$ cross sections calculated with mixing between proton and neutron channels (dashed curves) and without mixing (continuum curves). The absorptive potential W has been set equal zero. In the bottom of the figure are given the energies of the 1^- states

FIG. 14 , reproduced from Nuclear Physics, A. 132, 1969, 649.

A good agreement is observed with the Saclay (γ, n) data at least when one considers the 4 predicted peaks at $\simeq$ 17, 19.2, 20 and 22 MeV (arrows in Fig. 15).

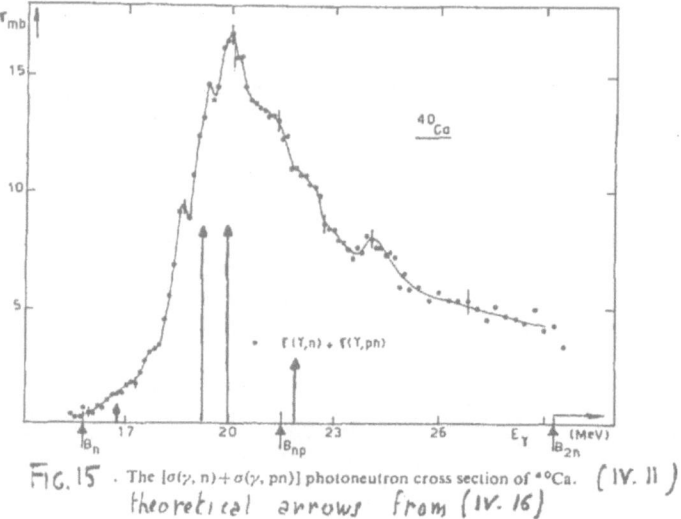

FIG. 15 . The $[\sigma(\gamma, n) + \sigma(\gamma, pn)]$ photoneutron cross section of ^{40}Ca. (IV. 11)
theoretical arrows from (IV. 16)

But the predicted peak cross section is about 5 times larger than the experimentally observed one and the predicted width of about 1 MeV for the central peaks is only one third the experimental one. The consideration of an absorptive potential well, of course, improves dramatically the predictions (Fig. 16).

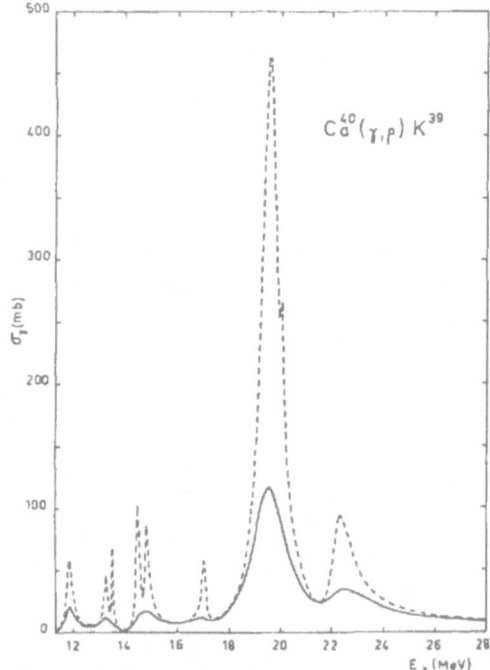

Fig. 16

in full curve the absorptive potential W (MeV) = 0.06 E_γ (MeV) −0.5

in dashed curve W = 0

The $\left(\gamma_i |\nu\right)$ predictions are compared with the $\left(\nu, \gamma_0\right)$ data obtained at Stony Brook by E.M. Diener et al. |IV.19|. Once again the predictions are better from the point of view of the energy positions of the peaks than from the points of view of absolute cross sections and widths. (Fig. 17)

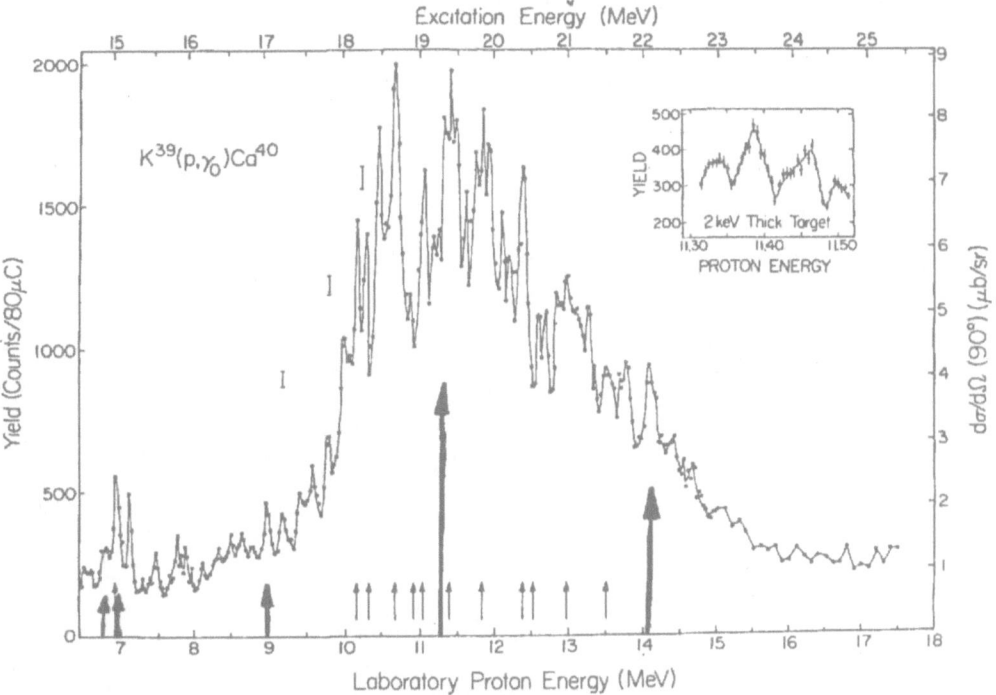

FIG. 17 : comparison of experimental results by Diener (IV.9) with theoretical predictions by Marangoni - Saruis (IV.16).

IV.F Consideration of n particles - n holes states in the fine structure of the GDR.

As an example of this problem let us consider the experimental GDR of light nuclei. One observes a gross resonance shape, with a width of between 5 and 10 MeV and containing a varying amount of structure. The number of distinguishable spikes is rather small for 1p shell nuclei, increases sharply for 2s-1d shell nuclei and tends to disappear in the overall shape of the GDR for A $>$ 40 nuclei. Such a behaviour has been interpreted as being due to the fact that f-wave particles, from e.g. ^{28}Si, emerge much less readily from the nucleus than the d-wave nucleons from ^{12}C. This favours the coupling of the 1p-1h states of the GDR to the 2p-2h states existing at the same energy as the GDR itself, and which are more numerous for s-d than for p-shell nuclei[20]. Moreover, in some cases, ground state deformation or isospin splitting can further complicate the observed structure   (Figs. 18, 19, 20).

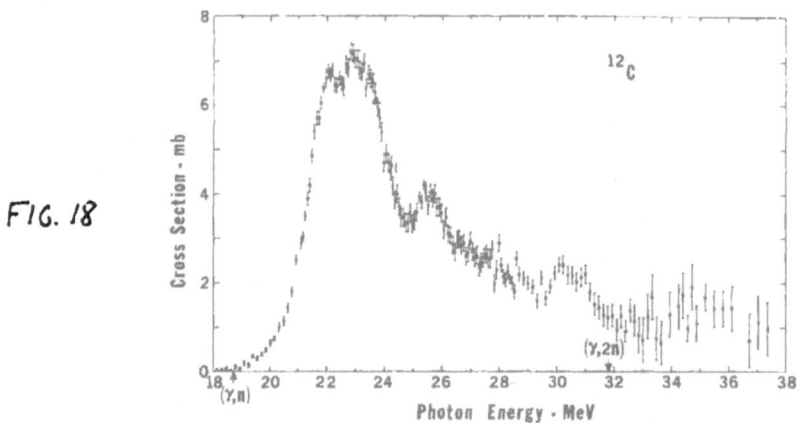

FIG. 18

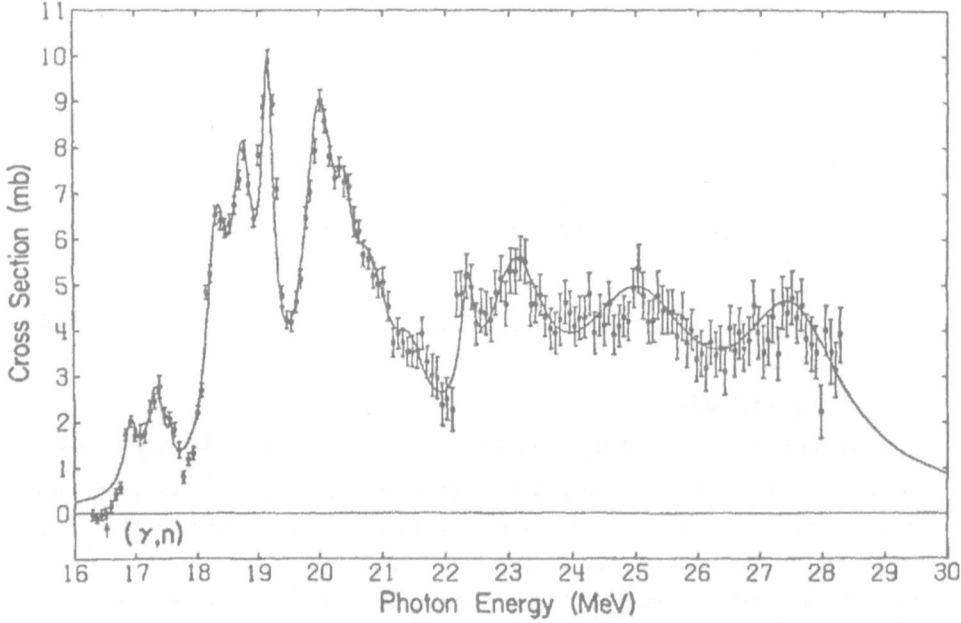

^{24}Mg total cross section $\sigma(\langle \gamma,n \rangle + (\gamma,pn))$ fitted with a calculated cross-section curve. The solid line is the sum of the 14 Lorentz lines whose parameters were adjusted to fit the data.

FIG 19   (Phys. Rev 4, 1971, 149)

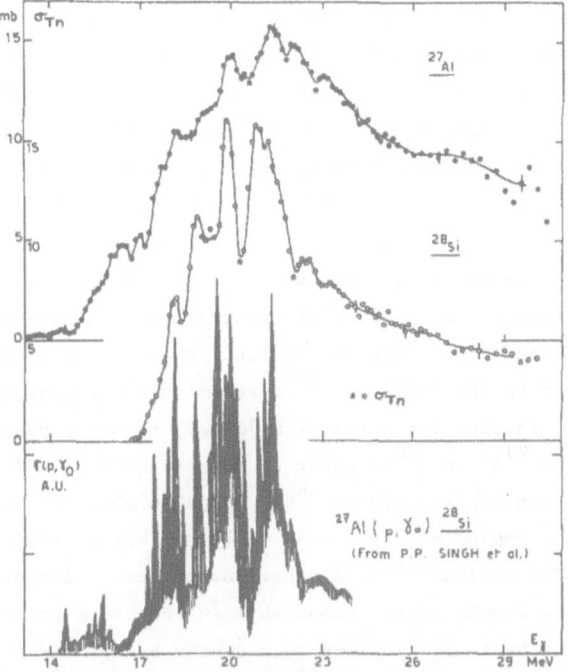

"Structure" observed in the σ_{Ta} curves of the $A = 4N$ and $A = 4N-1$ nuclei ^{28}Si and ^{27}Al respectively. High energy resolution curve $\sigma(p, \gamma_0)$ from ^{27}Al(p, γ_0)^{28}Si experiments is also shown for comparison.

FIG. 20 *(from ref IV. II)*

As can be seen from fig. 20, a reasonable overall agreement exists between the "structure" observed in the σ_{Tn} results of ^{27}Al and ^{28}Si, measured at Saclay with an energy resolution of $\Delta E \sim 180$ keV, and the corresponding high energy resolution data obtained from ^{27}Al$(p,\gamma_o)^{28}$Si. In particular, the "peaks" at 18.1, 18.9 and 19.85 MeV, observed in the rising part of the GDR of ^{28}Si, are practically coincident with the peaks at 18.2, 19.1 and 19.9 MeV observed in the rising part of the GDR of ^{27}Al. These peaks, observed in the σ_{Tn} curves with widths of ~ 400 keV, suggest that they are intermediate structure states since their widths are significantly greater than those of the underlying "compound nucleus" states. Therefore, if the numerous spikes seen in the ^{27}Al$(p,\gamma_o)^{28}$Si reaction are connected to 2p-2h configurations [20], lasting for a time $\sim \hbar/70$ keV, then the "peaks" observed in σ_{Tn} curves of ^{27}Al and ^{28}Si ought to correspond to simpler 1p-1h configurations lasting $\sim \hbar/400$ keV in ^{28}Si. One also observes that the peaks in the E = 17-20 MeV region of the A = 4N nucleus ^{28}Si, are narrower than those in the A = 4N-1 nucleus ^{27}Al, a behaviour one could actually have expected because these simple dipole components in the odd A nucleus ^{27}Al are somewhat smeared out by their coupling to a denser background of more complicated states.

Some particular theoretical treatments of such a coupling are summarized below.

IV.F.1. <u>Coupling of the 1p-1h dipole state to the first low energy 2^+ state.</u>

This is just a microscopic equivalent of the collective dynamic model in which the dipole oscillation is modulated by the low energy vibrations of the nuclear surface. As an example Kamimura |21| considered the dipole state $\psi_m(1^-)$ as constituted by simple 1p-1h states $\phi_m(1^-)$ and by more complicated states $[\phi_m(1^-) \otimes \phi(2^+)]_{1^-}$. The $\phi(2^+)$ state is made of 8 1p-1h states (with $0\hbar\omega$ or $2\hbar\omega$ excitation energy) in the ^{32}S case.

$$\psi_m(1^-) = \sum_n a_n \phi_n(1^-) + \sum_{n'} b_{n'}\left[\phi_{n'}(1^-) \otimes \phi(2^+)\right]_{1^-}$$

The obtained predictions for the dipole lines of ^{32}S are compared to the experimental data in Fig. 21 and show, of course, a much larger fragmentation than the 1p-1h predictions by Farris|9|, (fig. 9) in good agreement with the experimental data.

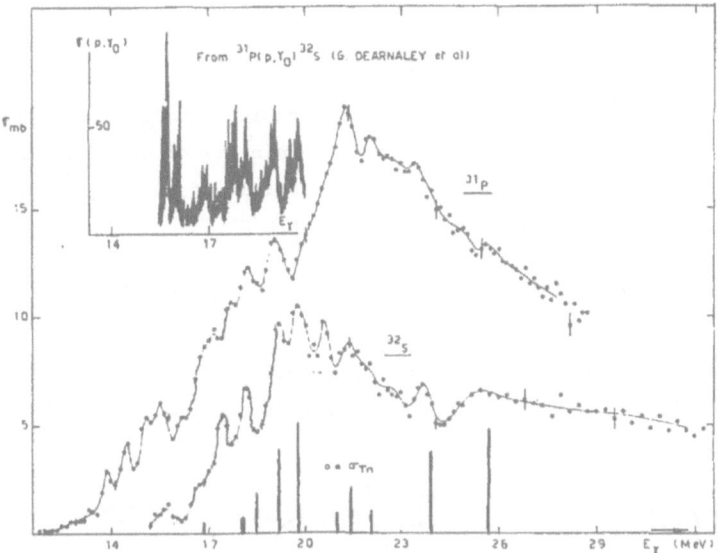

FIG: 21 "Structure" observed in the σ_{Ta} curves of the $A = 4N$ and $A = 4N-1$ nuclei ^{32}S and ^{31}P respectively. High energy resolution curve $\sigma(p, y_0)$ from $^{31}P(p, y_0)^{32}S$ experiments is also shown for comparison.

IV.F.2. The effect of the quasi-bound states.

In order to explain the rapidly varying structure observed in the continuum, Bohr considered some long-lived compound states. In this compound
nucleus model, the available excitation energy above some particle emission
threshold is shared amoung many nucleons. Therefore, the probability that
one nucleon can escape is very small, hence the long life of these states
and the narrowness of the associated resonant structure. V. Gillet
|22| showed that such a concept could be adapted to the shell-model where
the excited configurations can be classified according to their number of
particle-hole pairs, i.e. in a doubly closed shell nuclei, 1p-1h, 2p-2h,
3p-3h...np-nh. In such nuclei, the large gap between filled and unfilled
shell model orbitals allows only a few of these above configurations to
exist between, say the (γ,n) threshold and the GDR region. In this GDR
energy region some (np-nh) configurations where all nucleons are bound, can
be nevertheless excited through a coupling to the simple 1p-1h unbound configurations. Reversely, such discrete bound (np-nh) states will also decay
through their small coupling to 1p-1h unbound states only, which explains

their long life   and the associated fine structure, the density of which
is linked to the density of these ϱB states.

Therefore, in V. Gillet's approach the complete nuclear wave function
$\psi_i^n(E)$ at energy E, includes $|\alpha, \varepsilon\rangle$ Slater determinants of orthonormalized
single particle states (with one unbound particle of energy ε and with quan-
tuum number α), and also quasi-bound states $|\lambda\rangle$.

$$\psi^n(E) = \sum_\alpha \int d\varepsilon \, a_\alpha^{nE}(\varepsilon) |\alpha, \varepsilon\rangle + \sum_\lambda a_\lambda^{n,E} |\lambda\rangle$$

The mixing coefficients a_α and a_λ are obtained by resolving the set of
equations $H\psi^n = E\psi^n$. Then one can write the absorption rate of γ with
energy E, by means of the 1-body operator $\mathcal{O}$ as .

$$T(E) = \frac{2\pi}{\hbar} \sum_n \left| \sum_\alpha \int d\varepsilon \langle \alpha\varepsilon | \mathcal{O} | 0 \rangle a_\alpha^{nE}(\varepsilon) \right|^2$$

V. Gillet pointed out that whereas the matrix elements are only func-
tions of the 1p-1h configurations, the effect of the ϱB states $|\lambda\rangle$ is however
felt through the a_α coefficients which were obtained by solving equations in
which the $|\lambda\rangle$ states had to be considered. Such an effect in a (γ, n) or
(γ, p) experiment is summarized on Fig. 22.

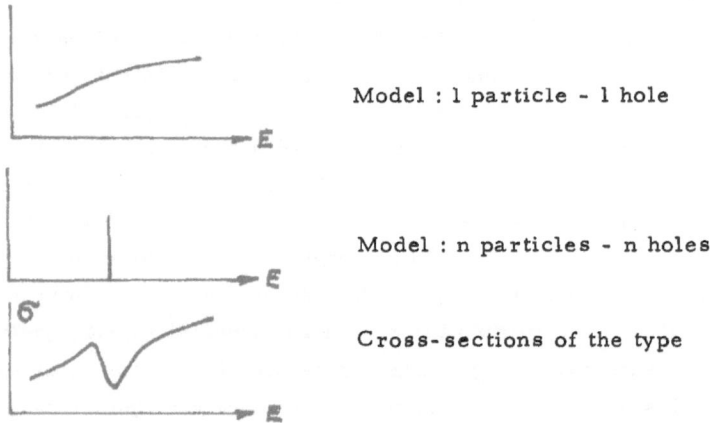

Model : 1 particle - 1 hole

Model : n particles - n holes

Cross-sections of the type

Fig. 22 : Effect of the mixing of an np-nh quasibound state with
the 1 p - 1 h continuum on cross-sections via a one-
body operator : (γ, γ') , (α, α') , (p, γ) , (γ, n) , ...

reproduced from ref. IV. 22

In the case of γ absorption by ^{16}O, the observed cross-section, say in the (γ,n) channel |11|, seems to show a strong evidence in favor of such QB state effects. Fig. 23 shows some dips observed at 21.1 MeV ($\vec{A}$), 22.5 MeV ($\vec{B}$), 24.9 MeV ($\vec{C}$) and possibly 26.1 MeV $\vec{D}$, which were not predicted by simple 1p-1h computations.

FIG. 23

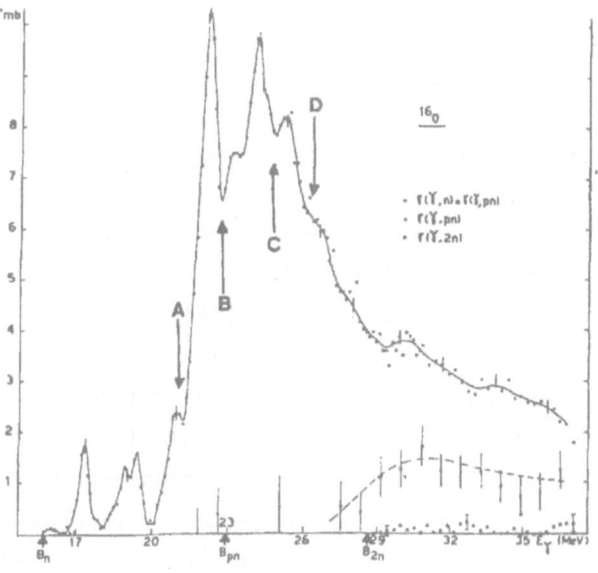

Partial photonuclear cross sections [$\sigma(\gamma, n) + \sigma(\gamma, pn)$], $\sigma(\gamma, pn)$ and $\sigma(\gamma, 2n)$ of ^{16}O   (IV. 11)

Actually the experiments ^{12}C (α,γ) ^{16}O|23|, ^{14}N(d,γ) ^{16}O|24| and ^{13}C(^{3}He,γ)^{16}O|25| show peaks respectively at energies corresponding to the following excitation energies in ^{16}O : 21 MeV, 22.7 MeV and 25.2 + 26 MeV respectively (Fig. 24).

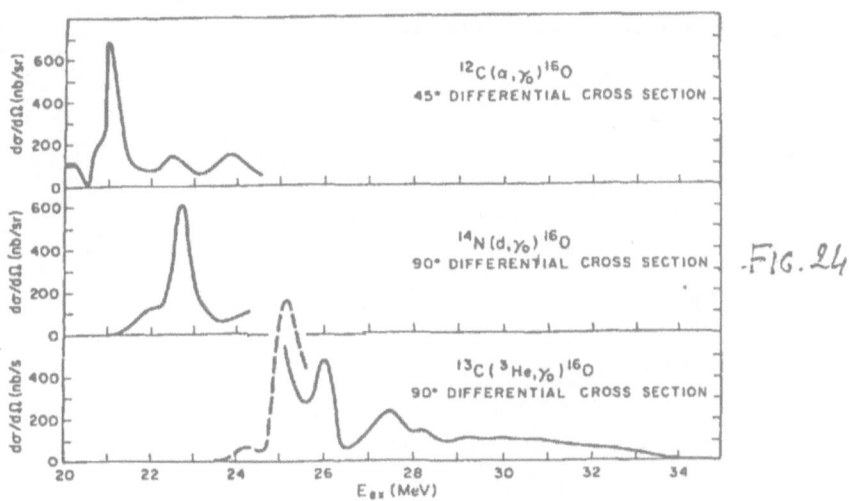

FIG. 24

Although the (γ, n) cross section is produced by a 1-body o-
perator, which does not couple np-nh states directly to the ground state,
the coupling of such np-nh quasibound states to the 1p-1h continuum could
thus explain the observed structure A, B, C, D, which should be connected :

A to a $(4p-4h)_{1^-, \; T=1}$ Q.B. state

B to a $(2p-2h)_{1^-, \; T=1}$ Q.B. state

C and D to a $(3p-3h)_{1^-, \; T=1}$ QB state.

IV.F.3. <u>A 3p-3h model for the fine structure observed in the ^{16}O GDR.</u>

Another type of intermediate structure, namely the 3p-3h states, was
considered to be mostly responsible for the observed fine stucture of the G. D. R.
by Shaking and Wong. C. Shakin and W. Wong [26] tried to specify the concept
of doorway states in their study of the fine structure (which they call " intermediate structure ") in the G. D. R. of a doubly closed shell nucleus such as ^{16}O.
In what follows, their approach is summarized mostly because the use of the
projector technique is certainly clarifying from a pedagogical point of view.
However, the various projectors are associated with several types of states,
whose classification, according to a given hierarchy of complications, is very
similar to the one already used by C. Bloch IV. 37 .

As the starting point, they admit that the usual linear combination
of $|1p-1h\rangle$ states may be considered as providing a good description of the
entrance "dooway state D" for a dipole electromagnetic excitation and can
explain the observed "gross structure". (Fig 25).

FIG. 25

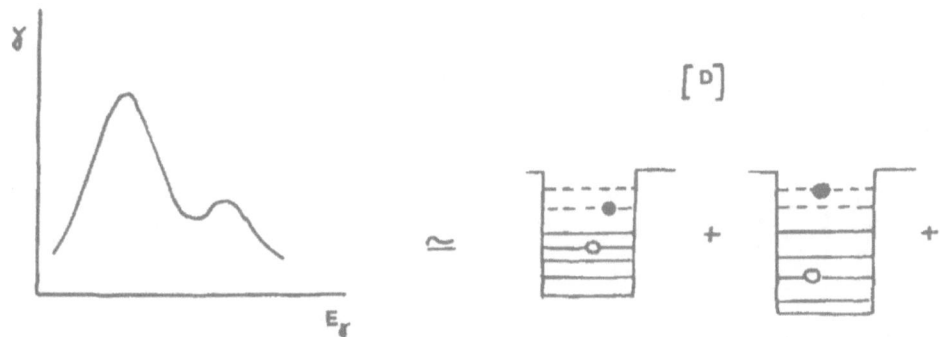

But a better experimental energy resolution (say $\Delta E_\gamma \simeq$ 100 keV) allows to see some "intermediate structure" which can be then associated to some more complicated "secondary doorway states" q (Fig. 26) (for example 3p-3h states).

FIG. 26

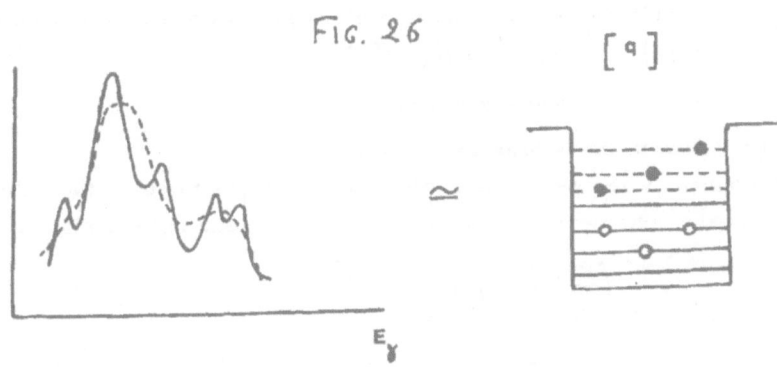

A very good experimental résolution (such as the one obtained in the reverse reactions (p,γ_o) (α,γ_o) indŭced with char-ged particles of well defined energy, should have to be connected with very complicated np-nh states X for which one cannot hope to achieve a microscopic analysis easily (Fig. 27).

FIG. 27

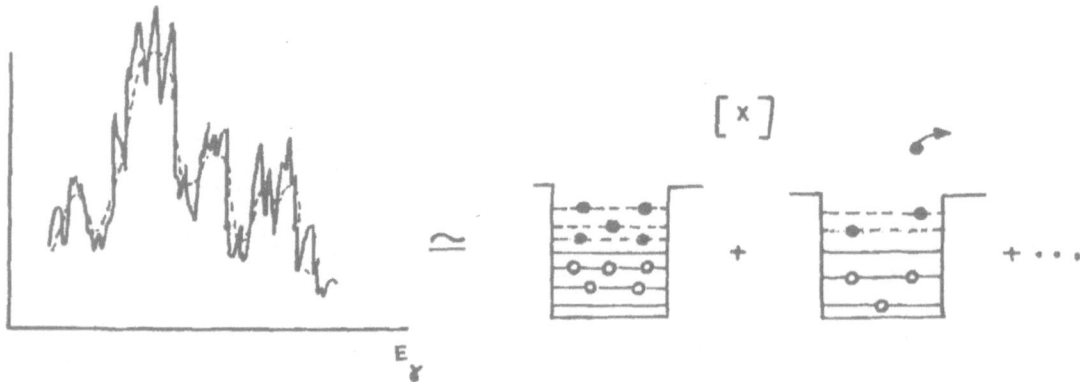

Thus, C. Shakin and W. Wang are led to considered separate Hilbert spaces (each one associated to a given projection operator P, D, Q).

1) P space : which contains the continuum channel ψ_c that one wants to study : (γn_0), (γ, n_0)...This channel will be reached by applying the electromagnétique interaction H_γ to the target nucleus $|\phi_0\rangle$ with the transition matrix element $T_c(\gamma) = \langle \psi_c | H_\gamma | \phi_0 \rangle$

2) D space : of primary doorway states D.

3) Q space : sum of two subspaces q and x.

Of course, when using the projection operators to solve the Schröodinger equation at the considered energy E, with the nuclear halmitonian H

$$(E - H) | \psi_c \rangle = 0$$

one has

$$P + Q + D = 1$$

$$| \psi_c \rangle = P | \psi_c \rangle + D | \psi_c \rangle + Q | \psi_c \rangle$$

$$\langle \psi_c | Q H_\gamma | \phi_0 \rangle = 0$$

(Doorway assumption   for the entrance channel). Finally, C. Shakin and W. Wang showed that one can write $T_c(\gamma)$ as a sum of a direct term and a resonant term :

$$T_c(\gamma) = <\Psi_c|H_\gamma|\Phi_o> + \sum_D \frac{<\Psi_c|H|D><D|H_\gamma|\Phi_o>}{E - E_D - \Delta_D(E) + \frac{i}{2}[\Gamma_D(E) + I]}$$

where the observed intermediate structure is produced by the rapid energy dependance of Δ_D and $\Gamma_D(E)$:

$$\Delta_D(E) - \frac{i}{2}\Gamma_D(E) \simeq \sum_Q \frac{|<D|H|Q>|^2}{E - E_Q + i\frac{I}{2}}$$

(I = averaging energy interval).

As a particular application of the above theory C. Shakin and W. Wang chose the $^{16}O(\gamma,n_o)^{15}O_{GS}$ channel.

They considered for their secondary doorway space q, 3p-3h states (3 bosons states) with $J_\pi = 1^-, T_o1$ built from the following low lying states.

J = 3⁻	T = 0	at 6.1 MeV
J = 1⁻	T = 0	7.1 MeV
J = 3⁻	T = 1	13.12 MeV
J = 2⁻	T = 1	12.7 MeV
J = 1⁻	T = 1	17.43 MeV

The 3p-3h states of the q space, are then constructed by first coupling two $|J_1 T_1>$ states   to an intermediate 2p-2h $|J_{12} T_{12}>$ state, which is then coupled to a $|J_3 T_3>$ state

Although the sum of the unperturbed energies is then larger than the energy of the dipole GDR states, their energies are brought down by pp and hh interactions (cf Table 4). Fig. 28a and 28b give and idea of the corresponding theoretical predictions :

EXTRACTED From

EXTRACTED FROM PHYSICAL REVIEW LETTERS 26, 15 (1971)

Table **4** The energy spectrum, configurations, and coupling-matrix elements to the dipole states, for the three-boson states.

Energy (MeV)	Adjusted Energy (MeV)	Configuration $(J_1T_1)(J_2T_2)J_{12}(J_3T_3)$	Coupling Matrix Element to Dipole State at 22.3 MeV	24.3 MeV
26.7	26.6	(10) (10) 2 (11)	0.014 MeV	0.0069 MeV
26.4	26.6	(10) (10) 2 (31)	0.0033	0.0016
25.9	25.7	(10) (10) 2 (21)	0.042	0.013
25.8	25.7	(30) (10) 4 (31)	-0.061	-0.030
25.6	25.7	(30) (10) 3 (31)	0.19	0.092
25.1	25.0	(30) (10) 3 (21)	-0.14	-0.07
25.0	25.0	(30) (10) 2 (11)	-0.047	-0.02
24.7	24.85	(30) (10) 2 (31)	0.23	0.11
24.2	24.85	(30) (30) 0 (11)[a]	0.40	0.20
24.1	24.85	(30) (10) 2 (21)	-0.22	-0.11
23.3	22.8	(30) (30) 4 (31)	0.0033	0.0017
23.3	22.8	(30) (30) 2 (11)	0.098	0.049
23.1	22.8	(10) (10) 0 (11)	0.29	0.14
23.0	22.8	(30) (30) 2 (31)	0.090	0.045
22.5	22.8	(30) (30) 2 (21)	0.13	0.066
20.2	21.25	(30) (30) 0 (11)	0.48	0.24

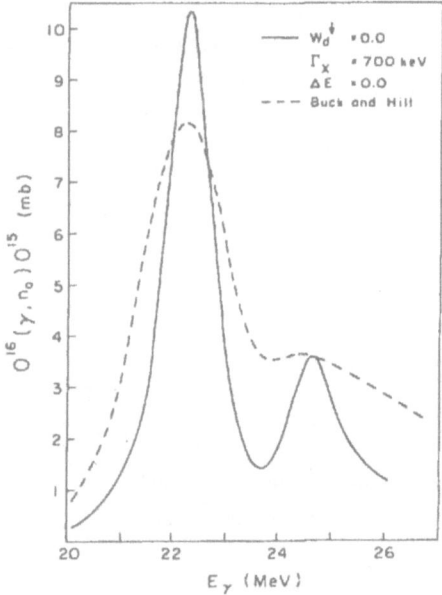

The doorway-state gross-structure calculation (solid line), which is obtained from our T matrix, , by setting $W_d^{\downarrow}(E)=0$. The quantity Γ_x (chosen to be 700 keV) is used as an energy-averaging parameter in the T matrix. The cross section is compared with that of the complex optical-model calculation of Buck and Hill (dashed line).

FIG. 28a

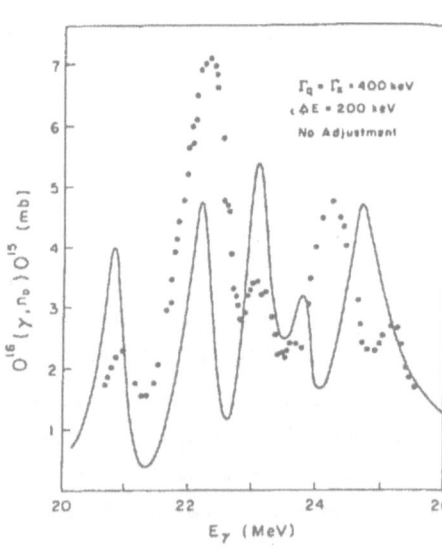

The intermediate structure as obtained without adjustments in the secondary-doorway energies and their couplings to the doorways is shown. We note that the strength at 22.3 MeV is shifted too much by the coupling and the energies of the intermediate resonances are not correctly reproduced. The experimental data are taken from Caldwell et al.,

FIG. 28b

IV.F.4. The fine structure of the GDR as a consequence of the "nuclear coexistence".

The previous examples, which try to explain the fine structure observed in the GDR of a doubly closed shell nucleus ^{16}O, showed that whereas everybody agrees that such a fine structure is due to the admixture of the main dipole doorway states ($\sim$ gross structure at 22.5 and 25 MeV) with other neighbour compound resonance states, the main problem lies in the identification of these compound nuclear states in the GDR region and in the mechanism coupling these states to the dipole strength.

A. Goswani and R. Graves |27| considered the phenomenum of "nuclear coexistence", i.e the fact that a deformed coexistent state (mostly 4p-4h) exists in ^{16}O characterized by the rotationnal sequence 0^+, 2^+, 4^+ built on the $0'^+$ state at 6.05 MeV. One can write :

$$|\psi(0^+)> = \alpha|\phi_o> + \beta|\phi_o'>$$

$$|\psi(0^{+'})> = -\beta|\phi_o> + \alpha|\phi_o'>$$

$$\alpha^2 + \beta^2 = 1 \qquad \left(\frac{\beta}{\alpha}\right)^2 = 0.11$$

Then the classical $|p.h>$ states coupled to $J^\pi = 1^-$ and $T = 1$ can be built either on the GS 0^+ state or on the $0'^+$ state. Considering then :

$$|h> = \{a_\mu^+ a_\ell\}_{JT} |\phi_o>$$

$$|\tilde{h}> = \frac{1}{\sqrt{N_k}} \{a_\mu^+ a_\ell\}_{JT} |\phi_o'>$$

Goswani and Graves showed that the dipole states $|\lambda>$ can be written :

$$|\lambda> = \sum X_k^{o\lambda} |h> + \sum' X_k^{o'\lambda} |\tilde{h}>$$

$$X_k^{o\lambda} = <\lambda|h>$$

$$X_k^{o'\lambda} = <\lambda|\tilde{h}>$$

where the dipole strengh D of the dipole state $|\lambda\rangle$ can be written as

$$|D|^2 = |\sum_k D_k \langle \lambda | a_\mu^\dagger a_\ell | \psi_0 \rangle|^2$$

$$= |\alpha \sum_k D_k \chi_k^{o\lambda} + \beta \sum_k D_k \sqrt{N_k} \chi_k^{o\lambda}|^2$$

$$D_k = \langle h | \frac{\tau_3}{2} r Y_{10} | \rho \rangle$$

Results of the calculation (Fig. 29) clearly show a splitting of the dipole-states obtained from the classical 1p-1h model. This leads to a better agreement (Fig. 30) with the (γ,n) data $|11|$.

FIG. 29 from ref.[IV.27]

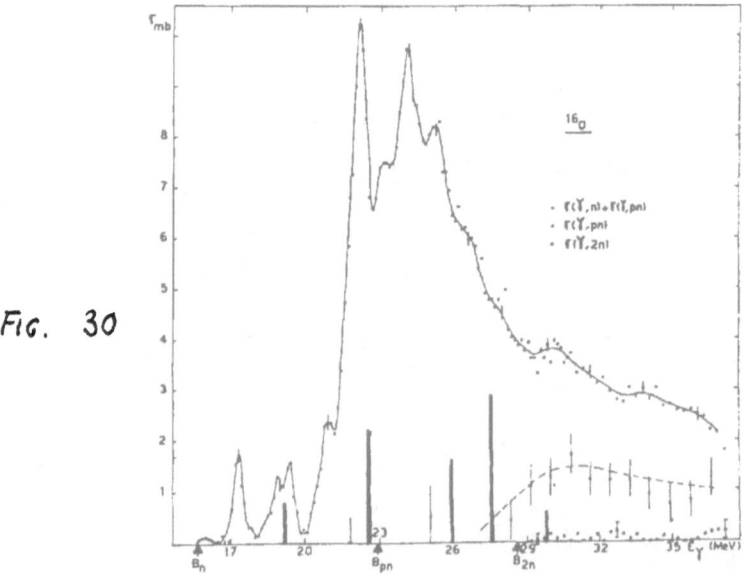

Dipole states of ^{16}O. (a) shows the dipole states in the particle-hole model; (b) shows the results of the present calculation. The percentage dipole strength is also indicated for each state.

FIG. 30

Partial photonuclear cross sections $[\sigma(\gamma,n)+\sigma(\gamma,pn)]$, $\sigma(\gamma,pn)$ and $\sigma(\gamma,2n)$ of ^{16}O

IV. G- A microscopic description of the damping width.

So far, one has only considered the influence of np-nh states on the splitting of the GDR. The consideration of the 1p-1h approximation led to the evaluation of the escape width $\Gamma^\uparrow$. But the microscopic study of 2p-2h states, especially in heavy nuclei, can lead to an evaluation of the damping width $\Gamma^\downarrow$ associated with the doorway dipole | 1p-1h ⟩ state $|\varphi_0\rangle$. Suppose that the $|\varphi_0\rangle$ state can decay via the two body force into the (2p - 2h) background states $|\varphi_i\rangle$. If one diagonalizes the hamiltonian $H = H_0 + V$ in the $\left\{ \varphi_0, \varphi_1, \varphi_2 \cdots \varphi_n \right\}$ space such that :

$$\langle \varphi_i | H | \varphi_j \rangle = \varepsilon_i \, \delta_{ij} \qquad i,j = 1, 2, \cdots n$$

$$H | \psi_j \rangle = \lambda_j | \psi_j \rangle$$

$$| \psi_j \rangle = a_{j0} | \varphi_0 \rangle + \sum_{i=1}^{n} a_{ji} | \varphi_i \rangle$$

then, the various $\langle \varphi_i | H | \psi_j \rangle = \lambda_j \langle \varphi_i | \psi_j \rangle$ equations give the various a_{ji} coefficients and the percentage of the doorway state $|\varphi_0\rangle$ in the eigensta-te $| \psi_j(E) \rangle$ is given by :

$$a_{j0}^2 = \cfrac{1}{1 + \sum_{i=1}^{n} \cfrac{|\langle \varphi_i | H | \varphi_0 \rangle|^2}{(\lambda_j - \varepsilon_i)^2}}$$

Davidson |28| pointed out that this quantity a_{j0} measure the spreading of the doorway state $|\varphi_0\rangle$ and that the distribution of a_{j0}^2 with energy will show some sort of gross resonance effect. By assuming a constant spacing "d" between the energies ε_i of the |2p 2h ⟩ φ_i states and a constant $|\langle \varphi_i | H | \varphi_0 \rangle|^2$ one can find a Breit-Wigner shape around E_0 : (Fig. 31)

$$a^2_{f_0}(E_f) = \frac{d}{2\pi} \ \frac{\Gamma^{\downarrow}}{(E_f - E_0)^2 + (\frac{\Gamma^{\downarrow}}{2})^2 + \frac{d\Gamma^{\downarrow}}{2\pi}}$$

with the damping width $\Gamma^{\downarrow}$:

$$\Gamma^{\downarrow} = \frac{2\pi}{d} \ |<\phi_i | V | \phi_0>|^2$$

or

$$\Gamma^{\downarrow} = 2\pi \ |<2p \cdot 2h | V | 1p \cdot 1h>|^2 \rho (2p \cdot 2h)$$

Davidson computed $\Gamma^{\downarrow} = 3$ MeV in ^{208}Pb

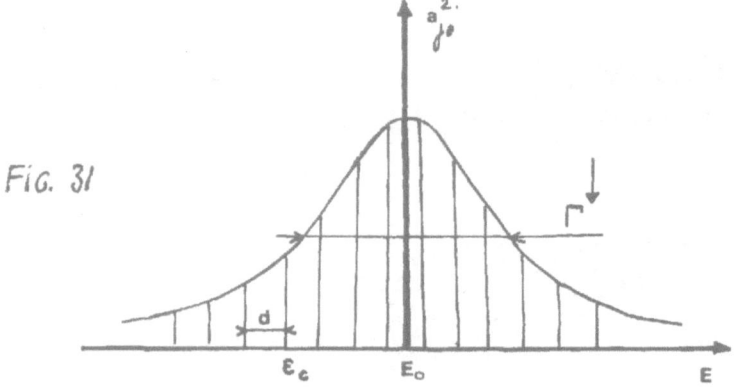

FIG. 31

(one can notice that if $d \simeq 20$keV and $\Gamma^{\downarrow} \simeq$ a few MeV, $\frac{d\Gamma^{\downarrow}}{2\pi} \ll \Gamma^{\downarrow}$).

E.D. Mshelia et al |29| computed effectively the density of (2p-2h) states (approximately 35 000 such states in the energy range 5 to 20 MeV of excitation energy in ^{208}Pb.

(Fig. 32) shows that at least qualitatively the observed width (Fig. 33)
Γ = 4.05 MeV is fairly well reproduced.

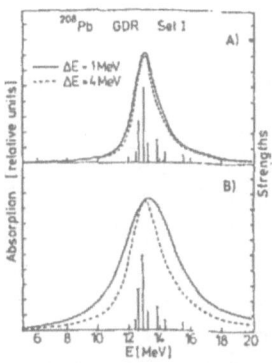 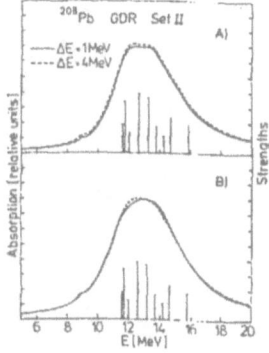

FIG. 32 : from ref. IV.29.

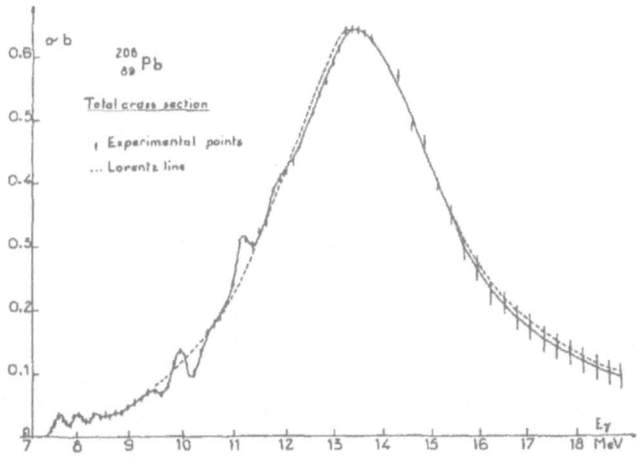

Total photonuclear cross section $\sigma_{\gamma,T}(E)$ of ^{208}Pb and best Lorentz line fit ·

FIG. 33 : from ref. IV.5.

Finally the computed variation of the density $P(2p-2h)_1^-$ shows (Fig. 34) a quasi parabolic variation of the damping width $\Gamma^{\downarrow}$ versus E :

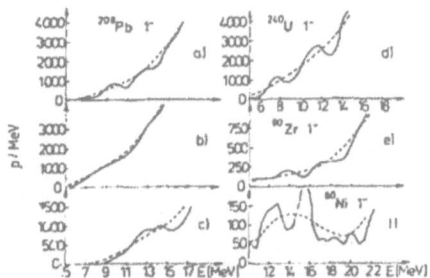

Energy dependence of the densities of 1^- 2p-2h states: a–c) for ^{208}Pb calculated with the single-particle energy sets I, II, and III respectively; d–f) for the nuclei ^{240}U, ^{90}Zr and ^{60}Ni respectively all calculated with the set I only.

FIG. 34 : reproduced from ref. IV.29.

IV.H - THE MICROSCOPIC MODELS AND THE SUM RULES.

Although this topic will be extensively treated in other lectures, a brief summary is nevertheless given below for the main systematic experimental data.

IV. H. 1 - The energy weighted sum rule σ_0.

Since one can write for the cross section of the dipole photoabsorption

$$\sigma(E) \simeq \frac{4\pi^2 e^2}{\hbar c} E \, |< \psi_f \,|\, D \,|\, 0 >|^2 \, \rho(E)$$

one sees that the integrated cross section

$$\int^{E_M} \sigma_{\gamma, abs}(E).dE = \sigma_0(E_M)$$

is equivalent to what is classically known as "the energy weighted sum rule" associated with the electric multipole operator $Q_{T\lambda\mu}$ [30]

$$\int_{EW}^{T\lambda} = \sum_m (E_m - E_0) |< m | \, Q_{T\lambda 0} \,| 0 >|^2$$

$$= \frac{1}{2} < 0 | \left[Q_{T\lambda 0} \,, \left[H, \, Q_{T\lambda 0} \right] \right] | 0 >$$

Hence

$$\sigma_0 = \int \sigma(\gamma, abs) \, dE = \frac{4\pi^2 e^2}{\hbar c} < 0 | \left[D_x, \left[H, D_x \right] \right] | 0 >$$

if the dipole operator

$$D_x = \sum_{i=1}^{A} \frac{\tau_i^3}{2} x_i$$

Then if one writes for the nuclear hamiltonian

$$H = \sum_i \frac{p_i^2}{2m} + \sum_{k<\ell} V_{k\ell}$$

one knows that the first term in H gives the Thomas-Reiche-Kuhn contribution $\frac{2\pi^2 e^2 \hbar}{Mc} \frac{NZ}{A} = 60 \frac{NZ}{A}$ MeV-mb and the nucleon-nucleon interaction term gives an enhancement factor K

$$\int_0^\infty \sigma(\gamma, abs)\, dE = \frac{2\pi^2 e^2 \hbar}{Mc} \cdot \frac{NZ}{A} \cdot (1 + K)$$

$$K = \frac{A}{NZ} \frac{M}{\hbar^2} \langle 0 | [[Y, D], D] | 0 \rangle$$

In table 5, B. Berman summarized Livermore and Saclay data and showed that an integration up to $E_M \simeq 30$ MeV gives already $\simeq 1.10$ times the TRK prediction. Hence an enhancement factor $K \simeq 0.10$ if $E_M = 30$ MeV and $\simeq 0.8$ if $E_M = 140$ MeV [31] has to be considered. W. T. Weng et al recently showed that this was mostly due to 2-body correlations due to the tensor force [32].

Nucleus	$E_{\gamma max}$ (MeV)	$\dfrac{\sigma_{int}(\gamma, tot)}{60NZ/A}$	$\sigma_{-1} A^{-4/3}$ (mb)	$\dfrac{\sigma_{-1}}{0.00225 \, A^{5/3}}$ (mb-MeV^{-1})	$\dfrac{\sigma_{int}[(\gamma, 2n) + (\gamma, 3n)]}{\sigma_{int}(\gamma, tot)}$	Reference
^{90}Zr	27.6	0.795	0.147	0.83	0.092	Berman et al., 1967
	25.9	0.945	0.175	1.00	0.039	Leprêtre et al., 1971
^{91}Zr	30.0	0.820	0.160	0.98	0.181	Berman et al., 1967
^{92}Zr	27.8	0.804	0.154	0.93	0.414	Berman et al., 1967
^{93}Nb	24.3	0.967	0.186	1.12	0.209	Leprêtre et al., 1971
^{94}Zr	31.1	0.813	0.160	1.01	0.547	Berman et al., 1967
^{107}Ag	29.5	0.858	0.155	0.89	0.194	Berman et al., 1969a
^{115}In	31.1	1.111	0.202	1.17	0.278	Fultz et al., 1969
^{116}Sn	29.6	0.978	0.175	0.99	0.248	Fultz et al., 1969
^{117}Sn	31.1	1.102	0.199	1.16	0.271	Fultz et al., 1969
^{118}Sn	30.8	1.072	0.190	1.07	0.297	Fultz et al., 1969
^{119}Sn	31.1	1.145	0.202	1.17	0.334	Fultz et al., 1969
^{120}Sn	29.9	1.185	0.209	1.19	0.330	Fultz et al., 1969
^{124}Sn	31.1	1.123	0.200	1.16	0.361	Fultz et al., 1969
^{127}I	29.5	0.933	0.164	0.93	0.256	Bramblett et al., 1966b
	24.9	1.074	0.201	1.18	0.196	Bergère et al., 1969
^{133}Cs	29.5	1.026	0.182	1.04	0.257	Berman et al., 1969a
^{138}Ba	27.1	1.022	0.183	1.05	0.242	Berman et al., 1970c
^{139}La	24.3	0.980	0.177	1.02	0.147	Beil et al., 71
^{141}Pr	29.8	1.001	0.175	0.97	0.167	Bramblett et al., 1966b
	16.9	0.691	0.138	0.85		Beil et al., 1971
	18.1	0.678ᵃ	0.128ᵃ	0.75ᵃ		Young, 1972
^{142}Nd	20.2	0.901	0.170	1.00	0.024	Carlos et al., 1971
^{143}Nd	19.8	0.910	0.176	1.08	0.094	Carlos et al., 1971
^{144}Nd	20.2	0.896	0.170	1.01	0.299	Carlos et al., 1971
^{145}Nd	20.2	0.965	0.193	1.26	0.323	Carlos et al., 1971
^{146}Nd	20.2	0.905	0.173	1.05	0.347	Carlos et al., 1971
^{148}Nd	18.8	0.795	0.155	0.97	0.491	Carlos et al., 1971
^{150}Nd	20.2	0.931	0.178	1.09	0.416	Carlos et al., 1971
^{153}Eu	28.9	1.022	0.181	1.03	0.311	Berman et al., 1969b
^{159}Tb	28.0	0.997	0.175	1.00	0.386	Bramblett et al., 1964
	27.4	1.109	0.198	1.15	0.243	Bergère et al., 1968
^{160}Gd	29.5	1.099	0.195	1.14	0.448	Berman et al., 1969b
^{165}Ho	28.9	1.057	0.183	1.04	0.312	Berman et al., 1969b
	26.8	1.202	0.215	1.24	0.272	Bergère et al., 1968
^{175}Lu	23.0	0.990	0.177	1.02	0.253	Bergère et al., 1969
^{181}Ta	24.6	0.835	0.146	0.82	0.404	Bramblett et al., 1963
	25.2	1.142	0.201	1.14	0.269	Bergère et al., 1968
^{186}W	28.6	1.123	0.191	1.06	0.449	Berman et al., 1969b
^{197}Au	24.7	1.045	0.179	0.98	0.262	Fultz et al., 1962b
	21.7	1.080	0.190	1.06	0.156	Veyssière et al., 1970
^{208}Pb	26.4	0.982	0.167	0.93	0.183	Harvey et al., 1964

TABLE 5 (from ref. V. 36)

Fig. 35 shows a slightly different presentation of the Livermore and Saclay data which was recently given by B. Berman [36] . There one plotted, in TRK units, the whole area below the Lorentz lines fitting the experimental GDR. This area should represent the El absorption extrapolated to the meson threhold, with the exceptions of the El overtones and of the high energy contributions such as the quasideuteron effect

$$\sigma_0' = \int_0^\infty \sigma_{LORENTZ}(E)dE = \frac{\pi}{2}\sigma_0\Gamma_0 \qquad \text{for spherical nuclei}$$

$$\sigma_0' = \int_0^\infty \sigma_{LORENTZ}(E)dE = \frac{\pi}{2}\left(\sigma_1\Gamma_1 + \sigma_2\Gamma_2\right) \text{for deformed nuclei}$$

For A > 100 such an extrapolated value of the σ_0 sum rule $\simeq$ 1.25 TRK. (Table 5$_b$)

Of course, for medium nuclei, Fig. 35 cannot be used since it takes into account only the photoneutron (and not the photoproton) channels.

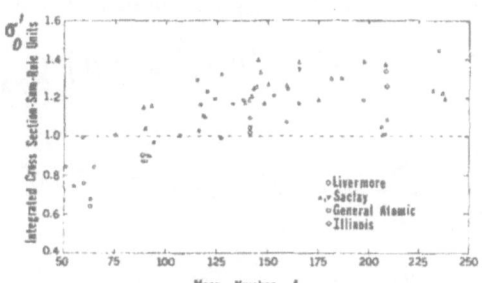

Extrapolated integrated cross sections derived from the Lorentz parameters in units of 60NZ/A MeV-mb.

FIG. 35 : reproduced from Rev. Mod. Phys. 47, 3, 1975.

TABLE 5_b

nucleus	σ_0 (MeV·b)	σ'_0 (MeV·b)	$0.06\frac{NZ}{A}$	$\frac{\sigma_0 A}{0.06\,NZ}$	$\frac{\sigma'_0 A}{0.06\,NZ}$
$_{53}$I	2.02±0.14	2.30±0.12	1.85	1.09±0.07	1.24±0.07
$_{58}$Ce	2.13±0.15	2.53±0.13	2.04	1.05±0.07	1.24±0.07
$_{62}$Sm	2.48±0.17	2.92±0.14	2.18	1.14±0.07	1.34±0.07
$_{68}$Er	2.70±0.19	3.04±0.16	2.42	1.12±0.07	1.26±0.07
$_{71}$Lu	2.65±0.18	2.96±0.16	2.53	1.05±0.07	1.17±0.07
average values				1.09±0.07	1.25±0.07

IV. H. 2 - The bremsstrahlung weighted sum rule σ_{-1}

The $\dfrac{1}{E}$ shape of bremsstrahlung spectra explains the name of the sum rule

$$\sigma_{-1} = \int \sigma(\gamma, abs)\, \frac{dE}{E}$$

which is actually the classical non energy weighted sum rule $S_{NEW}^{T\lambda}$

$$S_{NEW}^{T\lambda} = \sum_n |\langle n | Q_{T\lambda 0} | 0\rangle|^2 = \langle 0 | Q_{T\lambda 0}^2 | 0\rangle$$

Here

$$\sigma_{-1} = \frac{4\pi^2}{3\hbar c} \sum_n \langle 0 | D | n\rangle \langle n | D | 0\rangle$$

$$D = e \sum_{i=1}^A \frac{I_i^3}{2} r_i$$

r_i position of the ith nucleon with respect to the nuclear center of mass.

Levinger [33] showed that

$$\sigma_{-1} = \frac{4\pi^2}{3} \frac{e^2}{\hbar c} \frac{NZ}{A} \langle r^2\rangle$$

which gives

$$\sigma_{-1} = 0.36\, A^{4/3}\ mb \quad \text{for an harmonic oscillator shell model}$$

$$\sigma_{-1} = 0.30\, A^{4/3}\ mb \quad \text{for a finite square potential well}$$

Table 6 summarizes Saclay data for which the average value is at least 40% smaller than the above predictions

$$\sigma_{-1}\,(E_\eta = 30\,MeV) \simeq (0.20 \pm 0.02)\,A^{\frac{4}{3}}\,mb$$

- Table 6 -

Z^A	$_{50}Sn$	$^{127}_{53}I$	$_{56}Ba$	$^{139}_{57}La$	$_{58}Ce$	$_{62}Sm$	$^{159}_{65}Tb$	$^{165}_{67}Ho$	$_{68}Er$	$_{71}Lu$	$_{73}Ta$	$^{197}_{79}Au$
σ_{-1} (mb)	120±10	129±10	143±12	146±10	140±12	167±14	172±12	194±14	186±15	182±15	206±17	238±20
$\sigma_{-1}\cdot A^{-4/3}$	0,20±0,02	0,20±0,02	0,20±0,02	0,20±0,02	0,19±0,02	0,21±0,02	0,20±0,02	0,21±0,02	0,20±0,02	0,19±0,02	0,20±0,02	0,21±0,02

This reduction of the shell model prediction was attributed to two-body correlations by Lane and Melkian [34]. However Weng et al [35] pointed out that, since σ_{-1} involves only matrix elements of z^ℓ, this emphasizes large separation of the nucleon pairs and should decrease the importance of short range correlations in σ_{-1}.

IV. H. 3- The σ_{-2} sum-rule.

The sum rule $\sigma_{-2} = \int \sigma(\gamma, abs)\,\dfrac{dE}{E^2}$

is connected to the polarizability α of the nucleus, i.e. the dipole electric moment induced in the nucleus by an unity electric field

$$\sigma_{-2} = \frac{2\pi^2}{\hbar c}\alpha \quad , \quad \alpha = \frac{e^2 R^2 A}{40\,K}$$

(K symmetry term of the Weizsacker mass formula, Migdal [33] predicted

$$\sigma_{-2} = 2.25\,A^{\frac{5}{3}}\,\mu b.\,MeV^{-1}$$

Table 7 shows that most of the Saclay data lies at about $2.7 \, A^{5/3} \mu b. Mev^{-1}$, hence $\bar{K} \simeq 20 \, Mev$.

Table 7

nucleus	$_{50}Sn$	$^{137}_{53}I$	$_{56}Ba$	$^{139}_{57}La$	$_{58}Ce$	$_{62}Sm$	$^{159}_{65}Tb$	$^{165}_{67}Ho$	$_{68}Er$	$_{71}Lu$	$_{73}Ta$	$^{197}_{79}Au$
$\sigma_2 \, A^{-5/3}$ (MeV^{-1})	$27,2 \, 10^{-4}$	$26,8 \, 10^{-4}$	$26,2 \, 10^{-4}$	$25,7 \, 10^{-4}$	$25,1 \, 10^{-4}$	$27,7 \, 10^{-4}$	$25,7 \, 10^{-4}$	$28,2 \, 10^{-4}$	$26,8 \, 10^{-4}$	$23,5 \, 10^{-4}$	$25,9 \, 10^{-4}$	$25,8 \, 10^{-4}$
K_M (MeV)	$19 \pm 1,5$	$19,5 \pm 1,5$	$19,7 \pm 1,5$	$20 \pm 1,5$	$21 \pm 1,5$	$18,7 \pm 1,5$	$20 \pm 1,5$	$18,5 \pm 1,5$	$19,5 \pm 1,5$	$22 \pm 1,5$	$20 \pm 1,5$	$20 \pm 1,5$

average value ⟶ $K_M = 20$ MeV ± 1,5 MeV.

REFERENCES OF CHAPTER IV

[IV. 1] G. Brown and M. Bolsterli
Phys. Rev. Letters 3, 472 (1959)

[IV. 2] G. Brown
Unified theory of nuclear models, North-Holland, Amsterdam

[IV. 3] V. Gillet and N. Vinh-Mau
Nucl. Physics 54, 321-351 (1964)

[IV. 4] V. Gillet, A. M. Green and E. A. Sanderson
Nucl. Physics 88, 321-343 (1966)

[IV. 5] A. Veyssière, H. Beil, R. Bergère, P. Carlos and A. Leprêtre
Nucl. Physics A 159, 561-576 (1970)

[IV. 6] T. T. S. Kuo, J. Blomqvist and G. E. Brown
Phys. Letters 31 B, 93 (1970)

[IV. 7] S. M. Perez
Phys. Letters 33 B, 317 (1970)

[IV. 8] W. H. Bassichis and F. Scheck
Phys. Rev. 145, 771 (1966)

[IV. 9] S. A. Farris and J. M. Eisenberg
Nucl. Phys. 88, 241 (1966)

[IV. 10] J. Blomqvist and T. T. S. Kuo
Phys. Lett. 29 B, 544 (1969)

[IV. 11] A. Veyssière, H. Beil, R. Bergère, P. Carlos, A. Leprêtre
and A. de Miniac
Nucl. Phys. A 227, 513-540 (1974)

[IV. 12] S. S. M. Wong, D. J. Rowe and J. C. Parikh
Physics Letters 48 B, 403 (1974)

[IV. 13] G. F. Bertsch - Phys. Rev. letters 31, 121 (1973)
G. F. Bertsch and S. F. Tsai
Phys. Reports 18, 125-158 (1975)

[IV. 14] J. P. Blaizot to be published

[IV. 15] J. Raynal, M. A. Melkanoff and T. Sawada
Nucl. Phys. A 101, 369-407 (1967)

[IV. 16]   M. Marangoni and A. M. Saruis - Phys. Lett. 24 B , 218 (1967)

M. Marangoni and A. M. Saruis - Nucl. Phys. A132, 649-672 (1969)

M. Marangoni and A. M. Saruis - Nucl. Phys. A166, 397-412(1971)

[IV. 17]   J. Birkholz - Nucl. Phys. A 189 , 385-402 (1972)

[IV. 18]   R. F. Barrett, L. C. Biedenharn, M. Danos, P. P. Delsanto

W. Greiner and H. G. Wahsweiler

Rev. Mod. Phys. 45 , 44 (1973)

[IV. 19]   E. M. Diener, J. F. Amann and P. Paul

Physical Review C 7 , 695 (1973).

[IV. 20]   E. Hayward- Many body description of nuclear structure

ed. C. Block (Academic Press, NY 1966, p. 559)

[IV. 21]   M. Kamimura- Nuclear structure studies using electron

scattering and photoreaction, Sendai Conference 1972

M. Kamimura, K. Ikeda and A. Arima

Nucl. Phys. A 95 , 129-160 (1967)

[IV. 22]   V. Gillet, M. A. Melkanoff and J. Raynal

Nucl. Phys. A97 , 631-640 (1967)

V. Gillet - Dubna Conference 1968

[IV. 23]   K. A. Snover, E. G. Adelberger and D. R. Brown

Phys. Rev. Letters 32 , 1061 (1974)

[IV. 24]   M. Suffert- Asilomar Conference, March 1973

[IV. 25]   N. G. Puttaswamy and D. Kohler

Physics Letters 20 , 288 (1966)

H. D. Shay, R. E. Peschel, J. M. Long and D. A. Bromley

Phys. Rev. C 9 , 76 (1973)

[IV. 26]   W. L. Wang and C. M. Shakin

Phys. Rev. C 5, 1898 (1972)

C. M. Shakin - Asilomar Conf. March 1973

[IV. 27]   A. Goswami and R. D. Graves

Physics Letters 39 B , 499 (1972)

[IV. 28]   A. M. Davidson - Nucl. Phys. A 180 , 208-216 (1972)

[IV. 29]   E. D. Mshelia, K. Roos and W. Greiner

Nucl. Phys. A 212, 157-181 (1973)

[IV. 30]   A. M. Lane - Nuclear Theory (W. A. Benjamin, Inc., 1964, New-York, Amsterdam)

[IV. 31]   J. Ahrens and all - Nuclear structure studies using electron scattering and photoreaction - Sendai conference 1972, p. 213

[IV. 32]   W. T. Weng, T. T. S. Kuo and G. E. Brown
            Physics Letters 46 B , 329 (1973)

[IV. 33]   J. S. Levinger - Nuclear photodisintegration . Oxford University Press, 1960

[IV. 34]   A. M. Lane and A. Z. Mekjian
            Physics Letters 43 B , 105 (1973)
            A. M. Lane and A. Z. Mekjian
            Phys. Rev. C 8 , 1981 (1973)

[IV. 35]   W. T. Weng, T. T. S. Kuo and K. F. Ratcliff
            Phys. Letters 52 B , 5 (1974)

[IV. 36]   B. L. Berman and S. C. Fultz
            Rev. Mod. Phys. 47, 3 , 713 (1975)

[IV. 37]   C. Bloch and V. Gillet - Phys. Lett. 16, 62 (1965)
                                     Phys. Lett. 18, 58 (1965)
            C. Bloch, in many body description of nuclear structure and reaction - Academic Press 1966 .

CHAPTER V

THE DECAY CHANNELS FROM THE E1 GIANT STATES -

In the previous chapters, most of the reported data referred to a study of the doorway state through which the giant E1 resonance is excited. However many things can also be learnt by studying the decay channels from the GDR states and in particular the competition between them.

Several examples of such competitions will be extensively dealt with in other lectures and, therefore, will just be quoted here. Let us quote for instance :

1- The study of isospin impurities in light self conjugate nuclei by measuring the ratio of (γ, p) to (γ, n) cross sections leading to mirror levels in the residual nuclei [1], [2] .

2- Similarly the (γ, α) reaction in self conjugate (T = 0) nuclei can lead to a measure of the isospin mixing [1] .

3 - In T $\neq$ 0 nuclei, an isospin splitting of the GDR states into two collective resonances occurs ; one with isospin T ($T_<$) and the other with isospin T + 1 ($T_>$). The selection rules for the particle (γ, p) and (γ, n) decays from these states can help to identify the isospin of E1 resonances [3, 4] . One can also try to study the $T_>$ part of the GDR through a careful study of the energy spectrum of the emitted photo-neutrons [5] .

4- In even-even nuclei, a careful comparison of the elastic photon scattering from GDR states, with the inelastic (γ, γ') scattering towards the first and the second 2^+ levels should provide a very sensitive test of the dynamic collective models [6] .

5- Similar studies of elastic and inelastic scattering of plane
polarized 15. 1 MeV photons, as performed by E. Hayward [7] ,
should provide one with the same type of information.

Only a few typical examples, dealing with a study of the competition
between G. D. R. decay channels, will therefore be reviewed in this chapter.
Special attention will be paid to those studies which became possible only by
using the " monochromatic " photon beams developed at Livermore and
Saclay.

V. A - Competition between the (γ , n) and (γ , 2n) decay modes in heavy nuclei.

If one supposes that all electric dipole absorptions E1 by a nucleus
lead to the formation of a compound nucleus before the evaporation of one
or two neutrons can occur, then several approaches exist which allow one
to extract the nuclear temperature parameter Θ or the nuclear level
density parameter ˚a˝for the target minus-one-neutron nucleus from the
measured partial cross sections of the target nucleus.

The principle of such a determination is summarized in fig. 1 where
one assumes that the target nucleus $_Z A_N$ has been excited at energy E by
absorption of a photon.

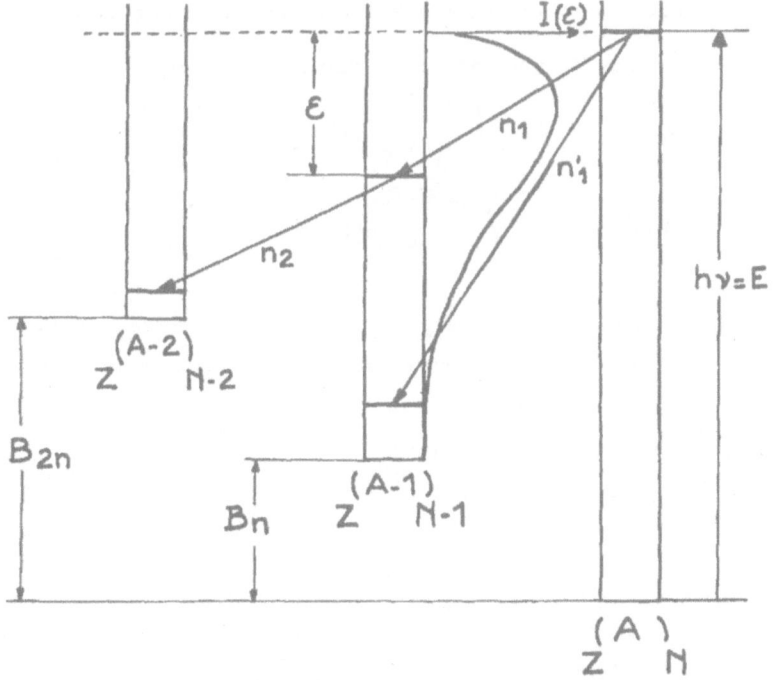

Fig.1

As a first step towards deexcitation a neutron is evaporated with energy ε from the compound state at energy E created in the nucleus $_Z A_N$. Following Blatt and Weisskopf [8] one can assume that the energy spectrum of such neutrons will be approximately:

$$I_n(\varepsilon)\, d\varepsilon = C^t . \varepsilon . \rho(U) \qquad (1)$$

where U is the effective excitation energy of the (A-1) nucleus, $U = E - B_n - \varepsilon - \delta$, ε is the kinetic energy of the emitted neutron, δ = pairing energy of the (A - 1) nucleus, ρ (U) is the level density formula still to be determined and B_n is the threshold value for the (γ, n) reaction.

One now assumes that, if the level reached in the daugther nucleus $_Z (A-1)_{N-1}$ after emission of one neutron n_1 by $_Z A_N$, is high enough to allow the emission of a second neutron n_2 towards the nucleus $_Z (A-2)_{N-2}$, then this second process must occur and a (γ, 2n) reaction will be observed. On the other hand if the first neutron n_1' has been emitted by the nucleus $_Z A_N^*$ with too large an energy ε , the excitation U of the nucleus $_Z (A-1)_{N-1}$ will not allow the emission of a second neutron. The nucleus $_Z (A-1)_{N-1}$ will decay by a γ cascade only and a reaction (γ, n) will be observed. Hence we can write the following general expression

as a function of the energy E of the absorbed photons :

$$\frac{\sigma_{\gamma, 2n}(E)}{\sigma_{\gamma, n}(E) + \sigma_{\gamma, 2n}(E)} = \frac{\int_{\varepsilon=0}^{E-B_{2n}} \varepsilon\, \rho(U)\, d\varepsilon}{\int_{\varepsilon=0}^{\varepsilon = E-B_n-\delta} \varepsilon\, \rho(U)\, d\varepsilon} \qquad (2)$$

If one now assumes that the level density form of the (A-1) nucleus at the excitation energy U is of the type described by Blatt and Weisskopf then

$$I_n(\varepsilon)\, d\varepsilon = C . \varepsilon \, \exp\left(\frac{-\varepsilon}{\Theta}\right) d\varepsilon \qquad (3)$$

where $\textcircled{H}$ is the nuclear temperature of the nucleus $_Z(A-1)_{N-1}$ at the energy $E-B_n$. If one assumes $\textcircled{H} \simeq$ constant, then eq (2) can be rewritten as follows :

$$\frac{\sigma_{\gamma,2n}(E)}{\sigma_{\gamma,n}(E) + \sigma_{\gamma,2n}(E)} = 1 - \left[1 + \frac{E - B_{2n}}{\textcircled{H}}\right] exp\left[-\frac{E - B_{2n}}{\textcircled{H}}\right] (4)$$

If we suppose that the level density in the (A-1) nucleus will be better represented by a Fermi gas type formula such as

$$\rho(U) = \frac{C}{U^2} exp\, 2\sqrt{aU} \tag{5}$$

then we can use expression (2) as it stands.

Both formulae, however, predict the complete disappearance of the $\sigma(\gamma,n)$ curve a few MeV above the $(\gamma,2n)$ threshold B_{2n}, for reasonable values of $\textcircled{H}$ and a . This prediction was verified by the results of Harvey [9] and Fultz [10] for ^{208}Pb and ^{197}Au respectively.

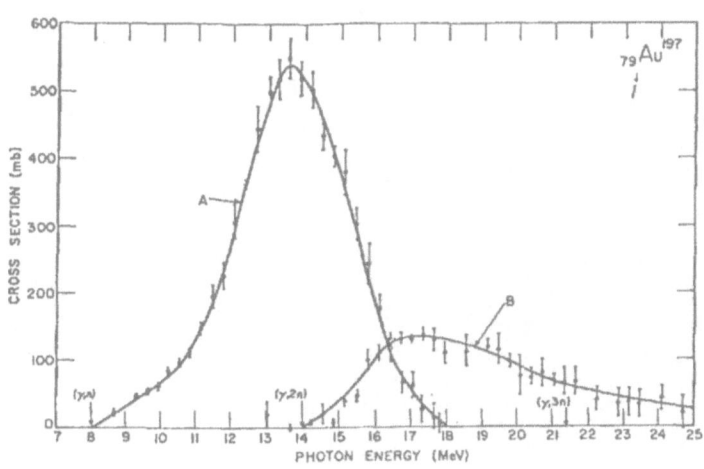

Partial cross-section curves for gold. Curve A consists of $\sigma(\gamma,n) + \sigma(\gamma,np)$. Curve B consists of $\sigma(\gamma,2n) + 3\sigma(\gamma,3n)$.

FIG. 2 : from ref. [V. 10]

Fig. 2 shows the Livermore data for ^{197}Au, where the above analysis leads to the value a = 17 MeV^{-1}.

However, the hypothesis that all neutrons are emitted only once the nucleus $_Z A_N$ has reached an equilibrum, typical of the compound nucleus, was not confirmed by specific studies of the spectrum of the photoneutrons emitted after absorption of monoenergetic photons by the nucleus $_Z A_N$. Some recent experiments by Calarco [11] on the photoneutron spectrum of ^{208}Pb (for the energy interval 12 MeV $\leqslant$ E $\leqslant$ 17 MeV) have shown that the fraction of " direct " neutrons is a constant and equal to n_D = 0. 14 $\pm$ 0. 03; what is called here " direct neutrons " is the fraction of photoneutron emitted by nucleus $_Z A_N$ in a somehow " direct process ", before any equilibrium of the compound nucleus type has been reached. Fig. 3 [12] shows how this percentage n_b of " direct " photoneutrons appears as a high energy bump (12 % of all the photoneutrons) in the energy spectrum of the photoneutrons from ^{209}Bi (γ , n) which is represented by a maxwellian straight line only in its lower energy part (Fig. 3).

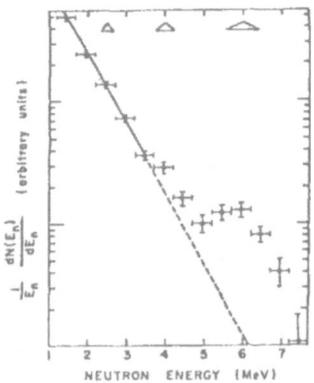

Fig. 3
Semilogarithmic plot of the spectrum (V. 12)

One admits that the low-energy part of the neutron spectrum can be completely represented by any of the evaporation formulae and that the residue of high-energy neutrons, which cannot thus be fitted, pertains to the so called direct interaction neutrons. In our specific case we assume that x % of all nuclear photon absorptions by the electric dipole process will be followed by the emission of such a direct high energy neutron. Such a direct neutron emission towards the low-lying energy levels of the (A-1) nucleus makes the emission of a second neutron, constituting a $(\gamma, 2n)$ reaction impossible. This was confirmed by the Saclay study on ^{208}Pb and ^{197}Au [13] which did not show a σ (γ, n) cross section falling to zero a few MeV above the B_{2n} threshold. (Fig. 4)

FIG. 4.

reproduced from ref. V.13

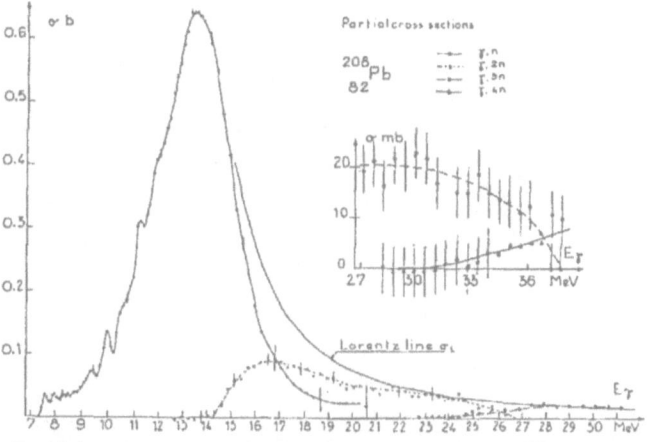

Partial photoneutron cross sections $\sigma_{\gamma, n}$, $\sigma_{\gamma, 2n}$, $\sigma_{\gamma, 3n}$, and $\sigma_{\gamma, 4n}$ of ^{208}Pb. We also show the descending part of the unique Lorentz line giving the best fit to the experimental $\sigma_{\gamma, T}(E)$ curve.

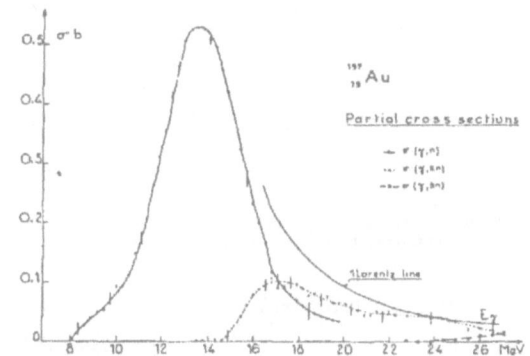

Partial photoneutron cross sections $\sigma_{\gamma, n}$, $\sigma_{\gamma, 2n}$ and $\sigma_{\gamma, 3n}$ of ^{197}Au. We also show the descending part of the unique Lorentz line

Assuming therefore x $\simeq C^r$ in the region covering a few MeV above B_{2n} , one has there

$$\frac{\sigma_{\gamma,2n}(E)}{\sigma_{\gamma,n}(E) + \sigma_{\gamma,2n}(E)} = (1-x) \cdot \frac{\int_{\varepsilon=0}^{\varepsilon=E-B_{2n}} \varepsilon \cdot \rho(U) \cdot d\varepsilon}{\int_{\varepsilon=0}^{\varepsilon=E-B_n-\delta} \varepsilon \, \rho(U) \, d\varepsilon} \qquad (6)$$

hence a very good theoretical fit of the experimental σ (γ , 2n) curve can be obtained (Fig. 5).

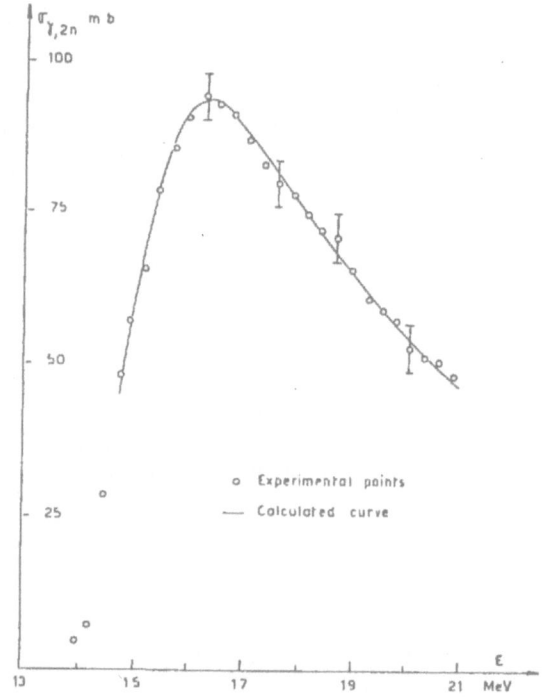

Actual fit between our experimental $\sigma_{\gamma,2n}(E)$ curve for [208]Pb and the analytical expression given in formula 6 for $x = 0.26$ and $a = 7.4$ MeV[-1].

FIG. 5 : from ref V. 13

A cruder way to get this percentage x of direct (or " non statistical ") neutrons is provided by plotting (Fig. 6) the experimental photoneutron multiplicity M (E) defined as follows

$$M(E) = \frac{\sigma_{\gamma,n} + 2\sigma_{\gamma,2n}}{\sigma_{\gamma,n} + \sigma_{\gamma,2n}} \qquad (7)$$

When using this approach, it can be shown that above the $(\gamma,2n)$ threshold
B_{2n} the value of M (E) tends towards its asymptotic form :

$$M_A = \frac{x_A \sigma_T + 2(1-x_A)\sigma_T}{x_A \sigma_T + (1-x_A)\sigma_T} = 2 - x_A \qquad (8)$$

(with $\quad \sigma_T \simeq \sigma(\gamma,n) + \sigma(\gamma,2n)$

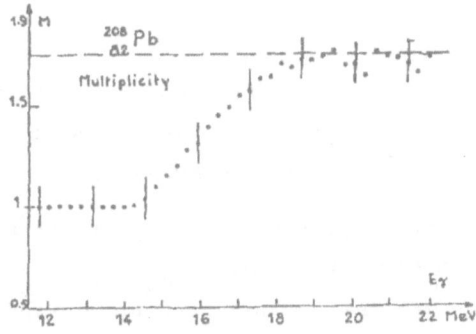

Neutron multiplicity M of ^{208}Pb as a function of incident photon energy.

FIG. 6 : from ref. V. 13.

The results obtained for x in the interval $B_{2n} \leqslant E \leqslant B_{2n} + 5\,\text{MeV}$
and x_A for $E \geqslant B_{2n} + 5$ MeV are in excellent agreement for ^{208}Pb as
well as for ^{197}Au. This leads us to conclude that the contribution of
" direct " neutrons $n_D = x_A / (2 - x_A)$ is approximately 15% and 20%
for ^{208}Pb and ^{197}Au respectively.

As can be seen from table I our values of n_D are again in good agreement with the n'_D and n''_D values (obtained from photoneutron spectra) given by Calarco [11]   (12 MeV $\leqslant$ E $\leqslant$ 17 MeV) and Mutchler [14]   (E = 14 MeV).

Table I

(from ref V.13)

"Direct" neutron contributions and level density parameters for ^{208}Pb and ^{197}Au obtained from the present work (undashed symbols) compared to values given by other authors (dashed symbols), where x or x_A is the fraction of the total cross section responsible for the emission of a "direct" neutron, n_D is the percentage of "direct" neutrons and a is the level density parameters for the target minus one neutron nucleus

	x	x_A	n_D	n'_D	n''_D	a (MeV^{-1})	a' (MeV^{-1})	a'' (MeV^{-1})
^{208}Pb	0.26±0.05	0.27±0.05	0.15±0.04	0.14±0.03	0.18±0.03	7.4±1	11.7	8.17
^{197}Au	0.31±0.05	0.35±0.05	0.20±0.04		0.14±0.04	12±2	16.6	19.13

However, our a values are substantially lower than the a' values found by Buccino [15]   using (n, n') reactions and the a" values given by Facchini et al [16]   and calculated from slow neutron resonance data.

If the introduction of a percentage x of " direct " neutrons provides a good fit of the observed branching ratio $\dfrac{\sigma(\gamma, 2n)}{\sigma(\gamma, n)}$, it seems desirable to get a deeper insight into the reaction mechanism responsible for the emission of those direct photoneutrons.

Griffin [17] and Blann [18]   have recently introduced the concept that, for (p, n) and (α, n) reactions, neutrons might be emitted before the nucleus could attain a statistical equilibrium. These so-called " precompound " neutrons would then have an energy spectrum richer in high-energy neutrons than an evaporation spectrum, and its precise shape would depend on the initial number n of excitons. (1 exciton is either a particle elevated above the fermi level, or a hole below the fermi level).

They considered a time dependent approach of the formation of the

compound nucleus. In fig. 7 for example a proton is captured in the potential well of a target nucleus where it brings an excitation energy E, concentrated initially on a 1 p state. Then, successively, the two-body interaction
scatters this excitation energy into a 2p-1h state, then a 3p-2h, then a
4p-3h state .... (Fig. 7) until finally one reaches the compound state with
an average number ñ of excitons

$$\overline{m} = \sqrt{g\,E}$$
$$g = \frac{2}{\pi}\,a \simeq 0.6\ a\ \text{Mev}^{-1}$$

hence for ^{208}Pb: $a \simeq 8\ \text{MeV}^{-1}$, and E = 15 to 20 MeV
in the region where the competition $(\gamma, n)/(\gamma, 2n)$ is
observed,

$\bar{n} = 8$ to 10 (i. e. 4p-4h to 5p - 5h states)

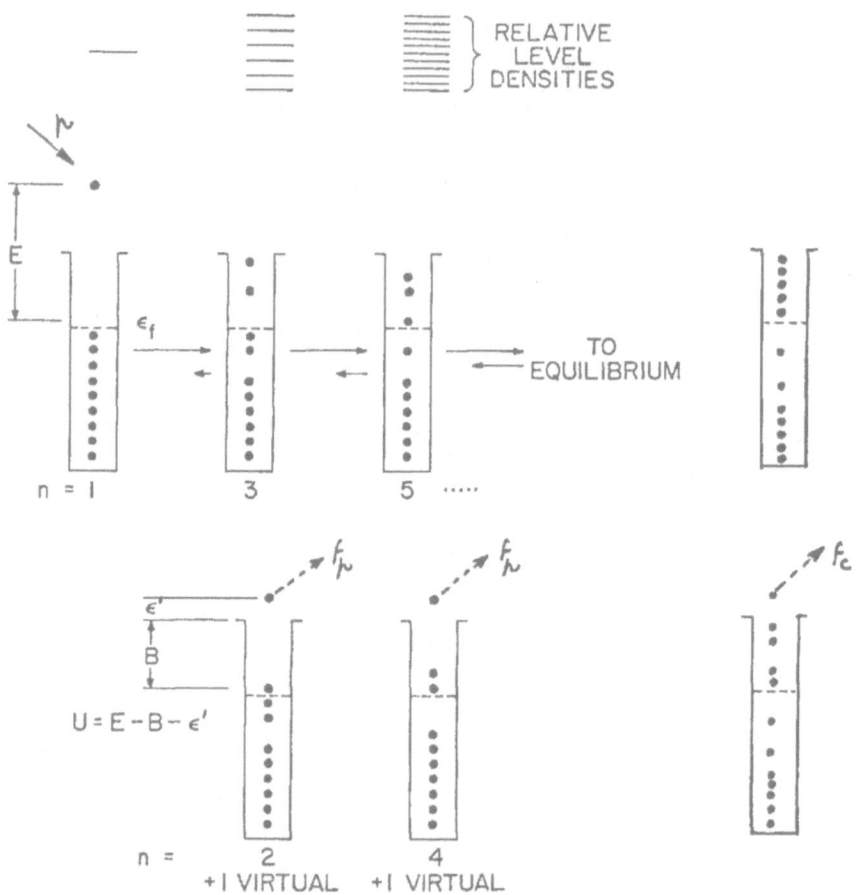

Pictorial representations of the ideas inherent in the exciton model. The series of two-body interactions leading toward an equilibrium distribution is illustrated, as well as the possibility of the configuration at each exciton number of having one or more particles in unbound states.

FIG. 7. : reproduced from ref. [V.19]

These " precompound states " now have a probability for the emission of a neutron (f_p). But, since the available energy is less " dispersed " among the excitons than in the final compound nucleus configuration, the ensuing neutron energy spectrum from such " precompound states " will be richer in high energy neutrons than a typical Maxwellian spectrum, characterizing the neutrons evaporated from the statistical equilibrium of the compound state, with an emission probability (f_c).

in some simple cases the initial number of excitons can be known :

2 particle- 1h in reactions (pn)

5 particle - 1h in reactions (α, n)

Then a good agreement was obtained between the observed neutron spectra and theoretical fits by a compound part W_c (E) and a precompound part W_p (E) as seen in fig. 8 [20] and fig. 9 [21] .

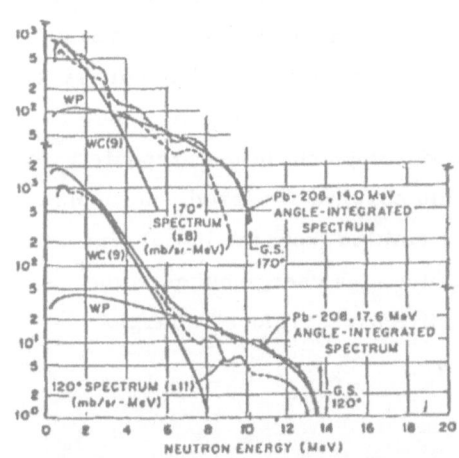

Angle-integrated experimental neutron production cross sections, $\sigma(E_i) = \sum_j \sigma(E_i,\theta_j)\Delta\Omega_j$, and cross sections measured at $\theta_j = 120°$ and $170°$. The $\sigma(E_i)$ are fitted with a precompound component, WP, and a compound component, $WC(g)$, calculated from Griffin's theory

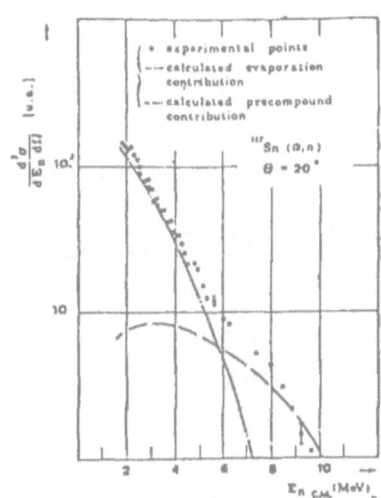

Precompound and evaporation analysis in the ^{117}Sn(α, n) reaction

FIG 8 : ^{208}Pb(p,n)

from ref. V.20

Fig 9 : ^{117}Sn (α,n)

from ref. V. 21

Unfortunately in the case of a photoneutron decay from the G. D. R, the definition of the " exciton configuration " of the doorway state is not simple since one has to deal with a coherent combination of elementary $1p$-$1h >$ states. Recently V. K. Luk'yanov [22] proposed to take into account the collective nature of the $1p$-$1h >$ doorway states by considering a modified density $\tilde{\rho}(1p$-$1h)$ of the initial $1p$-$1h$ states with respect to the classical density $\rho(1p$-$1h)$

where

$$\tilde{\rho}_{1p\text{-}1h} = \frac{\sigma_{\gamma\,exp}}{\sigma_{\gamma\text{-}1part}}\; \rho_{1p\text{-}1h} \gg \rho_{1p\text{-}1h}$$

$\sigma_{\gamma\,exp}$ = total photoabsorption cross section .

If one tries then to apply a classical program to compute the energy spectra of precompound neutrons starting from the above initial conditions, Luk'yanov observed that to consider an increase of the ($1p$-$1h$) density is equivalent to saying that the system lives for a comparatively long time in the initial $1p$-$1h$ states. This hinders the development of the preequilibrium decay process; as a result one should observe a relative enhancement of the channel in which a neutron is directly emitted from the initial dipole state into the continuum. The importance of the high energy bump on the energy spectrum of the emitted photoneutrons must therefore be larger than in the case of (p, n) neutrons (Fig. 10).

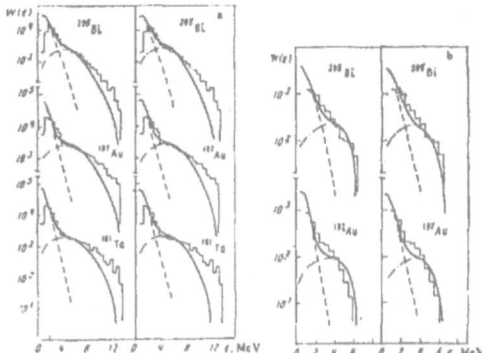

. Photoneutron spectra at the bremsstrahlung energy (a)
$E_{\gamma max}$ = 20 MeV and (b) $E_{\gamma max}$ = 14 MeV. On the left are the MPD
curves and on the right, the MEM curves. The solid lines depict the
total spectra, the dashed lines, the equilibrium spectrum, and the dashed-
dotted lines, the purely pre-equilibrium spectra.

FIG. 10 : reproduced from ref[V.22]

A qualitative verification is possible. With approximately the
same liquid scintillator detectors, the following competitions have been
studied (Fig. 11 and 12)

$$\frac{\sigma(p, 2n)}{\sigma(p, n)}$$ at Livermore [23]

$$\frac{\sigma(\gamma, 2n)}{\sigma(\gamma, n)}$$ at Saclay [13] [24]

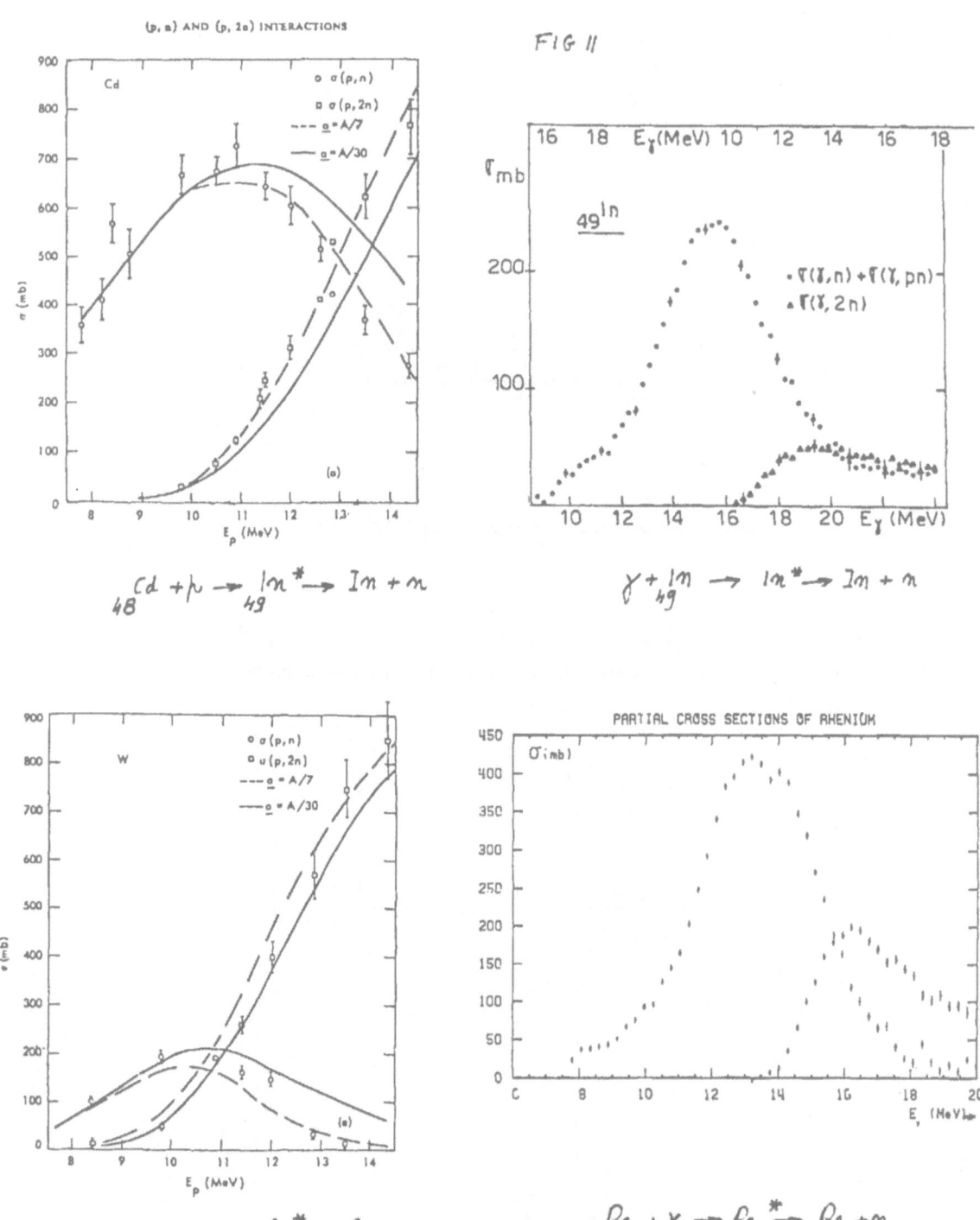

(p, n) AND (p, 2n) INTERACTIONS

Cd
○ σ(p,n)
□ σ(p,2n)
– – a = A/7
—— a = A/30

(a)

$$_{48}Cd + p \longrightarrow _{49}In^* \longrightarrow In + n$$

FIG 11

49^{In}

• Γ(γ,n) + Γ(γ,pn)
▲ Γ(γ,2n)

$$\gamma + _{49}In \longrightarrow In^* \longrightarrow In + n$$

W
○ σ(p,n)
□ u(p,2n)
– – a = A/7
—— a = A/30

(a)

$$_{74}W + p \longrightarrow _{75}Re^* \longrightarrow Re + n$$

PARTIAL CROSS SECTIONS OF RHENIUM

$$_{75}Re + \gamma \longrightarrow _{75}Re^* \longrightarrow Re + n$$

FIG. 12

For a given E_n energie, one knows the excited nucleus and its excitation energie U in the Livermore experiment. One can then make a comparison with the photo cross sections obtained with photons E_γ = U on these nuclei at Saclay. Table 2 shows that for a given excitation of the same nucleus, the ratio of the 2 n channel to the 1 n channel is always larger in the proton case showing therefore that the photoneutron spectrum is richer in higher energy components (fig. 13)

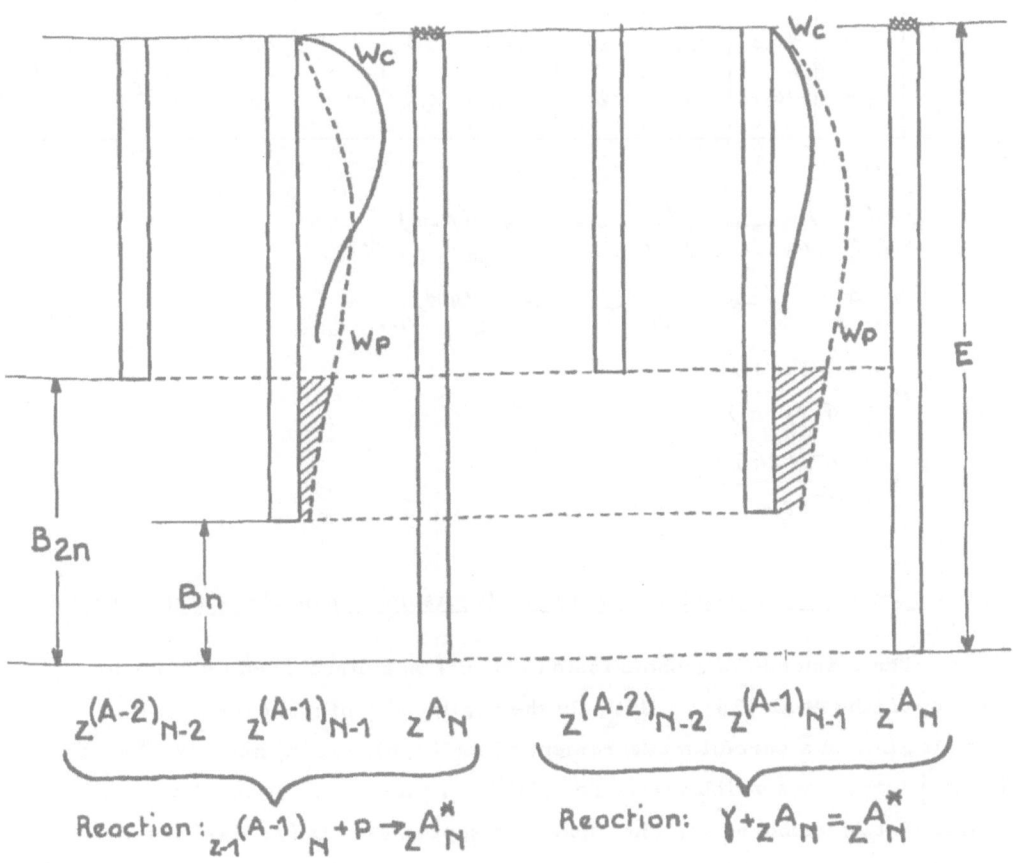

Fig.13

Table 2

$\begin{array}{cc}(A\text{-}1)\\ Z\text{-}1 & N\end{array}$	$\begin{array}{cc}A\\ Z & N\end{array}$	$B_p(A)$	$B_{2n}(A)$	E_p	$\sigma(P,n)$	$\sigma(P,2n)$	R_p	E_γ	$\sigma(\gamma,n)$	$\sigma(\gamma,2n)$	R_γ	$\dfrac{R_p}{R_\gamma}$
Cd_{48}	$In\,^{113}_{49}\,_{115}$	6,5	16,8	14,5	280	760	2,7	21,1	$35^{\pm3}$	$42^{\pm2}$	1,2	2,25
W_{74}	$Re\,^{185}_{75}\,_{187}$	5,7	14	12	150	400	2,6	17,7	$100^{\pm10}$	$120^{\pm10}$	1,2	2,15
Ag_{47}	$Cd\,^{108}_{48}\,_{110}$	8,5	17,7	14	200	850	4,2	22,5	$24^{\pm4}$	$40^{\pm2}$	1,7	2,5
$Nb\,^{93}_{41}$	$Mo\,^{94}_{42}$	8,5	17,8	14,5	250	700	2,8	23	$15^{\pm4}$	$30^{\pm5}$	2	1,4
$Zr\,^{90}_{40}$	$Nb\,^{91}_{41}$	5,1	21,8	14	200	300	1,5	19,1	$74^{\pm5}$	$48^{\pm2}$	0,65	2,3
$Au\,^{197}_{79}$	$Hg\,^{198}_{80}$	7,1	15,3	12,6	60	500	8,3	19,7	$22^{\pm5}$	$70^{\pm5}$	3,2	2,6
$Ta\,^{181}_{73}$	$W\,^{182}_{74}$	7,1	14,7	12	80	500	6,2	19,1	$18^{\pm3}$	$100^{\pm5}$	5,5	1,2

$$p + \underset{Z\text{-}1\quad N}{(A\text{-}1)} \longrightarrow \underset{Z\quad N}{A^*} \longrightarrow n + \underset{Z\qquad N\text{-}1}{(A\text{-}1)}$$

$$\gamma + \underset{Z\ N}{A} \longrightarrow \underset{Z\ N}{A^*} \longrightarrow n + \underset{Z\quad N\text{-}1}{(A\text{-}1)}$$

$$R_p = \frac{\sigma(p,2n)}{\sigma(p,n)}$$

$$R_\gamma = \frac{\sigma(\gamma,2n)}{\sigma(\gamma,n)}$$

V.B - Competition between the (γ,n) and $(\gamma,\text{fission})$ channels in fissile nuclei -

The existence of a photofission channel was discovered as early as 1941 by Haxby et al [25] . But only the availability of monochromatic γ-rays allowed a careful measurement of the (γ,n), $(\gamma,2n)$ and $[(\gamma,f) + (\gamma,nf)]$ processes which are in competition in the GDR region. For example at Livermore [26] one measured separately $\sigma(\gamma,\text{fission})$ with a fission chamber and later simultaneously

$$\sigma(\gamma, N) = \sigma(\gamma, n) + 2\sigma(\gamma, 2n) + 3\sigma(\gamma, 3n) + \bar{\nu}\,\sigma(\gamma, F)$$

and the multiplicity μ of neutron emission

$$\mu = \frac{\sigma(\gamma, N)}{\sigma(\gamma, tot)}$$

obtained by measuring separately the channels with $1, 2, 3 \ldots$ neutrons.

FIG. 14

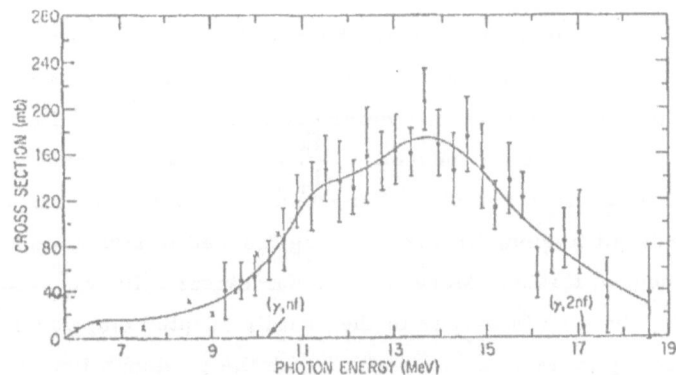

Photofission cross section of U^{235} as measured by BOWMAN, AUCHAM-PAUGH, AND FULTZ. Black dots represent data taken with photons from annihilation of positrons in flight. The crosses represent an extension of the measurement to lower energy with a bremsstrahlung beam.

FIG 15

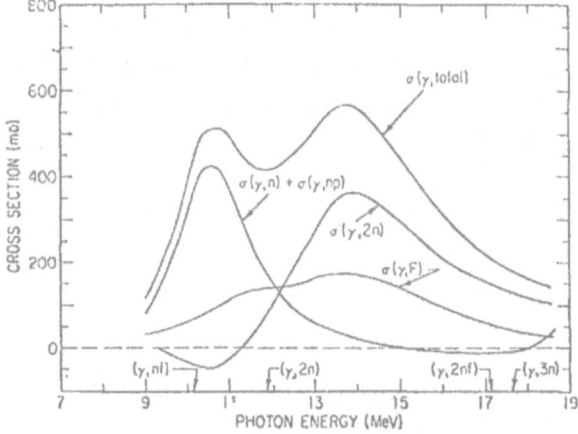

Summary of analysis by BOWMAN, AUCHAMPAUGH, AND FULTZ of the total photodisintegration cross section for U^{235} into its components. The negative cross sections arise from inaccuracies in the analysis.

FIG. 14 and FIG. 15 are reproduced from ref. V.26.

From these data, and writing

$$\sigma(\gamma, n) = \sigma(\gamma, tot) \frac{\Gamma_m}{\Gamma}$$

$$\sigma(\gamma, F) = \sigma(\gamma, tot) \frac{\Gamma_F}{\Gamma}$$

they get the ratio $\dfrac{\Gamma_n}{\Gamma_f} = \dfrac{\sigma(\gamma, n)}{\sigma(\gamma, F)}$ if the energy is below the (γ, 2n) and (γ, nf) threshold .

They were thus the first to observe the characteristic splitting of the GDR of a fissile nucleus into two components, a phenomenon observed for other permanently deformed nuclei as well . However they found that the photon-induced $\dfrac{\Gamma_n}{\Gamma_f}$ ratio was strongly energy dependent, a result in complete disagreement with data obtained from neutron-induced fission, bremsstrahlung induced fission and charged-particle-induced fission. Moreover the numerical value of $\dfrac{\Gamma_n}{\Gamma_f}$ turned out to be well in excess of the usually adopted empirical relationship connecting these $\dfrac{\Gamma_m}{\Gamma_f}$ values with the fissionability parameter Z^2/A.

By measuring separately, for each energy E of the incoming photons, the branching ratio towards the 1n, 2n, 3n .... 7n channels and by using the law (known from neutron induced fission)giving the average number of fission neutrons $\overline{\gamma}(E) = aE + b$ and their probability repartition $P_{\overline{\gamma}}(\gamma)$, one can get the true partial cross sections $\sigma(\gamma, n)$, $\sigma(\gamma, 2n)$, $\sigma(\gamma, F)$ shown in Fig. 16 [27] .

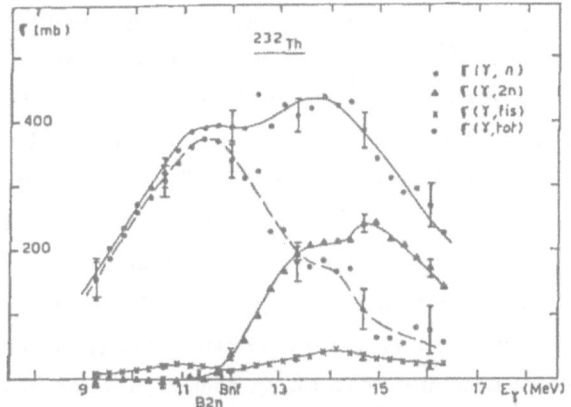

Partial and total photonuclear cross sections $\sigma(\gamma, n)$, $\sigma(\gamma, 2n)$, $\sigma(\gamma, F)$ and
$\sigma_{tot} = \sigma(\gamma, n) + \sigma(\gamma, 2n) + \sigma(\gamma, F)$ of $^{232}_{90}$Th.

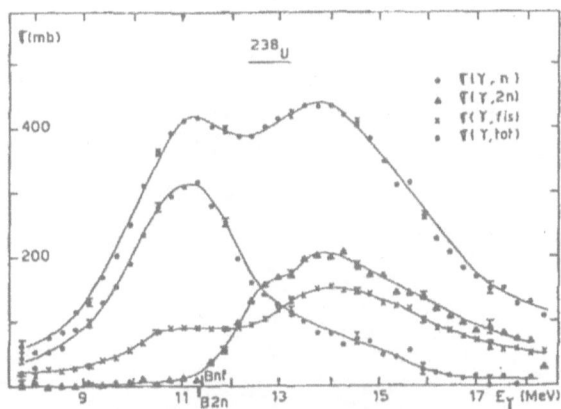

FIG. 16 : reproduced
from ref. V.27

Partial and total photonuclear cross sections $\sigma(\gamma, n)$, $\sigma(\gamma, 2n)$, $\sigma(\gamma, F)$ and $\sigma_{tot} = \sigma(\gamma, n) \div$
$\sigma(\gamma, 2n) + \sigma(\gamma, F)$ of $^{238}_{92}$U.

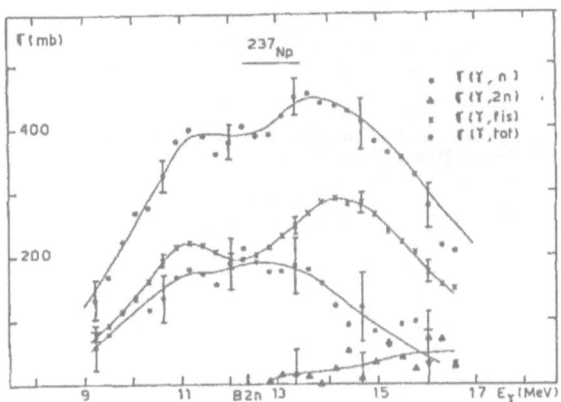

Partial and total photonuclear cross sections $\sigma(\gamma, n)$, $\sigma(\gamma, 2n)$, $\sigma(\gamma, F)$ and
$\sigma_{tot} = \sigma(\gamma, n) \div \sigma(\gamma, 2n) \div \sigma(\gamma, F)$ of $^{237}_{93}$Np.

V. B. 1- <u>Comparison between total absorption and photofission cross sections.</u>

In fig. 17 we present the $\dfrac{\sigma(\gamma, F)}{\sigma(\gamma, tot)}$ ratios as a function of the excitation energy E for the three nuclei concerned. The behaviour of these ratios closely resembles the results obtained in fast-neutron-induced-fission experiments (fig. 18 and 19) where the $\sigma(n, f)$ curve, after an initial sharp rise, flattens out over several MeV until a new rise sets in at about $E_n \simeq$ 5 to 7 MeV. The second rise is attributed to the fact that the excitation energy of the target nucleus is high enough to permit evaporation of one neutron without reducing the excitation energy of the residual nucleus below its own fission threshold. In such a case the system gets a second chance to undergo fission by the (n, nf) reaction.

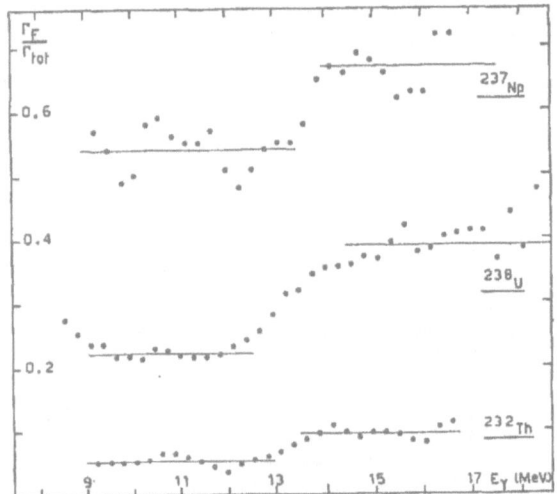

Behaviour of the ratio $\sigma(\gamma, F)/\sigma_{tot}$ as a function of the excitation energy E of the fissioning nuclei $^{232}_{90}$Th, $^{238}_{92}$U and $^{237}_{93}$Np.

FIG. 17

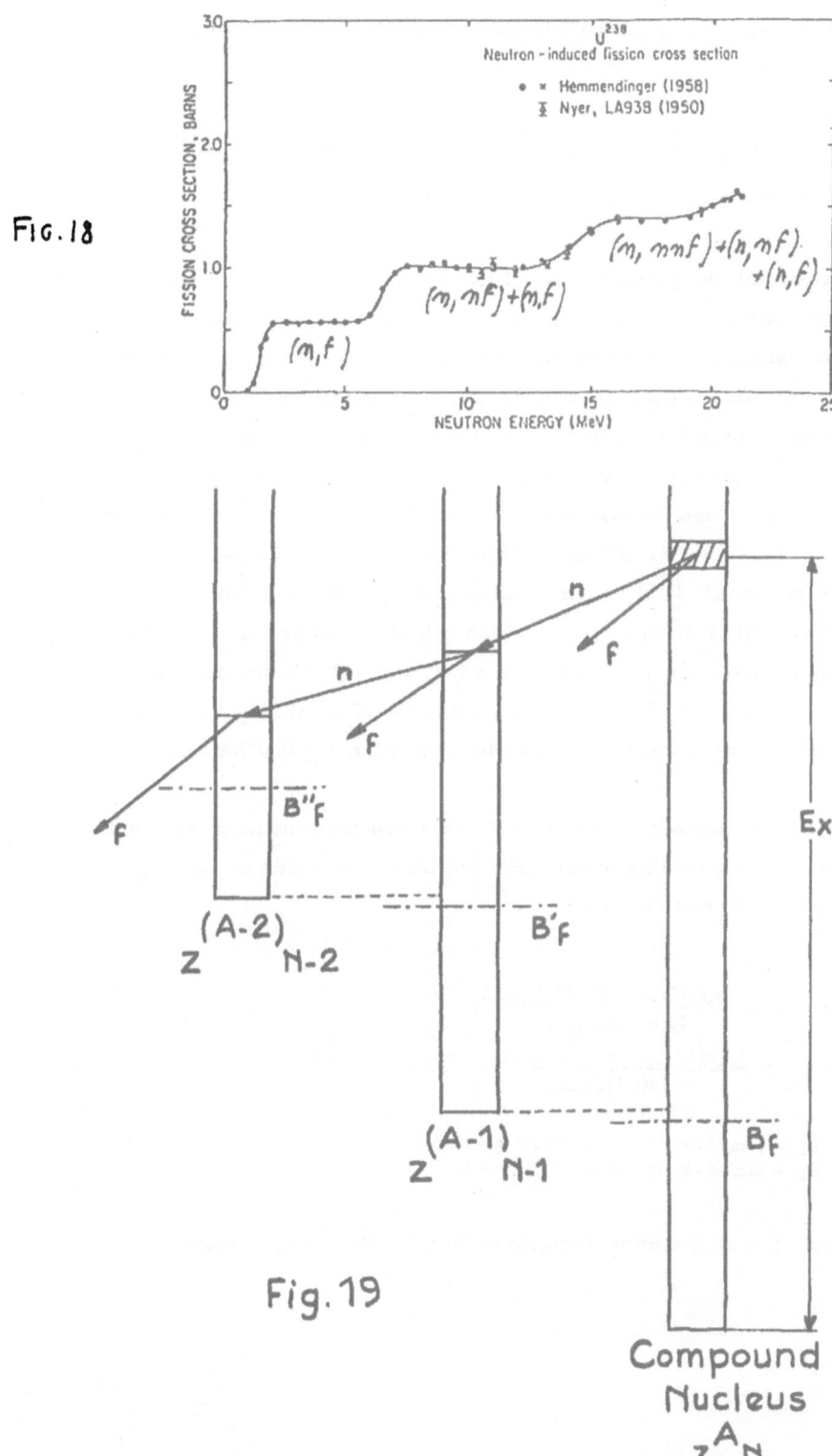

FIG. 18

Fig. 19

The same reasoning adapted to photofission   then allows us
to write :

$$\sigma(\gamma, F) = \sigma(\gamma, f)_A \qquad \text{(a)}$$

$$\sigma(\gamma, F) = \sigma(\gamma, f)_A + \sigma(\gamma, nf)_{A-1} \quad \text{(b)}$$

where (a) is valid for the photofission of the A- nucleus and corresponds
to the first flat portion of our curves, and where the second term in (b)
represents the "second chance photofission" of the (A-1) nucleus and
corresponds to the flat-topped step in fig. 17. One notes that the (γ, nf)
threshold values of 11.8 MeV and 11.2 MeV, given by Gindler et al [29]
for ^{232}Th and ^{238}U respectively, do in fact agree with the onset of the
step in our curves. If one evaluates the $(\gamma, 2n)$ threshold of ^{238}U $\simeq$ 11.3 MeV and by
taking 5.7 $\pm$ 0.2 MeV for the fission barrier value of ^{236}U as recom-
mended by Fraser et al [30] , one obtains for the " third -chance
photofission threshold" a value of 17.1 MeV which again agrees with the
position of the second step in fig. 17. The experimental second-chance
photofission threshold for ^{237}Np obtained from fig. 17 is 12.1 $\pm$ 0.4 MeV
which would then correspond to a fission barrier of 5.4 $\pm$ 0.4 MeV in
^{236}Np.

One can also compare the ratio P_n obtained from neutron-induced
fission with the corresponding expression for photon-induced fission P_γ
for an identical target nucleus A :

$$P_n = \frac{[\sigma(n, f)_{A+1} + \sigma(n, nf)_A]_{E_n > B'_{nf}}}{[\sigma(n, f)_{A+1}]_{E_n < B'_{nf}}},$$

$$P_\gamma = \frac{[\sigma(\gamma, f)_A + \sigma(\gamma, nf)_{A-1}]_{E > B_{nf}}}{[\sigma(\gamma, f)_A]_{E < B_{nf}}},$$

where

B'_{nf} = second-chance neutron-fission threshold,
B_{nf} = second-chance photofission threshold.

Results of such a comparison, given in table 3 show an interesting resemblance and this in spite of the fact that a " one-neutron difference " between the two formulae exists.

Table 3

Comparison of P_n and P_γ defined in the text	for ^{232}Th, ^{238}U and ^{237}Np		
	^{232}Th	^{238}U	^{237}Np
P_n	2.0	1.8	1.4
P_γ	1.9±0.2	1.76±0.1	1.25±0.3

V. B. 2 - Comparison between fission and neutron exit channels .

Experimental results further enable one to evaluate the fundamental neutron-to-fission width ratio $\frac{\Gamma_n}{\Gamma_f}$ (^{238}U). The raw experimental data $\sigma(\gamma,n)/\sigma(\gamma,f)$, (without the $x \simeq 15\%$ direct-interaction correction discussed previously), has been taken to be equal to the $\frac{\Gamma_n}{\Gamma_f}$ ratio shown in fig. 20 for E $\leqslant$ B$_{2n}$. One finds that results are a reasonable extension of the data published by Mafra [31] for photons below 9 MeV and tend to a constant asymptotic value of $\frac{\Gamma_n}{\Gamma_f}$ = 3.45 ± 0.2 for 9.5 MeV $\leqslant$ E $\leqslant$ 11.5 MeV, in agreement with the statistical-model interpretation. However, this result is in complete disagreement with a value of $\frac{\Gamma_n}{\Gamma_f}$ = 6 published by Bowman [26] for the photofission of ^{235}U at E = 10 MeV.

One can also evaluate the $\frac{\Gamma_n}{\Gamma_f}$ ratio for ^{237}U as follows : if one considers first the experimentally constant $\frac{\Gamma_n}{\Gamma_f}$ ratio for ^{238}U above 9.5 MeV and if one further assumes this experimental ratio to remain constant over the whole GDR region, one can then extrapolate a value of $\frac{\sigma(\gamma,f)}{\sigma(\gamma,tot)}$ = 0.22 above the B$_{2n}$ = 11.4 MeV and B$_{nf}$ = 11.3 MeV thresholds and write :

$$\sigma(\gamma,nf) = \sigma(\gamma,F) - 0.22\,\sigma_{TOT}$$

$$\frac{\Gamma_n}{\Gamma_f}(^{237}U) = \frac{\sigma(\gamma,2n)}{\sigma(\gamma,nf)}$$

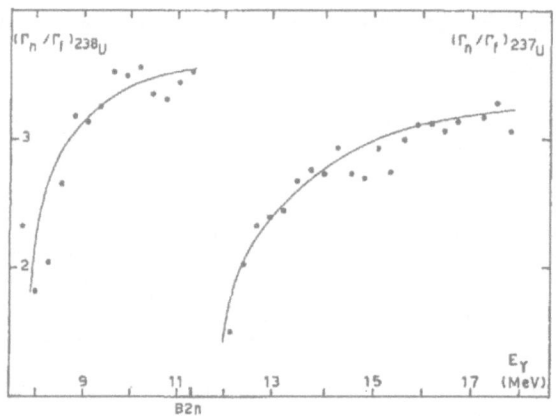

Fig. 20 - The Γ_n/Γ_f ratios versus the excitation energy E measured for ^{238}U and calculated for ^{237}U.

Table 4

The photoneutron-to-photofission width ratios Γ_n/Γ_f extracted from our experimental data

	^{232}Th	^{231}Th	^{238}U	^{237}U	^{237}Np
Γ_n/Γ_f	18.5±3	14.5±3	3.45±0.2	3.10±0.2	0.9±0.2

This ratio, also plotted in fig. 20 has a behaviour similar to the (Γ_n/Γ_f) ^{238}U curve but with an asymptotic value of 3.1 ± 0.2. Following the same procedure one can also evaluate the Γ_n/Γ_f ratios for ^{232}Th, ^{231}Th and ^{237}Np but the big statistical errors and the small $\sigma(\gamma,2n)$ cross section measured for ^{237}Np made the evaluation of $\frac{\Gamma_n}{\Gamma_f}$ for ^{236}Np impossible. The different Γ_n/Γ_f values obtained are shown in table 4 and are also plotted against the fissionability parameter (Z^2/A) in fig. 21.

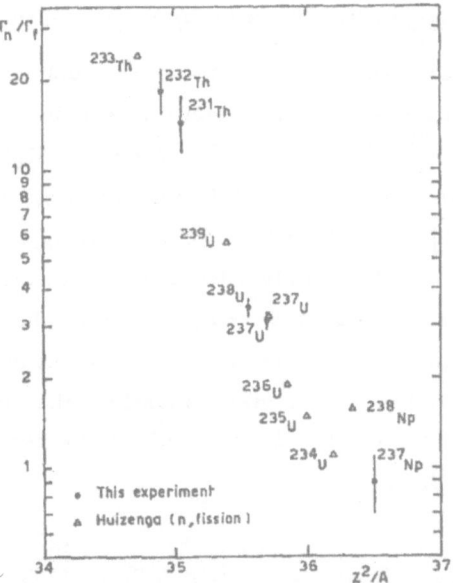

Photoneutron-to-photofission width ratios Γ_n/Γ_f as a function of the fissionability parameter Z^2/A obtained from table 4 together with neutron-induced width ratios Γ_n/Γ_γ as published by Huizenga *et al.*

FIG. 21

Therefore, the $\dfrac{\Gamma_n}{\Gamma_f}$ values, found either in neutron or in photon induced fission, are not significantly different for a given nucleus although the angular momenta of the compound nucleus are much more restricted in a photon induced experiment   since mostly E1 and possibly E2 transitions happen.

V. C - <u>The statistical competition between the (γ, n) and (γ, p) channels.</u>

The comparison between specific channels, such as (γ, p_0) and (γ, n_0), can yield specific information about the isospin impurities in T = 0 nuclei [1] and the $\frac{\sigma(\gamma, p)}{\sigma(\gamma, n)}$ branching ratio from the higher energy part of the GDR may also yield nuclear information about the isospin of the dipole states [3, 4]. However, it is worth noticing that, in the lower energy part of the GDR, namely a few MeV above the respective thresholds for proton (B_p) and neutron (B_n) emission, the observed $\frac{\sigma(\gamma, p)}{\sigma(\gamma, n)}$ ratio seems mostly connected to the number of open proton and neutron channels and therefore yields mostly rather trivial information of a statistical nature about the density of levels in the residual nuclei $_{Z-1}(A-1)_N$ and $_Z(A-1)_{N-1}$

V. C. 1 - <u>A ≅ 90 nuclei</u>.

Let us consider for example, the curves $\sigma_{Tn} = \sigma(\gamma, n) + \sigma(\gamma, 2n)$ for the Mo isotopes [32] (Fig. 22) and the corresponding integrated photoneutrons cross sections defined by the following expression.

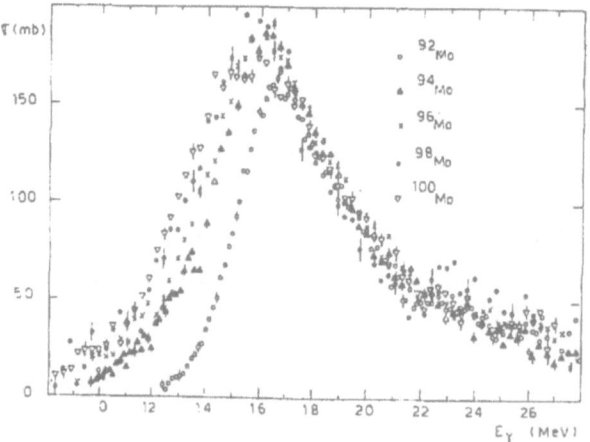

Experimental total photoneutron cross section $\sigma = \sigma(\gamma, n) + \sigma(\gamma, pn) + \sigma(\gamma, 2n)$ for Mo isotopes.

FIG. 22 from ref. [V. 32].

$$\sigma_{on} = \int_{B_n}^{E_M} \sigma_{T,n}(E) \; dE$$

$$\sigma_{-1n} = \int_{B_n}^{E_M} \sigma_{T,n}(E)/E \; dE$$

$$\sigma_{-2n} = \int_{B_n}^{E_M} \sigma_{T,n}(E)/E^2 \; dE$$

where E_M constitutes the different upper integration limits.

Let us use the Thomas, Reiche and Kuhn classical sum rule for the $_Z A_N$ nucleus namely $\dfrac{0.06 \; NZ}{A}$ MeV. b. As can be seen from Table 5, the only nuclei for which one obtains approximately the previously observed (cf. Chapter IV) $\dfrac{\sigma_{on}}{0.06 \; NZA^{-1}} \approx 1.1 \pm 0.05$ values are ^{96}Mo, ^{98}Mo and ^{100}Mo .

- Table 5 -

Nucleus	^{92}Mo	^{94}Mo	^{96}Mo	^{98}Mo	^{100}Mo
E_M (MeV)	29	29	29.5	29	28.5
σ_{on} (MeV b)	1.10	1.37	1.53	1.60	1.58
$\dfrac{\sigma_{on}}{0.06 \; NZA^{-1}}$	0.804	0.98	1.08	1.11	1.08

The sudden drop in the value of this ratio observed for ^{94}Mo, and even more so for ^{92}Mo, seems to indicate that the experimental $\sigma_{T,n}(E)$ values alone do not really represent the total photon absorption cross section $\sigma_{TOT}(E)$, a fact which would also explain why it is impossible to fit a single Lorentz line to $\sigma_{T,n}(E)$ values for ^{92}Mo since such a Lorentz line characterizes the form of the total photon absorption curve $\sigma_{TOT}(E)$ and not $\sigma_{T,n}(E)$. It seems reasonable to suppose that the missing strength $[\sigma_{TOT}(E) - \sigma_{T,n}(E)]$ is made up essentially of the non-observed (γ, p) exit channel contribution. Using experimental proton yield curves, obtained from (e, e'p) reactions, K. Shoda et al [33] were able to deduce the total photoproton cross sections $\sigma_{T,p}(E)$ of the N = 50 nuclei ^{88}Sr, ^{89}Y, ^{90}Zr and ^{92}Mo (Fig. 23).

FIG. 23.

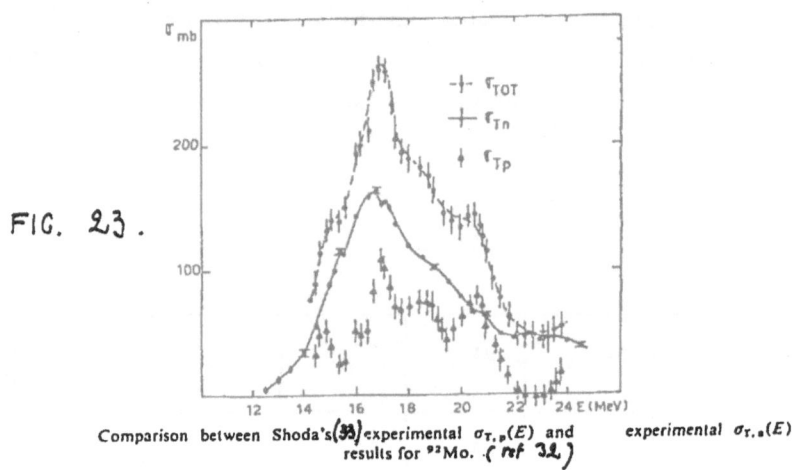

Comparison between Shoda's(33) experimental $\sigma_{T,p}(E)$ and   experimental $\sigma_{T,n}(E)$ results for ^{92}Mo. (ref 32)

Table 6 presents integrated photoproton cross sections together with photoneutron cross sections for some $88 \leqslant A \leqslant 100$ nuclei. The following definitions were used :

$$\sigma_{on} = \int^{26 \text{ MeV}} \sigma_{T,n}(E) \ dE$$

$$\sigma_{op} = \int^{26 \text{ MeV}} \left[\sigma_{T,p}(E)\right] \ dE = \int^{26 \text{ MeV}} \left[\sigma(\gamma,p)+\sigma(\gamma,np)\right] dE$$

where it should be noted that the integration limits in the evaluation of σ_{on} are not the same as those used in table 5.

- Table 6 -   from ref. [V. 32]

Integrated photoproton and photoneutron cross sections as defined in the text and graphically presented in fig. 13

Target nucleus	^{92}Mo	^{94}Mo	^{90}Zr	^{89}Y	^{93}Nb	^{96}Mo	^{88}Sr	^{98}Mo	^{100}Mo
Isospin T	4	5	5	5.5	5.5	6	6	7	8
$(B_n - B_p)$ (MeV)	5.2	1.2	3.6	4.4	2.8	−0.1	−0.5	−1.2	−2.3
σ_{on} (MeV · b)	1.03 ±0.07	1.32±0.07	1.28±0.07	1.36 ±0.07	1.39±0.07	1.46±0.07	1.42 ±0.07	1.53 ±0.07	1.54±0.07
$\frac{\sigma_{on} A}{0.06 \ NZ}$	0.75 ±0.05	0.95±0.05	0.95 ±0.05	1.04 ±0.05	1.01±0.05	1.03±0.05	1.09 ±0.05	1.06 ±0.05	1.06±0.05
σ_{op} (MeV · b)	0.660±0.06 [a])		0.21 ±0.02 [a])	0.150±0.02 [a])			0.130±0.02 [a])	0.140±0.015 [b])	
$\frac{\sigma_{op} A}{0.06 \ NZ}$	0.47 ±0.05		0.16 ±0.02	0.11 ±0.015			0.10 ±0.015	0.10 ±0.02	
$\frac{(\sigma_{on}+\sigma_{op})A}{0.06 \ NZ}$	1.22 ±0.07		1.11 ±0.05	1.15 ±0.05			1.19 ±0.05	1.16 ±0.05	

[a]) Values taken from ref. 33
[b]) Value obtained from ref. 34 after normalization to our σ_{on} value for ^{92}Mo.

In fig. 24 (a) and (b), σ_{on} and σ_{op} are plotted against isospin T and it is interesting to note that both curves display a monotonic behaviour. Moreover, one observes that a certain lack of strength in the σ_{on} results, represented by curve (a), is compensated by an increase in the corresponding σ_{op} results, represented by curve (b). Therefore, since the σ_{op}/σ_{on} ratio decreases as T increases and as (B_n - B_p) decreases, can a connection between this branching ratio and the behaviour of (B_n-B_p) be established ?

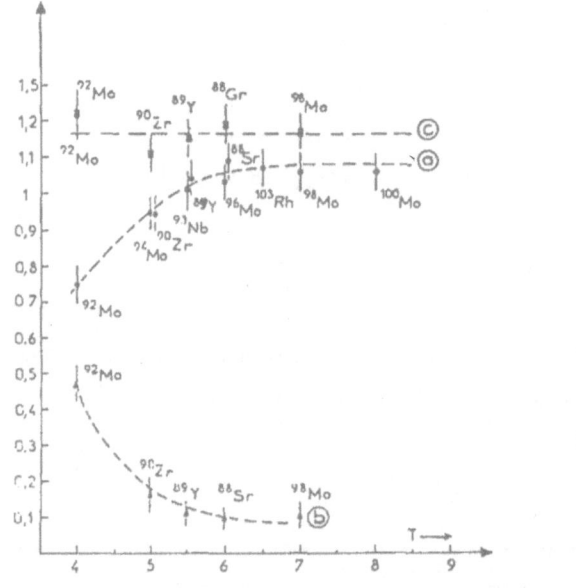

Graphical representation of integrated photoproton and photoneutron cross sections
where $a = \sigma_{0n} A/0.06\ NZ$; $b = \sigma_{0p} A/0.06\ NZ$;
$c = (\sigma_{0n} + \sigma_{0p}) A/0.06\ NZ$.

FIG. 24 from ref. [V.32]

Let us make the approximate assumption that all neutron and protons emerging from the $_Z A_N$ nucleus are " boiled-off " from the compound nucleus at an excitation energy $E_x = E$. The relative neutron (and proton) intensity distribution can be written [8] as a function of the channel energy ε;

$$I_N(\varepsilon) = \text{constant} \cdot \varepsilon \cdot \sigma_c(\alpha) \cdot W'(E_x - B_m - \varepsilon) \quad (10)$$

$$I_P(\varepsilon) = \text{constant} \cdot \varepsilon \cdot \sigma_c(\beta) \cdot W''(E_x - B_p - \varepsilon) \quad (11)$$

ε = kinetic energy of outgoing particle,

$\sigma_c(\alpha)$ and $\sigma_c(\beta)$ = cross sections for formation of the compound nucleus via channels α and β associated with the (γ, n) and (γ, p) processes respectively,

W' = level density in $_Z(A-1)_{N-1}$ nucleus

W'' = level density in $_{Z-1}(A-1)_N$ nucleus.

But available neutron and proton penetrabilities, as a function of ℓ and ε, now indicate that, for nuclei in the $A \simeq 90$ region, only neutrons with energies $\varepsilon \gtrsim 1$ MeV [or $\varepsilon \gtrsim 5$ MeV for protons] contribute appreciably to the $\varepsilon \sigma_c(\alpha)$ and $\varepsilon \sigma_c(\beta)$ terms which in turn will determine the neutron and proton contributions.

One can then define a maximum possible excitation energy $E'_x = E-B_n - 1$ MeV in the residual $_Z(A-1)_{N-1}$ nucleus after the emission of a single neutron from the target A and similarly one can define a maximum possible excitation energy $E''_x = E-B_n-5$ MeV in the $_{Z-1}(A-1)_N$ residual nucleus after the emission of a single proton from A. (fig. 25)

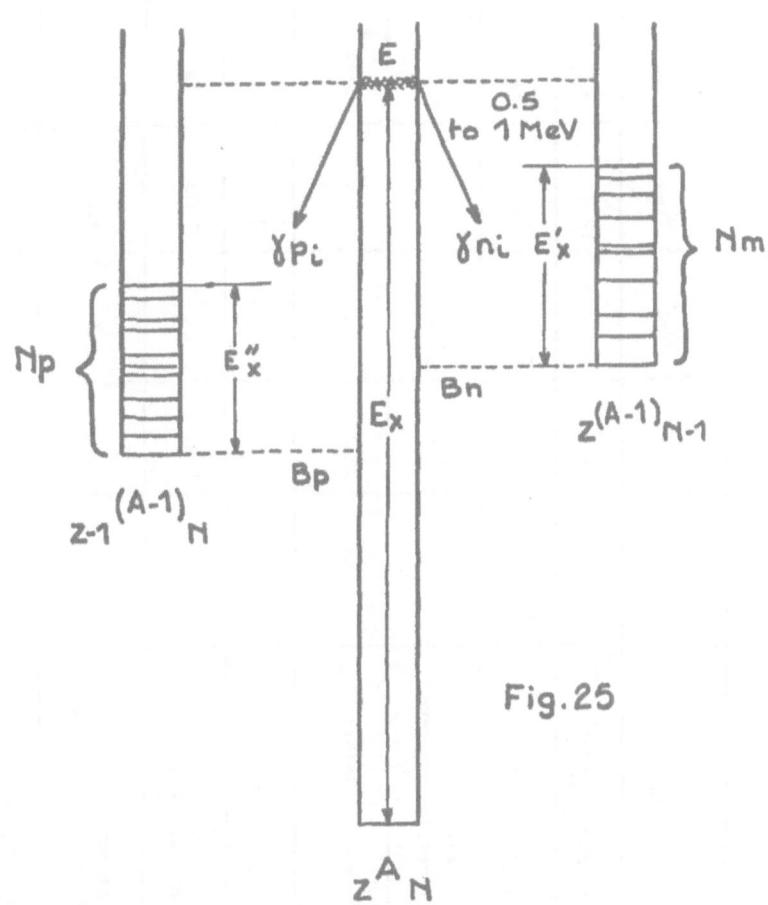

Fig. 25

Let us now make the very simple assumption that the $\sigma(\gamma,p)/\sigma(\gamma,n)$ ratio is mainly controlled by the relative level densities in the residual nuclei and therefore should vary as N_p / N_n where N_n and N_p are the actual number of available levels below the excitation energy E'_x and E''_x in the $_Z(A-1)_{N-1}$ and $_{Z-1}(A-1)_N$ nuclei respectively. (table 7)

TABLE 7

	88Sr	89Y	90Zr	92Mo	94Mo	96Mo
Target nucleus	^{88}Sr	^{89}Y	^{90}Zr	^{92}Mo	^{94}Mo	^{96}Mo
Threshold B_n (MeV)	11.1	11.5	12	12.7	9.7	9.2
Residual nucleus	^{87}Sr	^{88}Y	^{89}Zr	^{91}Mo	^{93}Mo	^{95}Mo
Maximum E'_x in residual nucleus	4.6	4.2	3.7	3	6	6.5
N_n = Nb of levels $< E'_x$	34	$>29^*$	15	14	$>34^*$	$>42^*$
Threshold B_p	10.6	7.1	8.4	7.5	8.5	9.3
Residual nucleus	^{87}Rb	^{88}Sr	^{89}Y	^{91}Nb	^{93}Nb	^{95}Nb
Maximum E''_x in residual nucleus	1.1	4.6	3.3	4.2	3.2	2.4
N_p = Number of levels $< E''_x$	3	8	10	27	$>21^{**}$	$>7^{**}$
$B_n - B_p$	0.5	4.4	3.6	5.2	1.2	-0.1
N_p/N_n	0.09	<0.28	0.66	1.95	<0.62	<0.16

α channel $\left\{ \begin{array}{l} (\sigma,n) \text{ channel} \\ (E_n \geqslant 1 \text{ MeV}) \end{array} \right.$

β channel $\left\{ \begin{array}{l} (\sigma,p) \text{ channel} \\ (E_p \geqslant 5 \text{ MeV}) \end{array} \right.$

Since no experimental (γ, p) data exist for most Mo isotopes, one can nonetheless try and check the above assumptions against some other N = 50 nuclei for which the required experimental $\sigma_{T,n}$ and $\sigma_{T,p}$ data are available. In table 8 are presented previous experimental $\sigma_{T,n}(E)$ data together with Shoda's $\sigma_{T,p}(E)$ results at E = 16.7 MeV, an energy which corresponds roughly to the position of the GDR maximum for nuclei in this mass region. One observes that the $\sigma_{T,p}/\sigma_{T,n}$ ratio at E = 16.7 MeV decreases as T increases and that the N_p/N_n ratio, also shown in Table 8 decreases in the same way. It seems therefore reasonable to conclude that the $\frac{\sigma(\gamma,p)}{\sigma(\gamma,n)}$ ratio is primarily a function of N_p/N_n (number of open channels), at least in the lower energy half of the GDR.

Table 8

Total photoneutron cross section $\sigma_{T,n}$, total photoproton cross section $\sigma_{T,p}$ and total photo-absorption cross section $\sigma_{T\,max} = \sigma_{T,n} + \sigma_{T,p}$ at $E = 16.7$ MeV for nuclei shown

Nucleus	^{92}Mo	^{90}Zr	^{89}Y	^{88}Sr
T	4	5	5.5	6
$\sigma_{T,n}$ (16.7 MeV) (mb)	160±10	210±10	225±10	210±10
$\sigma_{T,p}$ (16.7 MeV) (mb)	95±10	28± 5	15± 5	10± 5
$\sigma_{T\,max}$ (16.7 MeV) (mb)	255±20	238± 5	240±15	220±15
$(B_n - B_p)$ (MeV)	5.2	3.6	4.4	0.5
N_p/N_n	1.95	0.66	< 0.28	0.09

The number of actually available (N_n) and (N_p) levels in the residual nucleus after emission of a single neutron or proton from the target nucleus A excited at 16.7 MeV

V. C. 2 - The case of ^{58}Ni and ^{60}Ni -

It is well known that the ratio

$$\frac{\int \sigma_{\gamma,p}(E)\,dE}{\int \sigma_{\gamma,n}(E)\,dE}$$

when the integration is carried out over the GDR region, varies drastically from ^{58}Ni to ^{60}Ni [35, 36] , as shown in table 9 [37]

Table 9

	$\int^{25} \sigma(\gamma,n)\,dE$	$\int^{25} \sigma(\gamma,p)\,dE$	$\int^{25} \sigma_{abs}(\gamma)\,dE$
Ni⁵⁸	185±3 (28%)	465∓3 (72%)	650
Ni⁶⁰	482±12 (74%)	168∓12 (26%)	650

Recent experimental (γ, n) data, obtained at Livermore, allow a cleaner comparison between the σ (γ, n) curves and the σ (γ, p) data obtained at Moscow and Tohoku [36] (Fig. 26).

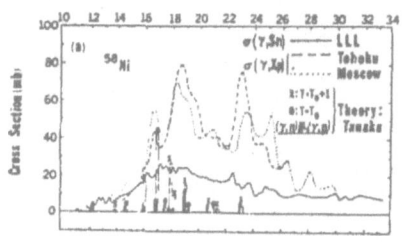 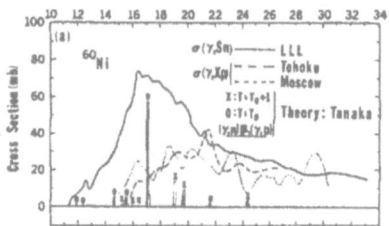

FIG. 26 : reproduced from ref. [V. 36]

This fact has been closely connected to a distribution of the dipole strength, in nuclei with ground state isospin T, into dipole states with isospin $T_< = T$, decaying mostly through the (γ, n) channel, and dipole states with isospin $T_> = T+1$, decaying mostly through the (γ, p) channel. The ratio of the corresponding integrated cross sections $\frac{I(T_>)}{I(T_<)}$ has been predicted by Fallieros and Goulard [38] as

$$\frac{I(T_>)}{I(T_<)} = \frac{T^{-1} - \frac{3}{2} A^{-\frac{2}{3}}}{1 + \frac{3}{2} A^{-\frac{2}{3}}} \qquad (12)$$

$$\frac{I(T_>)}{I(T_<)} = \begin{cases} 0.8 & \text{for } ^{58}Ni \ (T=1) \\ 0.36 & \text{for } ^{60}Ni \ (T=2) \end{cases}$$

The simple identification $\sigma (\gamma , p) \rightleftarrows T_>$ and $\sigma (\gamma , n) \rightleftarrows T_<$ follows qualitatively the above variation $\frac{I(T_>)}{I(T_<)}$ but not quantitatively since the (γ , p) mode is far more dominant in ^{58}Ni than one could expect from the $\frac{I(T_>)}{I(T_<)}$ variation.

Once again, one can try to consider the statistical effect of the respective number of (γ , pi) and (γ , ni) open channel. With the same notation as in V. C. 1, one has at $E_X = 17$ MeV .

$$^{60}Ni \begin{cases} B_m = 11.4 \; MeV \\ B_p = 9.5 \; MeV \end{cases} \quad \begin{aligned} &E_x' \cong 17 - B_m - 1 = 4.6 \; MeV \; \text{in} \; ^{59}Ni \rightarrow (N_m) \\ &E_x'' \simeq 17 - B_p - 5 = 2.5 \; MeV \; \text{in} \; ^{59}Co \rightarrow (N_p) \end{aligned}$$

$$^{58}Ni \begin{cases} B_m = 12.2 \; MeV \\ B_p = 8.2 \; MeV \end{cases} \quad \begin{aligned} &E_x' \cong 17 - B_m - 1 = 3.8 \; MeV \; \text{in} \; ^{57}Ni \rightarrow (N_m') \\ &E_x'' \cong 17 - B_p - 5 = 3.8 \; MeV \; \text{in} \; ^{57}Co \rightarrow (N_p') \end{aligned}$$

Hence one finds :

$$\text{for} \; ^{60}Ni \quad \frac{N_m}{N_p} = \frac{60}{15} = 4$$

$$\text{for} \; ^{58}Ni \quad \frac{N_m'}{N_p'} = \frac{9}{28} = 0.3$$

It is really surprising to observe that at $E_x \simeq 17$ MeV the experimental ratio $\dfrac{\sigma(\gamma,n)}{\sigma(\gamma,p)}$ is practically equal to the computed ratio $\dfrac{N_n}{N_p}$

$$^{60}\text{Ni} \qquad \frac{\sigma(\gamma,n)}{\sigma(\gamma,p)} \simeq \frac{70 \; mb}{20 \; mb} \simeq 3.5$$

$$^{58}\text{Ni} \qquad \frac{\sigma(\gamma,n)}{\sigma(\gamma,p)} \simeq \frac{25 \; mb}{50 \; mb} = 0.5$$

Therefore one can conclude that, very likely, any quantitative connection between $T_<$ ($T_>$) dipole strengths and the (γ,n) [or (γ,p)] strengths must be attempted only when the statistical factors weighting the (γ,n) and (γ,p) channels have been taken into account properly.

V. C. 3 - _s-d shell nuclei_ -

In their study of the isospin mixing in the GDR of self conjugate nuclei [39] Mahaux and Saruis pointed out that, as soon as numerous (γ,n_i) and (γ,p_i) channels toward excited states in the residual nuclei begin to open up, one can expect a certain predominance of the $\sigma(\gamma,p)$ over the $\sigma(\gamma,n)$ cross section, since a lower B_p threshold value and a greater number of available proton channels at the energy of the GDR, will then favour the (γ,p_i) channels. The total photon absorption cross sections σ_{tot} of ^{28}Si, ^{40}Ca and ^{32}S were measured by Ahrens et al [40] and Wyckoff et al [41] , and one can therefore compare the Saclay $\sigma(\gamma,n)$ data [42] measured around the maximum of the GDR situated at $E \simeq 20$ MeV, with the corresponding $\sigma(\gamma,p)$ cross sections, assumed equal to $[\sigma_{tot} - \sigma(\gamma,n)]$, since directly measured $\sigma(\gamma,p)$ data are seldom available.

Assuming pure E1 absorption and since most of the available low-energy levels, in the appropriate A = 4N−1 nuclei have positive parities, the penetrability factors P ($\ell = 1$) become greater than 0.5 for protons with energies E_p greater than 3.5, 3.5 and 4 MeV, emitted from ^{28}Si,

^{32}S and ^{40}Ca respectively. If one further admits that all neutrons with energies $E_n > 0.5$ MeV will indeed escape from the target nucleus, one can then evaluate the (N_p/N_n) ratio, where N_p and N_n represent the number of actually available levels in the appropriate residual nucleus pertaining to the (γ, p) and (γ, n) channels for $E_p > 3.5$ MeV and $E_n > 0.5$ MeV respectively.

Table 10

B_p and B_n threshold values together with the ratios of available levels (N_p/N_n) in residual nuclei corresponding to (γ, p) and (γ, n) reactions on ^{28}Si, ^{32}S and ^{40}Ca

	^{28}Si	^{32}S	^{40}Ca
B_p (MeV)	11.6	9	8.3
B_n (MeV)	17.2	15.1	15.7
N_p/N_n	3.5	3.6	5
$\sigma(\gamma, p)/\sigma(\gamma, n)$ at $E \approx 20$ MeV	2.5	4.6	4.6

As can be seen from table 10, these calculated N_p / N_n values are found to be of the same order of magnitude as the ratio of the evaluated $\sigma(\gamma, p)$ to the measured $\sigma(\gamma, n)$ value, thus confirming that the latter ratio might be a function of the relative positions of the B_p and B_n thresholds. Such an analysis seems further supported by the fact that the experimental $\sigma(\gamma, p)/\sigma(\gamma, n)$ ratio of ^{40}Ar is much smaller than unity and expresses the fact that the $B_n = 9.9$ MeV neutron threshold lies below the corresponding $B_p = 12.5$ MeV proton threshold in the ^{40}Ar target nucleus.

In this ^{40}A case, one observes that around the maximum of the GDR at 17.5 MeV, one has approximately $\sigma(\gamma, n) \gtrsim 30$ mb whereas Ehhalt et al [43] obtained $\sigma(\gamma, p) = 5$ mb only. These results seem also to agree with the calculated number of available neutron levels in the residual ^{39}Ar nucleus namely $N_n > 30$ whereas the available number of proton levels N_p in the ^{39}Cl residual nucleus turns out to be only a few units.

    A similar effect can be observed in the respective behaviour of
the (γ, pn) and $(\gamma, 2n)$ channels where the former is strongly enhanced
whenever the B_{pn} threshold lies well below the corresponding B_{2n} value.
Let us make the simple assumption that only neutrons or protons with
energies corresponding to a penetrability factor exceeding 0.5 can possibly
bly emerge from the target nucleus. Then, for all excitation energies
above $E \simeq \left(B_{pn} + 3 \right)$ MeV, the successive emission of a proton and

Table 11

B_{pn} and B_{2n} threshold values for ^{16}O, ^{31}P, ^{32}S, ^{39}K and ^{40}Ar					
	^{16}O	^{31}P	^{32}S	^{39}K	^{40}Ar
B_{pn} (MeV)	23	17.9	21.2	18.2	20.6
B_{2n} (MeV)	28.9	23.6	28.6	25.1	16.4

a neutron (or first a neutron, then a proton) will be favoured as long
as this E-value remains well below the B_{2n} threshold.

    Table 11 presents the B_{pn} and B_{2n} threshold values of ^{16}O, ^{31}P,
^{32}S and ^{39}K, all nuclei for which it could be experimentally shown that :

    a) $\sigma (\gamma, pn)$ increases for $E > \left(B_{pn} + 3 \right)$ MeV.

    b) The $\sigma (\gamma, 2n)$ contribution remains negligible when compared
to $\sigma (\gamma, pn)$ for at least 4 to 5 MeV above B_{2n}.

    c) $\sigma (\gamma, pn) \simeq \frac{1}{2} \sigma_{Tn} \simeq \sigma (\gamma, n)$ for E = 30-35 MeV in ^{16}O
and for E = 26-30 MeV in ^{31}P, ^{32}S and ^{39}K.

    On the other hand, the B_{2n} and B_{pn} threshold values for ^{40}Ar,
also shown in table 11, indicate that the former value of 16.4 MeV lies
well below the latter B_{pn} = 20.6 MeV, a fact borne out by our experi-
mentally observed values $\sigma (\gamma, 2n)$ = 22 mb and $\sigma \left[(\gamma, n) + (\gamma, pn) \right]$ =
9 mb at E = 22 MeV (Fig. 27 and 28).

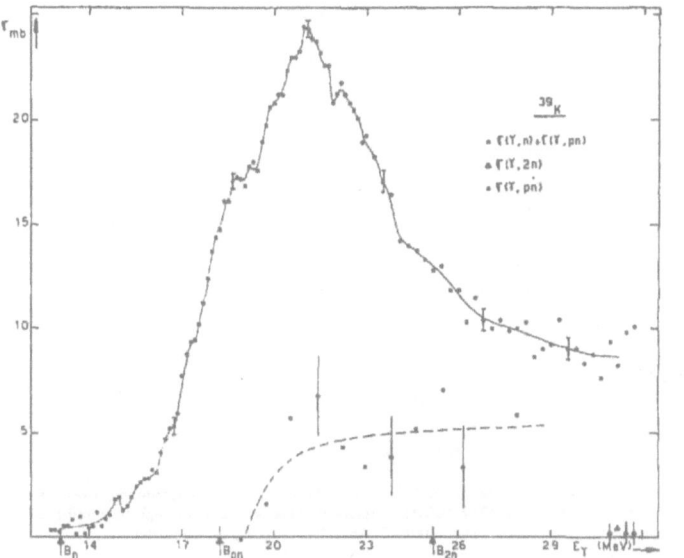

FIG. 27
from ref. V.42

'. Partial photoneutron cross sections $[\sigma(\gamma, n) + \sigma(\gamma, pn)]$, $\sigma(\gamma, pn)$ and $\sigma(\gamma, 2n)$ of ^{39}K.

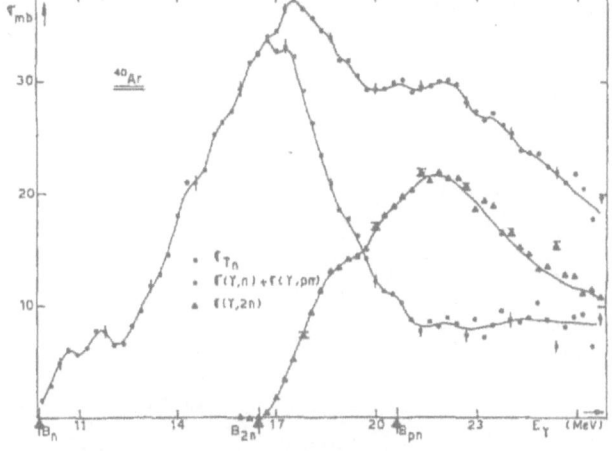

FIG. 28
from ref. V.42

Photoneutron cross sections σ_{Tn}, $[\sigma(\gamma, n) + \sigma(\gamma, pn)]$ and $\sigma(\gamma, 2n)$ of ^{40}Ar.

Another way to look at this possible statistical effect on the $\dfrac{\sigma(\gamma, n)}{\sigma(\gamma, p)}$ ratio is to look at the integrated photoneutron cross section σ_o^n presented in fig. 29, as a fractional value of the TRK sum rule $0.06 \dfrac{NZ}{A}$ MeV barn.

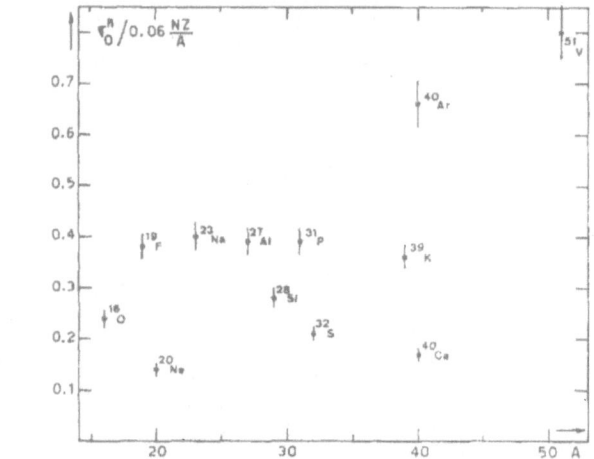

Ratio of experimental integrated photoneutron cross section σ_0^n over the Thomas, Reiche and Kuhn sum rule [0.06 NZ/A]. Numerical values and upper integration limits E_M are taken from table 3. Also $\Delta\sigma_0^n = \pm 7\%$ for all nuclei.

FIG. 29

from ref. V.42

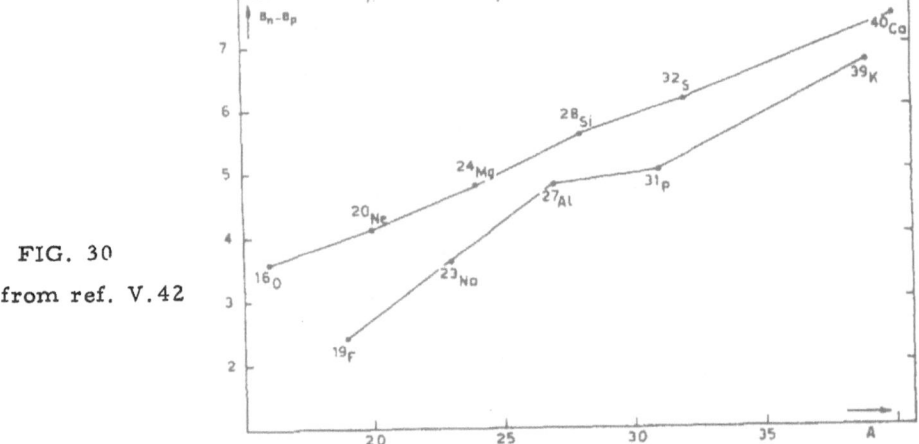

Threshold differences [$B_n - B_p$] for $A = 4N$ and the neighbouring $A = 4N - 1$ nuclei.

FIG. 30

from ref. V.42

One notes that all $A = 4N$, $T = 0$ nuclei show roughly the same 0.2 value. It can also be seen that this ratio decreases steadily when going from ^{28}Si to ^{32}S and to ^{40}Ca, a behaviour reminiscent of the one already observed for the corresponding $\sigma(\gamma, n)/\sigma(\gamma, p)$ or N_n/N_p ratios discussed above.

On the other hand the A=4N-1, $T = \frac{1}{2}$ nuclei show an average
ratio value of 0.38. These general traits can at least be qualitatively
understood if one considers the following :

(i) The difference $B_n - B_p$ between the neutron and proton
thresholds is always greater in A = 4N than in the A = 4N-1 neighbour
nuclei as can be seen in fig. 30. It follows that the experimental
$\sigma (\gamma, n)/ \sigma (\gamma, p)$ ratio, and hence also the corresponding
$\sigma_0^m /(0.06 \; NZ/A)$ values, ought to be smaller for the A = 4N
nuclei.

(ii) The above $\sigma_0^m /(0.06 \; NZ/A)$ ratio, for A = 4N nuclei, is
expected to decrease even more if one also takes into account the following experimental observation made by Shoda [44] .

He observed that, for doubly even nuclei, the (γ, p_0) channel is
strongly enhanced whereas for odd-A nuclei, the protons were preferentially
emitted towards the excited levels of the residual nucleus. But since these
latter, low-energy protons, have a lower penetrability than the corresponding
P_0 protons in A = 4 N nuclei, the total $\sigma (\gamma, p)$ cross section should decrease
and hence the ration $\sigma_0^m /(0.06 \; NZ/A)$ should increase for odd-A nuclei.

(iii) In A = 4N nuclei, as for example in ^{28}Si, one notes that both
the (γ, p) and (γ, n) channels lead to odd-A residual nuclei. This is not
the case for A = 4N-1 nuclei as can be seen for example in ^{27}Al where
the (γ, p) and (γ, n) channels lead to the doubly even $^{26}_{12}Mg^{14}$ and to the
doubly odd $^{26}_{13}Al$ nuclei respectively. For a target nucleus which has
absorbed a photon of energy E, one is then led to compare the densities
of the available neutron and proton levels in the appropriate residual
nucleus.

This, in turn, means one must evaluate an expression of the type

$$\rho(u) = C . u^{-2} \exp[2\sqrt{au}]$$

where

$$u = E - B_n \quad \text{for} \quad ^{26}_{13}Al$$

$$u = E - B_p - 2\Delta \quad \text{for} \quad ^{26}_{12}Mg$$

But the introduction of the pairing energy Δ means that one is now dealing with an " effective threshold " for proton emission $\left(B_p + 2 \Delta\right)$, a fact which enhances the neutron channel in A = 4N-1 nuclei.

Moreover, one should note that in T $>$ 1 nuclei, the (B_n-B_p) difference becomes rather small for ^{45}Sc (T $=\frac{3}{2}$) and becomes negative for ^{40}Ar (T=2) and $^{51}V(T =\frac{5}{2})$ which, as can be seen from fig. 29, increase the relative importance of the (γ, n) channel even further.

Finally, fig. 31 represents a graphical summary of the evolution of the above discussed $\sigma_0^m/(0.06\ NZ/A)$ ratio as a function of T where a certain correlation is clearly apparent.

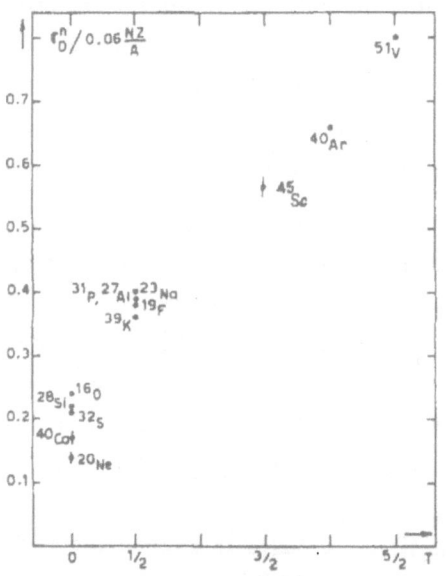

FIG. 31 from

ref. V. 42

The $[\sigma_0{}^n/(0.06\ NZ/A)]$ ratio as a function of isospin T. Possible overall errors of $\pm 7\%$ are to be applied to all nuclei shown.

V. D - <u>The branching ratio towards the various levels in the residual nucleus</u> .

The comparison of the (γ, n_0) (γ, n_1)...(γ, n_i) channels from a given dipole state (fig. 32) has not been often studied so far but provides a powerful tool to understand some nuclear structure problems.

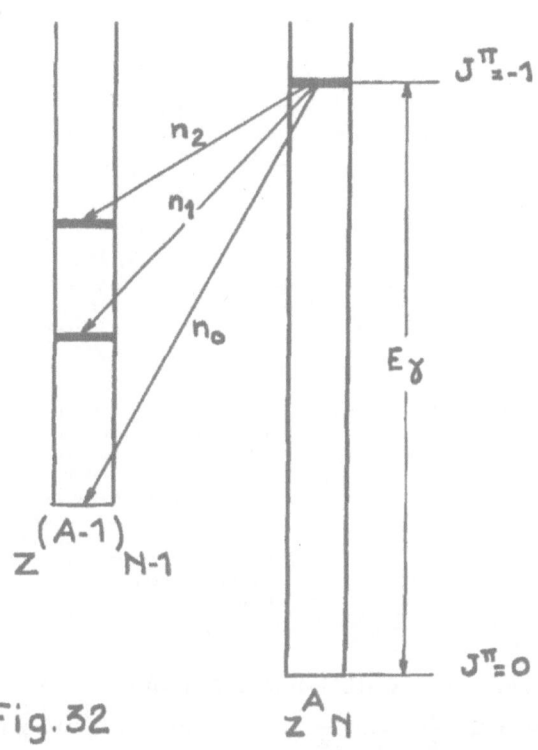

Fig. 32

Just as an example let us quote the case of ^{16}O where the levels in the residual nuclei ^{15}O and ^{15}N are well separated in energy (Fig. 33).

At Livermore, a combination of an anlysis of the neutron energy and of the γ' spectra in the reaction $(\gamma, n\gamma')$ [45] allowed to get the data of fig. 34

FIG. 33

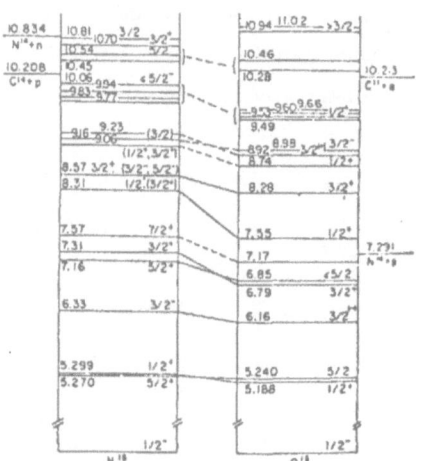

Energy-level diagrams for N^{15} and O^{15}.

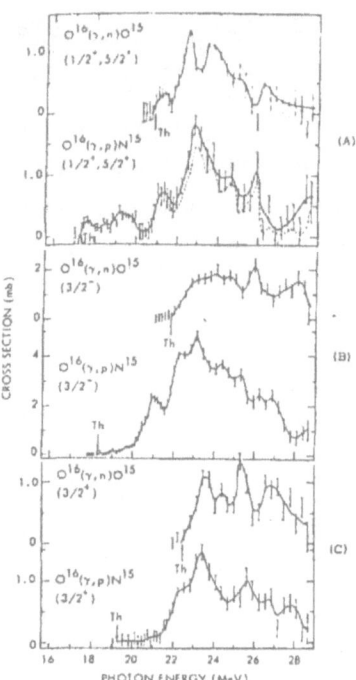

FIG. 34 : reproduced from ⟶

ref. V.45

O^{16}(γ, $n\gamma'$) and (γ, $p\gamma'$) mirror-level, final-state cross sections. (a) Top: O^{15}($\frac{1}{2}^+$, $\frac{5}{2}^+$ unresolved) 5.2-MeV, final-state cross section. Bottom: N^{15}($\frac{1}{2}^+$, $\frac{5}{2}^+$ unresolved) 5.3-MeV, final-state cross section. Dashed line shows effect of subtracting 9.22-MeV level cascades (Ref. 2). (b) Top: O^{15}($\frac{3}{2}^-$) 6.18-MeV, final-state cross section. Bottom: N^{15}($\frac{3}{2}^-$) 6.33-MeV, final-state cross section. (c) Top: O^{15}($\frac{3}{2}^+$) 6.79-MeV, final-state cross section. Bottom: N^{15}($\frac{3}{2}^+$) 7.30-MeV, final-state cross section.

With high resolution Ge(Li) diodes Baglin and Thomson clearly confirmed [46] that the $\frac{1}{2}^+$ and $\frac{5}{2}^+$ levels are appreciably populated in ^{16}O (γ, $n\gamma'$) and ^{16}O (γ, $p\gamma'$) reactions (Fig. 35).

FIG. 35 :
reproduced from
ref. V. 46

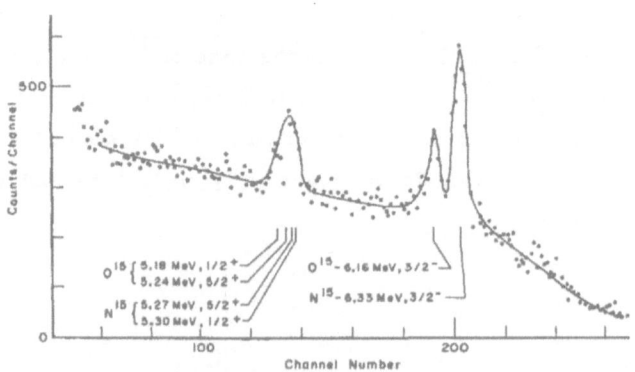

Typical spectrum from germanium detector.

Thus one explains the strong population of the $\frac{3-}{2}$ state ($\simeq$ 6 MeV) and of the $\frac{1-}{2}$ ground state by the creation of a 1p-1h dipole state in ^{16}O where the emission of a proton (or neutron) leaves a hole in the $p\frac{1}{2}$ or $p\frac{3}{2}$ shell.

But the above data showed that 25% of the cross section populates the positive parity state $\frac{1}{2}^+$ and $\frac{5}{2}^+$ at 5.2 MeV. This implies an important non pure 1p-1h excitation of the GDR in ^{16}O.

BIBLIOGRAPHY (Chapter V)

V. 1   S. S. Hanna - Asilomar, mars 1973.

V. 2   C. P. Wu, F. W. K. Firk and T. W. Phillips
            Physcal Rev. Letters $\underline{20}$, 1182 (1968).

V. 3   P. Paul- Proceedings, Asilomar, mars 1973.

V. 4   K. Shoda, University of Melbourne
            Reports Um-P-74/22.

V. 5   C. P. Wu, F. W. K. Firk and B. L. Berman
            Phys. Lett. $\underline{32\,B}$, 675 (1970).

V. 6   H. Arenhövel - Asilomar, mars 1973.

V. 7   E. Hayward, W. C. Barber and J. J. Mc Carthy
            Phys. Review $\underline{C10}$, 2652 (1974)

V. 8   Blatt and Weisskopf - Theoretical nuclear physics
            (John Wiley - New-York, London, Sydney)

V. 9   R. R. Harvey, J. T. Caldwell, R. L. Bramblett and S. C. Fultz
            Phys. Rev. $\underline{136}$, B 126 (1964).

V. 10   S. C. Fultz, R. L. Bramblett, J. T. Caldwell and N. A. Kerr
            Phys. Rev. $\underline{127}$, 1273 (1962)

V. 11   J. R. Calarco - University of Illinois (Thesis 1969).

V. 12   F. T. Kuchnin, P. Axel, L. Griegee, D. M. Drake, A. O. Hanson
            and D. C. Sutton
            Phys. Rev. $\underline{161}$, 1236 (1967).

V. 13   A. Veyssière, H. Beil, R. Bergère, P. Carlos and A. Leprêtre
            Nucl. Phys. $\underline{A\,159}$, 561-576 (1970).

V. 14   G. S. Mutchler - M. I. T. Thesis 1966.

V. 15   S. G. Buccino, C. E. Hollandsworth, H. W. Lewis and
            P. R. Bevington
            Nucl. Phys. $\underline{60}$, 17 (1964)

V. 16   U. Facchini and E. Saetta-Menichella Energia Nucleare
            $\underline{15}$, 54 (1968).

V. 17   J. J. Griffin - Phys. Rev. Lett. $\underline{17}$, 478 (1966).

V. 18   M. Blann and F. M. Lanzafame - Nucl. Phys. $\underline{A142}$, 559 (1970).

V. 19   C. K. Cline and M. Blann
        Nucl. Phys. A 172 , 225-259 (1971).

V. 20   V. V. Verbinski and W. R. Burrus
        Phys. Rev. 177, 1671 (1969).

V. 21   A. Alevra et al. - Nucl. Phys. A209 , 557 -571 (1973).

V. 22   V. K. Lik'yanov, V. A. Seliverstov and V. D. Toneev
        Sov. J. Nucl. Phys. 21 , 508 (1975)

V. 23   G. Chodil et al. - Nucl. Phys. A93 , 648-672 (1967)
        R. G. Thomas and W. Bartolini - Nucl. Phys. A106 , 323-336 (1968).

V. 24   A. Leprêtre et al. - Nucl. Phys. A175 , 609-628 (1971).
        A. Leprêtre et al. - Nucl. Phys. A219 , 39-60 (1974).

V. 25   R. O. Haxby, W. E. Shoupp, W. E. Stephens and W. A. Wells
        Phys. Rev. 59 , 57 (1941).

V. 26   C. P. Bowman, G. F. Auchampaugh and S. S. Fultz
        Phys. Rev. 133 , B676 (1964).

V. 27   A. Veyssière, H. Beil, R. Bergère, P. Carlos, A. Leprêtre
        and K. Kernback
        Nucl. Phys. A 199 , 45-64 (1973).

V. 28   M. Soleilhac, J. Frehaut and J. Gauriau
        J. Nucl. Energ. 23 , 257 (1969).

V. 29   J. E. Gindler, J. R. Huizenga and R. A. Schmitt
        Phys. Rev. 104 , 425 (1956).

V. 30   J. S. Fraser and J. C. D. Milton
        Nuclear fission in Ann. Rev. Nucl. Sci. vol. 16 (1966) .

V. 31   O. Y. Mafra, S. Kuniyoski and J. Gioldemberg
        Nucl. Phys. A 186 , 110 (1972).

V. 32   H. Beil - Nucl. Phys. A227, 427 (1974).

V. 33   K. Shoda - Nucl. Phys. A239 , 397 (1975).

V. 34   R. W. Gellie - Nucl. Phys. 60 , 343 (1964).

V. 35   K. Min and T. White - Phys. Rev. Lett. 21, 16 , 1200 (1968).

V. 36   S. C. Fultz et al. - Phys. Rev.C10,2 , 608 (1974) .

V. 37   Y. Tanaka - Prog. Theor. Phys. 46, 3 , 787 (1971).

V. 38   S. Fallieros and B. Goulard
        Nucl. Phys. A 147 , 593 (1970)

V. 39   C. Mahaux   and A. M. Saruis - Nucl. Phys. A138 , 481 (1969).

V. 40   J. Ahrens et al. Proceed. Asilomar, Mars 1973.

V. 41   J. M. Wyckoff et al.
         Phys. Rev. 137 , B576 (1965).

V. 42   A. Veyssière - Nucl. Phys. A227, 513 (1974) .

V. 43   D. Ehhalt et al. - Z. Phyz. 187 , 210 (1965).

V. 44   K. Shoda - Nucl. Phys. 72 , 305 (1965).

V. 45   J. T. Caldwell - Phys. Rev. Lett. 19, 8 , 447 (1967).

V. 46   J. Baglin and M. Thomson - Conf. Nucl. Struct. Tokyo (1967).

PHOTONUCLEAR REACTIONS ABOVE THE GIANT

DIPOLE RESONANCE : A SURVEY

Giovanni Ricco
Istituto di Scienze Fisiche
Università di Genova

Viale Benedetto XV, 5-16132 Genova

I- Introduction

The main purpose of these lectures is to give a general introduction to medium energy photonuclear reactions. I shall therefore
present a rapid survey of the existing experimental data from immediately above the Giant Dipole Resonance (G.D.R.) up to the real isobar
production region with a brief discussion on the involved physical
problems. The subject will therefore be treated at a phenomenological
level, more detailed theoretical discussions on the specific arguments
will be in fact given in other lectures (1) (2).

In a preliminary discussion of photonuclear reactions at inter
mediate energies three main questions arise:

1) What is the relative importance of photon absorption processes above
the G.D.R.   The answer has been given by recent experiments: the total
absorption cross section in O^{16}, measured by the Mainz group (3) up to
140 MeV, is reported in fig. 1. It is evident that the excitation of
the G.D.R. does not exhaust the integrated cross section: the contributions from threshold to 30 MeV and from 30 to 140 MeV being roughly
equal.

2) What is the photon-nucleus interaction responsible of the observed
absorption cross sections at high energies.

3) What physical informations, concerning nuclear forces and nuclear
structure, can be obtained from the analysis of photonuclear experiments at intermediate energies.

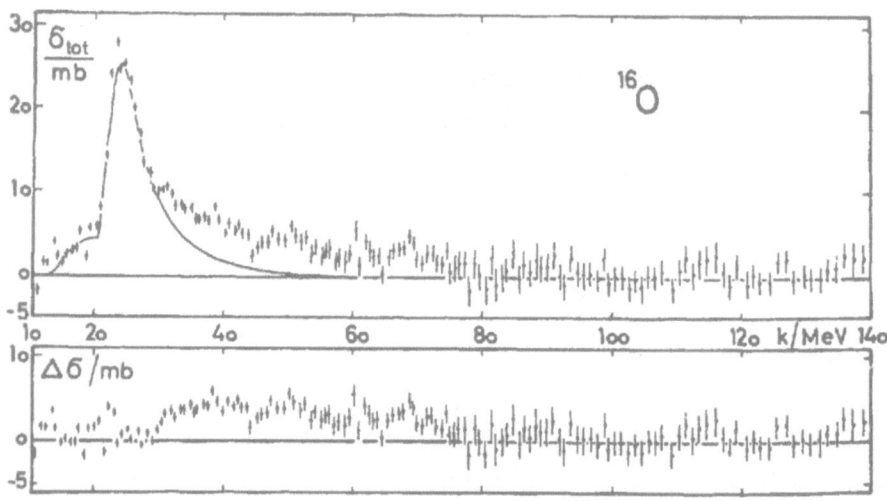

Fig. 1   Total absorption cross section in O^{16}

Before attempting a detailed discussion on the last two questions one should point out that the present stage of this field of
research often allows only tentative speculations rather than definite
conclusions. The main source of uncertainty must be probably ascribed
to the still unsufficient reliability of the experimental data. The
degree of precision of the available data is still in fact rather poor,
when compared to other branches of experimental physics. The worse
handicap is the use of continuous bremsstrahlung beams which considera
bly reduces the reliability of the measured cross sections and introduces ambiguities in the data analysis. The main effort of many laboratories in the last few years has been therefore devoted to the develope
ment of monochromatic photon facilities or of monochromatization techniques on the existing beams. A rapid survey of the obtained results
is presented in figs. 2 and 3.

The simplest method to identify the γ ray energy in a continuous bremsstrahlung spectrum ϕ (k, kmax) is the well known "photon
difference" method reported in fig. 2a.

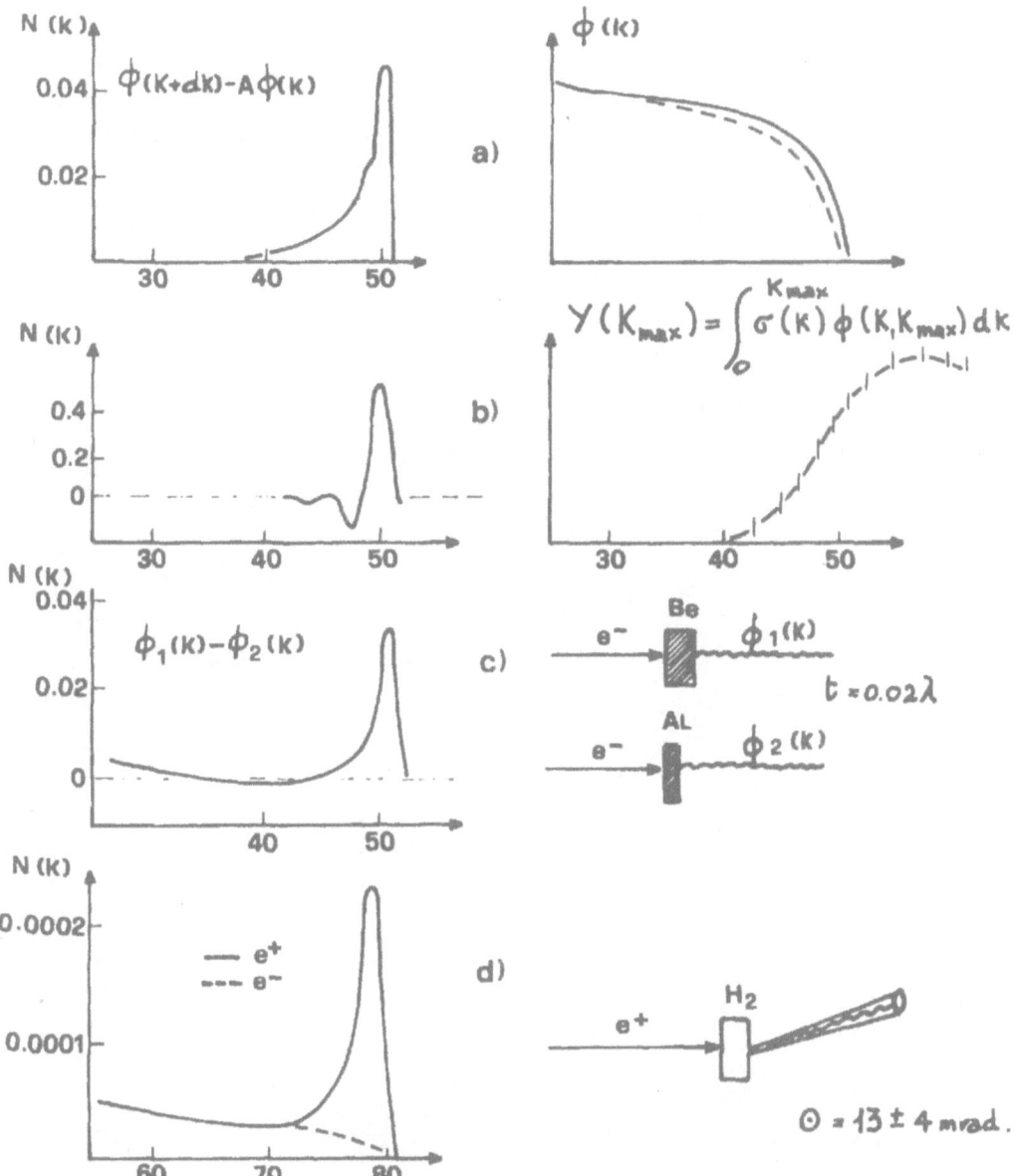

Fig. 2   Available sources of quasi monochromatic γ ray beams. On the vertical axis the number of photons per incident electron is reported.

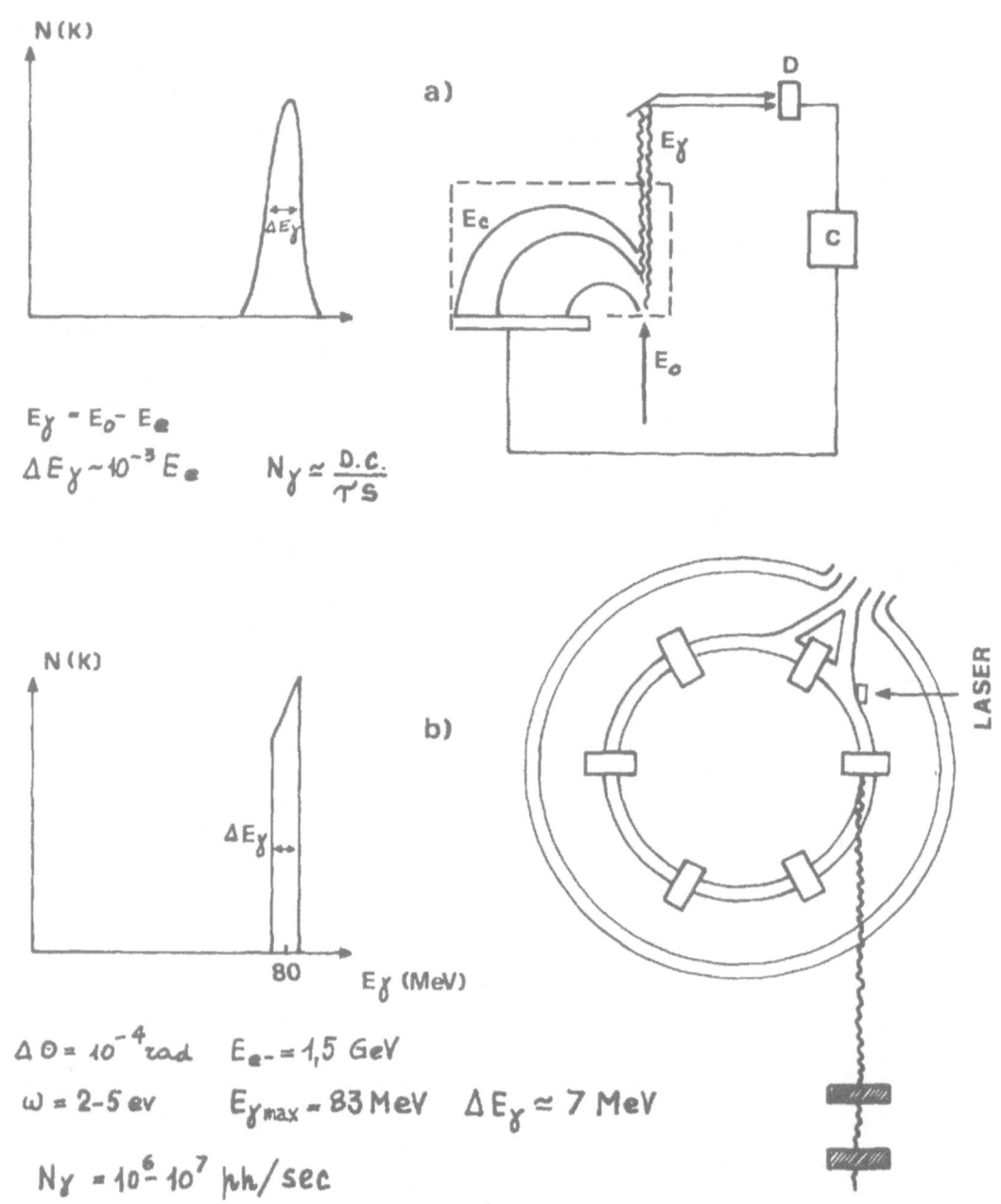

$E_\gamma = E_0 - E_e$

$\Delta E_\gamma \sim 10^{-3} E_e$ $N_\gamma \simeq \dfrac{D.C.}{\tau s}$

$\Delta \theta = 10^{-4}$ rad $E_{e^-} = 1,5$ GeV

$\omega = 2-5$ ev $E_{\gamma max} \sim 83$ MeV $\Delta E_\gamma \simeq 7$ MeV

$N_\gamma = 10^6 - 10^7$ ph/sec

Fig. 3   Future facilities for the production of monochromatic γ ray beams.

Two bremsstrahlung spectra of slightly different end point
energies k_{max} and $k_{max} + dk$ are subtracted through a linear combination

$$N(k) = \phi(k, k_{max} + dk) - A\phi(k, k_{max})$$

where the constant A is optimized to give a minimum low energy tail in
the resulting difference. The photon spectrum $N(k)$ shows a rather
asymmetric shape with a slow low energy tail and poor resolution.
Moreover this method must trust theoretical bremsstrahlung cross section
calculations near the end point energy to get the optimum A value.
A slight improvement can be obtained by varying the bremsstrahlung
energy k_{max} in close steps. The yield curve

$$[1] \qquad Y(k_{max}) = \int_{0}^{K\,max} \sigma(k)\phi(k, k_{max})\,dk$$

corresponding to a defined nuclear cross section $\sigma(k)$ is measured
as a function of the bremsstrahlung energy k_{max}. The Volterra integral
equation $[1]$ can be inverted by numerical techniques (4) to get the
experimental cross section $\sigma(k)$. The equivalent photon spectrum used
in this method is reported in fig. 2b. The photon peak is narrower but
exhibits oscillations in the low energy side, which can be reduced only
by decreasing the step intervals and increasing the statistical precision. The method is therefore rather massive and its reliability is
strongly affected on one side by the propagation of statistical errors
in the matrix inversion and on the other side by sistematic errors due
to the sensitivity to the detailed shape of the bremsstrahlung spectrum
close to the end point energy.

An interesting modification of the photon difference method
has been recently developed by the Glasgow group (5). If the photon
spectra from two different Z radiators, having the same thickness
in radiation length units, are alternatively subtracted, the lower
energy contributions $(k \leqslant 0.9\ k_{max})$ are approximately cancelled while
a peak is produced (fig. 2c) near the end point energy by virtue of
the different collision losses in the two radiators. A FWHM resolution
as good as 2 MeV at 80 MeV can be obtained but a small low energy
tail is still left.

In fig. 2d is shown the energy distribution of photons produced
by annihilation in flight of high energy positrons on a H_2 radiator (6)
The spectrum shows, besides the usual bremsstrahlung tail, the high
energy annihilation peak. If the radiation produced by an electron beam
under identical conditions is subtracted the difference spectrum should
show the peak contribution only, due to the charge independence of
electron and positron bremsstrahlung cross sections. Using this technique
the subtraction is performed in a more reliable way because of the
enhanced difference between the two spectra around the end point energy
but the available intensity per positron is decreased as much as two
order of magnitudes (fig. 2d). Annihilation photon beams are now availa
ble in Livermore and Saclay up to 100 MeV and in Frascati from 80 to
300 MeV.

All the discussed methods are basically different approaches
to a subtraction technique between continuous spectra, a further effort
is therefore required to obtain realmonochromatic photon spectra i.e.
without any low energy tail. Two facilities of this kind have been
proposed and are presently under construction. The first one is based
on the well known "tagged photon" method where the photon energy is
identified by the coincidence between the final electron and the
reaction product (fig. 3a). If the final electron energy E_e is accurately measured we have $E_\gamma = E_o - E_e$ and a FWHM resolution of the
order of $10^{-3} E_e$ but the available intensity is limited by the accidental coincidence rate to $I_\gamma = D.C./(S\tau)$ where S is the true to random
counts ratio, τ the coincidence resolving time and D.C. the machine
duty cycle. In average experimental conditions ($T = 2\ 10^{-9}$ s $\simeq 50$)
an intensity of the order of 10^7 photons/sec would require a D.C.
close to one. This method can therefore be applied only on continuous
beam electron accelerators still under development (7).

The second facility is based on the inverse Compton scattering.
From the 180° collision of an high energy electron beam with the low
energy optical photons from a power laser high energy photons can be
obtained. Being this a two body reaction the photons collected in a
small collimation angle around the forward direction are fairly mono-

chromatic (8). Moreover in these conditions the polarization of the laser beam is conserved and the final high energy photons are also polarized. This method will be applied to the 1.5 GeV electron beam of the Adone storage ring in Frascati and the parameters of this facility are reported in fig. 3b. The estimated intensity should be of the order of 10^7 photons/sec with a 9% FWHM at 80MeV and a 95% polarization.

II- Giant Multipole Resonances

We shall now begin to analyse the available data in the excitation energy region immediately above the G.D.R., say in the range 20 to 50 MeV. The existence of E1 and E2 collective states at such an high excitation energy has been debated for a long time. Both collective models (9) and many body calculations (10) have predicted the possibility of electromagnetic transitions to states lying higher than the G.D.R.

If we look again to the total absorption cross section in O^{16}(fig. 1), some fluctuations may be noted in the experimental cross section above 30 MeV but the statistical accuracy does not afford any definite conclusion. A clear evidence of quadrupole states lying above the G.D.R. has been found only recently (11) in inelastic electron scattering experiments. Fig. 4 shows the spectrum of electrons inelastically scattered from Zr^{90} at different momentum transfers, the γn cross section is also plotted in the upper part of the figure for comparison. The Giant Resonance region observed by electron scattering exhibits a more complex structure than reported by photonuclear experiments, in particular

a) Narrow structures appear on the low energy side of the G.D.R.

b) A large bump is evident at excitation energy around 27 MeV i.e. about 1.5 times the G.D.R. At the relatively high momentum transfer of this experiment the multipolarity can be deduced by comparison of the q dependence of the inelastic (ee') cross section with theoretical calculations: the E2 assignement seems to be consistent with the experimental data in fig. 4.

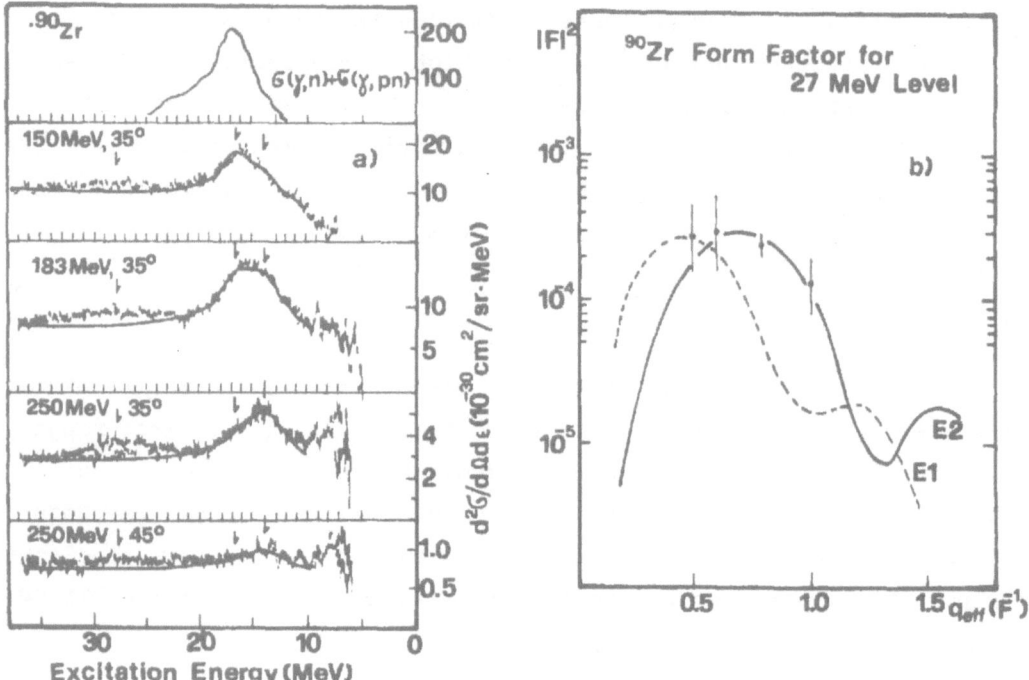

Fig. 4   a) The spectrum of electrons inelastically scattered from Zr[90]
             at different momentum transfers
          b) Inelastic from factor for the 27 MeV level in Zr[90]

Unfortunately E2 and E0 multipoles are undistinguishable using this method. These high lying E2 or E0 transitions are quite general and have been sistematically (12) observed in many nuclei (fig. 5), the average excitation energy being $E(E2) \simeq 120/A^{1/3}$. Isovector E2 transitions just around this energy value were already predicted by collective models (9) , this would support the E2 assignment.

The discovery of these isovector quadrupole resonances is of course of great interest for medium energy photonuclear reactions because it shows the existence of E2 collective states lying at high excitation energy which might be important to explain the observed absorption cross section above the G.D.R.

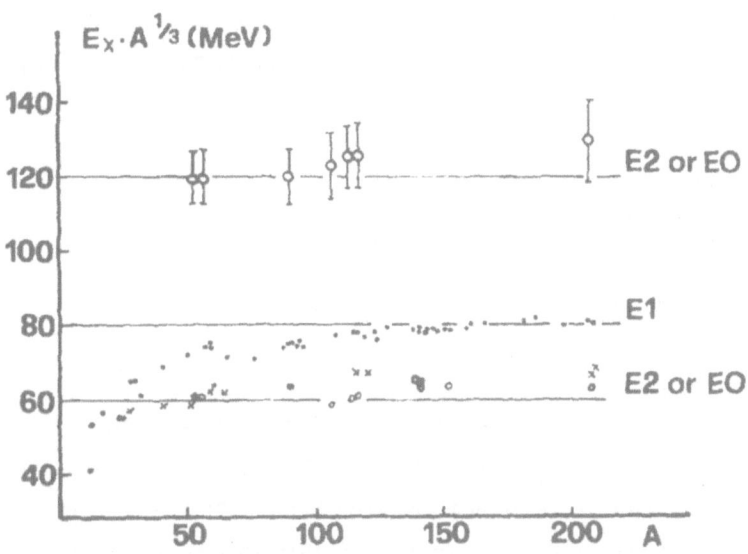

Fig. 5   Centroid energies (MeV) of the observed Giant Multipole Resonances
multiplied by $A^{1/3}$

A recent calculation (13) of the total absorption cross section
in O^{16}, taking into account the existence of the quadrupole resonances
and including all electric multipoles upto E3 in the transition amplitudes,
shows (fig. 6) that the E1 and E2 contributions become comparable
already at 80 MeV . From an experimental point of view the observation
of structures in the total cross sections above 20 MeV is not therefore
sufficient but some effort must be devoted to get the unambiguous separation of the partial contributions of the different multipoles. At
intermediate energies the measured single multipole cross sections
(S.M.C.S.) would be in fact a much more significative test of photon-
-nucleus interaction models. A possible approach to this problem is
provided by the systematic measurement of photonucleon differential
cross sections $\frac{d\sigma}{d\Omega}$ at various photon energies.

Using unpolarized γ rays the photonucleon angular distribution
is given by

$$d\sigma/d\Omega = A_o(E)\left[1 + \sum_{k=1}^{4} A_k(E)\, P_k(\cos\theta)\right] \qquad A_o = \sigma(E)/4\pi$$

where the coefficients A_k are related to the multipole transition amplitudes EL and ML or to their interference terms (EL, ML)

$$A_1 \Rightarrow (E1, M1) + (E1, E2)$$
$$A_2 \Rightarrow (E1)^2 + (M1)^2 + (E2)^2$$
$$A_3 \Rightarrow (E1, E2)$$
$$A_4 \Rightarrow (E2)^2$$

The connection between the coefficients A_k and the S.M.C.S. σ(Ek) is not at all straightforward. It is of course always possible in the framework of an interaction model to calculate the A_k coefficients and compare with the measured values but the inverse problem is extremely difficult and not yet solved in a completely model independent formalism. In order to illustrate by an example I shall report some analysis performed in the G.D.R. region where most work has been done. If we assume the 1p – 1h model to be reasonably valid for the Giant Resonance of O^{16} the A_k coefficients can be expressed as a function of the 1p – 1h complex transition amplitudes. In table I the main 1p – 1h proton transition corresponding to E1 or E2 absorption in the Giant Resonance of O^{16} and the relative S matrix elements are reported using a simplified notation

TABLE I

Multipolarity	Configuration	S
E1	$p_{1/2}^{-1} - d_{3/2}$ $p_{1/2}^{-1} - 2s_{1/2}$	$d_{3/2}\, e^{i\delta_d}$ $s_{1/2}\, e^{i\delta_s}$
E2	$p_{1/2}^{-1} - 2p_{3/2}$ $p_{1/2}^{-1} - f_{5/2}$	$p_{3/2}\, e^{i\delta_p}$ $f_{5/2}\, e^{i\delta_f}$

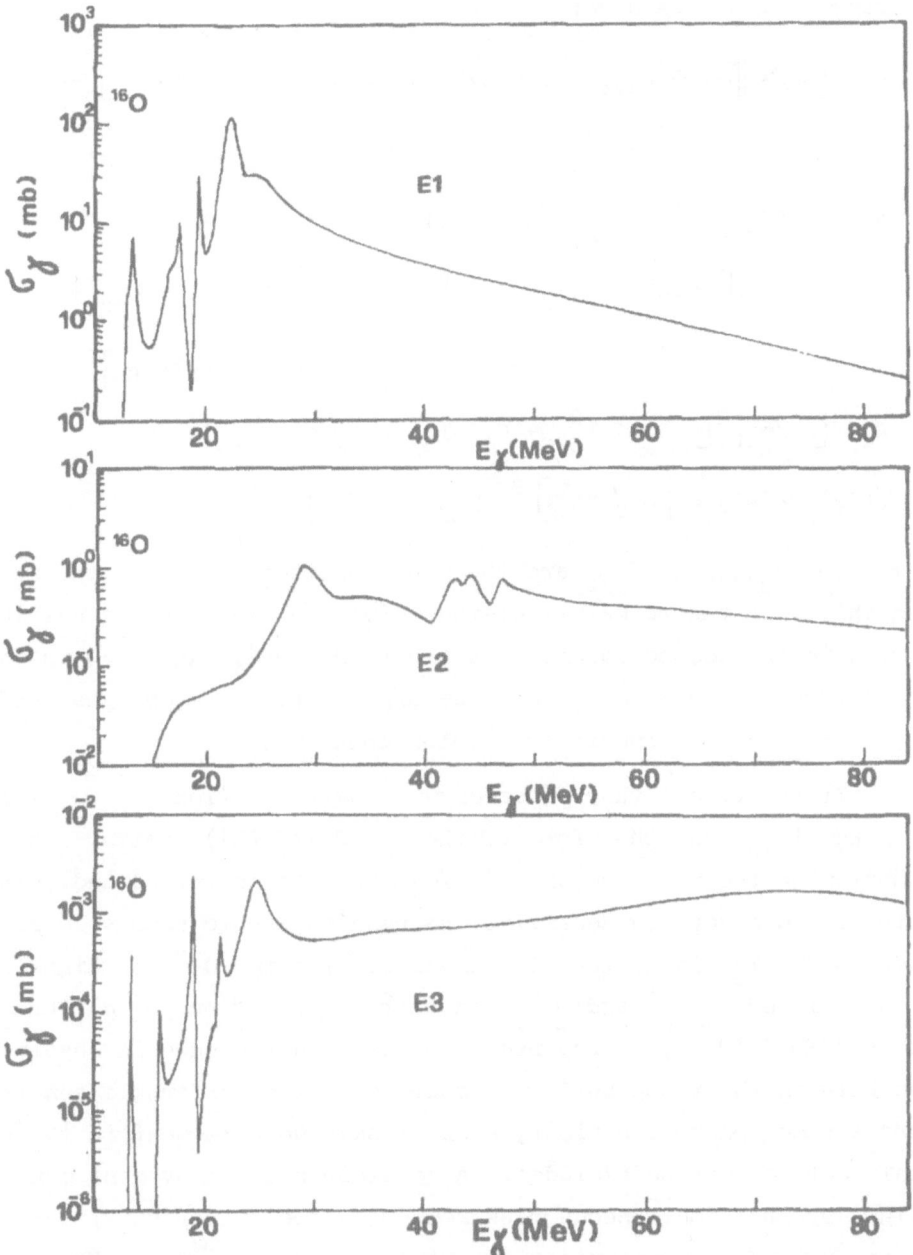

Fig. 6   Computed (13) dipole, quadrupole and octupole absorption cross
         section in 0^16

Following ref. 14 we have:

$$A_1 = \left[\text{Qk}/\alpha\right]\left[-1.09 p_{3/2} s_{1/2} \cos\Delta_{ps} + 0.15 p_{3/2} d_{3/2} \cos\Delta_{pd} -\right.$$
$$\left. - 1.1 f_{5/2} d_{3/2} \cos\Delta_{fd}\right]$$

$$A_2 = -0.5 d_{3/2}^2 + 1.41 s_{1/2} d_{3/2} \cos\Delta_{sd}$$

$$A_3 = \left[\text{Qk}/\alpha\right]\left[-0.22 f_{5/2} s_{1/2} \cos\Delta_{fs} - 0.93 p_{3/2} d_{3/2} \cos\Delta_{pd} +\right.$$
$$\left. + 0.5 f_{5/2} d_{3/2} \cos\Delta_{fd}\right]$$

$$A_4 = \left[\text{Qk}/\alpha\right]^2\left[-0.114 f_{5/2}^2 + 0.56 f_{5/2} p_{3/2} \cos\Delta_{fp}\right]$$

$$\sigma(E2)/\sigma(E1) = \left[\text{Qk}/\sqrt{5}\alpha\right]^2\left[p_{3/2}^2 + f_{5/2}^2\right]$$

where $s_{1/2}$ $d_{3/2}$ $p_{3/2}$ $f_{5/2}$ are the real amplitudes and δ_ℓ the real phase shifts of the S matrix elements for the 1p - 1h transitions reported in the second column of table I, $\Delta_{\ell\ell'} = \delta_\ell - \delta_{\ell'}$, Q is the ratio between the quadrupole and the dipole effective charges and $\alpha^{-1} = 1.6F$ is the harmonic oscillator constant.

It is evident that in order to get any evaluation of the amplitudes $p_{3/2}$ and $f_{5/2}$ and therefore of the $\sigma(E2)/\sigma(E1)$ ratio, some assumption on the relative phases $\Delta_{\ell\ell'}$ has to be introduced. The problem is to check how sensitive the results are to such specific assumptions. In fig. 7 the E2 cross section in O^{16} obtained from the analysis of γp_o and γn_o angular distributions by different authors (14) (15) , is reported. The large difference in absolute value between the two experiments confirms the considerable sensitivity to the assumed phase behaviour, even if some desagreement in the experi- mental data cannot be excluded. A considarable improvement can be obtained by the simultaneous measurement of the differential cross section $d\sigma/d\Omega$ and polarization $P(\theta)$ or, equivalently, by the inverse polarized proton capture experiments.
If the proton is polarized, the analyzing power is given by

$$A(E,\Theta) = \frac{\frac{d\sigma^{\uparrow}}{d\Omega} - \frac{d\sigma^{\downarrow}}{d\Omega}}{\frac{d\sigma}{d\Omega} + \frac{d\sigma}{d\Omega}} = \frac{1}{\frac{d\sigma}{d\Omega}} \left(\sum_{K=1}^{4} B_k(E) \sin k\Theta \right)$$

where the B_k coefficients have a dependence from the transition ampli‐ tudes very similar to that shown for the A_k coefficients, namely:

$$B_1 \Rightarrow (E1, M1) + (E1, E2)$$

$$B_2 \Rightarrow (E1)^2 + (M1)^2 + (E2)^2$$

$$B_3 \Rightarrow (E1, E2)$$

$$B_4 \Rightarrow (E2)^2$$

In terms of the 1p – 1h model, for each B_k the same ampli‐ tudes and phases are involved than for the corresponding A_k , for example for E1 absoption in O^{16} we have

$$B_2 = -1.06 s_{1/2} d_{3/2} \sin \Delta_{sd}$$

If polarized and unpolarized proton data are simultaneously analyzed, for the same number of free parameters the number of equations is about doubled: the ambiguities are therefore strongly reduced and a unique physically consistent solution can be obtained. The E2(γP_o) cross section in O^{16} from polarized proton data (16) is also plotted in fig. 6, the desagreement with photonuclear experiments is still considerable. The extension of this technique to higher proton energies would be certainly desirable for the investigation of S.M.C.S. above the G.D.R., but the available facilities are presently limited to proton energies up to about 20 MeV.

An almost equivalent method seems to be the measurement of phodisintegration cross sections from polarized photon beams. Using linearly polarized (P = 1) photons we can define an analyzing power:

$$A(E) = \frac{\frac{d\sigma^{\perp}}{d\Omega} - \frac{d\sigma^{\prime\prime}}{d\Omega}}{\frac{d\sigma^{\perp}}{d\Omega} + \frac{d\sigma^{\prime\prime}}{d\Omega}} = \frac{\sin^2\Theta}{\frac{d\sigma}{d\Omega}} \left[1.5 B_2 + 0.94 A_3 \cos\Theta - 0.63 A_4 (7\cos^2\Theta - 1) \right]$$

The sensitivity of the measurement to angular distribution coefficients like A_3 and A_4, which depend from the E2 transition amplitudes, is here strongly enhanced. Moreover a wise choice of the scattering angles Θ can afford the separate measurement of each coefficient.

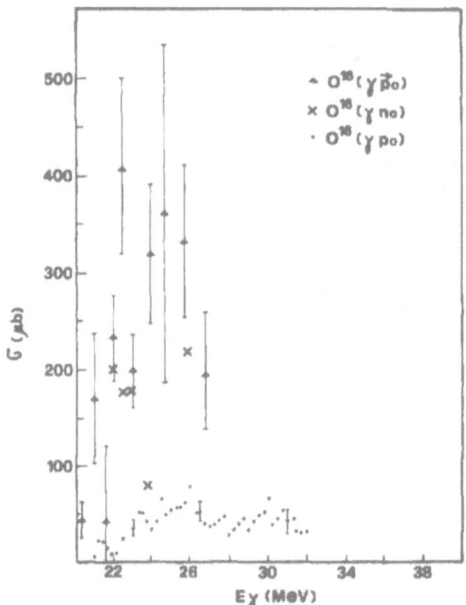

Fig. 7   The γp_0 and γn_0 E2 cross section in O^{16} as measured in ref. (14) (15) (16)

Polarized photon beams are therefore going to be a very promising facili ty for a quantitative investigation of the relative importance of the different multipoles with increasing γ ray energy . Unfortunately, even if good accuracy data may be soon available, the extraction of S.M.C.S. from the angular distribution A_k and B_k coefficients remains up to now a model dependent problem. From an experimental point of view it would be therefore very interesting the development of some model independent approach similar, for example, to the analysis of tran sition charge densities as obtained from inelastic electron scattering cross sections (17).

As a conclusive remark I would like to point out that our knowledge of high lying collective states is still presently very poor and
photonuclear reactions might become, if properly developed, an important
tool for the investigation of the physical nature of these resonances.

III- The γn and γp reactions

Above 50 MeV the total absorption cross section (fig. 1)
remains fairly constant. The simplest process to investigate the involved
reaction mechanisms in this energy region is provided by the two body
(γp) and (γn) reactions. The cross section $d\sigma/d\Omega$ for emission of
protons in a definite energy interval $E_p \pm \Delta E_p$ at 45°, measured by
the Genova - Torino group (18), is plotted in fig. 8 as a function of
the γ ray energy (left hand curves): a pronounced peak is observed
corresponding to the ground state (γp_o) transition.

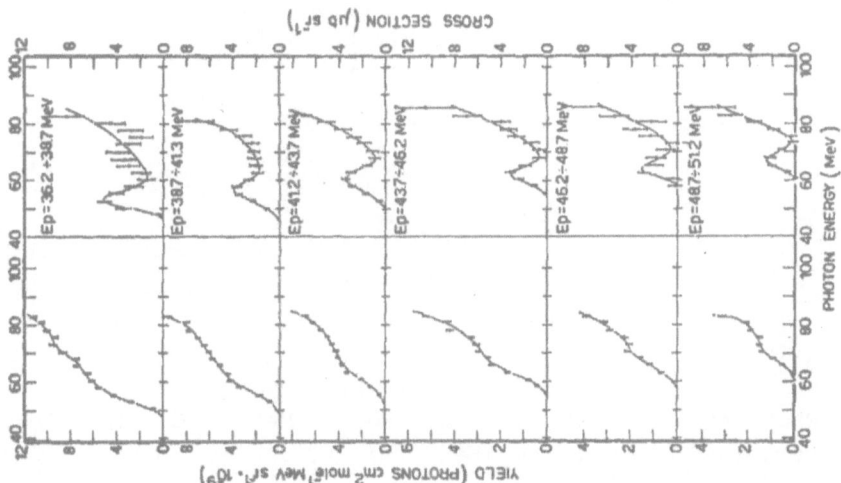

Fig. 8   Proton yields and cross sections at 45° , relative to the
          proton energy interval ΔE_p = 2.5 MeV   in Lithium

These measurements have been improved by the Glasgow group (19)
using the monochromatization technique previously described.

The data for Li^6 Li^7 and C^{12} are reported in fig. 9, the γ ray energy is now determined inside 2MeV FWHM and the differential cross section is plotted as a function of the proton energy.

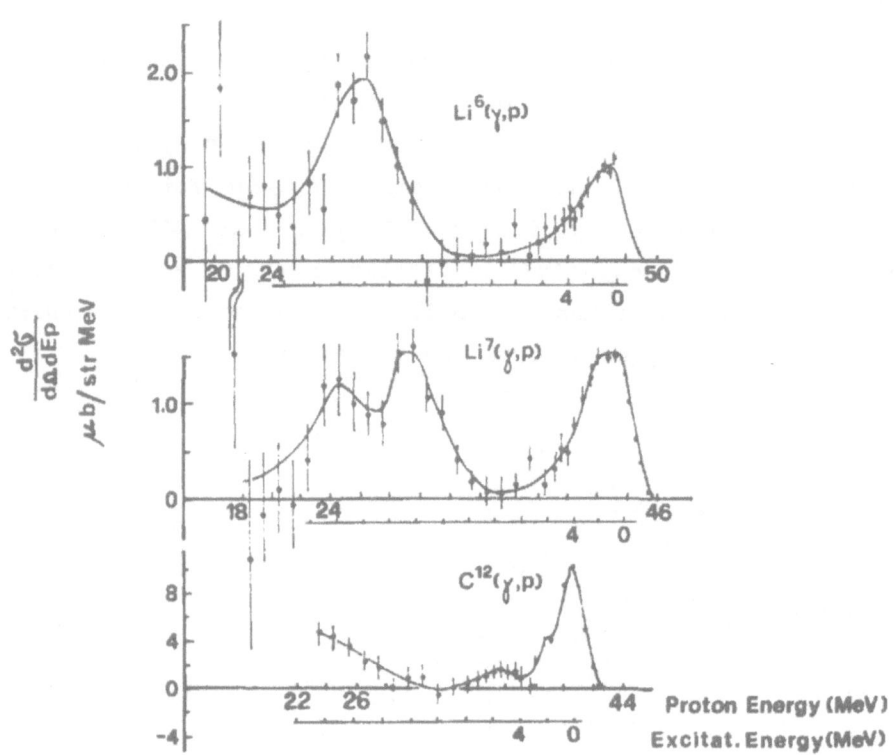

Fig. 9   Photoproton spectra for E_γ = 60 ± 1MeV Θ = 45° from
ref. (19)   a) $Li^6(\gamma p)He^5$ b) $Li^7(\gamma p)He^6$ c) $C^{12}(\gamma p)B^{11}$

The ground state transition peak is clearly evident in all the investigated nuclei, a second large bump is also observed in the Li isotopes at proton energies roughly corresponding to the excitation of $1s^{-1}$ hole states in these nuclei. All these results suggested the simple interpretation of the γ p reaction as a sort of nuclear photo-electric effect where the proton is directly picked out from single particle states leaving the residual nucleus in the corresponding hole state (20) . This process is very interesting in the nucleus because of the particular kinematics of photonuclear reactions. The absorbed photon transfers in fact to the proton high energy but relatively low momentum, the direct interaction is therefore allowed by momentum conservation only with nucleons in a high momentum state. The γ p cross section should be therefore related to the high momentum components in the single particle wave functions.

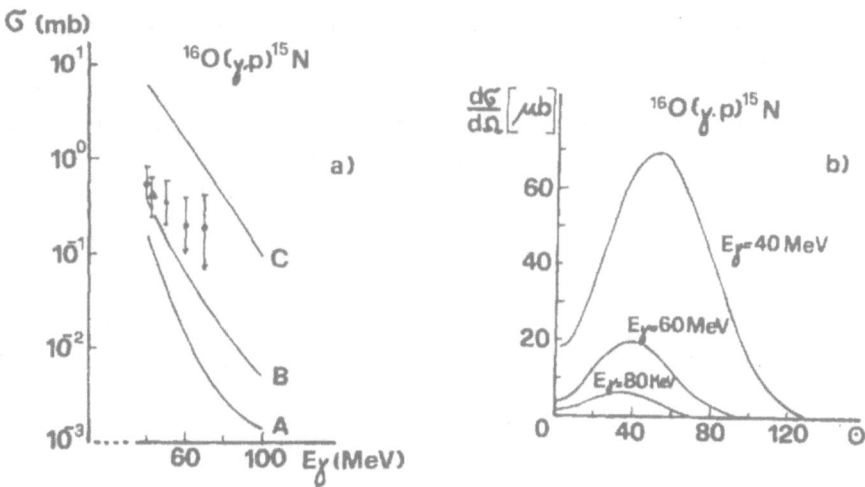

Fig. 10   a) Integrated cross sections for $O^{16}(\gamma p)N^{15}$ in the independent particle model. The bound state is calculated in
a W - S potential, the scattering state belongs to this
same potential in curve A, to an energy dependent W - S
potential in curve B and is a plane wave in curve C.
b) The calculated angular distributions for the same reaction.

The preliminary analysis (18) seemed in fact to confirm this scheme but these conclusions were sistematically criticized by the Bochum group (21) In fig. 10 is reported the direct γP_o cross section for $^1P_{1/2}$ protons in O^{16}, computed using various average potential wells for the initial and final states. A reasonable agreement is obtained only using plane waves (B A) for the final proton state, more realistic wave functions give cross sections about one order of magnitude too low. The E1 - E2 interference produces in all cases a forward peaked proton angular distribution, as experimentally observed. A further check of the direct interaction mechanism is provided by the measurements of the (γn) cross section, performed by groups in Mainz (22) and Saskatoon (23). In both experiments the ground state γn_o transition has been observed and the comparison between the magnitude and angular dependence of γp_o and γn_o cross sections leads to the following conclusions:

a) The two cross sections have comparable magnitude over a wide energy range (fig. 11)

b) Both angular distributions are forward peaked, the asymmetry being larger for γp_o data (fig. 12)

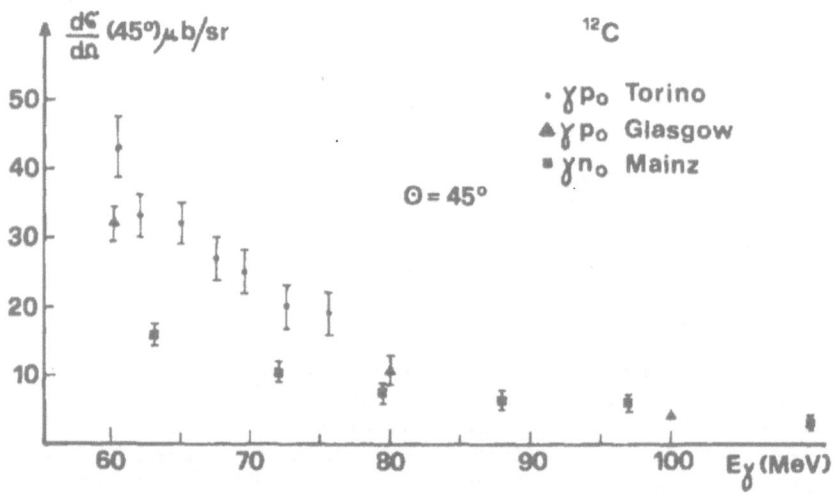

Fig. 11 γp and γn cross sections $d\sigma/d\Omega$ at $\Theta = 45°$ as a function of the γ ray energy.

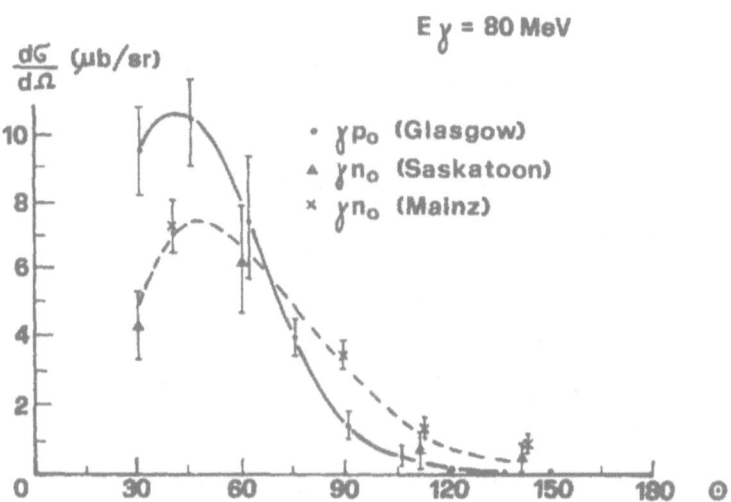

Fig. 12 γp and γn angular distributions at E_γ = 80 MeV.

These results are in definite desagreement with the assumption
of a purely direct interaction model. In this picture neutron emission
should in fact be allowed only by the center of mass motion terms, describing the absorption of the photon by the recoiling mass A - 1
nucleus (20) , and the $\sigma(\gamma n_o)/\sigma(\gamma p_o)$ ratio is expected to be low.
Moreover the E2 effective charge is for the neutron extremely small:

$$q_{eff}(E2) = Z/A^2$$

The E1 - E2 interference is therefore negligible and the
neutron angular distribution should be roughly symmetric around 90°.
Short range correlations, i.e. the effect of the short distance repulsive
neutron - proton interaction may increase the absolute cross section,
introducing larger high momentum components in the single mucleon wave
functions, and produce more symmetric γn and γp angular distributions. The result of a quantitative calculation performed by
Weise and Huber (24) is reported in fig. 13.

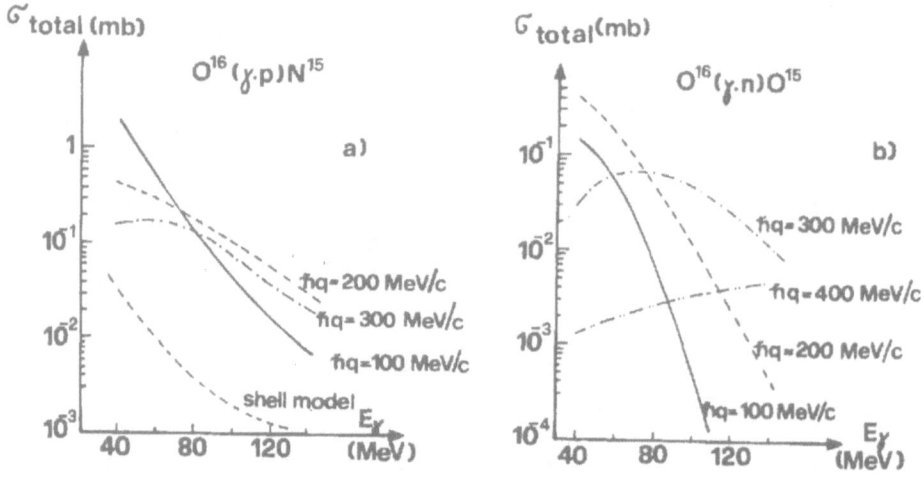

Fig. 13 Computed (24) total γp and γn cross sections in O^{16} for various values of the excanged momentum $\hbar q$

Unfortunately in this calculation the exchanged nucleon momentum $\hbar q$, related to the range of the repulsive core, is treated as a free parameter and a good fit is obtained with $\hbar q \simeq 250$ MeV/c a value which has been criticized by many authors (21) (25) as unrealistic. In fact the nuclear wave functions, computed using this figure, heal too slowly (21) and no fit can be obtained to other data (25), like the elastic electron scattering form factors, also sensitive to the short distance behaviour of the wave functions. Calculations performed with more realistic parameters seem to indicate a negligible contribution of short range correlations to the photonuclear cross sections up to γ ray energies of about 100 MeV (26).

All these analysis were performed assuming the validity of the impulse approximation also at relatively low transferred energy. Since 1964 G.E. Brown (27) had shown that radiative proton capture cross sections in medium heavy nuclei cannot be explained above the Giant Dipole Resonance without introducing a semidirect interaction mechanism through the intermediate excitation of G.D.R. states. This model has been extended by different groups (28) (13) (29) to include the excitation of the E2 giant resonances and applied to the analysis of γ p and γ n reactions above 20 MeV. The main assumptions are illustrated in fig. 14: besides diagram a) which describe the usual impulse approximation, the effect of residual interactions has been added by diagrams like b) and c) which include the intermediate excitation of particle - hole pairs in the initial and final state respectively. It is easy to show that diagrams b) and c) correct the theory in the right sense. The electromagnetic vertex describes now the interaction with a collective state while the nuclear vertex is intrinsically charge independent, the contribution to proton and neutron emission is therefore pratically the same. Moreover, if both E1 and E2 intermediate particle - hole states are included, the inter-ference terms will shift the nucleon angular distribution in the forward direction, the shift being very similar for protons and neutrons.

The contribution of the semidirect process is expected to be relevant at photon energies E_γ still comparable with the energy E_R of the intermediate resonance (fig. 14), when $E_\gamma \gg E_R$ the impulse approximation (diagram a)) becomes dominant. Quantitative calculations have been performed by three different groups and I shall briefly review their results. A preliminary evaluation of the importance of the final state interaction (graph. b) in the reaction $Pb^{209}(\gamma n)Pb^{208}$ has been attempted by Carbone, Cenni, Malvano and Molinari. (29) Their tool was to explain an old experiment, performed at the Torino laboratory (30) , where a strong forward asymmetry was observed in the photoneutron angular distribution from Bi^{209} at γ ray energies above 20 MeV. Close to the G.D.R. diagram c) is negligible, the calcu-lation has therefore retained only diagrams a) and b) and the E1 and E2 intermediate resonances known from the experiments. The computed differentialcross section is reported in fig. 15b, compared with the prediction of the pure impulse approximation (fig. 15a): the final state interaction produces a strong forward shift in the neutron angular dis-tribution in substantial agreement with the experimental data (fig. 15c) The γ p and γ n cross sections in light nuclei have been calculated up to 100 MeV by the Bochum group(28) including the initial and final state interaction with the intermediate excitation to the giant E1 and E2 resonances.

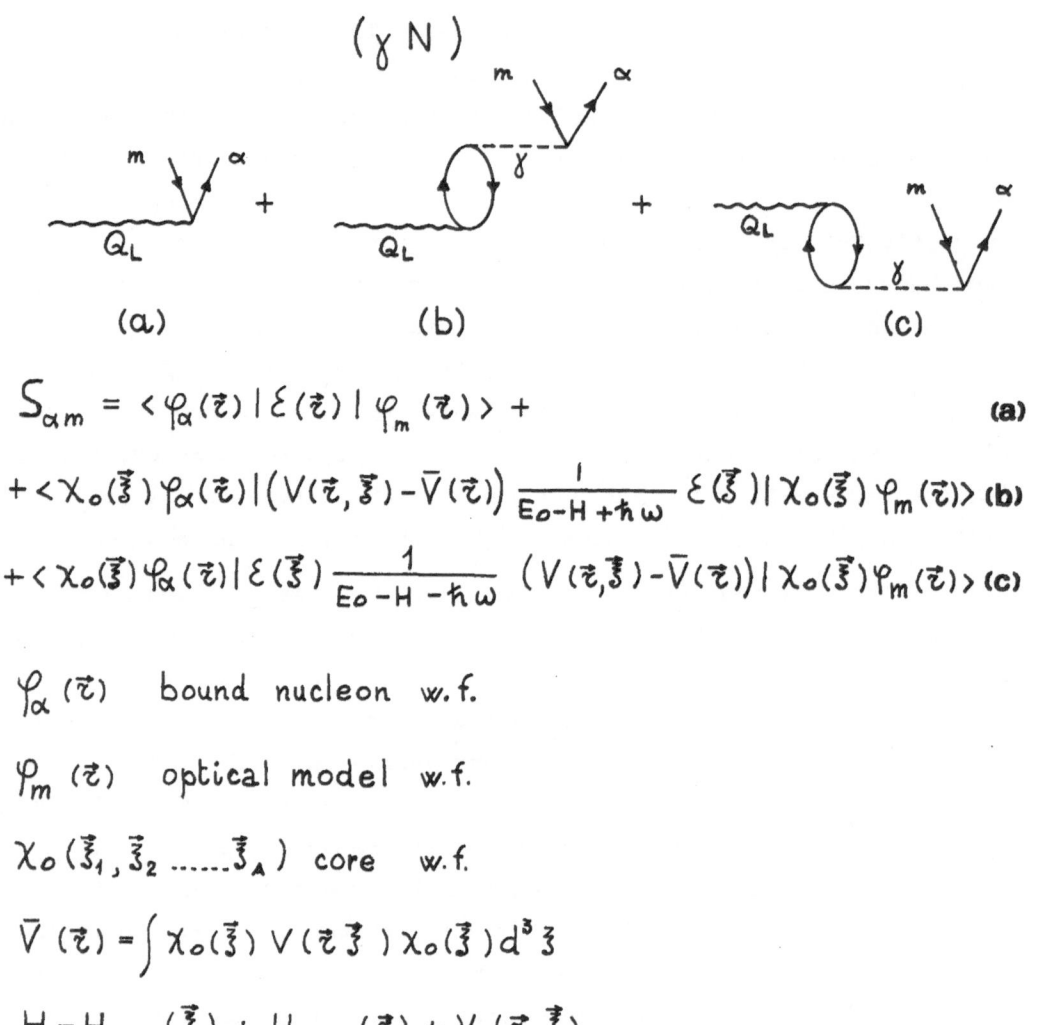

(γN)

(a)   (b)   (c)

$$S_{\alpha m} = \langle \varphi_\alpha(\vec{z}) | \mathcal{E}(\vec{z}) | \varphi_m(\vec{z}) \rangle + \quad \textbf{(a)}$$

$$+ \langle \chi_0(\vec{3}) \varphi_\alpha(\vec{z}) | (V(\vec{z},\vec{3}) - \bar{V}(\vec{z})) \frac{1}{E_0 - H + \hbar\omega} \mathcal{E}(\vec{3}) | \chi_0(\vec{3}) \varphi_m(\vec{z}) \rangle \; \textbf{(b)}$$

$$+ \langle \chi_0(\vec{3}) \varphi_\alpha(\vec{z}) | \mathcal{E}(\vec{3}) \frac{1}{E_0 - H - \hbar\omega} (V(\vec{z},\vec{3}) - \bar{V}(\vec{z})) | \chi_0(\vec{3}) \varphi_m(\vec{z}) \rangle \; \textbf{(c)}$$

$\varphi_\alpha(\vec{z})$ bound nucleon w.f.

$\varphi_m(\vec{z})$ optical model w.f.

$\chi_0(\vec{3}_1, \vec{3}_2 \ldots \ldots \vec{3}_A)$ core w.f.

$\bar{V}(\vec{z}) = \int \chi_0(\vec{3}) V(\vec{z}\,\vec{3}) \chi_0(\vec{3}) d^3 \vec{3}$

$H = H_{core}(\vec{3}) + H_{part}(\vec{z}) + V(\vec{z},\vec{3})$

Fig. 14

The static part of the electromagnetic interaction with the exchanged pion current and the effect of short range correlations have also been taken into account. In fig. 16a is shown the separate contribution of the different terms to the γ p cross section, in fig. 16b is plotted the neutron angular distribution at 63 MeV, a detailed description of this work will be given elsewhere (1). In this approach a particular emphasis is given to the role of the excanged meson currents.

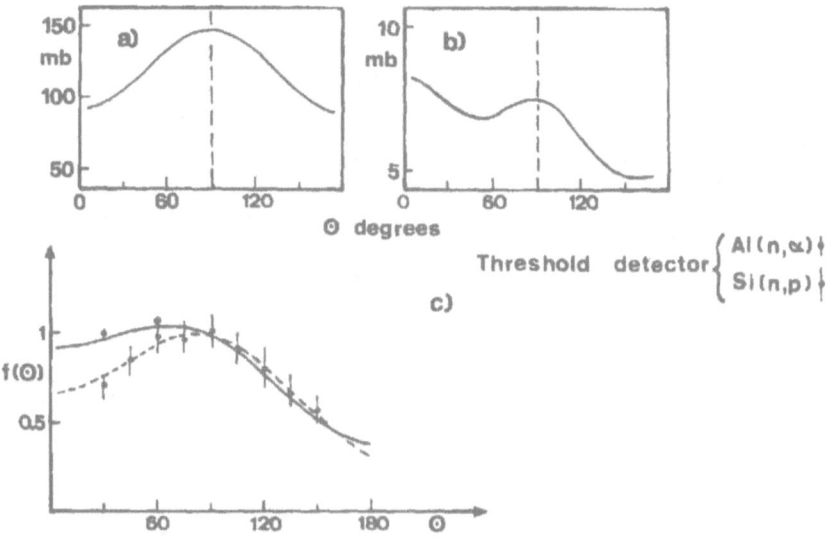

Fig. 15   The Pb^{209} (γn) Pb^{208} cross section at E$_\gamma$ ≈ 25 MeV   a) pure
impulse approximation   b) impulse approximation   plus final
state interactions   c) Experimental angular distribution
from ref.(29)

The improvement of the fits with respect to the preliminary
impulse approximation calculations is   considerable ∶ absolute cross
sections as well as energy and angular dependence seem to be
reproduced.

Very similar physical assumptions are followed in the   paper
of Marangoni, Ottaviani and Saruis(13) : the photodisintegration cross
section of   O^{16} is computed up to   80 MeV including the final state
interaction (graph. b) and summing up all the possible E1   and   E2
1 particle - 1 hole intermediate states.   The contribution of the charge
exchange potential is also accounted through a zero range interaction.
The most attractive features of this work are the correct treatment
of continuum states and the complete calculation of the photodisintegration cross section   from threshold up to   80 MeV.

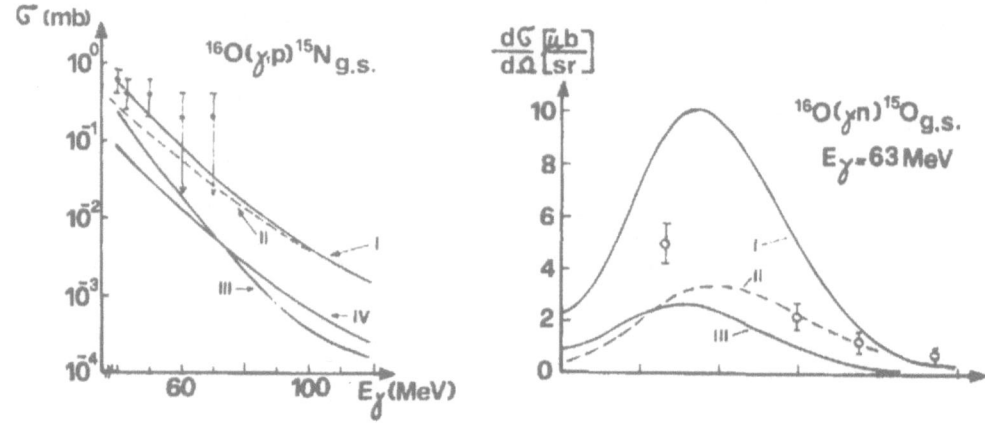

Fig. 16 a) Total cross section of the reaction $O^{16}(\gamma p_0)N^{15}$
b) Angular distribution of the reaction $O^{16}(\gamma n_0)O^{15}$
I = total transition matrix   II = shell model + exchange
current contributions   III = shell model + initial and
final state interactions   IV = pure shell model

The high energy part is therefore evaluated using in the residual
interaction the same parameters which give a satisfactory fit to the
Giant Resonance region. The numerical results are plotted in fig. 17:
again the most important features of the experimental data seem to be
correctly reproduced.

We might therefore conclude that the inclusion of the semidirect
process has greatly improved our knowledge of the single nucleon photo
emission at intermediate energies. Unfortunately this new picture is,
to my opinion, less attractive than the old impulse approximation.
The most interesting feature of the nuclear photoelectric effect was
in fact a fairly direct connection between the high energy γ p or
γ n cross sections and the high momentum tail of the nucleon momentum
distribution, an important aspect of nuclear structure where the experimental information is still lacking .

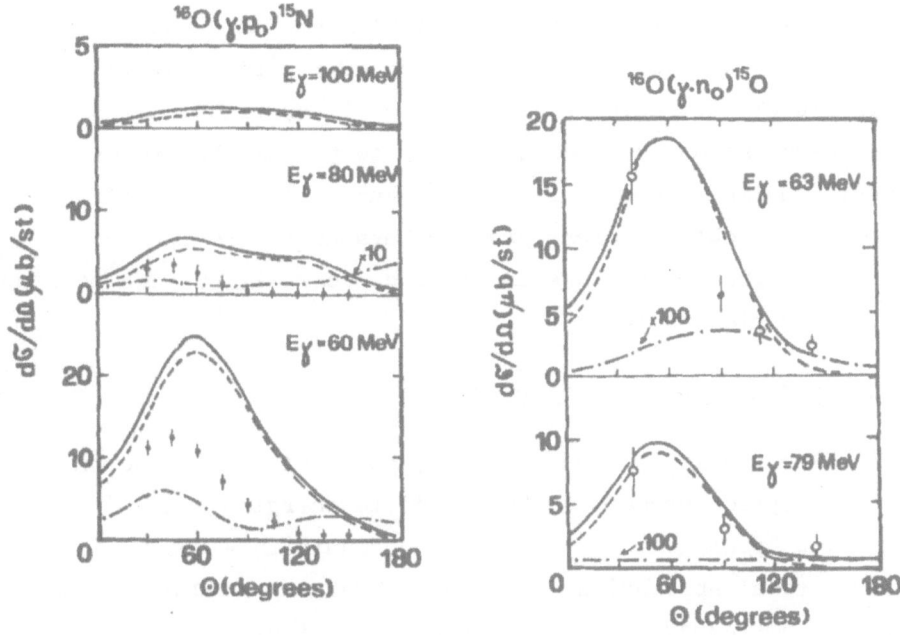

Fig. 17   Computed (13) differential cross sections for the reactions $O^{16}(\gamma p_o)N^{15}$ and $O^{16}(\gamma n_o)O^{15}$. Solid and dotted curves correspond to different choices of the $1f_{7/2}$ binding energy. The dash dot curve is the pure shell model result.

The strong perturbations introduced by the final state inter-actions mask the observation of the physical properties of the initial nuclear state. Nevertheless there are in the investigation of the photonucleon emission at intermediate energies possibilities, still experimentally unexplored, to obtain important nuclear structure infor-mations. In the Glasgow experiment (19) some evidence has been found of excitation of the deep $1s^{-1}$ hole states (fig. 9), this effect should be more carefully investigated using the now available quasi monochromatic γ ray facilities. Photon data on the properties of deeply bound nucleons like binding energy, widths and momentum distribution, would be a desi-derable complement to the well known results of quasi elastic scat-tering experiments (31). A preliminary evaluation of the importance

of distorsion effects in these particular channels would be of course
necessary.   Moreover the contribution of the final state interactions
should rapidly decrease, as shown in fig. 14, with increasing photon
energy becoming almost negligible for away from the highest E2 resonances. The measurement of the γ p and γ n cross sections above
100 MeV   might therefore provide less distorted informations on
fundamental properties like short range correlations and exchange currents in complex nuclei.

IV- Quasi deuteron absorption

The single nucleon emission, does not seem to be the dominant
photon absoption process at intermediate energies: in O^{16} the integrated γ p plus γ n cross section accounts only about 17% of the
observed total absorption between 30 MeV and 170 MeV. Competitive
many body disintegration channels are important and have been identified
in experiments using visualizing techniques, typically diffusion chambers
or stream chambers. The results of measurements performed be Gorbunov
and coworkers (32) (33) are reported in table II. The two body γ np
reaction accounts for about 22% of the total integrated absorption
cross section above 30 MeV in O^{16}; a similar figure is also obtained
for C^{12} if we identify, following Taran (33) , the γ npα reaction
as a particular γ np channel leaving the residual B^{10} nucleus in
a highly excited state according to:

$$\gamma + C^{12} \Rightarrow B^{10*} + n + p \Rightarrow Li^6 + \alpha + n + p$$

Since 1956 Levinger (34) assumed the two body absorption process
to be the most important reaction channel at photon energies above
100 MeV. At these energies the γ ray can penetrate the nucleus and
interact with a two nucleon cluster; assuming the interaction to be
predominantly dipole absorption, only neutron - proton pairs can contribute to the cross section, of these the ones in S^3 state,"quasi deu
terons", are expected to be responsible of the majority of the interactions.

TABLE II

O^{16}

REACTION	INTEGRATED CROSS SECTION (MeV – mb)		
	$E_\gamma \leq 30$ MeV	$E_\gamma \leq 170$ MeV	$30 \leq E_\gamma \leq 170$ MeV
(γp)	100 ± 4	117 ± 5	17 ± 6
(γn)	65 ± 2	86 ± 3	21 ± 3
(γpn)	9 ± 1	60 ± 5	51 ± 5
$(\gamma \alpha)$	4	4	–
Remaining Reactions	23	$141 \begin{smallmatrix} +\ 0 \\ -\ 16 \end{smallmatrix}$	$118 \begin{smallmatrix} +\ 0 \\ -\ 16 \end{smallmatrix}$
TOTAL	201	$408 \begin{smallmatrix} +\ 10 \\ -\ 30 \end{smallmatrix}$	225 ± 20

C^{12}

	$E_\gamma \leq 40$ MeV	$E_\gamma \leq 170$ MeV	$40 \leq E_\gamma \leq 170$ MeV
(γpn)	11.5	42 ± 5	30.5
$(\gamma pn\alpha)$	0.4	22.6 ± 4.5	22.2
$(\gamma p\alpha)$	4.9	10.2 ± 1	5.3
$(\gamma n\alpha)$	4.1	8.6 ± 1.7	4.5
$(\gamma p\, t)$	1.6	9.4 ± 1.2	7.8
(γnH_e^3)	1.0	5.6 ± 1.2	4.6

If the state of a n - p pair in the nucleus is assumed to approach
at sufficiently short distance the wave function of the free deuteron,
the cross section for the photodisintegration process can be written

$$\sigma_{np}(E_\gamma) = L \frac{NZ}{A} \; \sigma_D(E_\gamma) \qquad\qquad E_\gamma \geqslant 100 \text{ MeV}$$

where σ_D is the deuteron photodisintegration cross section, NZ/A
the number of quasi deuteron pairs per unit volume in the nucleus and
L a proportionality constant which is related to the probability to
find the two nucleons close enough to afford the absorption. As we
shall see later on using the Wilson model, this interaction distance
should be of the order of the pion Compton wavelength.

The most unambiguous experimental test of the two body process
is provided by the detection of neutron proton coincidences at photon
energies 200 ÷ 300 MeV. Experiments (35) (36) (37) have been performed in various laboratories and the more significative results can
be summarized as follows: a) The differential γ np cross section
exhibits an angular dependence very similar to the free deuteron photodisintegration cross section in the same kinematical conditions (fig.18)
The angular correlation in complex nuclei is obviously smeared out by
the motion of the n - p pairs in the nucleus.   b) The ratio
$\sigma_{np}(E_\gamma)/\sigma_D(E_\gamma)$ results roughly proportional to NZ/A as predicted
by the quasi deuteron model, but strong desagreements still exist among
the values of the constant L reported by the different experiments
(table III). The value $L \simeq 10$, obtained by the Glasgow group (35)
is supported by the estimates from total neutron (38) , and activation
(39) cross section measurements performed at γ ray energies above
100 MeV.

The success of the semiphenomenological model of Levinger
stimulated more accurate theoretical and experimental work. In 1960
Gottfried (40) gave a rather extensive theory of the two body photodisintegration. In this approach the photon nucleus interaction
Hamiltonian is assumed to be dominated by two body operators, the ground
state of the nucleus is described by a correlated wave function of the
Jastrow type and the Born Approximation is used for the final nucleon waves.

TABLE III

Experimental L values

He	Li	C	O	Ca	Experiment
	9.6±2.3	12.4±3	10.3±2.6	8.7±2.1	n-p coincidences (Glasgow) (35)
6.3±1	4.2±0.6				n-p coincidences (Illinois)(37)
10 – 12					$S^{32}(\gamma\,np)$ (Illinois) (39) $(\gamma\,nT)$ (Orsay) (38)

$^2H(\gamma np)$

$^{16}O(\gamma np)$

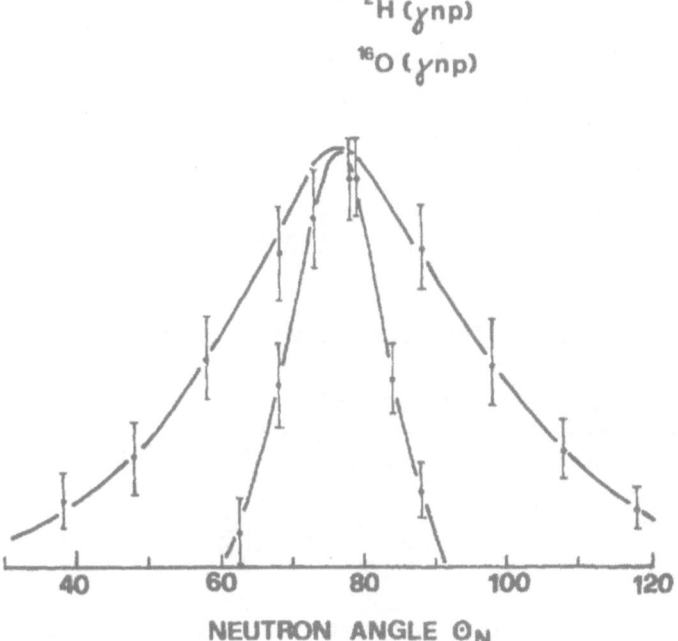

NEUTRON ANGLE Θ_N

Fig. 18 Comparison of the neutron – proton angular correlations in H^2 and O^{16} (36)

The differential cross section

$$[2] \quad d\sigma = (2\pi)^{-4} F(P) S_{fi} \, \delta(E_f - E_i) \, d^3k_p \, d^3k_n$$

is given, for closed shell nuclei, by the product of three factors: the available phase space, the probability to find the $n - p$ pair with central momentum $\vec{P} = \vec{k}_p + \vec{k}_n - \vec{\omega}$ in the Slater determinant and the probability S_{fi} that two particles in a state of relative motion given by the short range correlation function perform a transition to the final scattering state. If we now assume the $n - p$ pair wave function to be, at relative distances much less than the nuclear radius, identical to the free deuteron wave function in the same range we get the Levinger result

$$S_{fi} = 3 \gamma^3 D_{fi}$$

where γ is a proportionality constant and the transition probability D_{fi} is related to the free deuteron photodisintegration cross section in the centre of momentum frame *

$$D_{fi} = \frac{4\pi}{k_P^* E_P^*} \left(\frac{d\sigma_D}{d\Omega} \right)^*$$

In closed form the "quasi deuteron" cross section is generally written:

$$\frac{d\sigma_{pn}}{d\Omega} = L \frac{NZ}{A} CF(P) \left(\frac{d\sigma_D}{d\Omega} \right)^* J \, B(E_\gamma)(1 - \beta_D \cos\Theta_D) d^3k_p \, d\varepsilon_n$$

$$E_\gamma = \varepsilon_n + \varepsilon_p + \bar{\varepsilon}$$

where C is a normalization constant which has to be introduced in order to get for the integrated cross section the same expression given by Levinger, $B(E_\gamma) (1 - \beta_D \cos\Theta_D)$ the incident photon spectrum corrected for the Doppler shift (Θ_D is the angle between $\vec{\omega}$ and the quasi deuteron momentum $\vec{P}$) and J the Jacobian of the transformation to the centre of momentum frame (defined by $k_p + k_n = 0$).

The missing energy $\bar{\mathcal{E}}$ should be determined by complete kinematic experiments, tentative evaluations seem to indicate $\bar{\mathcal{E}} \simeq 30$ MeV.

The form factor $F(P)$ is given by:

$$F(P) = \sum_{nn'\ell\ell'} \sum_{\rho=|\ell-\ell'|}^{\ell+\ell'} (2\ell+1)(2\ell'+1) \left| <11'00\,\rho 0> \right|^2$$

$$\left| \int_0^\infty R_{n\ell}(r) R_{n'\ell'}(r) J_\rho(Pr) r^2 \, dr \right|^2$$

where $R_{n\ell}(r)$ is the radial part of the single particle wave function $\psi_{n\ell m}(r)$ and $j_\rho(Pr)$ is the spherical Bessel function of order ρ $F(P)$ describes therefore the shell model (Slater determinant) pair momentum distribution, all the short range dynamical effects being included in the transition probability S_{fi}

In the expression $[2]$ the form factor $F(P)$ is a rapidly varying function of P while the squared matrix element S_{fi} mainly depends on the γ ray energy ω and on the relative final momentum $\vec{k} = 1/2(\vec{k}_p - \vec{k}_n)$. An important consequence is that the angular distribution yields informations on the shell model part $F(P)$ of the pair correlation function, while the more interesting short range correlations mainly influence the energy dependence and absolute magnitude (strength constant L) of the two body photodisintegration cross section. The measured pair momentum distribution $F(P)$, obtained from Glasgow (35) and Bonn (41)data, is reported in fig. 19. The continuous curve rapresents a shell model calculation with a realistic h.o. parameter value: the theoretical distribution is in both experiments considerably wider than the experimental one. The h.o. parameter can always be adjusted, for each shell, in order to fit the data but the physical meaning becomes then obscure and, moreover, it is extremely difficult to find a unique recipe for the different experiments. Such discrepancy does not seem to be due to some model dependence: calculations (40) performed using various average potential wells give in fact very similar results. Final state interactions are instead expected to influence the angular correlation function leading to a distorted pair momentum distribution (40)

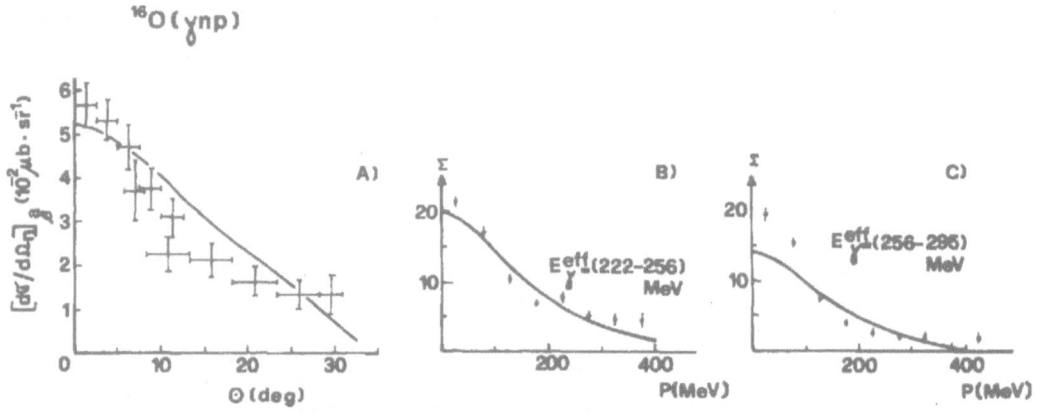

Fig. 19   Differential cross sections for the reaction $O^{16}(\gamma np)$
A) ref. (35)   B), C) ref. (41)   . The curves are shell
model calculations with a realistic h.o. parameter.

This effect has not been completely disregarded in the data analysis
but roughly simulated by correction factors. The real part of the
optical potential introduces refraction effects only important at
very forward angles but the imaginary part gives an absorption correction as large as 70%. It seems therefore very important to handle
the distortion of the outgoing waves in a more sophisticated way, also
including the possible transitions to intermediate excited states,
before attempting any conclusion about the dynamical state of the n - p
pair in the nucleus .

        Before concluding this subject I would like to point out the
physical interest of this reaction for the investigation of nuclear
structure. All the short distance effects like short range correlations
and exchange currents, already considered in the single nucleon cross
sections, should be more clearly observed in electromagnetic matrix
elements where a two body density is involved. The importance of final
state interaction requires experiments where the detected nucleon
energies are above at least 100 MeV. Moreover if monochromatic or

quasi monochromatic γ rays are available in coincidence experiments, the direct measurement of the residual excitation energy in the nucleus becomes possible. As an example (42) of the expected results the excitation energy distribution of two hole states in B^{10}, following π^- capture in C^{12}, is shown in fig. 20: peaks are observed at $\sim$ 0 MeV, corresponding to p^{-2} states, and $\sim$ 40 MeV corresponding to S^{-2} states.

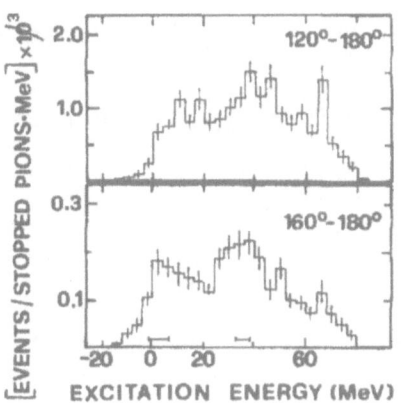

Fig. 20   Energy distribution of residual states in B^{10} from π^- capture in C^{12}

The "quasi deuteron" is not the only many body absorption mechanism observed at intermediate γ ray energies. From this point of view the data on C^{12} (33) reported in table II are particularly significative: reactions 1) and 2) can be interpreted as two nucleon absorption, reactions 3) and 4) as one nucleon absorption, the α particle following in all cases from the decay of excited states in the residual nucleus. Reactions 5) and 6) might be ascribed to the interaction of the photon with quasi α clusters in the nucleus, but the experimental data are still too poor to afford any conclusion. A more detailed experiment has been performed by a group in Bonn (41) to measure the photon – deuteron angular correlation from the γ pd reaction in O^{16} at γ ray energies in the range 200 – 300 MeV. In close similarity with Gottfried's expression $\begin{bmatrix} 2 \end{bmatrix}$, the γ pd angular correlation $\sum(P, E_\gamma)$ is assumed to be given by the product of two factors

$$\sum(P, E_\gamma) \propto F(P) \frac{d\sigma^*}{d\Omega}(E_\gamma)$$

where $\vec{P} = \vec{k}_p + \vec{k}_d - \vec{\omega}$ If angular correlation measurements are performed at various γ ray energies E_γ , an empirical $F(P)$ function, fairly independent from E_γ , can be obtained (fig. 21 a), b)). The ratio $\sum/F(P)$ is reported in fig. 21 c) as a function of E_γ and compared with the free He^3 photodisintegration cross section:

$$\frac{d\sigma}{d\Omega}(\gamma + He^3 \Rightarrow p + d)$$

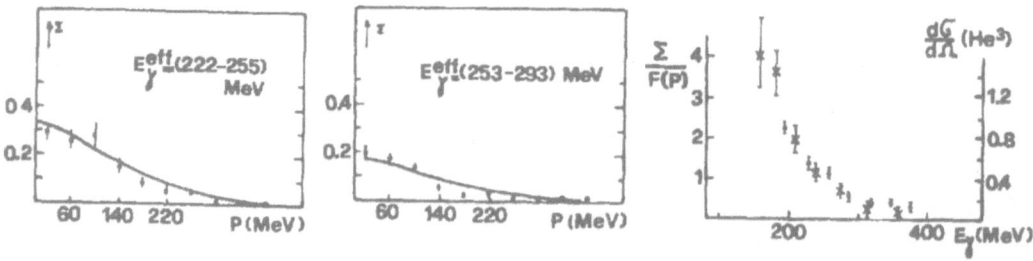

Fig. 21   a), b) Differential cross section for the γ pd process in O^{16}
c) comparison between the $\sum/F(P)$ ratio(.)and the He^3 two body photodisintegration cross section.(×)

The similar energy dependence confirms the validity of this simple approach.   Photonuclear reactions at intermediate energies seem therefore to be a very promising method for the experimental investigation of many body correlation effects in nuclear structure.

V- <u>The photoproduction of pions</u>

Real pion photoproduction from nuclei may be described in the framework of the impulse approximation (I.A.).   The nucleus is considered, in the classical nuclear physics picture, as an ensemble of neutrons and protons interacting through a two body force without changing their intrinsic properties. The production matrix element M_π is written as the sum of one body transition amplitudes between the initial $|J_i M_i\rangle$ and final $|J_f M_f\rangle$ nuclear states. At γ ray energies close to the production threshold (S wave pions) we have:

$$[3] \quad M_\pi = \langle J_f M_f | \sum_{J=1}^{A} \tau_J^\pm \; \vec{\sigma}_J \cdot \vec{\varepsilon} \; e^{-i\vec{k}\cdot\vec{\tau}_J} | J_i M_i \rangle$$

where τ_J $\vec{\sigma}_J$ an the nucleon isospin and spin operators $\vec{\varepsilon}$, $\vec{k}$ the photon polarization and momentum. The validity of the I.A. [3] is difficult to establish in complex nuclei because of theoretical uncertainties introduced by the assumptions on nuclear wave functions and final state interactions (D.W.I.A.) in [3].

Donnelly and Walecka (43) have proposed an interesting test of validity of the I.A. based on the comparison of [3] with similar weak and electromagnetic transition matrix elements connecting the same initial and final states or their analogs. Namely:
The Gamow – Teller β decay matrix element

$$[4] \quad M_\beta = \langle J_f M_f | \sum_{J=1}^{A} \vec{\sigma}_J \; \tau_J^\pm | J_i M_i \rangle$$

The Axial vector muon capture matrix element

$$[5] \quad M_\mu = \langle J_f M_f | \sum_{J=1}^{A} \vec{\sigma}_J \; \tau_J^- \; e^{-i\vec{v}\cdot\vec{\tau}_J} | J_i M_i \rangle$$

The electron scattering M1 spin flip matrix element

$$[6] \quad M_e = \langle J_f M_f | \sum_{J=1}^{A} (\mu_p - \mu_n) \; \vec{\sigma}_J \; \tau_J^3 \; e^{-i\vec{q}\cdot\vec{\tau}_J} | J_i M_i \rangle$$

The reduced matrix element of any of the multipole operators $0_{JT}(q)$ discussed above may be written (43).

$$\langle J_f \| O_{JT}(q) \| J_i \rangle = \sum_{a a'} \langle a \| O_{JT}(q) \| a' \rangle \psi_{JT}^{fi}(a, a')$$

where a, a' run over a complete set of single particle quantum numbers (usually nlj), the quantities $\langle a \| O_{JT}(q) \| a' \rangle$ are the single particle reduced matrix elements, reasonably well known, and the complexity of the nuclear many body problem is buried in the one body transition density matrix elements $\psi_{JT}^{fi}(aa')$. The coefficients ψ_{JT}^{fi} are generally adjusted to fit the q dependence of the elastic and inelastic electron scattering cross sections for M1 transitions [6] to the final $| J_f M_f \rangle$ states and the obtained set is used to compute the matrix elements [3] [4] [5] between the same $| J_i M_i \rangle$, $J_f M_f$ states or their analogs. The experimental cross section for the reaction $C^{12}(\gamma \pi^-) N^{12}$ (44) around threshold is reported in fig. 22 along with the theoretical predictions (DWIA). The calculation of Koch (44) (curve k) includes only S wave pion waves distorted by a Kissingler optical potential and a transition density matrix parametrized with shell model wave functions. The calculation of Nagl and Uberall (curves 1 and 2) (45) includes distorted S, p and d waves and a transition density parametrized by a Helm model: the difference between the curves 1 and 2 is due to different choices of the pion momentum in the interaction Hamiltonian. In all cases the density matrix has been fitted to the inelastic electron scattering data to the 1^+ 15.1 MeV state in C^{12}. All the calculations agree with the measured cross section at threshold but the observed slope seems to be larger. Further experimental and theoretical work must certainly be done before any definite conclusion may be attempted, but from the present results large deviations from the DWIA do not seem to be expectable.

The measurement of pion photoproduction near threshold is there fore, when involving transitions between known bound states, a consistency test of the Impulse Approximation. Conversely, once the validity of the I.A. is clearly established, the experimental investi gation of the matrix element [3] can provide further spectroscopic informations on the final nuclear states.

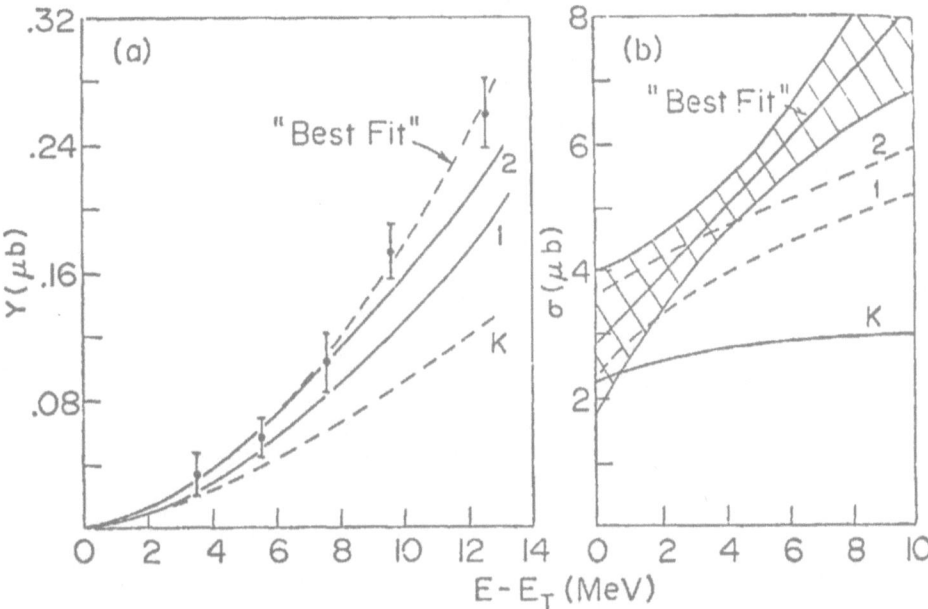

Fig. 22   Yield and b) Cross section versus energy above threshold
for the reaction $C^{12}(\gamma\pi^-)N^{12}$. The curves are discussed
in the text.

This aspect is particularly important at higher photon energies
($E_\gamma \geqslant 200$ MeV) when the quasi bound G.R. states can be reached. The
similarity of [3] with the photoabsorption matrix element

$$[7] \qquad M_\gamma \propto\; <J_f\, M_f \;|\; \sum_{J=1}^{A} (1 + \tau_3)\; \vec{\varepsilon}\cdot\vec{p}\; e^{i\vec{k}\cdot\vec{r}} \;|\, J_i M_i >$$

suggests the possibility, in pion production, of excitation of 1p – 1h
vibrational states very similar to the Giant Multipole Resonances.
The presence of the spin operator in [3] changes the involved selection
rules favouring the $\Delta S = 1$ transitions, the so called spin – isospin
vibrations.   These transitions have been observed in the inverse
π^- capture experiments (46) : the spectrum of capture γ rays from
the reactions $C^{12}(\pi^-\gamma)B^{12}$ and $B^{10}(\pi^-\gamma)Be^{10}$ is reported in fig. 23

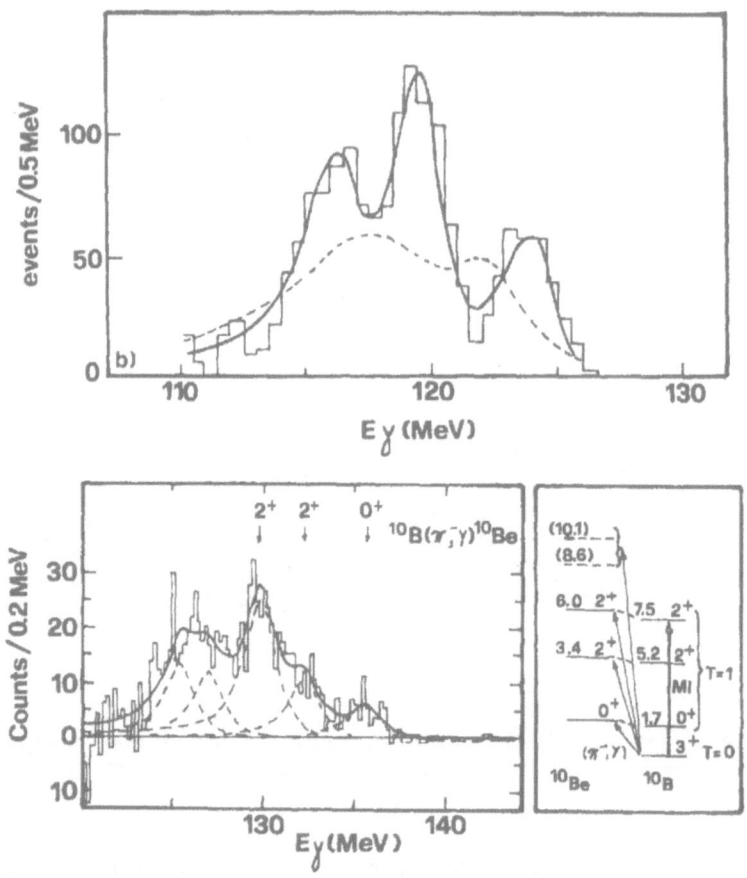

Fig. 23 Spectrum of γ rays from the π⁻ capture in C¹² and B¹⁰

In the first reaction peaks are observed corresponding to final states in B¹² which are generally interpreted as analogs of the 15.1 MeV M1, 19.9 MeV M2 and 23.7 MeV E1 Giant Resonances in C¹². In the second reaction only transitions to low lying states of B¹⁰ have been clearly resolved. In these experiments only the energies of the residual states are   identified, angular momenta and parities are tentatively assigned by comparison with theoretical calculations. The π photoproduction, using monochromatic γ rays, would have the advantage that the pion angular distribution is sensitive to the assignements of the involved nuclear states. The excitation strength of the state is in fact a function of the momentum transfer q , which varies with the π detection angle (fig. 24).

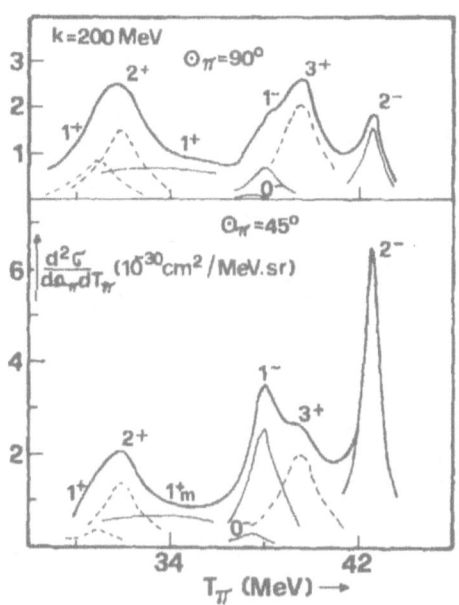

Fig. 24 Calculated (47) spectrum of pions from O^{16}, at $E_\gamma = 200$ MeV
$\Theta = 90°$ and $\Theta = 45°$

The only photonuclear experiment giving some evidence, even if indirect, of the excitation of spin – isospin Giant Resonances is the total absorption measurement in Li and Be (48) The cross section exhibits a steep change of slope at a photon energy about 20 – 30 MeV higher than the photoproduction threshold (fig. 25). This energy shift may be interpreted as due to the excitation of spin – isospin vibrations at 20 – 30 MeV in the residual nucleus.

At photon energies above 200 MeV the $J = 3/2$ $T = 3/2$ $N*(1236)$ resonance becomes the most important pion photoproduction channel. In the framework of the I.A. and neglecting final state interactions the total cross section $\sigma_{\gamma\pi^+}^z$ in complex nuclei may be described, in the resonance region, in terms of the cross section $\sigma_{\gamma\pi^+}^p$ with a free proton as target, namely (49)

$$[8] \qquad \sigma_{\gamma\pi^+}^z = z\,\sigma_{\gamma\pi^+}^p + \int\limits_{\vec{\nu}+\vec{k}\,|\leqslant d} \rho(\vec{k})\,d\vec{k}$$

$$\vec{k} = \vec{n} + \vec{\mu} - \vec{\nu} \qquad\qquad d = \sqrt{2M(\nu_0 - \mathcal{E} - \mu_0)}$$

where $\vec{\mu}, \vec{\nu}, \vec{n}$ are the momenta of the meson, photon and neutron respectively, $\vec{k}$ is the recoil momentum M, μ_0 are the neutron and meson rest masses, $\mathcal{E}$ the neutron binding energy and $\rho(k)$ the proton momentum distribution in the nucleus.

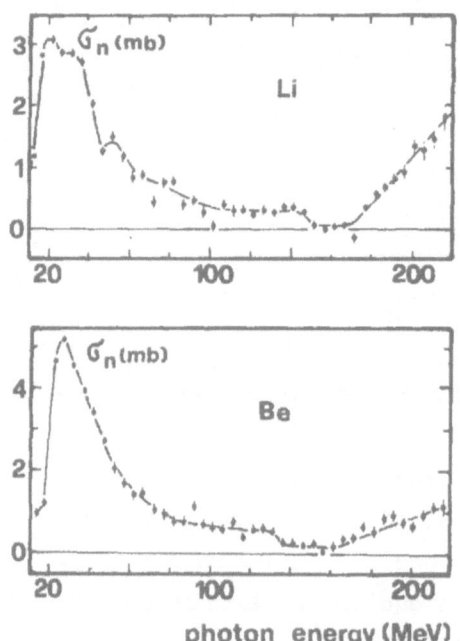

Fig. 25 Total absorption cross section in Li and Be.

The ratio $\sigma_{\gamma\pi^+}^z / \sigma_{\gamma\pi^+}^p$ is reduced because only a portion of the momentum distribution is energetically capable of contributing to the cross section. As the photon energy increases, the common volume increases so that eventually the proton momentum distribution is covered and the efficiency becomes unity. A more complete calculation has been performed by Laget (49) to include π - nucleus and nucleon-nucleus interactions in the final state, using proper optical potential wells. The comparison of the numerical results with experimental data (50) shows a reasonable agreement at least in the recoil momentum range $0 \leqslant k \leqslant 100$ MeV/c (fig. 30).

Other interesting effects, associated with the photoproduction of pions, are the importance of electromagnetic interactions with the exchanged meson current (M E C), the existence of isobar configurations (I C) in the nuclear wave functions and the interactions of the produced isobar with the rest of the nucleus.(2) All the mentioned effects are not obviously taken into account in the ordinary I. A. and D.W.I.A. calculations and might be therefore experimentally evidenced as deviations from the predicted I. A. results. This method will be reliable if reaction channels and kinematic conditions are chosen where the impulse approximation matrix element is strongly reduced. Some work has been performed in this direction following two different ways:

a) Investigation of reaction channels, above the π threshold, where no real π is produced in the final state since these purely nucleonic decay modes are expected to depend from more complicated many body interactions.

b) Photoproduction of pions on nucleons in a high momentum state, since the I. A. cross section [8] is proportional to the nucleon momentum distribution which becomes very small at high momenta.

A typical example of the first kind of experiments is given by the two body disintegration of H^2, He^3 and He^4 in the 3/2 3/2 resonance region. The influence of the MEC and IC on the energy dependence of the cross section for the reaction $H^2(\gamma p)n$ around the photoproduction threshold is shown in fig. 26 (51) Above 150 MeV the

total cross section for the same reaction, reported in fig. 27, shows
a large resonance peaked around 260 MeV, about 40 MeV lower than
the free N* centroid energy.

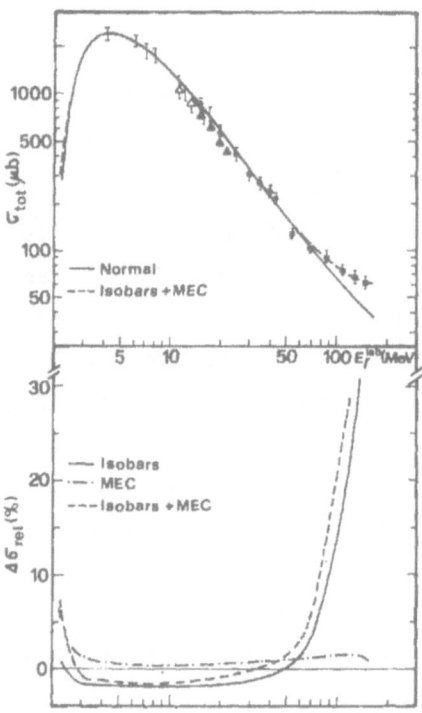

Fig. 26   Effect of the Meson Exchange Currents (MEC) and Isobar
          Configurations (I.C.) on the energy dependence of the
          deuteron photodisintegration cross section

This behaviour is not reproduced by the ordinary calculations which
neglect specific meson effects (Schiff curve). Since 1956 Wilson (52)
has explained this result by assuming a two body absorption process
where a pion is produced on one nucleon and riabsorbed by the other
we have

$$[9] \quad \sigma_d = \sigma(\pi)_{np} \, P \, \rho_{np} / \rho_\pi$$

where $\sigma(\pi)_{np}$ is the total cross section for production of mesons of any kind on the neutron and proton, P is the probability of riabsorption of the meson ($\sim$ 0.11), ρ_{np} the density of final states of the two nucleons and ρ_{π} the density of final states of the meson - - nucleon system. Close to the resonance mesons are produced in P waves, S waves being favoured at lower energies, but the phase space density factors in [9] enhance the S wave absorption.

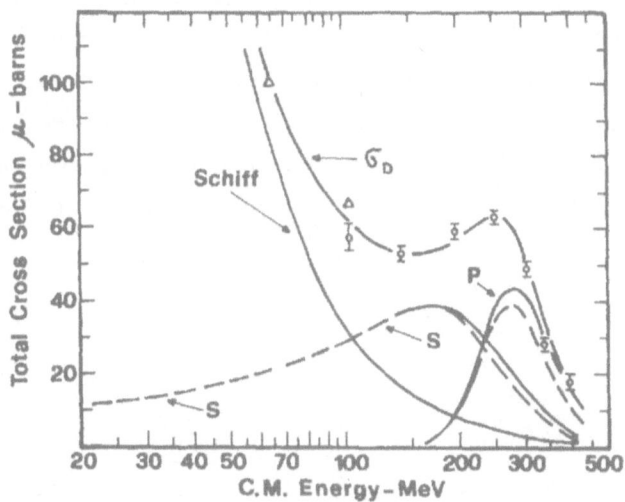

Fig. 27 Wilson (52) calculation of the total cross section of the
         disintegration of the deuteron. The curves marked S and P
         show the contribution which results from the reabsorption of
         S and P wave mesons, the curve marked Schiff shows the
         result of the calculation neglecting specific meson effects.

The final result, shown in fig. 27, is a resonant cross section shifted downward in energy by the S wave contribution in good agreement with the experimental data. The physical description of the high energy photodisintegration of the deuteron is of primary importance also for the investigation of "quasi deuteron" effects in nuclei. Systematic angular distribution measurements through the resonance region as well as more detailed calculations would therefore be highly desirable for a deeper understanding of these two body absorption mechanisms.

The cross sections for the reactions $He^3(\gamma\,d)$ and $He^4(\gamma\,t)$, measured in different laboratories(53)(54)(55) are plotted in figs. 28 and 29. Besides the poor agreement between the various data sets, the He^3 and He^4 two body photodisintegration cross sections exhibit an energy dependence very different from that observed in the deuteron. The two cross sections are in fact about two order of magnitude lower and do not show any clear resonant behaviour, even if a change of slope is evident above 150 MeV. The quenching of the effect of the N* when one goes from the deuterium to the helium case is probably related (55) to the presence in the final state of bound deuteron and triton. The cross section for the photodisintegration of He^3 and He^4 is roughly given by the deuteron photodisintegration cross section multiplied by the probability for the final deuteron and triton to remain bound.

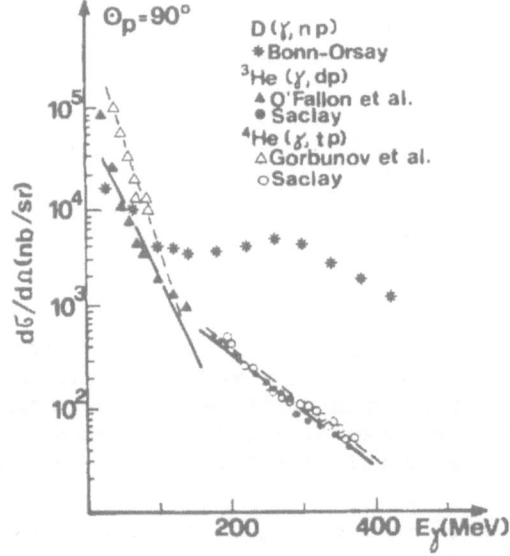

Fig. 28 Two body photodisintegration cross section for d, He^3, He^4 at 90°

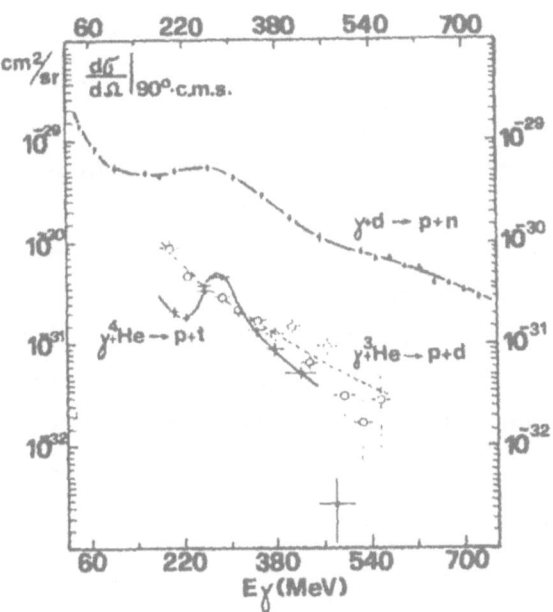

Fig. 29 Comparison of the experimental 90° c.m.s. differential
photodisintegration cross section in two bodies for d, He³,He⁴.
(O) He³ (Frascati (54)); (•) He⁴ (Frascati (54)) (∇) d (62)
(Δ) d (61)

A rough estimate of this last probability is given by the squared
electromagnetic form factor of the residual nuclei. With increasing
photon energies (or angle) the momentum transfered to the recoiling
nucleus increases and its probability to stay bound decreases washing
out any effect of the resonance. The cross section would probably show
again a resonant behaviour if all the final channels, including three
and four body photodisintegration, were added. Such experiments should
be soon possible using the new quasi monocromatic γ ray facilities.

An alternative approach to these problems, previously discussed,
is the selective investigation of photonuclear interactions with high
momentum nucleons. An interesting example of these experiments is
provided by the measurement of the $\gamma p \pi$ cross section in He⁴ (50)
In this experiment the initial momentum P of the nucleon, which is

equal in the I.A. to the momentum of the recoil mucleus, is determined
by the coincidence kinematics. Resonant pion production is therefore
studied on nucleons of defined momentum in the nucleus eliminating the
large averaging effect of the ground state momentum distribution. The
cross section is reported in fig. 30 as a function of the invariant
mass Q : at P = 50 MeV/c the data are fairly well reproduced by the
D.W.I.A. but at higher momenta (P $\geq$ 150 MeV/c) the resonance shows a
lower peak energy and a narrower width with respect to the quasi free
N* production. All these results, if confirmed, might support the
hypothesis of non negligible interactions between the N* and the rest
of the nucleus.

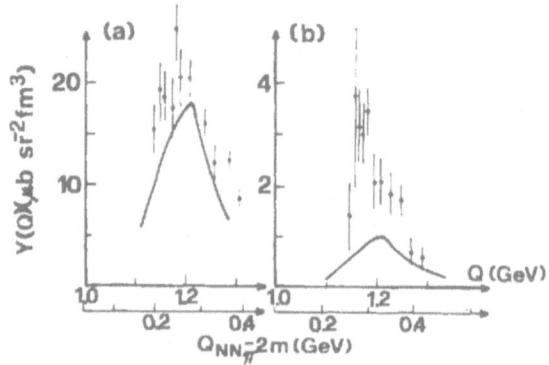

Fig. 30 Differential cross sections of the reaction $\gamma \, p \, \pi$ in He^4 as a
         function of the invariant mass   a) P = 50 MeV/c.
         b) P = 200 MeV/c. The curve shows the D.W.I.A. result.

VI - Sum rules

      No discussion on photonuclear reactions can end without mention
ing sum rules. A detailed description of the modern theoretical develop
ments in this field will be given in other lectures. I shall here
simply review the "old" sum rules and their comparison with the available experimental data. Sum rules have been developed following two
different lines:

A)   The T R K sum rule (56) gives the total absorption cross section, in the electric dipole approximation, following the technique used in atomic physics. The dipole oscillator strengths to the final states are summed up and the sum is evaluated, using closure, as the expectation value in the ground state of the double commutator $\left[\left[HD\right]\,D\right]$ between the nuclear Hamiltonian and the dipole operator D , namely

$$[10] \quad \sum = \int_0^\infty \sigma_{abs}^{E1}(E_\gamma)dE_\gamma = \frac{2\pi^2 e^2}{\hbar c}\left\{\left[\left[HD\right]D\right]\right\}_{oo} =$$

$$= -\frac{2\pi^2 e^2}{\hbar c}\left\{\left[\left[\sum_i \frac{P_i^2}{2m}\,D\right]D\right]\right\}_{oo} - \frac{m}{\hbar^2}\left\{\left[\left[\sum_{ij} V_{ij},D\right]D\right]\right\}_{oo}$$

If only the average central potential $V_o(r)$ is considered, the second term in $[10]$ vanishes and the first term corresponds to the "classic" sum rule $\sum_c = 60\ NZ/A$ mb MeV. Any nucleon – nucleon potential that does not commute with D may give a non zero contribution to the second commutator in $[10]$ (table IV). We can write in general

$$\sum = \sum_c (1 + k)$$

Any observed difference between the experimental integrated cross section and the classic sum rule $\sum_c$ may be interpreted as evidence of non negligible non central terms in the nuclear Hamiltonian. The experimental $\sum / \sum_c = 1 + k$ ratio, reported in fig. 31, gives an average $k \simeq 1$. Majorana exchange forces can account up to only about 0.4, the contribution of other non commuting interactions like for example the tensor part of nuclear forces (57) must therefore be accurately evaluated. We must also be aware that the upper integration limit of the experimental cross sections in fig.31 is of the order of 170 MeV while there are reasons to believe that a non negligible E1 strength lies at higher energies. Moreover the experimental k value includes also the non separable contribution of electric multipoles higher than E1. These ambiguities might be avoided following the GGT approach.

B)   The G G T (58) sum rule relates the Kramers – Kronig dispersion relation for the forward photon scattering amplitude from nuclei to the corresponding dispersion relation for the photon nucleon scattering amplitude. We have again

$$[11] \quad \sum = \int_{0}^{\mu} \sigma_{abs} \, (E) dE = \sum_{c} (1 + k)$$

$$k = \frac{A}{60NZ} \int_{\mu}^{\infty} \left(Z \, \sigma_{\pi}(p) + N \, \sigma_{\pi}(n) - \sigma(A) \right) \, dk$$

where $\sigma_{\pi}(p)$ and $\sigma_{\pi}(n)$ are the π production cross sections on the proton and the neutron respectively while $\sigma(A)$ is the total absorption cross section for the mass A nucleus.

TABLE IV

TERM	H	$\sum$ (mb. MeV) (63)
Central potential	$\sum\limits_{i=1}^{A} \dfrac{P_i^2}{2M} + V_o$	$60 \dfrac{NZ}{A}$
+		
Exchange (Majorana)	$\sum\limits_{i>j} x \, V_o^E \, P_{ij}^M$	60 NZ/A (1 + 0.8x)
+		
Spin – orbit	c 1 . s	60 NZ/A (1 + 0.8x)
+		
Nilsson – term	$- Dl^2$	60 NZ/A (1 + 0.8x)– negligible terms.
+		
Velocity dependent potential	$\sum\limits_{ij} \dfrac{\lambda}{M} (P^2 W(r) + W(r) P^2)$	60 NZ/A (1+0.8x+0.37)
+		
short range n – p correlations		60 NZ/A (1.67+0.78x + 0.06x^2)

x fraction of excange forces $\simeq$ 0.5

The integral in $\left[11\right]$ is now extended only up to the π meson production threshold μ and the cross section σ_{abs} includes all the electric multipoles. Nevertheless the numerical G G T estimate, reported in fig. 31 , gives k = 0.4 in desagreement with the data.

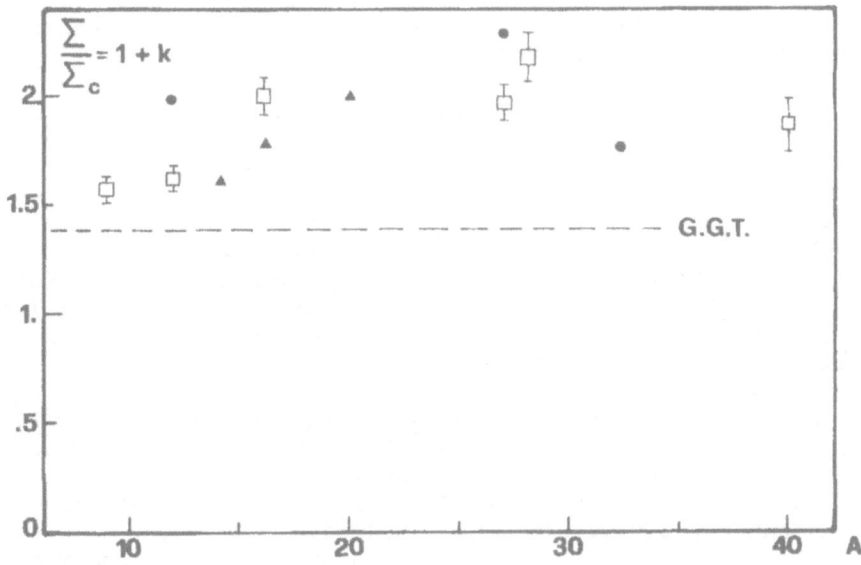

Fig. 31 The experimental Σ / Σ_c ratio as a function of the atomic mass number. ($\bullet$) γ nT Bishop et al (38); (∇)   Gorbunov et al. (60) ; ($\square$) Ziegler et al (3)   The dotted line gives the G G T sum rule.

The G G T calculation has been recently improved by Weise (59) to include the hadronic shadowing effect in the asymptotic value of σ(A) , leading to a much better agreement with the experiment.

R E F E R E N C E S

1) H.Hebach International School on Electro and Photonuclear Reactions Erice June 1976
2) H.Arenhövel International School on Electro and Photonuclear Reactions Erice June 1976

3) J.Ahrens, H.B.Eppler, H.Gimm, H.Gundrum, M.Kröning, P.Riehn, G.SitaRam, A.Zieger and B.Ziegler in Proceedings of the Int. Conference on Photonuclear Reactions and Applications Asilomar (1973) Edited by B.Berman p. 23.

4) A.Penfold and J.Leiss Phys. Rev. 114 (1959) 1332

5) J.L.Matthews and R.O.Owens Nucl. Instr. and Meth. 91 (1971) 37

6) G.P.Capitani, E.De Sanctis, S.Faini, C.Guaraldo, R.Scrimaglio, G.Ricco, M.Sanzone and A.Zucchiatti Lett. Nuovo Cimento 16(1976)453

7) J.S.Allen, P.Axel, A.O.Hanson, J.R.Harlan, R.A.Hoffswell, D.Jamnik, C.S.Robinson, D.C.Sutton, L.M.Young in Proceedings of the Int. Conference on Photonuclear Reactions and Applications Asilomar (1973) Edited by B.Berman p. 243

8) R.Malvano, C.Mancini and C.Schaerf Report LNF 67/48 Frascati (1967)

9) A.Bohr and B.R.Mottelson in Neutron Capture ray Spectroscopy (IAEA Vienna 1969)
M.Danos Nucl. Phys. 5 (1958) 23

10) J.Carver, D.C.Peaslee and R.B.Taylor Phys. Rev. 127 (1962) 2198

11) S.Fukuda and Y.Torizuka Phys. Rev. Lett. 29 (1972) 1109

12) Y.Torizuka, Y.Kojima, T.Saito, K.Itoh, A.Nakada, S.Mitsunobu, M.Nagao, K.Hosoyama, S.Fukuda and H.Miura Proceedings of the Int. Conference in Photonuclear Reactions and Applications Asilomar(1973) Edited by B.Berman p. 675
R.Satchler Phys. Reports 14C (1974) 98

13) M.Marangoni, P.L.Ottaviani and A.M.Saruis Report CNEN RT/FI (1976)

14) D.E.Frederick, R.J.Steward and R.C.Morrison Phys. Rev. 186(1969)992

15) J.W.Jury, J.S.Hewitt and K.G.McNeill Can. Journ. Phys. 48(1970)1635

16) S.S.Hanna, H.F.Glavish, R.Avida, J.R.Calarco, E.Kuhlmann and R.Lacanna Phys. Rev. Lett. 32 (1974) 114

17) B.Dreher, J.Friedrick, K.Merle, H.Rothhass and G.Lührs Nucl. Phys. 235A (1974) 219

18) G.Manuzio, G.Ricco, M.Sanzone and L.Ferrero Nucl. Phys.133A(1969)225
M.Sanzone, G.Ricco, S.Costa and L.Ferrero Nucl. Phys.153A(1970)401
E.Mancini, G.Ricco, M.Sanzone, S.Costa and L.Ferrero
Nuovo Cimento 15 (1973) 705

19) J.L.Matthews, D.J.Findlay, S.N.Gardiner and R.O.Owens Phys. Lett. 46B (1973) 186 and preprint Glasgow University (1976)

20) S.Shkyarevskii Sov. Phys. Jetp. 9 (1959) 1057

21) M.Fink, H.Hebach and H.Kümmel Nucl. Phys. 186(1972) 353

22) H.Schier and B.Schoch Nucl. Phys. 229A (1974) 93 and Lettere Nuovo Cimento 12(1975) 334

23) H.Miller, W.Buss and J.Rawlins Nucl. Phys. 163A (1971) 637

24) W.Weise and H.Huber Nucl. Phys. 162A (1971) 330

25) Ciofi degli Atti in The Nuclear Many Body Problem Editrice Compositori Bologna (1973) p. 365

26) A.Malecki and P.Picchi in Proceedings of the Int. Conference on Photonuclear Reactions and Applications Asilomar (1973) Edited by B.Berman p. 987

27) G.E. Brown Nucl. Phys. 57 (1964) 339

28) H.Hebach, A.Wortberg and M.Gari Nucl. Phys. 267A (1976) 425

29) M.Carbone, R.Cenni, R.Malvano and A.Molinari Nuovo Cimento 27 (1975) 60

30) O.Borello, F.Ferrero, R.Malvano and A.Molinari Nucl.Phys.31(1962)53

31) G.Jacob and Th.A.Maris Rev. Mod. Phys. 14 (1973) 6

32) A.N.Gorbunov and V.A.Osipova Sov. Phys. Jetp. 16 (1969) 27

33) G.G.Taran Sov. J. Nucl. Phys. 7 (1968) 301

34) J.S.Levinger Phys. Rev. 84 (1951) 43

35) J.Garvey, B.H.Patrick, J.C.Rutherglen and I.L.Smith Nucl. Phys.
    70 (1965) 241
    I.L.Smith, J.Garvey, J.C.Rutherglen and G.R.Brookes Nucl. Phys.
    1B (1967) 483

36) A.C.Odian, P.C.Stein, A.Wattemberg, B.T.Feld and R.Weinstein
    Phys. Rev. 102 (1956) 837

37) M.A.Barton and J.H.Smith Phys. Rev. 95 (1954) 573

38) G.Bishop, S.Costa, S.Ferroni, R.Malvano and G.Ricco Nuovo Cimento
    42 (1966) 1

39) J.R.Van Hise, R.A.Meyer and J.P.Hummel Phys Rev. 139 (1965) 554

40) K.Gottfried Nucl. Phys. 5 (1958) 557

41) H.Hartmann, H.Hoffmann, B.Mecking, G.Nöldeke in Proceedings of
    the Int. Conference on Photonuclear Reactions and Applications
    Asilomar (1973) p. 967

42) D.M.Lee, R.C.Minehart, S.E.Sobottka and K.O.Ziock Nucl. Phys.
    182A (1972) 20

43) T.W.Donnelly in Comptes Rendus of the Saclay Meeting on Electron
    Scattering at Intermediate Energies Saclay September 1975 p. 299

44) W.Bertozzi International School on Electro and Photonuclear
    Reactions Erice June 1976

45) A.Nagl and H.Überal Phys. Lett. 63B (1976) 291

46) H.W.Baer, J.A.Bistirlich, N. de Botton, S.Cooper, K.M.Crowe,
    P.Truöl and J.D.Vergados Phys. Rev. 12C (1975) 921

47) F.Kelly, L.Mc Donald and H.Überall Nucl. Phys. 139A (1969) 329

48) J.Ahrens, H.B.Eppler, H.Gimm, M.Kröning, P.Riehn, A.Zieger and
    B.Ziegler Phys. Lett. 52B (1974) 43

49) J.M.Laget Nucl. Phys. 194A (1972) 81
    M.Lax and H.Feshbach Phys. Rev. 81 (1951) 189

50) P.E.Argan, G.Audit, N.de Botton, J.M.Laget, J.Martin, C.Schuhl
    and G.Tamas Phys. Rev. Lett. 29 (1972) 1191

51) W.Fabian and H.Arenhövel Nucl. Phys. 258A (1976) 461

52) R.W.Wilson Phys. Rev. 104 (1956) 218

53) S.E.Kiergan,A.O.Hanson and L.J.Koester Phys. Rev. 8C (1973) 431

54) P.Picozza, C.Schaerf, R.Scrimaglio, G.Goggi, A.Piazzoli and
    D.Scannicchio Nucl. Phys. 157A (1970) 190

55) C.Tzara in Proceedings of the Int. Conference on Photonuclear
    Reactions and Applications Asilomar (1973) Edited by B.Berman
    p. 105

56) J.V.Noble Ann. of Phys. 67 (1971) 98

57) A.Arima, G.E.Brown, H.Hyuga and M.Ichimura Nucl. Phys. 205A(1973)27

58) M.Gell Mann, M.L.Goldberger and W.Thirring Phys. Rev. 96(1954)1612

59) W.Weise International School on Electro and Photonuclear Reactions
    Erice June 1976

60) A.N.Gorbunov, V.A.Dubrovina, V.A.Osipova, V.S.Silaeva and
    P.Cerenkov Sov. Phys. Jetp. 15 (1962) 520

61) R.Ching and C.Schaerf Phys. Rev. 141 (1966) 1320

62) C.Keck and A.V.Tollestrup Phys. Rev. 101 (1956) 360

63) L.Dohnert and O.Rojo Phys. Rev. <u>136B</u> (1964) 396

GIANT MULTIPOLE RESONANCES

Stanley S. Hanna

Department of Physics, Stanford University, Stanford, California 94305, USA[*]

I.   Introduction

In this series of lectures I want to survey and discuss the information on the
giant multipole resonances in nuclei. Since the giant electric dipole (El) resonance is thoroughly covered in the lectures of R. Bergère, I will emphasize the
other multipole resonances. However, as an introduction to the subject I would
also like to discuss some of the new developments which bear on the configurations
of the giant El resonances.

It is instructive to classify the giant multipole resonances according to the
basic oscillations of a nucleus [1,2], as is done in Fig. 1. The electric oscillations

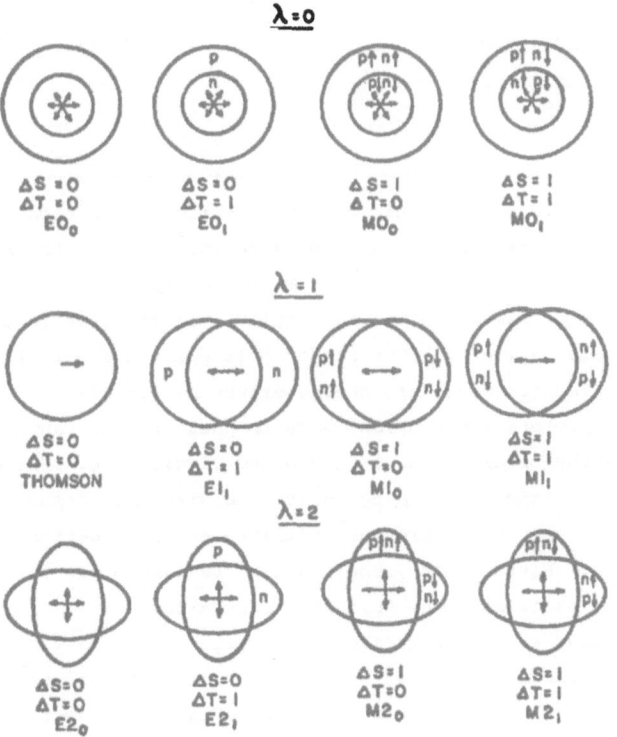

Fig. 1.   The giant multipole oscillations of a nucleus.

Table I. Quantum numbers and excitation energies of giant multipole resonances.

λ	ΔS	ΔT	π	$0^+ \to J^\pi$	Harm. Osc. Type	$\Delta n(\hbar\omega)$	Shell Model $\Delta n(\hbar\omega)$	$E(MeV/A^{1/3})$	Hydr. Model $E_{th}(MeV/A^{1/3})$	$E_{ex}(MeV/A^{1/3})$
0	0	0	+	0^+	EO_o	0,2	$\sqrt{2}$	60		
0	0	1	+	0^+	EO_1	0,2	4	170	151	173
0	1	0	-	0^-	MO_o	1				
0	1	1	-	0^-	MO_1	1				
1	0	0	-		Thomson					
1	0	1	-	1^-	$E1_1$	1	2	80	70	80
1	1	0	+	1^+	$M1_o$	0				
1	1	1	+	1^+	$M1_1$	0				
2	0	0	+	2^+	$E2_o$	0,2	$\sqrt{2}$	60		
2	0	1	+	2^+	$E2_1$	0,2	3	130	112	128
2	1	0	-	2^-	$M2_o$	1				
2	1	1	-	$.2^-$	$M2_1$	1				

are shown on the left and the magnetic ones on the right. In each case the isoscalar modes are shown in the left column and the isovector modes in the right column. The different multipoles are arranged in rows. The monopole mode is a spherically symmetric oscillation; the dipole vibration is axially symmetric; while the quadrupole oscillation has biaxial symmetry. The higher multipoles are not illustrated.

The isoscalar electric modes are characterized by oscillations of the nucleus as a whole in which protons and neutrons move in phase without any differentiation of spin. These are the oscillations of a charged liquid drop. The dipole mode corresponds to a translation and can be identified with the Thomson scattering. In the isovector electric modes the protons oscillate against neutrons, again without spin differentiation. Historically, the giant E1 resonance was first identified with this collective oscillation by Goldhaber and Teller.

The magnetic modes are characterized by oscillations involving spin rather than charge. In the isoscalar vibrations protons and neutrons with spin up oscillate against protons and neutrons with spin down, while in the isovector modes, protons with spin up oscillate against neutrons with spin up and neutrons with spin down against protons with spin down.

It is well-known that these oscillations can also be described by basic transitions in the microscopic shell model [3]. The quantum numbers associated with these transitions are given in Table I, along with the characteristic excitations in a

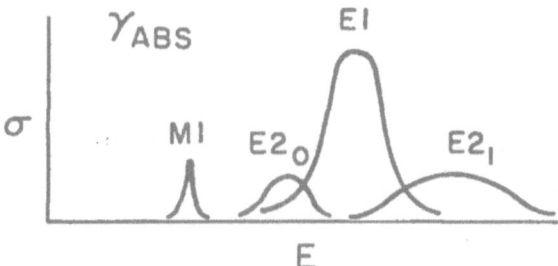

Fig. 2.   Idealized gamma-ray absorption curves showing the M1,
E1, E2$_o$(isoscalar, and E2$_1$(isovector) resonances.

harmonic oscillation model. Actual shell model calculations of the excitation energies are shown in some cases and compared with the results of the hydrodynamical
model. The relative locations of the dipole and quadrupole resonances are shown by
the idealized gamma absorption curves in Fig. 2.

II.   The Giant E1 Resonance

The giant E1 resonance has long been the object of intensive study. The three
important properties which characterize it are its systematic occurrence in all
nuclei, its great strength, and its localized nature [4].

(1)   In the medium and heavy nuclei the E1 resonance occurs at an energy of
about $77/A^{1/3}$ MeV. However, in the light nuclei, below ^{40}Ca, the energy of the
resonance falls off as shown in Fig. 3. We note here that if the giant E2 resonance
maintains a position of $63/A^{1/3}$ MeV in the light nuclei, then the E2 resonance will
cross the E1 resonance and lie above it in the lightest nuclei (as indicated by the
straight dashed line in Fig. 3).

(2)   The giant E1 resonance "exhausts" the classical E1 sum rule 60 $NZ/A^{1/3}$
mb·MeV. Actually it is now known that the total strength exceeds this sum rule and
this has been the object of recent study. However, this phenomenon will not concern
us here.

(3)   Perhaps the most impressive feature of the E1 resonance is its localized
nature, despite the fact that it occurs in the continuum where many decay channels
are open. From the lightest to the heaviest nuclei the width is given by $\Gamma/E \simeq 1/5$,
with several notable exceptions which can be attributed to the following causes:

(i)   nuclear deformation (well established),

(ii)   isospin splitting (well established in certain nuclei),

(iii)   excitation of deep hole states (not yet well established).

There are many other interesting and significant properties of the E1 resonance
which will not be discussed here [5,6]. Instead, we turn to the new information
that has been obtained on the configurations of the E1 resonances from a study of
the $(\vec{p},\gamma)$ reaction.

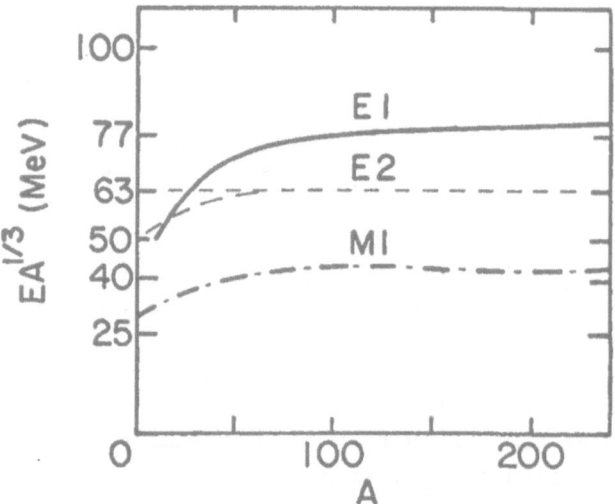

Fig. 3. Location of giant resonances. See text for
discussion of M1 and E2 resonances.

II.1. Configurations of the E1 resonance

The particle-hole model has quite successfully described many of the dominant features of the giant E1 resonance (GDR) in nuclei and in its simplest form provides naturally for the characteristic single-particle transitions of the type $\ell_j \rightarrow (\ell+1)_{j+1}$ (no spin flip) which carry large E1 strengths [3,7]. However, in the region of the GDR finer structure in the total cross section is often observed in capture reactions such as (p,γ_0) and in the inverse photonuclear reactions. A question that immediately arises is whether or not this structure is indicative of a change in nuclear configuration as one passes through the GDR, as has often been suggested. Not only have changes in the (1p-1h) configurations been proposed, but (2p-2h) or (3p-3h) configurations have also been invoked to explain the observed structure.

In contrast to the idea of a changing configuration is the remarkable constancy of the angular distributions observed throughout the GDR [8], including such well-defined levels as the E1 analogue states [9] which are often assigned to configurations different from the main GDR. To improve our understanding of the configurations in a GDR it is very helpful to know the relative phases as well as the amplitudes of the reaction matrix elements associated with the channels which form a GDR. For the proton channel, for example, these can be found only by combining angular distribution measurements from the polarized reaction $(\vec{p},\gamma_0)$ with unpolarized angular distribution data. In many cases unique solutions for the reaction amplitudes are obtained from such measurements.

We consider first the unpolarized (p,γ_0) experiment. It is well known that the cross-section can be expanded as follows

$$\sigma(E,\theta) = A_o(E)\left[1 + \sum_1^4 a_k(E)P_k(\theta)\right] \tag{1}$$

where $A_o(E)$ gives the resonance strength and the relationship between the coefficients a_k and the multipolarity of the radiation is given in Table II.

Table II. The dependence of the angular complexity of $\sigma(E,\theta)$ and $A(E,\theta)\sigma(E,\theta)$ on the multipolarity of the radiation.

Radiation	Unpolarized $\sigma(E,\theta)$	Polarized $A(E,\theta)\sigma(E,\theta)$
E1 or M1	a_2	b_2
E2	a_2, a_4	b_2, b_4
(E1,M1)	a_1	b_1
(E1,E2)	a_1, a_3	b_1, b_3
(M1,E2)	a_2	b_2

In all cases that have been investigated the M1 and E2 contribution may be important in $a_1(E)$, $a_3(E)$, and $a_4(E)$, but can be neglected in $a_2(E)$. Thus, we may fit the data with Eq. (1) and then isolate

$$\sigma(E,\theta) = A_o(E)[1 + a_2(E)P_2(\theta)] \tag{2}$$

where $A_o(E)$ and $a_2(E)$ carry the information on the E1 resonance in the (p,γ) reaction.

We now consider the polarized experiment, in which the gamma ray yield is sensitive to the degree of polarization perpendicular to the reaction plane. It is convenient to measure the analyzing power

$$A(E,\theta) = \frac{\sigma_\uparrow - \sigma_\downarrow}{\sigma_\uparrow + \sigma_\downarrow} \tag{3}$$

where $\sigma_\uparrow$ and $\sigma_\downarrow$ are the γ yields with p spin up and spin down, respectively.

The analyzing power can be expanded as follows:

$$A(E,\theta)\sigma(E,\theta) = A_o(E)\left[\sum_1^4 b_k(E)P_1^1(\theta)\right] \tag{4}$$

where the relationship between the coefficients b_k and the multipolarity of the radiation is given in Table II. Again, if we are interested in the E1 strength, we may neglect the M1 and E2 contribution to b_2 and extract from the polarization measurements

$$A(E,\theta)\sigma(E,\theta) = A_o(E)b_2(E)P_2^1(\theta) \tag{5}$$

where $b_2(E)$ carries the information on the E1 resonance. Thus, we obtain the three quantities

$$A_o(E), \quad a_2(E), \quad \text{and } b_2(E)$$

measured over the El giant resonance.

We may now pass from these three quantities to the configurations in the proton channel of the GDR. The proton configurations can, of course, be expressed in any desired coupling scheme. We may indicate this transformation formally by

$$
\begin{bmatrix}
A_o(E) \\
a_2(E) \\
b_2(E)
\end{bmatrix}
\longrightarrow
\begin{bmatrix}
\text{proton configuration} \\
\text{in jj, LS, or other} \\
\text{coupling scheme}
\end{bmatrix}
\tag{6}
$$

We illustrate this transformation below in Section II.2.

Finally, we must relate the configuration of the proton channel to the giant resonance itself. This is the task of theory, but it is clear that the observed proton configurations will severely restrict the allowed configurations of the GDR. We discuss a theoretical treatment of $^{16}O(\gamma, p_o)^{15}N$ below in Section II.3.

II.2. The El resonance in ^{16}O

In Fig. 4 are shown the data for $A_o(E)$, $a_1(E)$, $a_2(E)$ and $a_3(E)$ taken from the unpolarized experiment of O'Connell et al. [10]. It can be seen that the

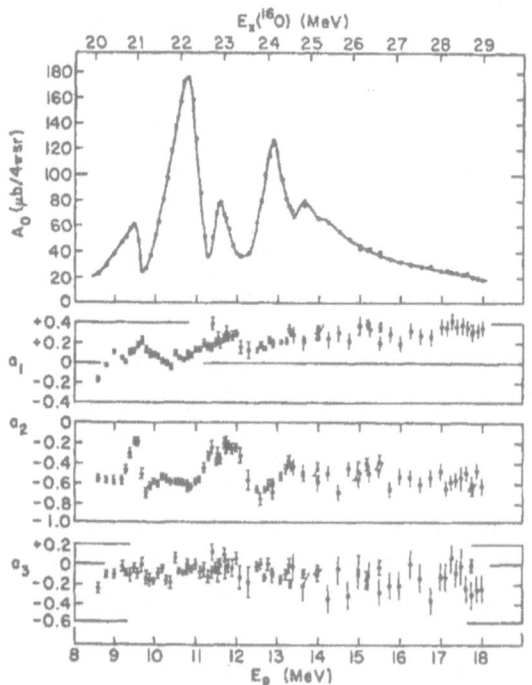

Fig. 4. Total yield and angular distribution coefficients in $^{15}N(p,\gamma_o)^{16}O$. Ref. 10.

coefficients a_1 and a_3 indicate the presence of E2 and possibly M1 radiation, but we now confine our attention to A_0 and a_2, as discussed above. It is seen that $a_2(E)$ is quite constant over the resonance at the "dipole value" of -0.6 except in three regions (E_x = 21, 23, and 24.5 MeV) where there is also marked fine structure in the total cross section $A_0(E)$.

It has been remarked before [11] that this structure in A_0 correlates well with resonances seen in the (d,γ), $(^3He,\gamma)$ and (α,γ) reactions which might indicate (n-particle, n-hole) configurations in ^{16}O. It is in fact possible to decompose the GDR empirically into two basic resonances (presumably the predicted 1p-1h states) at E_x = 22 and 24 MeV and three sharper levels at 21, 22.5, and 24.5 MeV (presumably np-nh states) which interfere in a characteristic manner with the two basic states [see Ref. 10]. It has also been shown [12] that this model can account for the structure seen in the a_2 curve (Fig. 4). Also, Shakin and Wang [13] have obtained theoretical agreement with the structure in A_0 using only 3p-3h states. This picture of interfering states is an attractive one and, although not firmly established, we shall adopt it in our discussion.

We now pass from the quantities A_0 and a_2 to the amplitudes of the proton channel in jj representation [see Eq. (6)] in the reaction $^{15}N(p,\gamma)^{16}O$. Only incident proton waves with (ℓ = 0, j = 1/2) and (ℓ = 2, j = 3/2) can combine with the $1/2^-$ ground state of ^{15}N to form a 1^- state in ^{16}O. Thus for E1 radiation we have the transition scheme $1/2^-(s_{1/2}, d_{3/2})1^-(E1)0^+$ which determines the angular distribution. The corresponding matrix elements may be written as

$$|s_{1/2}|e^{i\phi_s} \quad \text{and} \quad |d_{3/2}|e^{i\phi_d},$$

where $s_{1/2}$ and $d_{3/2}$ are the real amplitudes and ϕ_s and ϕ_d the real phases. From a straightforward calculation we obtain

$$a_2 = -0.5d_{3/2}^2 + \sqrt{2}|s_{1/2}||d_{3/2}|\cos(\phi_d - \phi_s) \tag{7}$$

$$1 = s_{1/2}^2 + d_{3/2}^2. \tag{8}$$

The normalization (8) eliminates A_0 from further consideration. Strictly speaking this analysis is valid for a direct or semi-direct capture process which is believed to dominate this reaction. In fact, as suggested above, the presence of "compound-like" amplitudes in the analysis may account for the observed interference effects.

We can now appreciate the problem of having only unpolarized results available: there are three unknown quantities, namely $s_{1/2}$, $d_{3/2}$ and $\phi_d - \phi_s$, but only two relationships (7) and (8) to determine them. Of course, the experimental value of a_2 severely restricts the amplitudes and phases but does not uniquely determine them. It is possible to plot the allowed solutions as curves in amplitude-phase space as done in Fig. 5. Since the expressions (7) and (8) are quadratic there are two equally acceptable solutions which are labeled I and II in Fig. 5. The solutions are shown for a_2 = -0.5 which is representative of the value throughout the whole

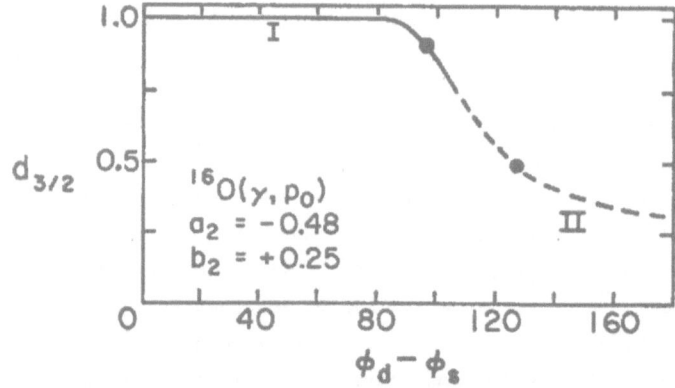

Fig. 5. Allowed solutions I (solid curve) and II (dashed curve)
for the p_o channel of $^{16}O(\gamma,p_o)^{15}N$. The curves give
values of $d^2_{3/2}$, $s^2_{1/2}$ (obtained with the help of Eq. 8)
and $\cos(\phi_d-\phi_s)$ allowed by $a_2 = -0.48$. For clarity the
pairing of solutions I with solutions II is not indicated in this plot, except for the dots which show the two
solutions produced by the additional condition
$b_2 = +0.25$.

GDR, except for the regions where there is fine structure. We note that there is
one solution (I) which is predominantly d-wave while the other (II) is predominantly
s-wave. The simple particle-hole model would of course prefer the former solution
[14-16].

We now turn to the polarized measurements on $^{15}N(p,\gamma)^{16}O$ carried out at
Stanford to see what light they can shed on the proton amplitudes. If the analyzing
power is measured as a function of angle at each energy then the quantity $b_2(E)$
can be obtained from Eq. (5). This new quantity can then be expressed in terms of
the amplitudes and phases

$$b_2 = \sqrt{2}/2|s_{1/2}||d_{3/2}|\sin(\phi_d-\phi_s) \qquad (9)$$

which gives a third relationship to go along with Eqs. (7) and (8). Thus, we have
three equations and three unknowns and unique solutions (I and II) can be obtained.

The measurements were carried out with a polarized beam in the setup shown in
Fig. 6. The spin direction for the protons could be set either up or down by select-
ing the appropriate rf transition in the polarized ion source. At a given energy
and angle the analyzing power $A(E,\theta)$, Eq. (3), was determined from measurements
made by frequently alternating runs with proton spin up with runs with proton spin
down. The values obtained for b_2 (see Eq. (5)] are shown in Fig. 7 along with the
curves for a_1, a_2 and the total yield A_o.

Throughout the main part of the GDR we see that b_2 is fairly constant at a
value of about 0.25. The constancy of both a_2 and b_2 means that the configuration in the proton channel remains constant throughout the GDR no matter what is
happening to the configuration of the GDR itself. This is a very remarkable result.
If we now impose the added condition Eq. (9) and adopt $b_2 = 0.25$ as representative

of the entire GDR we obtain the unique solutions I and II shown by the dots in Fig. 5. These are the characteristic solutions of the proton channel of the GDR.

It is of course interesting to see what causes the fluctuations in the coefficient a_2. This can be determined by obtaining the solutions at each experimental point. These solutions are shown in Fig. 7. The polarization results show that the

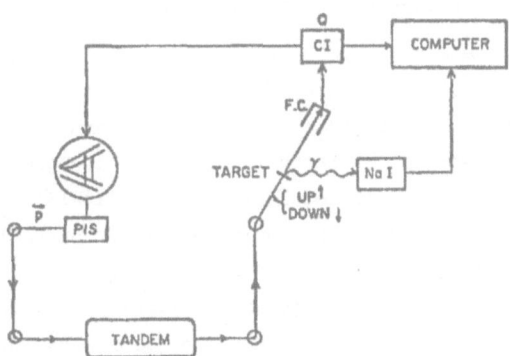

Fig. 6. Experimental arrangement used for making measurements with a polarized beam.

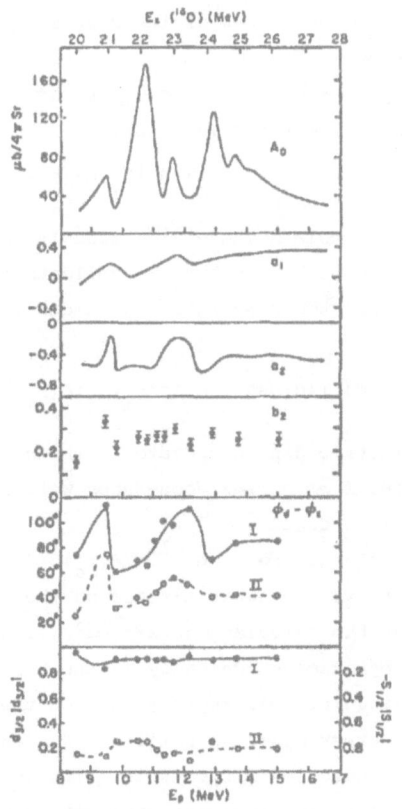

Fig. 7. Summary of E1 information on the GDR in $^{15}N(p,\gamma_0)^{16}O$. The two solutions for the proton channel are indicated by I and II.

fluctuations in a_2 are caused almost entirely by fluctuations in the phase difference $\phi_d - \phi_s$ rather than in the $s_{1/2}$ and $d_{3/2}$ amplitudes. It will be interesting to see if refinements in the theories of the GDR in ^{16}O can account for this phenomenon.

II.3. Analysis of the El resonance in $^{16}O(\gamma,p_o)^{15}N$

As we have seen, the giant dipole resonance of ^{16}O exhibits two dominant peaks (see Fig. 4) at excitation energies of 22.3 and 24.4 MeV. These two peaks carry a major part of the El strength and have been interpreted as collective single particle-hole excitations generated from a particle-hole interaction acting on unperturbed single-particle shell-model excitations [14-16]. In terms of this model the two peaks are predicted to have quite different particle-hole configurations, being dominantly $d_{5/2}p_{3/2}^{-1}$ at 22.3 MeV and $d_{3/2}p_{3/2}^{-1}$ at 24.4 MeV. On the other hand, we have seen that the angular-distribution and polarization measurements in the $^{15}N(p,\gamma_o)^{16}O$ reaction show that the $s_{1/2}$ and $d_{3/2}$ proton-capture matrix elements (the only ones allowed for El radiation by angular momentum and parity conservation) have remarkably constant relative amplitudes over both peaks. The following calculation was made to see if the simple shell-model description can account for such constancy.

The matrix elements $T_{\ell j}$ were determined by use of the doorway-state model of Feshbach, Kerman and Lemmer [17] which gives [13]:

$$T_{\ell j} = <\ell j|D_\gamma|0> + \sum_k <\ell j|V_{ph}|d_k><d_k|D_\gamma|0>(E - E_k + i\tfrac{1}{2}\Gamma_k)^{-1} \qquad (10)$$

The continuum nucleon and the hole state of the mass 15 target nucleus are described by $|\ell j>$. The doorway states $|d_k>$ are the two collective particle-hole configurations and E_k and Γ_k are their energies and widths. The particle-hole interaction V_{ph} was taken as

$$V_{ij} = -584.1(0.865 + 0.135\sigma_i \cdot \sigma_j)\delta(\vec{r}_i - \vec{r}_j).$$

The quantity D_γ is the electric dipole operator. The unperturbed single particle wave functions were generated from a real Wood-Saxon well adjusted to reproduce correctly the single-particle energies.

The results of the calculation for the $^{15}N(p,\gamma_o)^{16}O$ reaction are compared with the results discussed above (somewhat altered on the basis of new data) in Fig. 8. It is apparent that the calculations are consistent with Solution I and are able to reproduce the approximate constancy in the $s_{1/2}$ and $d_{3/2}$ amplitudes. Even the phase difference is quite well reproduced. On the right of Fig. 8 the comparison is made with the experimental quantities a_2 and b_2. It is seen that the remarkable constancy of these coefficients is quite well reproduced.

The success of this calculation indicates that it is quite possible to account for the experimentally observed properties of the GDR in ^{16}O within the framework

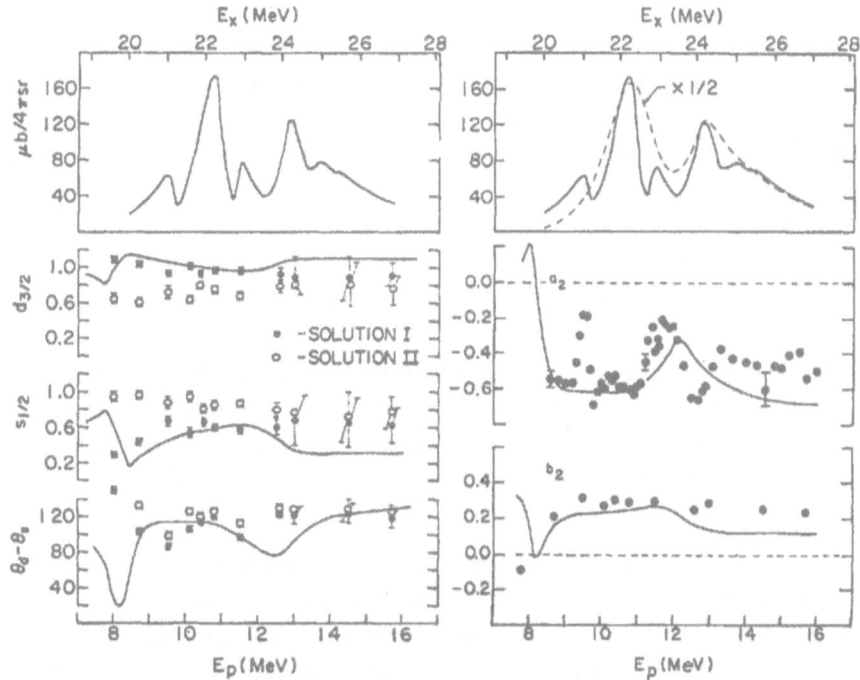

Fig. 8. Experimental data and theoretical fits for $^{15}N(p,\gamma_o)^{16}O$. The solid curves in the upper part of the figures are the experimental cross sections. The remaining solid curves are theoretical fits generated from Eq. (10). The broken curve (upper right) is the cross section generated from Eq. (10).

of the simple p-h model. However, we note that in this model the basic splitting of the GDR arises from a spin-orbit splitting since the dominant configurations are $p_{3/2}^{-1}d_{5/2}$ and $p_{3/2}^{-1}d_{3/2}$. We emphasize that the calculation does not uniquely establish such a spin-orbit splitting. It is still possible that the splitting arises from some other mechanism such as a deformation of the excited ^{16}O nucleus, or from interference with more complex configurations (in much the same way that the finer structure might arise from interference with np-nh configurations, see above). More complete data on the other particle channels of the GDR will contribute to our understanding of this problem.

III. The Giant M1 Resonance

Information on the giant M1 strength is now rather extensive and exists all the way from mass 6 to 208. The methods that have been used to study the M1 resonance can be summarized as follows:

(1) Capture reactions (X,γ) where X stands for a nucleon or nucleus. Early work was not directed specifically at locating and studying giant M1 strength. In

recent years the work, principally at Stanford, Argonne and Orsay, has investigated the M1 strength of $T_>$ and $T_{>>}$ levels of the light nuclei [5]. Some levels have also been studied by reactions of the type $(X, Y\gamma)$.

(2) Gamma-ray fluorescence (γ, γ'). These investigations of M1 strength, represented by the early studies at the National Bureau of Standards [18] and Illinois [19] are now being extensively pursued.

(3) Inelastic electron scattering at 180°. The use of 180° scattering to sort out magnetic from electric multipoles was pioneered at Stanford [20] and has recently been effectively continued at the Naval Research Laboratory [21] and at Darmstadt [22].

(4) The photoneutron process (γ, n) has been used at Livermore [23], Argonne [24], and Harwell [25] to give valuable information above the neutron threshold in heavy nuclei. Information comes also from the inverse (n, γ) reaction [26].

III.1. General properties

The basic M1 excitations are shown in Fig. 1; the isoscalar mode is a spin oscillation, while the isovector mode is a spin-isospin oscillation. As indicated in Table I the shell-model description involves non-parity changing excitations within a shell since the M1 matrix element vanishes for excitations to higher oscillator shells.

The basic magnetic dipole operator is given by

$$\mu = \mu_v + \mu_s = \frac{1}{2}(\mu_- + \frac{1}{2})\left[\sum_1^A \sigma_i^3 \tau_i^3 - \frac{1}{2}\sum_1^A j_i^3 \tau_i^3\right]_v$$
$$+ \frac{1}{2}(\mu_+ + \frac{1}{2})\left[\sum_1^A \sigma_i^3 + \frac{1}{2}\sum_1^A j_i^3\right]_s \tag{11}$$

where σ_i^3, j_i^3, and τ_i^3 are the third components of the spin, total angular momentum, and isospin operators, respectively, of the ith nucleon, $\mu_- = \mu_n - \mu_p = -4.7$, and $\mu_+ = \mu_n + \mu_p = 0.88$. Since the terms in $j_i^3 \tau_i^3$ and j_i^3 do not induce transitions in the j-j coupling model and since the terms in $\sigma_i^3 \tau_i^3$ and σ_i^3 are of comparable size, the fact that $\mu_- \gg \mu_+$ leads to the result that the isovector operator μ_v is usually much larger than the isoscalar operator μ_s (generalization of the Morpurgo rule). This result is nicely illustrated in ^{12}C where the gamma-ray excitation of the 12.7-MeV state (T = 0) is only about 1% of the 15.1-MeV state (T = 1). Thus, the giant isoscalar M1 resonance usually does not represent a significant gamma-ray absorption in nuclei.

If we consider only the isovector M1 absorption, we can classify the M1 giant resonances according to isospin (as can also be done for the isovector E1 resonances). The classification for self-conjugate, conjugate, and non-conjugate nuclei is shown schematically in Fig. 9.

A complete sum-rule treatment of the isospin geometry of the isovector M1

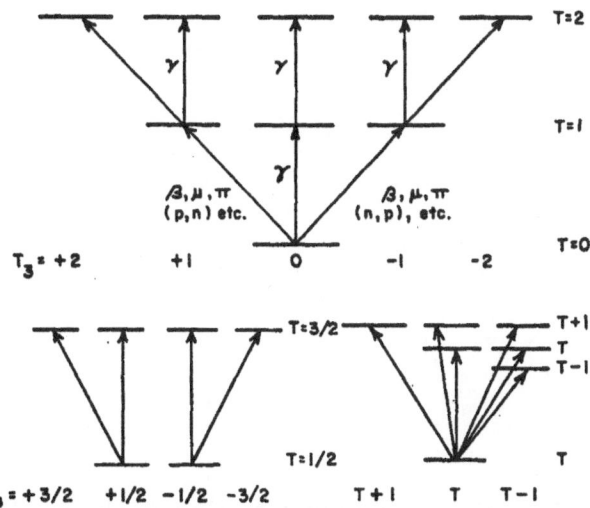

Fig. 9. Isospin classification of isovector excitations
in self-conjugate, conjugate, and non-conjugate
nuclei.

excitations has been given by Lipparini et al. [27].

In addition to studying the distribution of M1 strength, a basic objective is
of course to test the validity of the concept of isospin itself: the charge independence of the nuclear force. The pursuit of this goal leads to several specific
investigations such as the following:

(1)   Systematic study of Coulomb energies in nuclei.

(2)   Study of states of a given isospin $(T_>)$ as simple shell-model states.

(3)   Investigation of the interaction of a $T_>$ state with the background of
$T_<$ states. This investigation leads to a study of isospin mixing and the isospin
forbidden particle decays from the $T_>$ state.

(4)   Study of the allowed (and forbidden) electromagnetic transitions between
states of good isospin. Since the electromagnetic operator is known, this study
provides a very sensitive and basic test of the concept of isospin.

(5)   Comparison of the β-decays with the analogue γ-decays between states of
given isospin. Since the β-decay operator and electromagnetic operator are closely
related this comparison provides a sensitive test of the nuclear wave functions.

In the light nuclei many of the decay channels of the M1 resonances are closed
by isospin conservation. Three examples are shown in Fig. 10. A study of these
forbidden decay channels can provide important information on the amount of isospin
impurity in the states. In simple cases at least it is then possible to determine
whether the Coulomb force is sufficient to account for the observed impurity or
whether isospin breaking is required.

The important gamma-ray selection rules can be stated by writing the matrix

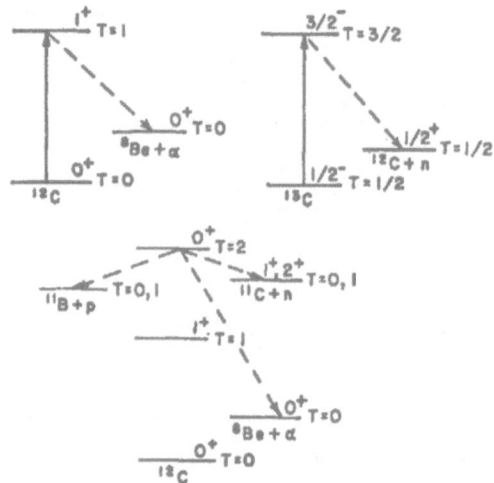

Fig. 10. Examples of isospin-forbidden decay
channels of M1 levels.

element of the M1 operator, Eq. (11), in the following form [28]:

$$M(T_3) = M_s + T_3 M_v \qquad \Delta T = 0$$

$$M(T_3) = \sqrt{(T_i^2 - T_3^2)} M_v \qquad \Delta T = \pm 1$$

where M_s and M_v are the scalar and vector matrix elements, $M(T_3)$ is the matrix element of the transition in the nucleus with isospin component T_3, and T_i is the isospin of the initial state.

From these selection rules we obtain the following basic electromagnetic selection rules:

(1) There should be no $\Delta T \geqslant 2$ transitions in nuclei,

(2) Mirror transitions with $\Delta T = \pm 1$ in conjugate nuclei should have identical transition probabilities.

(3) In nuclei with different values of T_3 the branching ratios for $\Delta T = \pm 1$ decays of analogue states should be identical.

We now turn to the comparison of the allowed β and γ analogue decays for which $\Delta T = 1$. It can be shown [5] that the reduced Gamow-Teller matrix element $B(GT)$ and the reduced M1 matrix element $B(M1)$ are connected by the relation

$$B(M1) = \left[A \; 1 + 0.11 \frac{<|\Sigma \ell|>}{<|\Sigma s|>} \right]^2 B(GT) \qquad (12)$$

where $<|\Sigma \ell|>$ and $<|\Sigma s|>$ are matrix elements of the orbital angular momentum operator and spin operator, respectively, and A is a constant which depends only on the isospin of the levels involved.

In evaluating the strengths of the M1 transitions we shall use the Kurath sum rule [29]:

$$61 \sum_i \Gamma_i / E_i^2 \; = \; -a<0| \sum \ell_j \cdot s_j |0> \qquad T = 0 \to T = 1 \qquad (13)$$

where Γ_i and E_i are the width and energy of the ith level, a is the spin-orbit coupling parameter and $<||>$ is the expectation value of the spin-orbit coupling in the ground state. One may insert the experimental values on the left of Eq. (13) and compare the sum with the right hand side which represents the total expected M1 strength. Other sum rules have been given by Lipparini et al. [27].

III.2. The light nuclei

Let us first consider the odd-odd nuclei which appear to fall in a rather special category. It is unfortunate that not many of these nuclei are available for study, as they are of considerable interest because of the presence of the unpaired neutron and proton in their ground state. The three examples shown in Fig. 11 all come from the 1p shell and illustrate three types of behavior. In ^{6}Li 90% of the M1 strength is concentrated in a single low-lying level: the spin- and isospin-flip transition of the "deuteron" type. In ^{10}B, in the middle of the shell, the strength is spread over levels rather widely spaced, but there is still a tendency toward concentration into a single level. In ^{14}N, the shell model predicts the M1 strength to be concentrated in a single level, which in nature becomes mixed with a neighboring level. The strengths of the transitions, given in terms of the M1 sum rule, have been derived from various measurements with gamma-ray, electron-scattering and capture reactions. The total sum rule strength for each nucleus is taken simply as the total M1 strength predicted by the shell model as calculated by Kurath [29].

Figure 12 surveys the M1 strength observed in other nuclei from A = 8 to 60. References are given in Table III. Where it is known, the isospin of the level is indicated by a solid line $(T_>)$ or a dashed line $(T_<)$. As noted above the isoscalar

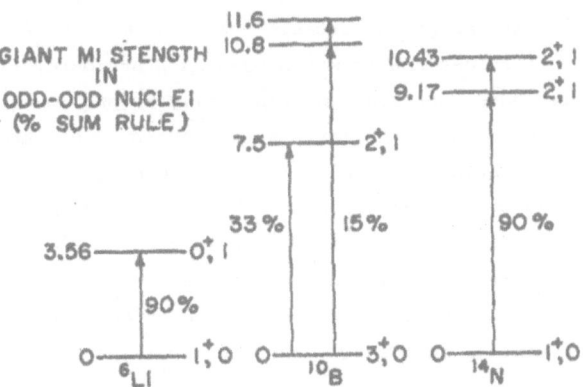

Fig. 11. The M1 strength in odd-odd nuclei. Refs. are in Table III.

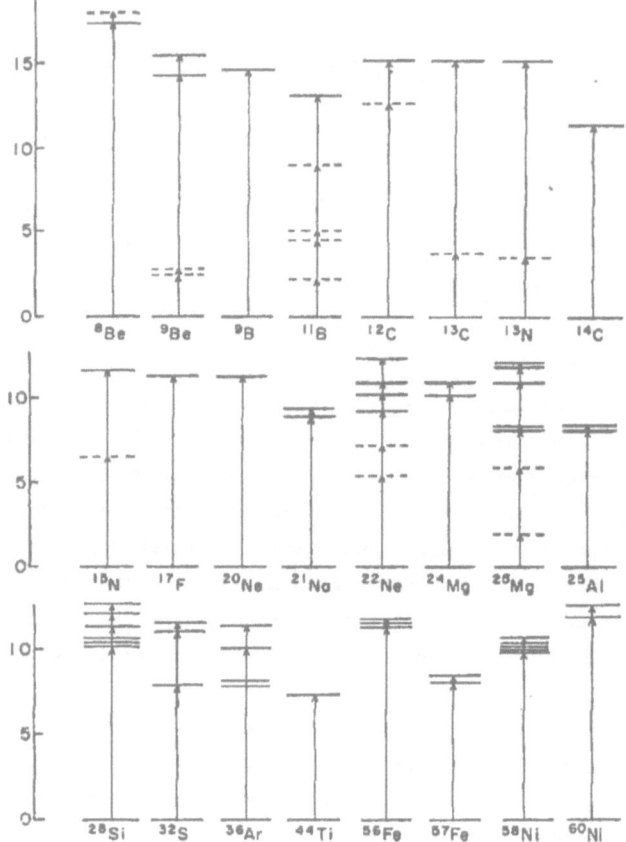

Fig. 12. The Ml strength in nuclei from A = 8 to 60.
References are in Table III.

strength is expected to be relatively very weak. We cite only some typical examples
and interesting features of the Ml transitions. The transition in ^{8}Be, observed
by the (p,γ) reaction in 1936, is the first giant resonance of any kind observed.
The resonance in ^{12}C was found in particle reactions many years ago and has since
been studied by many reactions and is often used as a standard. The transitions
in the odd mass nuclei have been studied principally by proton capture reactions
but also by gamma decay following nuclear reactions and by electron scattering.
The Ml levels in the 4N nuclei have been investigated by gamma fluorescence, electron scattering, and by gamma decay from the T = 2 levels above.
    A typical example of the feeding of Ml levels from a 0^{+}, T = 2 level above
is shown in Fig. 13 where the analogue transitions in ^{28}Al and ^{28}Si are compared. Although all the transitions have not yet been observed we note that the
branching ratios are similar in the two cases in accordance with the third selection

Table III. References on M1 strength.

Nucleus	References	Nucleus	References	Nucleus	References
^{6}Li	5,21,29,30	^{25}Mg	21,34	^{112}Cd	25
^{8}Be	5,29,30	^{25}Al	5,34	^{114}Cd	39
^{9}Be	5,21,30	^{26}Mg	21,34	^{116}In	40
^{9}B	5,30	^{28}Si	21,33,34,35	^{117}Sn	24,25
^{10}B	21,29,30,31	^{32}S	21,33,34,36	^{118}Sn	40
^{11}B	5,21,30	^{36}Ar	21,34,36	^{119}Sn	24,25
^{12}C	5,21,29,30	^{44}Ti	34	^{120}Sn	40
^{13}C	5,21,30	^{56}Fe		^{122}Sb	40
^{13}N	5,30	^{57}Fe	24	^{136}Ba	26
^{14}C	21	^{58}Ni	21,35,37	^{138}Ba	41
^{14}N	21,29,30	^{60}Ni	21,37	^{139}La	22
^{15}N	5,21,30	^{87}Sr	25	Ce	22
^{17}F	32	^{88}Sr	38	^{141}Pr	22
^{20}Ne	5,21,33	^{90}Zr	21	^{197}Au	21,42
^{21}Na	34	^{91}Zr	25	^{206}Pb	21
^{22}Ne	21	^{97}Mo	25	^{207}Pb	24
^{24}Mg	5,21,33,34	^{106}Pd	39	^{208}Pb	43-46

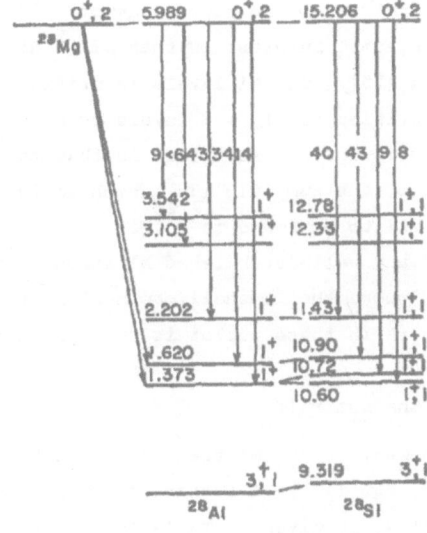

Fig. 13. Analogue M1 transitions in
the mass 28 system.

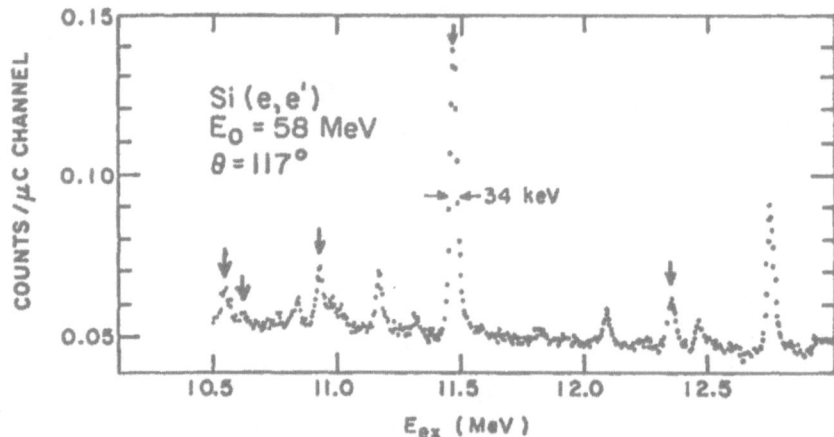

Fig. 14. Electron-excitation of M1 levels (labeled by
an arrow) in ^{28}Si. Ref. 35.

rule in Section III.1 above. We note the interesting fact that the lowest 1^+ level
in ^{28}Al is split in ^{28}Si, presumably by strong mixing between the $T = 1$ level
and a nearby 1^+, $T = 0$ level. The excitation of these M1 levels in ^{28}Si by
electron scattering is illustrated in the recent work from Darmstadt [35] shown in
Fig. 14. The electron scattering resolves the doublet and also excites the expected
analogues at 12.33 MeV and 12.79 MeV not yet observed in the gamma decay in ^{28}Si.
(The level at 12.79 MeV is not identified as 1^+ in Fig. 14, but is so identified
by Fagg et al. [21].) The example in Fig. 13 also shows the importance of studying the beta analogues and comparing them with the gamma decays. It is clear that
a more complete study is needed on the beta decay of ^{28}Mg.

Another example of special interest is that of ^{58}Ni studied at N.R.L. [21,37]
and recently at Darmstadt [35]. The M1 levels identified in the latter work are
shown in Fig. 15 and identified as $T_> = 2$ levels by comparison with analogue 1^+
levels in ^{58}Co. The particular interest here is that the total M1 strength in
these levels is only 12% of the sum rule [35] which would indicate that a large
amount of strength is still to be found in ^{58}Ni.

We note the lack of any well established M1 strength in the "closed-shell"
nuclei ^{16}O and ^{40}Ca, where the in-shell, spin-flip transitions are inhibited.
The location of M1 strength in these nuclei is an important experimental problem.

III.3. Comparison with the sum rule

The strengths of many of the M1 transitions in Fig. 12 have been compared
with the Kurath sum rule, Eq. (13). A summary of such a comparison based on
electron scattering results, is given in Table IV. In the s-d shell the spin-
orbit coupling parameter $a \simeq -2.0$ MeV. Several evaluations of the right hand side
of the sum rule are given:

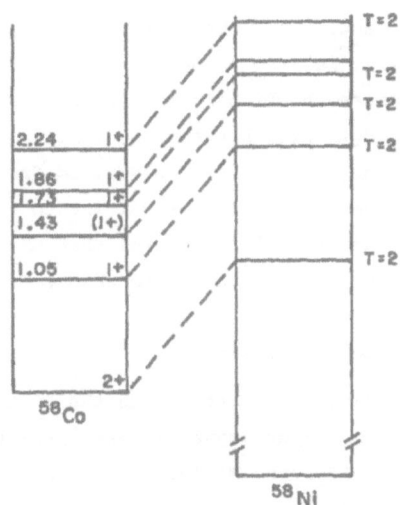

Fig. 15. Analogue M1 levels in the mass 58 system.

(1) a value calculated from a complete j-j shell-model picture,

(2) the value predicted from occupation numbers in the ground state obtained either from a large shell model calculation or from empirical evidence,

(3) values for either an oblate or a prolate deformation.

Table IV. Comparison of the measured M1 strength with the M1 sum rule.

Nucleus	$61\sum \Gamma_i / E_i^2$	$2<0|\sum \ell_j \cdot s_j|0>$		Prolate	Oblate
	Exp(a)	(b)	(c)	$\eta = +4$	$\eta = -4$
^{20}Ne	5.5	8.0	3.14		
^{22}Ne	4.7	12.0	8.24		
^{24}Mg	12.1	16.0		12.0	
^{26}Mg	19.6	20.0			
^{28}Si	20.5	24.0	11.4	8.6	15.2

(a) The first four values are calculated from data in Ref. [21]; the last value from Refs. [21] and [35].

(b) Calculated in the independent single-particle model.

(c) For ^{20,22}Ne the occupation numbers are calculated from the sd shell model [47]; for ^{28}Si empirical values are used [48].

We see that for ^{24}Mg and ^{26}Mg the observed strength is close to exhausting the extreme j-j value. For ^{20}Ne and ^{22}Ne the shell-model calculations indicate a strong reduction of the strength which is in better agreement with experiment. For ^{28}Si the measured strength lies between the limiting value and that obtained from occupation numbers in the ground state as derived from transfer reactions. We note, however, that the sum rule is very sensitive to these occupation numbers. If we consider a deformed model for ^{28}Si we see that the M1 strength favors either a small prolate deformation or a moderate oblate shape.

III.4. Comparison with analogue beta decays

Another important evaluation of the strengths of these M1 transitions can be made by comparing them to their analogue GT beta transitions by means of Eq. (12). Figure 16 shows the comparison for the M1 levels in the p and sd shells. We see that for several of the transitions the orbital contribution $<|\Sigma\ell|>$ is small, but that in many cases it plays an important role in the M1 strength (in most cases because the spin part $<|\Sigma s|>$ is small).

III.5. Comparison with shell model calculations

In recent years several large shell model calculations have been carried out in the 1p shell, the 2s-1d shell and elsewhere. It is of special interest

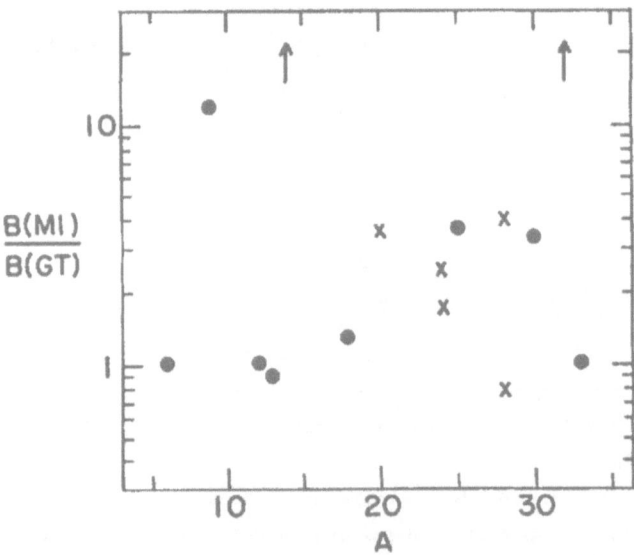

Fig. 16. Comparison of analogue gamma and beta transitions. The dots show $T = 1$ to $T = 0$ transitions, the crosses $T = 2$ to $T = 1$ transitions. The units are chosen so that $B(M1)/B(GT) = 1$ if $<|\Sigma\ell|> = 0$ in Eq. 12.

to compare the theoretical predictions of such calculations with the experimental results in this paper. Such a comparison is made in Figs. 17-20. The theoretical results are derived from the works of Wildenthal and collaborators [49,50].

In these figures the basic M1 transitions are illustrated for ^{20}Ne, ^{24}Mg, ^{28}Al, ^{28}Si, ^{32}P, and ^{32}S. Transitions from the 0^+, T = 2 to the 1^+, T = 1 to the 0^+, 2^+, T = 0 levels are considered. The theoretical predictions are given without brackets and compared to the experimental results within brackets. In general, the radiation width in eV is the quantity compared, but when the branching ratios appear to be more interesting they are given instead. The notation "N.O."

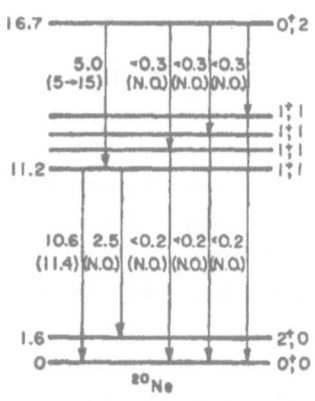

Fig. 17. Experimental (in brackets) and theoretical widths (in eV) in ^{20}Ne N.O. = not yet observed.

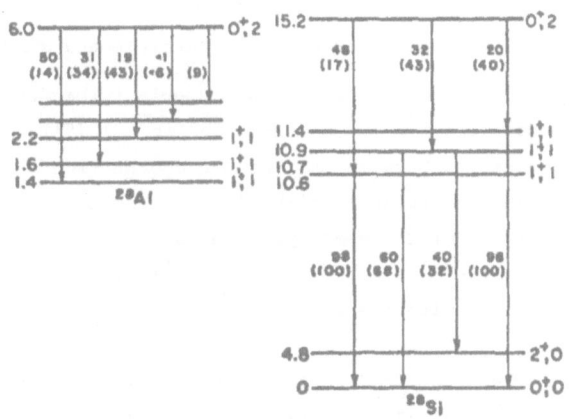

Fig. 19. Experimental (in brackets) and theoretical branching ratios in mass 28.

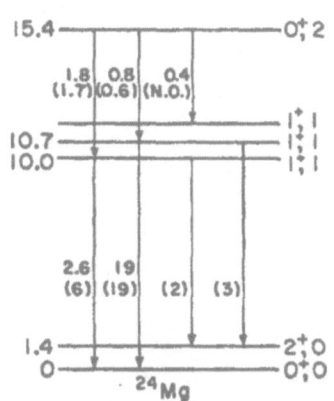

Fig. 18. Experimental (in brackets) and theoretical widths (in eV) in ^{24}Mg. N.O. = not yet observed.

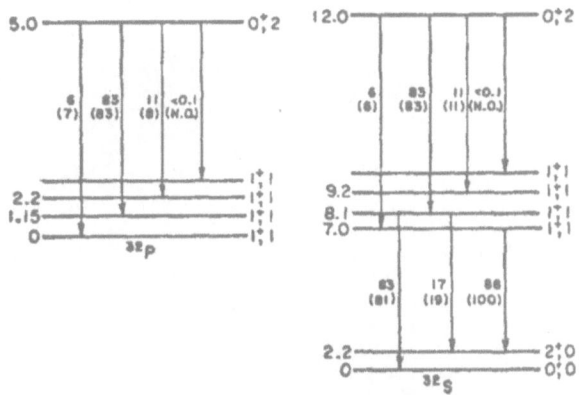

Fig. 20. Experimental (in brackets) and theoretical branching ratios in mass 32. N.O. = not yet observed.

stands for not observed and it can be seen that it corresponds usually to a theoretical prediction less than 0.5 eV (sometimes considerably less). In general the over-all experimental results are not yet very precise nor very extensive. Although there appear to be some discrepancies in the decay of the T = 2 level in mass 28, nevertheless, the general agreement is quite impressive expecially for the prominent transitions which have been well measured. This agreement is especially gratifying since it is just these dominant M1 transitions which should be well represented by the theory. It is to be hoped that the experimental situation will be refined so as to test the theory more closely.

III.6. Comparison of analogue particle and gamma decays

A good example of the isospin forbidden particle decay of M1 levels is the case of ^{13}C - ^{13}N shown in Fig. 21 [51]. In this case very large asymmetries are observed between the proton and neutron decays as can be seen in Table V.

The theoretical widths [52] in Table V were obtained by assuming mixing with $T_<$ states of the same spin and parity as predicted by shell model calculations. The agreement is reasonably good. It is found that the isotensor mixing must be comparable with the isovector mixing in order to produce the large observed asymmetries.

The M1 gamma decays of these mirror levels provide a good test of the rule (see Section III.1 above) that $\Delta T = 1$ gamma decays in mirror nuclei must be identical [5]. The most recent comparison [53] gives $\Gamma_{\gamma_0} = 24.5 \pm 1.5$ eV for ^{13}N and $\Gamma_{\gamma_0} = 23.3 \pm 2.7$ eV for ^{13}C. The agreement is very satisfactory but it would be desirable to decrease further the errors of both measurements. Examples of the selection rule requiring identical branching ratios for analogue $\Delta T = 1$ transitions in non-mirror nuclei are given in Figs. 19 and 20.

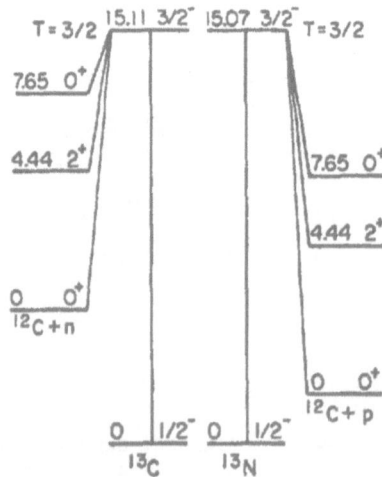

Fig. 21. Mirror decays from the lowest T = 3/2 levels of the mass-13 system.

Table V. Isospin forbidden widths of the lowest T = 3/2 levels of the mirror nuclei ^{13}C and ^{13}N compared with theoretical predictions. The errors in the experimental numbers are approximately ±25%.

	^{13}C		^{13}N	
	Γ_{n_o} (keV)	Γ_{n_1} (keV)	Γ_{p_o} (keV)	Γ_{p_1} (keV)
Exp	0.35	1.31	0.19	0.12
Theor	0.19	1.65	0.25	0.14

III.7. The heavy nuclei

A summary of M1 strength in the heavier nuclei is shown in Fig. 22. From A = 87 to 138 the work has been done principally with the (γ,n) reaction and many of the results are conflicting. In this figure, uncertainty or a lack of confirmation is indicated by single hatching, while better established cases are shown by cross hatching. "N.P." signifies that no particular peaking is seen in the M1 strength. A recent example of this work is the case of ^{138}Ba. From A = 139 to 197 the results are from electron scattering. A typical example is shown in Fig. 23 where the M1 resonance in Ce is the only one seen at 165°. Some M1 strength has also been observed in ^{141}Pr, ^{144}Nd, ^{205}Tl, ^{208}Pb, and ^{209}Bi by means of (γ,γ') experiments [54].

In the Pb region the results come principally from (γ,γ'), (γ,n), (X,X'), and (e,e') measurements. In early work considerable strength was reported [23] in the 7.8 MeV region by means of photoneutron measurements just above threshold. However, recent measurements [45] in which the neutron polarization was determined, found only one level at 7.99 MeV in the region above 7.0 MeV. Strength at 7.5 MeV has been identified in ^{207}Pb [24]. A resonance corresponding to this strength may also be present in electron scattering on ^{208}Pb and ^{206}Pb [55]. However, more recent measurements on ^{208}Pb have identified most of this strength as M2 [55]. The electron scattering also reveals possible M1 strength below the neutron threshold at about 7 MeV in ^{206}Pb and ^{208}Pb. Recent (p,p') and (d,d') measurements [56] have revealed only one 1^+ resonance at 7.06 MeV, other than the resonance at 7.99 MeV. Thus, at present in ^{208}Pb there are only two reasonably well established M1 levels as shown in Fig. 22.

A strong M1 level (5 eV) was reported in ^{208}Pb at 4.8 MeV in a resonance fluorescence experiment [43]. The assignment of this level was obtained by angular correlation and polarization measurements. However, it is very puzzling that this level has not been seen in electron scattering, and recently it was observed [44]

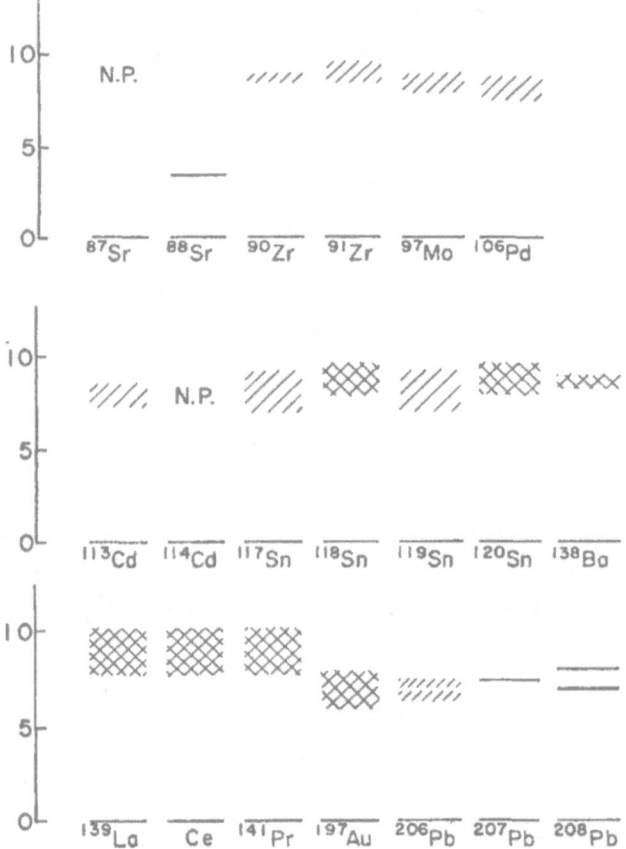

Fig. 22. The M1 strength in nuclei from A = 87 to 208.
References are in Table III.

in (α,α') scattering which indicates the level has natural parity, i.e., $J^\pi = 1^-$. In this case an angular distribution measurement confirms the E1 assignment [46]. It appears improbable therefore that any M1 strength has been seen at this low excitation energy in ^{208}Pb.

The presence of such a strong level at 4.8 MeV would represent a very large and significant spreading of the M1 strength in ^{208}Pb, which would be due presumably to the interaction of the $h_{11/2} \rightarrow h_{9/2}$ proton transition with the $i_{13/2} \rightarrow i_{11/2}$ neutron transition. On the other hand, if these transitions are identified with the two levels in the 7-8 MeV region, then the spreading is very small and theory could account more easily for the observed splitting.

A comparison of these two levels in ^{208}Pb with theory [57] is given in Table VI. We see that the splitting is well estimated by the calculation but that the intensities of the transitions are not, the upper transition being stronger than the lower one. Also, if the calculation exhausts essentially the M1 sum rule

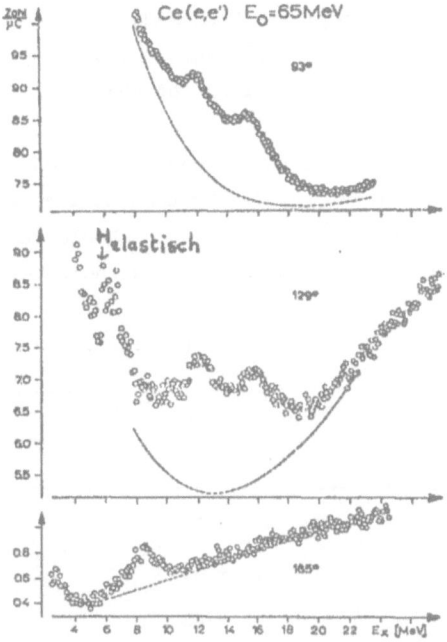

Fig. 23. Excitation of giant resonances in Ce by
inelastic electron scattering. Ref. [22].

then the experimental strengths are "too large".

It is premature, however, to make any conclusions about the M1 strength in ^{208}Pb. The experimental situation is far from complete or definitive. Detailed electron scattering measurements are in progress [35] and further (X,X') and (γ,γ') measurements are needed. The theory also may not yet be definitive and may need refinement on the basis of further experimental results.

Table VI. Comparison of two M1 levels in ^{208}Pb with theory (Ref. 56).

	Exp.		Theor. (Ref. 57)	
E_x (MeV)	7.06	8.00	7.50	8.31
B(M1)↓	5.6	3.3	1.9	3.7

III.8. Summary of giant M1 properties

In many cases shown in Figs. 12 and 22 the levels have been seen by several methods. In others, different levels are observed in different methods. It should also be emphasized that in many nuclei all the M1 strength has not been located. In most cases, no attempt has been made to assess critically the evidence, nor to evaluate the total reported M1 strength. Instead, these figures are presented to display the flavor and the extent of the M1 strength already observed in these nuclei.

It is instructive to associate the strong transitions in Figs. 12 and 22 with the strong spin-flip transitions provided by the shell model. We note from the sum rule discussed above that the major strength should reside in the spin-flip transitions of maximum ℓ. These transitions are indicated in Table VII.

Table VII. Strong M1 configurations

| Transition | A | |
	Proton	Neutron
$p_{3/2} \rightarrow p_{1/2}$	$5 \rightarrow 15$	$5 \rightarrow 15$
$d_{5/2} \rightarrow d_{3/2}$	$17 \rightarrow 41$	$17 \rightarrow 39$
$f_{7/2} \rightarrow f_{5/2}$	$41 \rightarrow 85$	$39 \rightarrow 67$
$g_{9/2} \rightarrow g_{7/2}$	$93 \rightarrow 139$	$73 \rightarrow 111$
$h_{11/2} \rightarrow h_{9/2}$	$197 \rightarrow 211$	$117 \rightarrow 149$
$i_{13/2} \rightarrow i_{11/2}$		$195 \rightarrow 211$

From the evidence presented in Figs. 12 and 22 we may extract the basic properties of the giant M1 resonances. These are summarized in Table VIII along with the E1 properties for comparison. The location of the M1 strength may not be as systematic as that of the E1 strength. This may be due in part to incomplete evidence, but may also arise from a real variation depending on the nuclear type

Table VIII. Summary of E1 and M1 properties

	E1 [a]	M1 [b]
E_x (MeV)	$77/A^{1/3}$	$45A^{1/3}$
Strength (MeV·mb)	$60 \, NZ/A^{1/3}$	$\simeq$ Sum Rule
Γ/E_x	0.2	$\simeq 0.2$

(a) Exceptions noted in text.

(b) See discussion in text.

or species (4N, 4N ± 1, etc.); the "anomalous" behavior of the odd-odd nuclei was noted above. In the heavy nuclei the M1 strength appears to be located at $45/A^{1/3}$ although the data are still very sparse. In the light nuclei the location of the strength appears to drop to about $35/A^{1/3}$ as shown in Fig. 3.

In several cases where the total M1 strength is known and the relevant proper-ties of the ground state are also known, the strength is found to practically exhaust the sum rule. This remark applies to a number of cases in the light nuclei. However, there are some puzzling exceptions such as in ^{60}Ni (see above). Much more systematic work needs to be done in the heavy nuclei. But the data accumulated so far suggest that a similar concentration of strength may be local-ized in the giant M1 resonance of the heavy nuclei.

The spreading of the M1 strength appears to be similar to that of the E1 strength although the data are still not complete enough for a quantitative comparison. In the light nuclei, the level density is low enough that the strength is usually concentrated in one or a few levels, but the spreading may turn out to be about the same as in the heavy nuclei (if allowance is made for the fact that there are not enough levels for the strength to spread over in the light nuclei). The spreading of the M1 strength, which in zero approximation is concentrated in pure shell-model configurations, is of considerable importance as it gives infor-mation on the interaction of these simple states with more complex configurations.

IV. The Giant Isoscalar E2 Resonance

The recent interest in the study of the E2 strength has stemmed from the observation in electron scattering [22] of compact isoscalar E2 resonances below the well-known E1 resonance and their identification in inelastic proton scatter-ing [58]. It appeared at first that these resonances might be of a different nature from the E2 strength seen earlier in (p,γ) work on the lighter nuclei where the strength was found in or above the E1 resonance. However, extensive work on capture reactions [59,60] and inelastic alpha scattering [61] has greatly clarified the picture. In the light nuclei it now seems that much of the E2 strength falls in or above the E1 resonance, but that in fact the E2 strength is spread downward over a wide region.

The existence of an isoscalar E2 resonance was predicted on quite general grounds by Blatt and Weisskopf [62], and by Bohr and Mottelson [63]. There have been several derivations of the excitation energy of this resonance based on the hydrodynamical model [63,64] and on a sum-rule approach [65-67]. All these cal-culations place the energy at (Table I)

$$\Delta E = \sqrt{2}\hbar\omega_o \simeq 60 \, A^{-1/3} \text{ MeV}$$

for a basic harmonic-oscillator excitation of $2\hbar\omega_o$. In addition, extensive shell-model calculations [68-70] based on 1p-1h excitations, i.e. $2\hbar\omega_o$ excitations

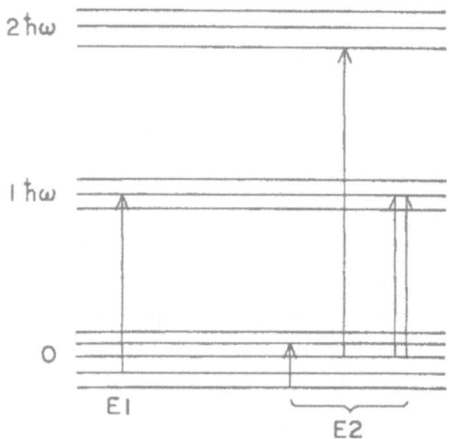

Fig. 24. Schematic E1 and E2 excitations in nuclei.

(see Fig. 24), have been carried out for spherical nuclei such as ^{16}O, ^{40}Ca, ^{90}Zr, and ^{208}Pb. These calculations also place the strength at about $60/A^{1/3}$ MeV. The interesting question is how much of the sum rule is to be found in the 1p-1h excitations and how much in other excitations such as those of the 2p-2h type (Fig. 24).

In this paper we use the Gell-Mann--Telegdi sum rule for the isoscalar E2 strength

$$\int (\sigma/E^2)\, dE = 0.255\,(Z^2/A) <R^2> \text{ }\mu b/MeV$$
$$= 0.22 \text{ } Z^2/A^{1/3} \text{ }\mu b/MeV \qquad (14)$$

where we have taken

$$<R^2> = 0.6(1.2A^{1/3})^2 \text{ fm}^2 \text{ .}$$

We may classify the methods that have been used to study E2 strength as follows:

(1) Capture reactions (X,γ). The first evidence for E2 strength in nuclei came from angular distributions of (p,γ) reactions [71]. The (p,γ) work has continued [10] and been made much more definitive by the use of polarized protons [60]. Inportant information has been obtained from the inverse (γ,p) process [72]. A great deal of evidence has also been accumulated from the (α,γ) reaction [59, 73-75].

(2) Inelastic scattering (e,e'). As mentioned above the first isoscalar E2 resonances in heavy nuclei were observed in (e,e') scattering [22] and the work has since been continued [76].

(3) Inelastic scattering by nuclear particles (X,X'). The isoscalar E2 resonances were identified in early results from (p,p') scattering appearing in the literature and then in current measurements [58]. Subsequent work has been carried out with (p,p') reactions [77,78], with $(^3He,^3He')$ reactions [79], with (d,d') reactions [80] and with (α,α') reactions [61].

(4) Indirect excitation through doorway states by (X,X') reactions. This method has been extensively used to obtain E2 (and other multipole strength) from

analysis of angular distributions of (p,p') reactions [81].

IV.1. The (α,γ_o) reaction

One of the most interesting features of the E2 strength observed in the light
nuclei is the relatively great strength of the α_o decay (where the barrier does
not prohibit it) as observed by the inverse (α,γ) capture reaction. Measured in
terms of their respective sum rules the α_o decay from the E2 excitations is about
ten times more probable than from the E1 resonances. As we shall see, however, the
E2 strength that decays by alphas is spread out over low excitation energies and
may indicate a "different kind" of E2 strength.

In the (α,γ) or (γ,α) process on 0-spin nuclei the E2 strength can be unambiguously separated from the E1 strength by means of angular distributions. This
technique has been used in several laboratories to obtain the E2 strength from 0-spin
nuclei over the range A = 16 to 60. References are given in Table IX and results
are summarized in Fig. 25. We emphasize that the E2 strength observed in these
experiments is only that seen in the α_o decay channel. Figure 25 includes not
only this "α_o strength", but also the total E2 strength measured in the bound and
low-lying discrete resonances (as extracted from various sources [34]). The arrows
mark the energy $63/A^{1/3}$ where compact isoscalar resonances have been observed in
medium and heavy nuclei. We see that there is little indication of such a resonance
in these data. This agrees with the other evidence (see below) that below [40]Ca
such a compact resonance does not exist. Instead the strength is more spread out
and from these data it appears that the α_o decay occurs well below the expected

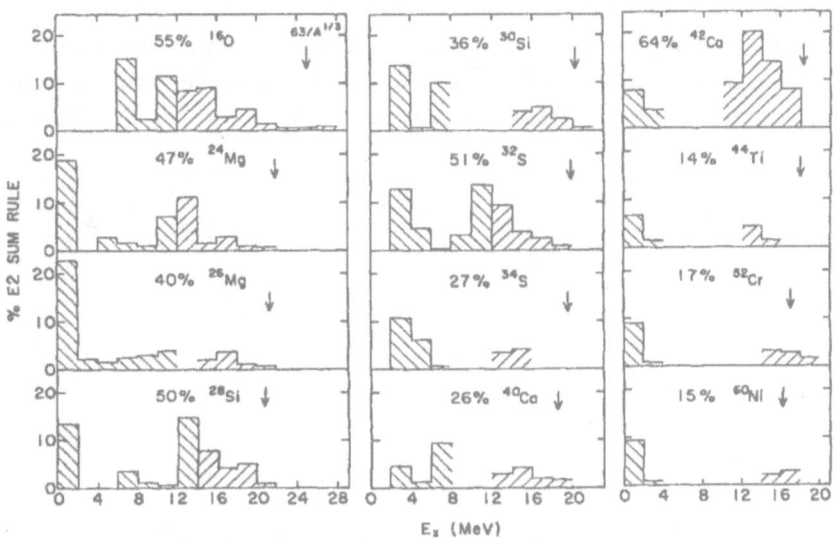

Fig. 25. Summary of E2 strength observed in low-lying resonances
and in (α,γ) reactions. References are in Table IX.

Table IX.   References on E2 strength from (α,γ) reactions.

Nucleus	References	Nucleus	References	Nucleus	References
^{16}O	75	^{32}S	59,73,74	^{44}Ti	82
^{24}Mg	59	^{34}S	59	^{52}Cr	74
^{26}Mg	59	^{40}Ca	74	^{60}Ni	74
^{28}Si	73,74	^{42}Ca	74		
^{30}Si	73	^{44}Ca	74		

location of the E2 resonance.   Above ^{40}Ca the (γ,α_o) data are not very extensive and the α_o channel may be weak because of increased competition from other decay channels and suppression by the Coulomb barrier.

In any case, the strength displayed in Fig. 25 is very large, averaging about 50% of the isoscalar E2 sum rule (wherever the data are reasonably complete), and is spread out over the region from the first 2^+ level up to $63/A^{1/3}$.   If one adds to the "α_o strength" any reasonable estimate of the E2 strength decaying into other channels, one concludes that a major portion of the isoscalar E2 sum rule is exhausted well below $63/A^{1/3}$.

IV.2.   The ($\vec{p},\gamma_o$) reaction

Evidence for E2 strength in the giant resonance region appeared when detailed angular distribution measurements became available in (p,γ) reactions [71].   The appearance of $P_3(\cos\theta)$ and in some cases $P_4(\cos\theta)$ terms in the angular distributions established the existence of E2 radiation interfering with the E1 resonance. Figure 4 shows values of a_3 (the coefficient of P_3) measured [10] throughout the

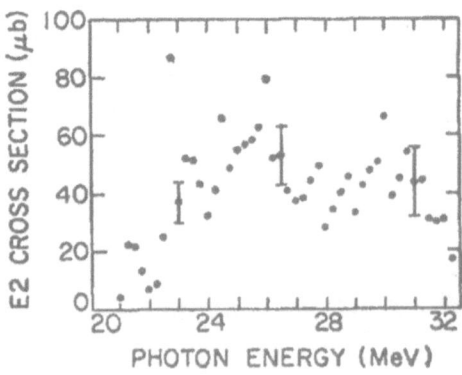

Fig. 26.   The E2 strength observed in ^{16}O(γ,p_o)^{15}N.   Ref. [72].

giant resonance of ^{16}O. The values range between -0.1 and -0.2. Unfortunately, one cannot extract the E2 strength unambiguously from a_3 alone, because the E2-E1 phase difference is not known. However, if one is willing to make plausible assumptions about this phase difference the E2 strength can be extracted [72], as shown in Fig. 26 for the reaction ^{16}O$(\gamma,p_o)^{15}$N. We observe the appearance of significant E2 strength between E_x = 20 and 33 MeV in ^{16}O. In several other nuclei from A = 4 to 40, E2 strength of this nature has been found in (p,γ) and (γ,p) measurements.

The breakthrough in these measurements came with the use of polarized protons in the (p,γ) measurements [60]. In a reaction such as ^{15}N$(\vec{p},\gamma)^{16}$O where the spin of ^{15}N is 1/2, specifying the proton spin makes it possible to remove the phase ambiguity and extract the E2 intensity from the polarized and unpolarized angular distributions.

In order to obtain a reliable measure of the E2 intensity it is necessary to make precise measurements of the angular distributions so as to extract accurate values of the coefficients in Eqs. (1) and (4) up to k = 4. The analysis is then expanded to include 2^+ resonances in ^{16}O and $p_{3/2}$ and $f_{5/2}$ waves in the reaction ^{15}N$(p,\gamma_o)^{16}$O. We now have the following matrix elements in the entrance channel:

$$|s_{1/2}|e^{i\phi_s}, \ |d_{3/2}|e^{i\phi_d}, \ |p_{3/2}|e^{i\phi_p}, \ \text{and} \ |f_{5/2}|e^{i\phi_f}$$

and the relationship between these quantities and the 8 experimental coefficients can be expressed in terms of 8 known functions along with the normalization condition

$$1 \ = \ 0.75(s_{1/2}^2 + d_{3/2}^2) + 1.25(p_{3/2}^2 + f_{5/2}^2). \tag{15}$$

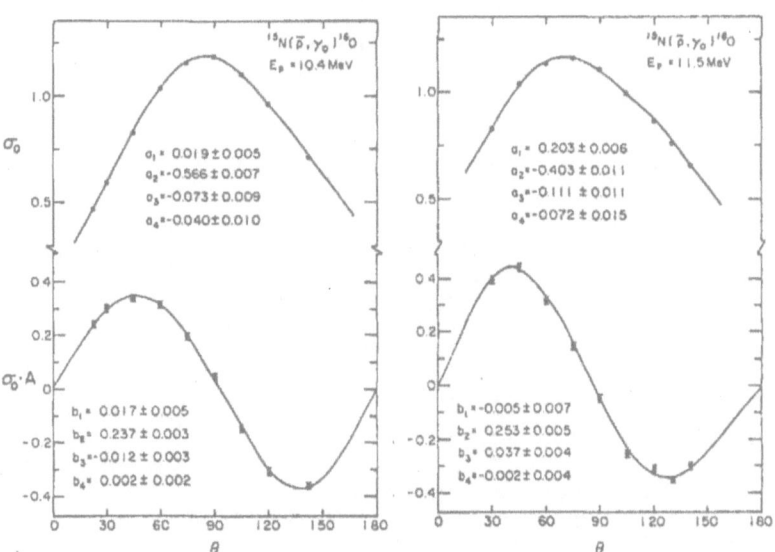

Fig. 27. Polarized and unpolarized angular distributions from ^{15}N(p,γ)^{16}O at E_p = 10.4 and 11.5 MeV.

We now have 9 equations and 7 quantities to be determined (4 amplitudes and 3 relative phases). However, for a complete solution it would be necessary to include M1 radiation in the analysis which would introduce 1^+ resonances and $p_{1/2}$ and $p_{3/2}$ waves in $^{15}N(p,\gamma_0)^{16}O$. Fortunately, the M1 radiation can be effectively eliminated from the analysis by omitting the coefficients a_1 and b_1 from consideration. We are then left with 7 equations and the 7 unknown quantities describing the E1 and E2 strengths. Once these have been determined, it is then possible to insert the known E1 and E2 parameters into the equations for a_1 and a_2 to test for the presence of M1 radiation. In the measurements made so far in the giant E1 regions of several nuclei the M1 contributions have been found to be very small.

Some typical angular distributions for $^{15}N(p,\gamma)^{16}O$ are shown in Fig. 27. These measurements illustrate the quality of the data that can be achieved by careful measurements in relatively short periods of time. On the basis of such measurements, carried out between $E_x = 19$ and 28 MeV and shown in Fig. 28, the $p_{3/2}$ and $f_{5/2}$ amplitudes and phases were extracted and the total E2 cross sections were obtained as shown in Fig. 29. Although there are still two solutions (I and II),

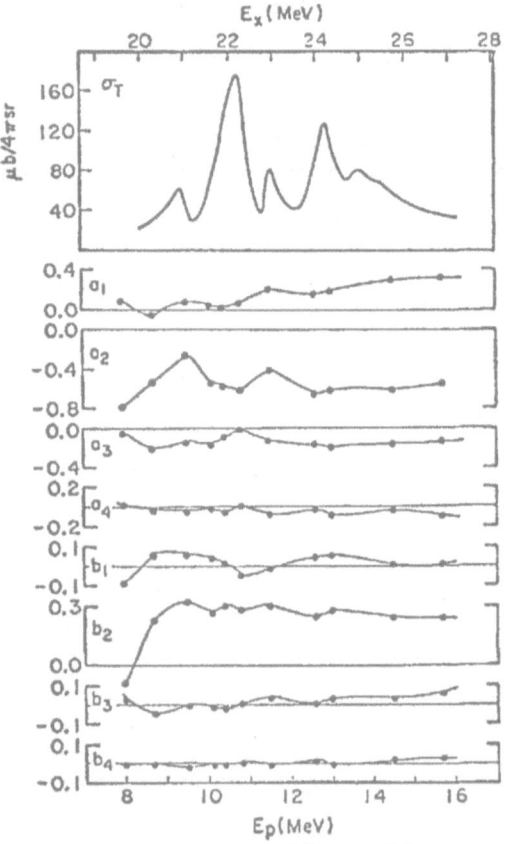

Fig. 28. Top: total cross section curve of $^{15}N(p,\gamma)^{16}O$. Below: the unpolarized coefficients $a_1 \ldots a_4$ and polarized coefficients $b_1 \ldots b_4$.

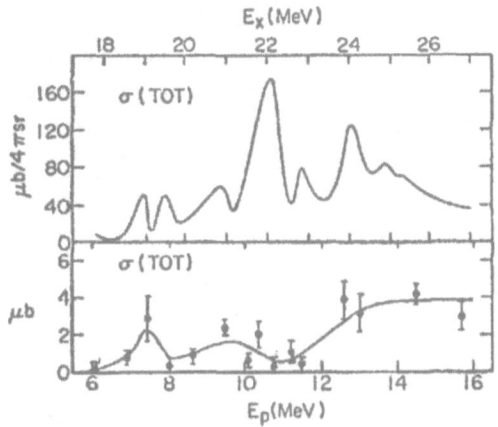

Fig. 29. The E2 total cross sections derived from the
coefficients in Fig. 28 for $^{15}N(\vec{p},\gamma_o)^{16}O$.

it is a fortunate circumstance that both solutions lead to essentially the same E2
cross sections. Although the absolute cross sections differ a little, the trend
of the E2 strength in Fig. 26 agrees quite well with that in Fig. 29.

In Fig. 30 we collect together all the E2 strength in ^{16}O determined by
gamma-ray reactions, taken from Figs. 25 and 29 with the latter curve extended to
higher energy by unpolarized (p,γ) work from Brookhaven [83]. As we expect to find
considerable strength in channels other than α_o and p_o, the total yield represented by Fig. 30 accounts for a very substantial portion of the isoscalar E2 sum
rule. It is possible, therefore, that some of the strength at higher energies is
isovector in character.

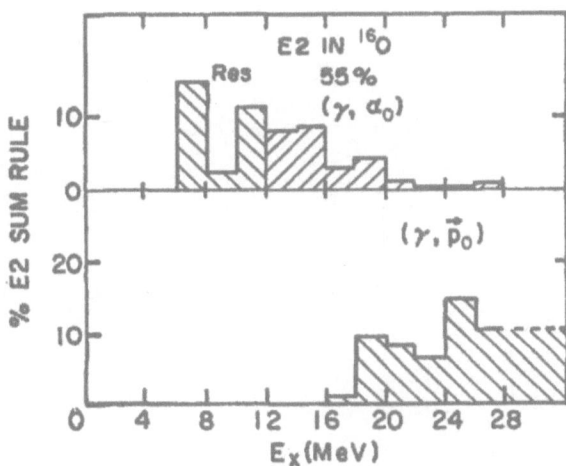

Fig. 30. Summary of E2 strength in ^{16}O from gamma reactions.

This suggestion would agree qualitatively with the theoretical calculations. We give in Figs. 31 and 32 the results of two calculations [68,69] of the E2 strength based on 1p-1h excitations of the $2\hbar\omega$ type (see Fig. 24). Both calculations place the major isoscalar strength in a compact peak at about 22 MeV and one of the calculations [68] shows strength that is chiefly isovector in character extending to higher energies. However, it is clear from Fig. 30 that the experimental

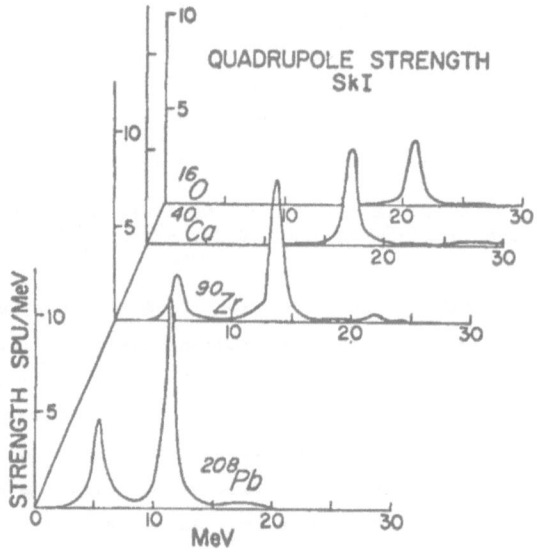

Fig. 31. Isoscalar E2 strength computed for ^{16}O, ^{40}Ca, ^{90}Zr, and ^{208}Pb. Reference [69].

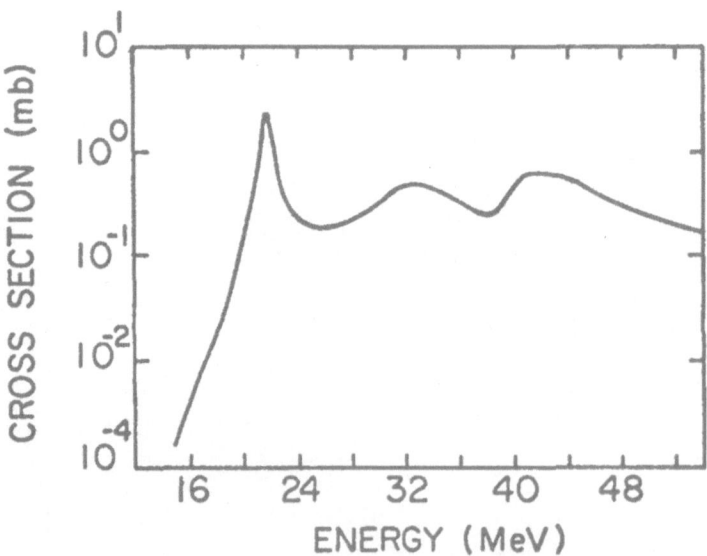

Fig. 32. Isoscalar and isovector E2 strength computed for ^{16}O. Reference [68].

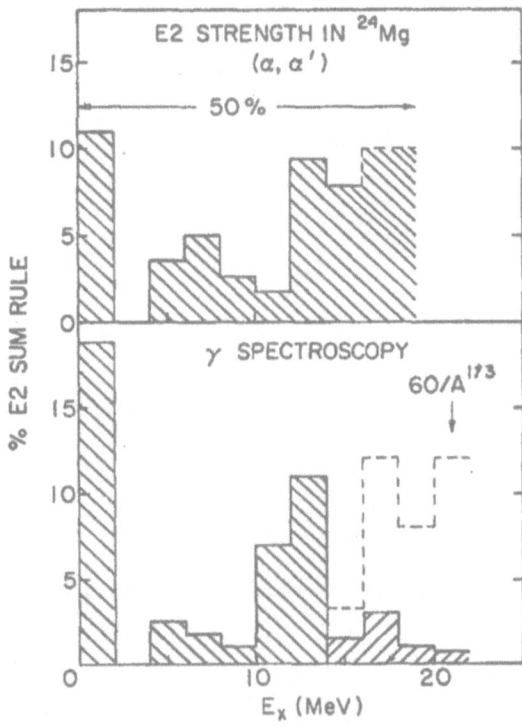

Fig. 33.   Comparison of the isoscalar E2 strength in ^{24}Mg from
gamma reactions with that from (α,α').   Reference [88].

isoscalar strength is more spread out than indicated by these 1p-1h calculations.
We suggest that this spreading is due to more complex excitations, such as 2p-2h
(see Fig. 24), 4p-4h, etc.   Indeed, an earlier calculation [84] of the E2 strength
in ^{16}O which included 2p-2h excitations showed a very pronounced spreading downward of the E2 levels.   Recently, more refined calculations [85-87] have confirmed
this spreading as shown in Fig. 40.

Based on the evidence in Fig. 25 we might expect the same picture as observed
in ^{16}O to hold up to ^{40}Ca, if not beyond.   This is confirmed by γ-ray measurements in ^{24}Mg (Fig. 33) and ^{32}S (Fig. 34).   In ^{24}Mg the E2 strength found
by γ spectroscopy in the bound levels [34] together with the α_o strength [59]
is compared with the strength observed by a detailed, high-resolution study [88]
of ^{24}Mg(α,α').   If a plausible correction for missing strength in other channels
is made (dashed line in lower histogram in Fig. 33), the two distributions are in
qualitative agreement and account for more than 50% of the isoscalar E2 sum rule.
In ^{32}S (Fig. 34) the picture is very similar to ^{16}O, except that the (γ,p_o) and
(γ,α_o) strengths [59] overlap more in ^{32}S.

The E2 strength in the p_o channel of ^{12}C has been investigated in recent
measurements [89] on ^{11}B$(\vec{p},\gamma_o)^{12}$C.   In this case the analysis is less definitive

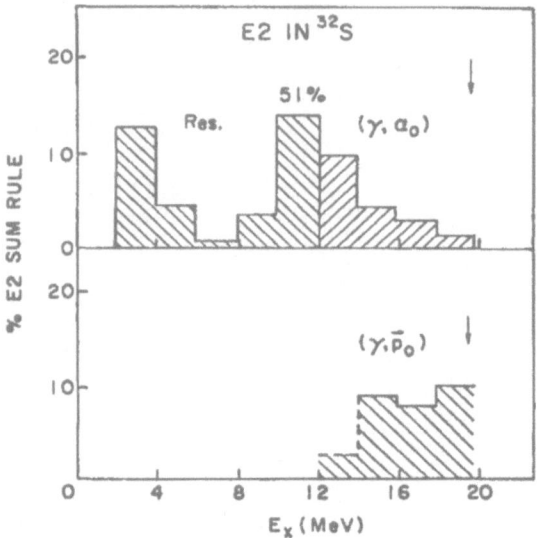

Fig. 34. Summary of E2 strength in ^{32}S from gamma reactions.

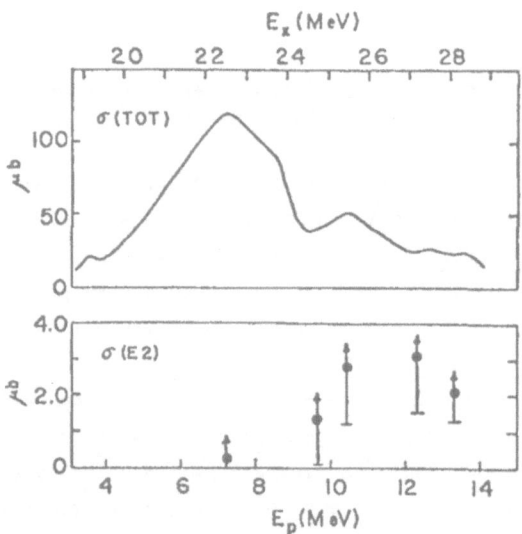

Fig. 35. Top: giant E1 resonance in ^{12}C observed with the ^{11}B(p,γ_0)^{12}C reaction.
Bottom: E2 strength derived from ^{11}B($\vec{p}$,γ_0)^{12}C.

than for ^{16}O and ^{32}S because the spin of the target nucleus ^{11}B is 3/2. Never-theless, lower limits and most probable values for the E2 strength can be found as shown in Fig. 35. The general picture in ^{12}C is very similar to that in ^{16}O. It has been shown recently [90] that much of the E2 strength seen in these (p,γ_0) reactions can in fact be accounted for by a semi-direct capture process which would not be included in the shell-model calculations of the E2 resonances and would contribute further to the spreading of the E2 strength.

IV.3. Inelastic scattering – the light nuclei

Examples of the isoscalar E2 resonances observed with the (p,p') reaction are shown in Fig. 36 [91] and with the (α,α') reaction in Fig. 37 [61]. A recent review of these resonances is given by Bertrand [91]. In this section we shall

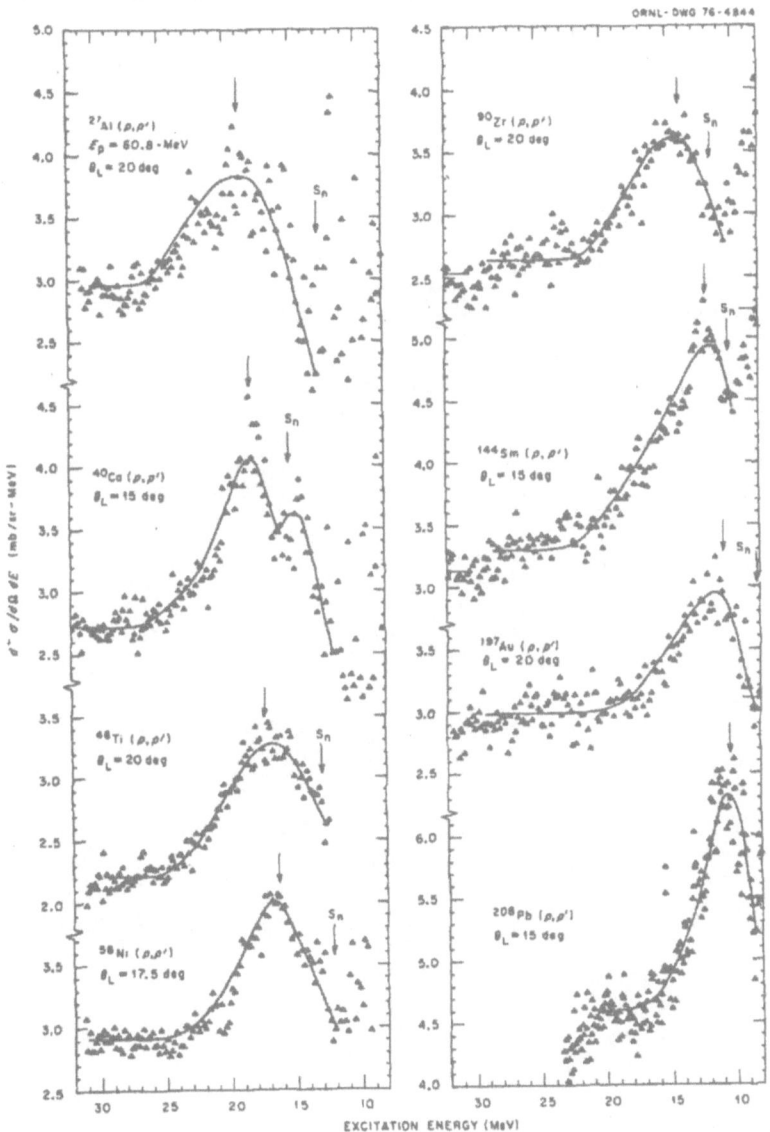

Fig. 36. Inelastic proton spectra for E_p = 61 MeV [91]. The neutron separation energy S_n and energy $60/A^{1/3}$ are marked by arrows.

be concerned chiefly with some of the problems concerning these resonances, especially in the light nuclei where the nucleus ^{16}O will be considered in some detail. In the heavy nuclei we shall concentrate chiefly on the important case of ^{208}Pb.

The early work [58] on (p,p') indicated a compact isoscalar resonance lying below the E1 resonance down to at least ^{27}Al. Such a resonance in ^{27}Al would be difficult to reconcile with the evidence presented above in Sections IV.1 and IV.2. The recent work [61] on (α,α') has greatly clarified the picture. The alpha reaction has the great advantage of selectively exciting isoscalar, electric multipoles only.

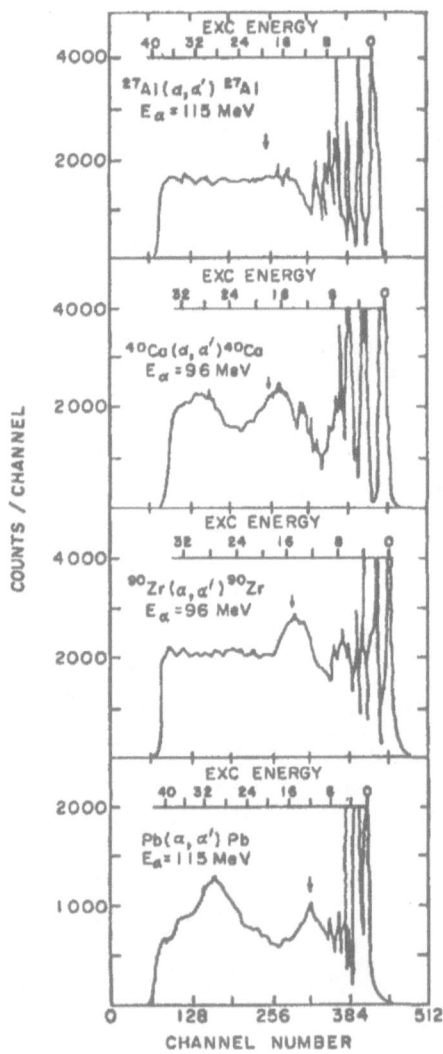

Fig. 37. Inelastic alpha excitation of ^{27}Al, ^{40}Ca, ^{90}Zr, and Pb. Reference [61].

In Fig. 37 we see a compact isoscalar resonance at $63/A^{1/3}$ in Pb, ^{90}Zr, and ^{40}Ca but in ^{27}Al the resonance has washed out into a broad shoulder indicative of a spreading out of the strength. This picture is maintained throughout the light nuclei such as ^{14}N, ^{16}O, ^{20}Ne, and ^{24}Mg [61]. The authors of this work make no attempt to draw a background under this shoulder in order to extract an E2 strength. It would appear that this evidence from (α,α') is qualitatively compatible with the evidence from gamma-ray reactions on light nuclei presented in Figs. 30, 33 and 34. This spreading of the strength in the light nuclei has now also been confirmed by recent measurements on the (p,p') reaction as shown in Fig. 36 [91].

Results [92] from ^{16}O(^{3}He, ^{3}He') are presented in Fig. 38. This reaction is much more sensitive to isoscalar E2 strength than to isovector strength. We see that the E2 strength is spread out over the 16 → 26 MeV region with a maximum at about 19 MeV. Thus, this isoscalar strength lies definitely below the GDR and the E2 strength seen in the $(\vec{p},\gamma_o)$ reaction (Figs. 26 and 29) and supports the suggestion that the latter yield may contain considerable isovector strength.

In a series of (α,α') measurements [93] at high energy, E_α = 146 MeV, it was found that the isoscalar E2 strength is displayed more clearly than in the work at 115 MeV [61]. The result for ^{16}O is shown in Fig. 39. That the strength above the dotted line is predominantly E2 is established by the angular distributions also shown in Fig. 39. Although differing in detail, the isoscalar strength from (α,α') agrees in its main features with the result from (^{3}He, ^{3}He') and appears to establish that the peak of the isoscalar E2 strength in ^{16}O is centered at about

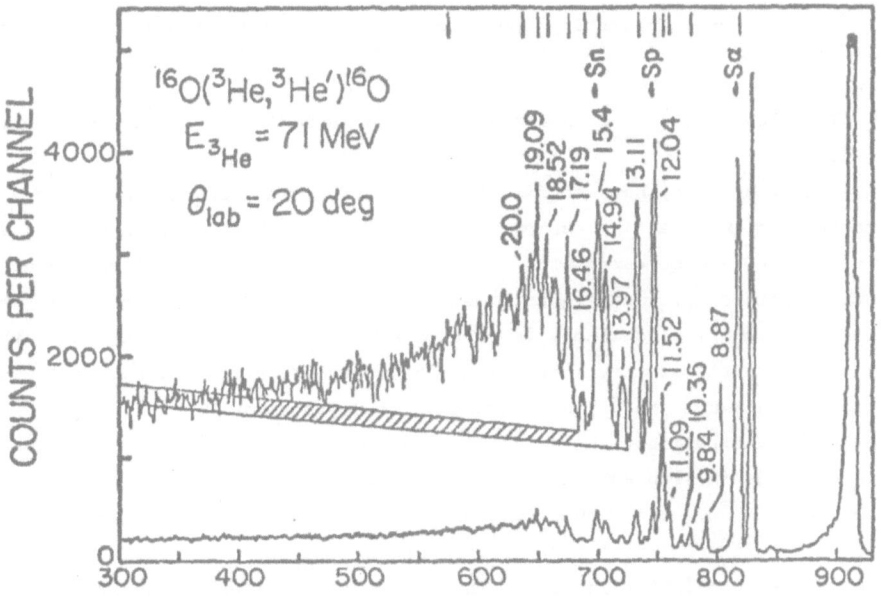

Fig. 38. Inelastic ^{3}He excitation of ^{16}O. Reference [92].

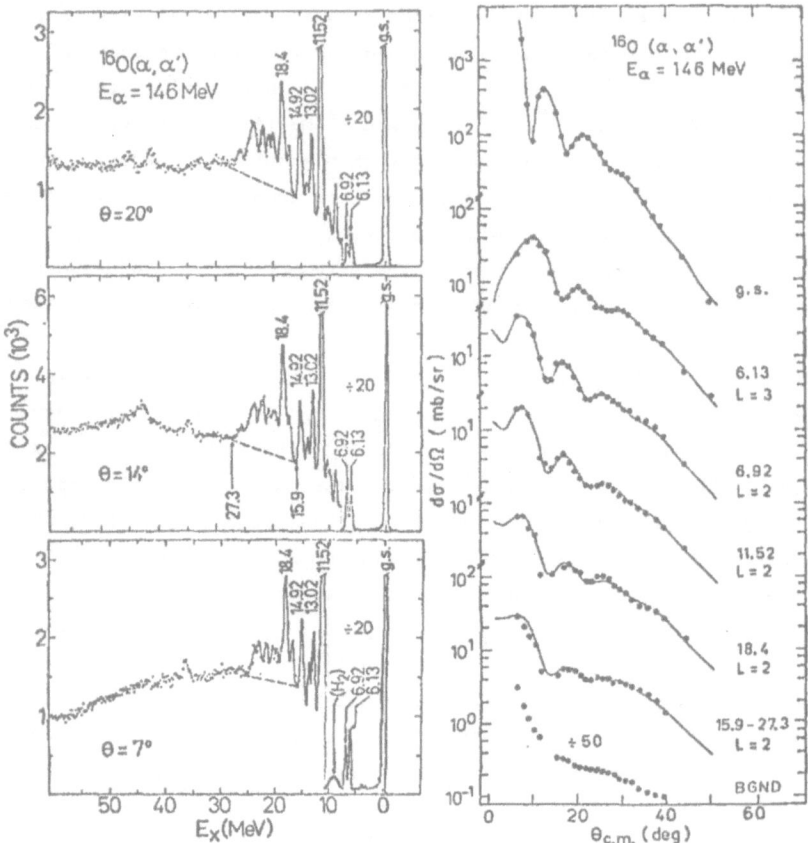

Fig. 39. Inelastic alpha excitation of ^{16}O and angular
distributions of 2^+ resonances. Reference [93].

19 MeV. This lowering of the isoscalar strength is indicated by the dotted line
in Fig. 3.

Harakeh et al. [94] have studied the inelastic (α, α') scattering at E_α = 104
MeV with results in essential agreement with Ref. 93. The results are shown at the
bottom of Fig. 40 and compared with the recent theoretical calculations [85-87]
that include more complex configurations. We see that the well established experimental E2 strength (solid lines) is grouped around E_x = 20 MeV. Although the
theoretical calculations vary in the location of the E2 strength, nevertheless,
they agree in showing considerable spreading of the strength.

In a recent experiment the Heidelberg-Jülich group [95] has observed the (α, α')
yield in coincidence with subsequent decay particles throughout the region of the
E2 resonance in ^{16}O. The result is shown in Fig. 41. As can be seen, under the
conditions of this experiment most of the "background" is eliminated and a strong
response is observed in the region of the E2 resonance. Further analysis of this
experiment should provide valuable information on the E2 resonance in ^{16}O.

315

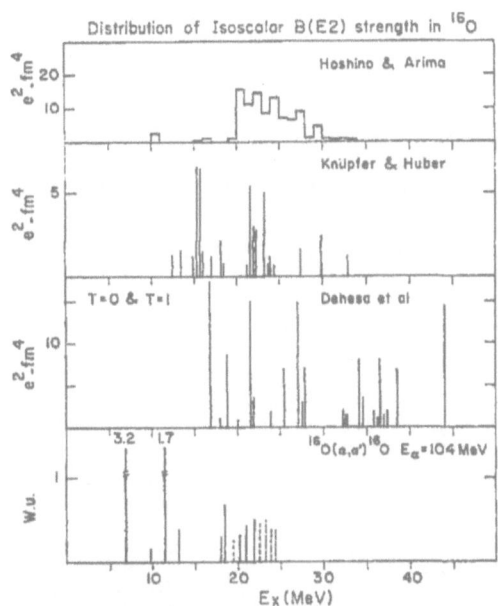

Fig. 40.   Comparison of experimental E2 strength with theoretical
predictions.   Dashed lines indicate doubtful E2 assign-
ments.   Reference [94].

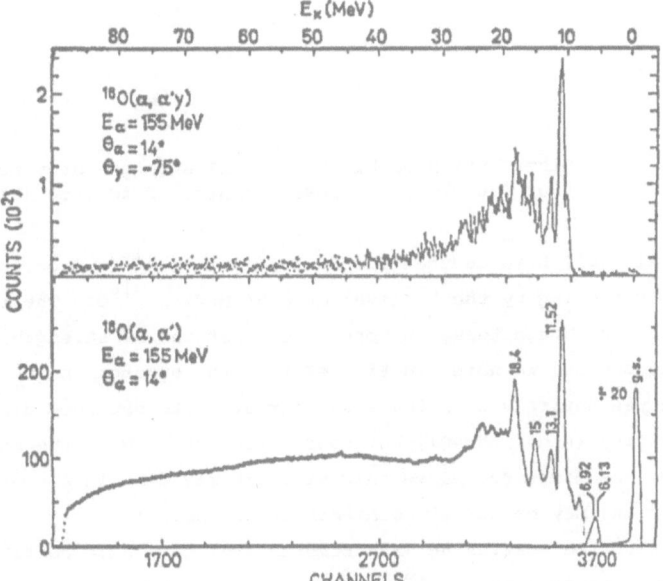

Fig. 41.   The ^{16}O$(\alpha,\alpha'\gamma)$ coincidence curve compared with
the ^{16}O(α,α') singles yield.   Reference [95].

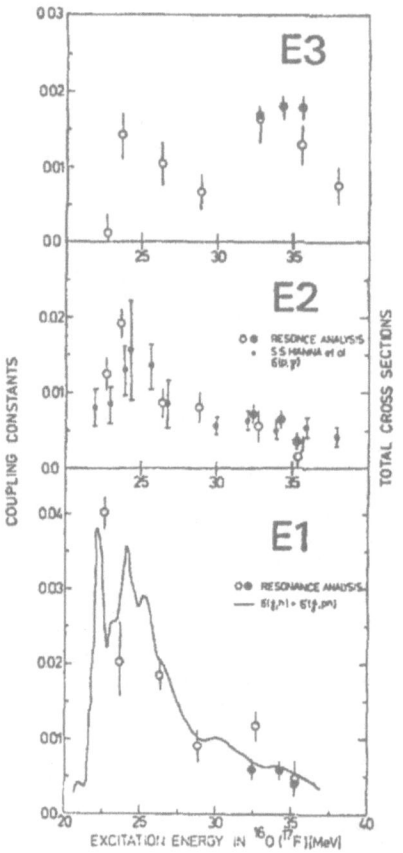

Fig. 42. Distribution of E1, E2, and E3 strength deduced
from $^{16}O(p,p')^{16}O^*$ measurements. Reference [81].

Geramb et al. [81] have deduced the E2 strength in ^{16}O from analysis of inelastic proton scattering by the 2^- level at 8.88 MeV in ^{16}O. Their results are
given in Fig. 42. Although these authors argue that the E2 strength they see is
predominantly isoscalar, we note in Fig. 42 that the strength peaks at 24 MeV and
thus does not agree entirely with the isoscalar strength obtained directly from
inelastic scattering (Figs. 38-40), but does agree better with the strength observed
in $^{15}N(\vec{p},\gamma)^{16}O$. As discussed above this strength may contain a considerable isovector component and may be largely semi-direct in character.

Recently, electron scattering measurements [96] have also confirmed the results
of the gamma-ray measurements on ^{16}O. In particular, the results show 43% of the
E2 sum rule below 20 MeV and only 20% between 20 and 30 MeV.

We now come to ^{40}Ca which could prove to be the pivotal case between the light
and heavy nuclei. Figure 43 shows the results [76] of inelastic electron excitation

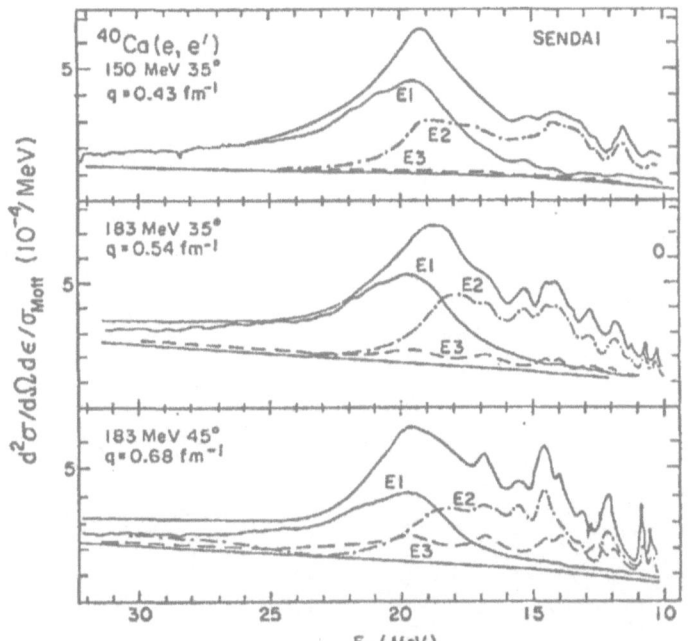

Fig. 43. Inelastic electron excitation of ^{40}Ca at different momentum transfers. Reference [76].

on ^{40}Ca. A background has been subtracted and a semi-empirical technique has been used to sort out the various multipoles. Actually, it is not possible to distinguish E2 from E0 strength. Thus, we see that the E2(E0) strength is rather uniformly spread between 10 and 20 MeV, in rather good agreement with the picture developed above for E2 strength in the lighter nuclei. Since the 1p-1h calculations (see Fig. 31) place the isoscalar E2 strength in a compact resonance at 17 MeV, an attempt was made [97] to divide the E2(E0) strength in Fig. 43 into an E2 resonance at 17.5 MeV and an E0 resonance at 14 MeV. This assignment of E0 strength is doubtful for at least three reasons:

(1) This strength in the region of 14 MeV, which appears in both (e,e') and (α,α') is probably seen strongly in (γ,α_0) [74]. Since E0 strength cannot be excited by gamma rays, the yield in this region is very probably E2.

(2) Results from particle-particle angular correlation measurements [98] indicate that the correlation throughout the region between 10 and 20 MeV is not isotropic. This result would rule out any major strength in spin-0 resonances.

(3) The theoretical calculations [68,69] make it seem doubtful that any major monopole strength would be found at such a low excitation energy.

IV.4. Inelastic scattering - the heavy nuclei

In the heavier nuclei the isoscalar E2 resonances have been the subject of

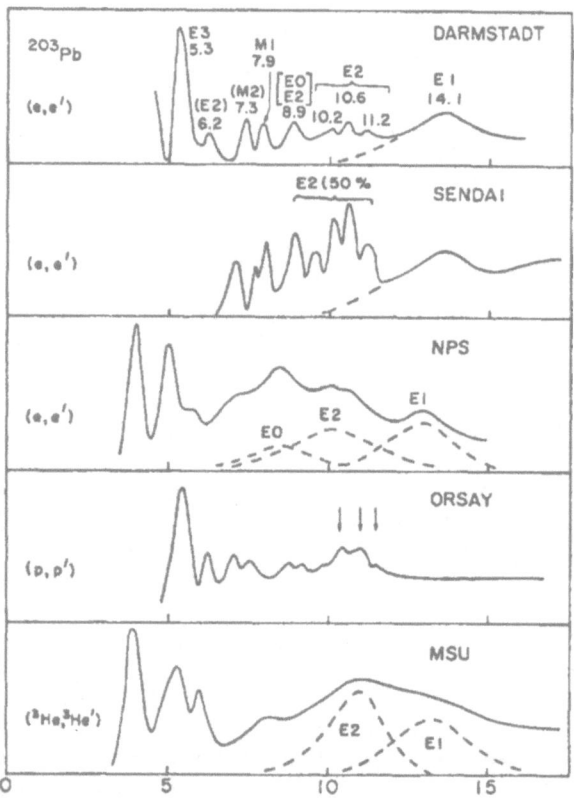

Fig. 44. Comparison of various results from inelastic excitation of ^{208}Pb. References [99, 107, 102, 77, 79], respectively, from top to bottom.

many papers and several reviews [91,99]. The analysis of these resonances has been based on the extensive work of Satchler [100]. We will not attempt to review the complete picture but instead will examine the E2 strength observed in ^{208}Pb.

Many spectra have been obtained on ^{208}Pb. Figures 44 and 45 are an attempt to compare some of the earlier work. It is clear that with good resolution, as in the (e,e') spectra, a great deal of structure is present in ^{208}Pb. The three peaks at 10.2, 10.6 and 11.2 MeV, seen in (p,p') and (e,e') are reported [99] as E2, as is the structure at 6.2 MeV [101]. Thus, the spreading of the E2 strength in Pb may be quite appreciable. The structure at 8.9 MeV could well be E2 but it has been argued [102] that this resonance is E0 on the basis that it is not observed in the (γ,n) yield, see Fig. 45, while the higher E2 resonances are probably seen as small wiggles on the dominant E1 cross section (see below). However, as was the case for ^{40}Ca discussed above, the calculations [68,69] suggest that the monopole strength should be higher than 9 MeV.

In view of the above arguments it is instructive to compare the (e,e') results with the (γ,n) measurements [103-105] made with monochromatic photons, as is done

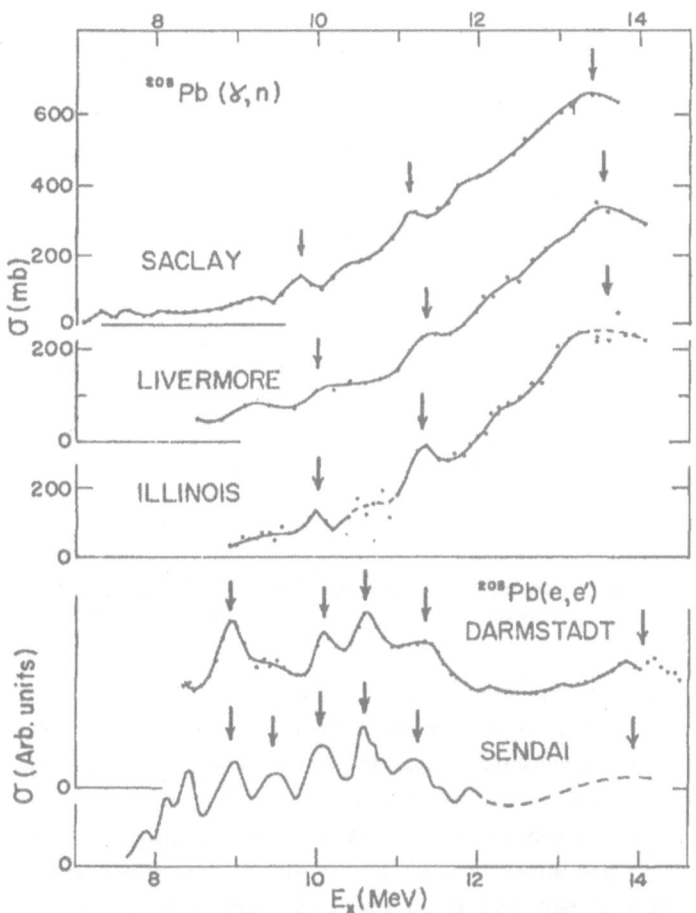

Fig. 45. Comparison of results from photonuclear excitation with monoenergetic photons and high-resolution inelastic electron scattering. References [103,104,105,99,107], respectively, from top to bottom.

in Fig. 45. The agreement among the (γ,n) curves is quite good, provided the Saclay curve is shifted up in energy by about 200 keV. It is seen that the two peaks at 10.1 and 11.4 MeV are quite well established in these curves. The strengths of these peaks are compatible with E2 radiation and coincide with the peaks seen in (e,e'). There is less evidence for the peaks at 10.6 and 8.9 MeV and the assignment of these peaks as E2 should not be considered as established at present. On the other hand, it has been recently shown [106] that the strength and breadth of the (e,e') peak at 8.9 MeV is compatible with the (γ,n) curves so that it is not necessary to assign this peak to E0. More recent work on this level is discussed below.

In Fig. 46 we show an attempt [108] to analyze the gross structure seen in a

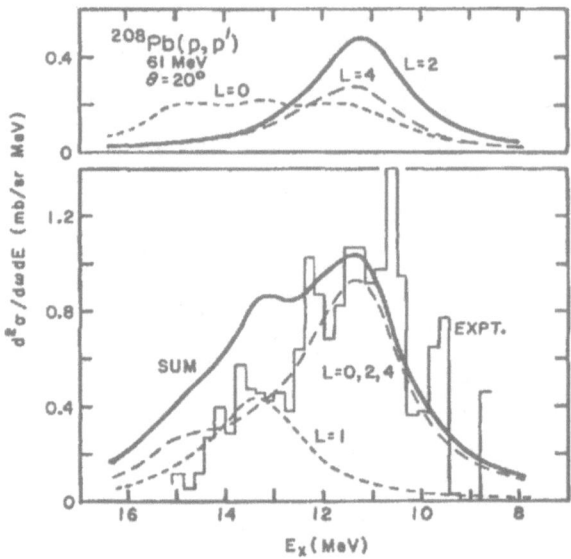

Fig. 46.   Fit to the gross structure in ^{208}Pb with
theoretical L = 0,1,2,4 multipoles [108].

^{208}Pb(p,p')^{208}Pb excitation curve obtained with low resolution. The experimental
response function is fitted with a superposition of L = 0,1,2, and 4 multipoles,
all of which contribute substantial amounts to the cross section. This mixture of
multipoles leads to a featureless angular distribution [108].

It is interesting to compare this low-resolution spectrum with one taken under
high resolution [109] and shown in Fig. 47. A single monopole state is identified
in this spectrum at 9.11 MeV, but it is found to carry only 13% of the sum rule.
No other monopole strength is identified. Although the authors state that there is
considerable E2 strength in this spectrum it seems difficult to confirm the states
at 10.2, 10.6, and 11.2 MeV identified in earlier work (see above). Although structure appears at 10.3 MeV much of it is attributed to octupole strength. The region
around 10.6 MeV appears to be obscured by a ^{12}C impurity. Structure appears at
11.2 MeV but seems to have the wrong angular distribution to be identified with
quadrupole radiation.

A high resolution study has also been made on the ^{208}Pb(γ,n) reaction [110].
The results of the analysis of the (γ,n$_o$) yield are shown in Fig. 48. We see again
that in the region of the E2 strength there is an abundance of fine structure with
very little correlation with the structure seen in the low-resolution work summarized
in Figs. 44 and 45. The main point made by the authors of this work is that the
strength of the structure seen in the 9.0 MeV region is such that an assignment of
E2 is to be preferred over the assignment of E0. A more general conclusion from

these high-resolution studies is that no definitive conclusions can be drawn regarding the distribution of multipole strength in ^{208}Pb below the E1 resonance until the multipolarities of all the fine structure peaks have been established. Thus, analyses of the type shown in Fig. 46 and that made recently by Pitthan, Buskirk and Walcher [111] should be re-examined in the light of these high-resolution studies. In the work of Pitthan et al. [111] an attempt was made to explain the low-resolution (e,e') results in terms of $T_>$ fine-structure peaks superimposed on broad $T_<$ resonances.

There is clearly much work that needs to be done to obtain a clear picture of the location, distribution and strength of the isoscalar E2 resonance in ^{208}Pb. We remark here that no other heavy nucleus has been scrutinized as intensively as has ^{208}Pb, so that it is possible that the multipole distribution is equally complex in other nuclei.

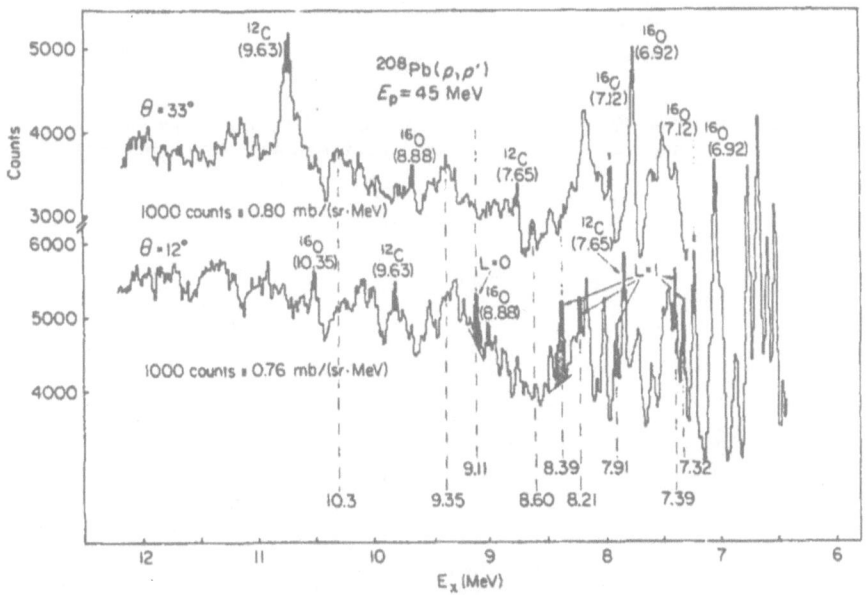

Fig. 47. High resolution study of ^{208}Pb with the (p,p') reaction. Reference [109].

IV.5. The E2 strength in deformed nuclei

The search for splitting of the isoscalar E2 resonance in deformed nuclei has recently been one of the active areas of research. In view of the beautiful splitting in the E1 resonance (Ref. 112 and Fig. 52) it was thought there might be a similar splitting in the E2 resonance. Such a splitting has been looked for in

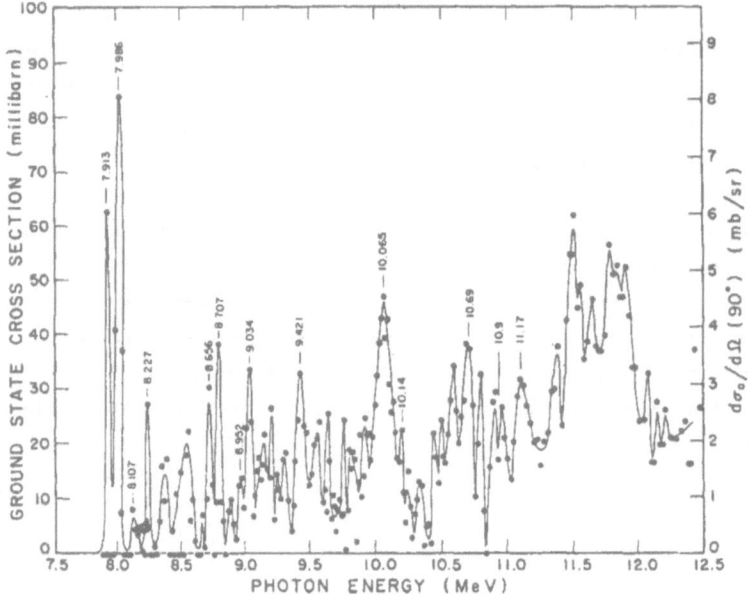

Fig. 48. The (γ, n_o) cross section on ^{205}Pb obtained by
neutron time-of-flight spectroscopy. Reference [110].

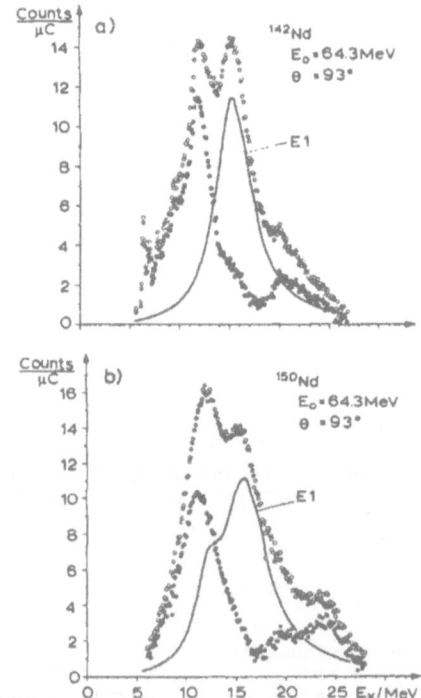

Fig. 49. The (e,e') spectrum (open circles) from ^{142,150}Nd after subtraction of
the radiation tail and background. The solid dots and the curve show
the decomposition into an E2 and an E1 cross section. Reference [106].

the Sm isotopes with 67-MeV protons [113], 80-MeV ^{3}He particles [114], and 115-MeV alphas [115], in the Nd isotopes with 50- and 64.3-MeV electrons [116], and in ^{165}Ho with 75 and 105 MeV electrons [117].

The investigations with electrons and alphas agree in showing a broadening of the E2 resonance considerably less than that observed in the E1 resonances. The electron scattering results [116] on Nd are shown in Fig. 49. After decomposition of the curves a width of 2.9 MeV is found for the E2 resonance in the spherical nucleus ^{142}Nd and 5.0 MeV for the deformed nucleus ^{150}Nd. A similar but somewhat smaller broadening was observed [115] in the Sm isotopes with alpha scattering.

These results have produced a flurry of theoretical activity [115,118,119]. The authors of Ref. 115 find that the usual quadrupole-quadrupole interaction would predict a splitting of 6 MeV but a renormalization of the interaction would produce a splitting of only 2 MeV in better agreement with experiment. The other theoretical treatments [118,119] predict similar broadenings.

IV.6. Summary of properties of isoscalar E2 resonances

In the heavy nuclei the E2 resonances are found at $63/A^{1/3}$, but as was discussed in Section IV.2 there is a tendency for the energy to decrease in the lighter nuclei. This behavior is shown in Fig. 3.

The measured widths of the E2 resonances are summarized in Fig. 50. It is apparent that there is a great deal of scatter in these measurements. This is due

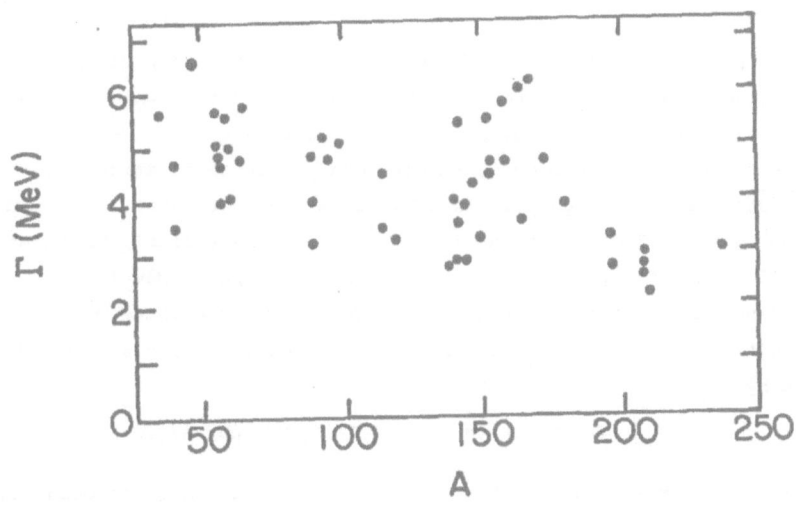

Fig. 50. Summary of the widths of isoscalar E2 resonances
obtained from inelastic (X,X') and (e,e') experiments.

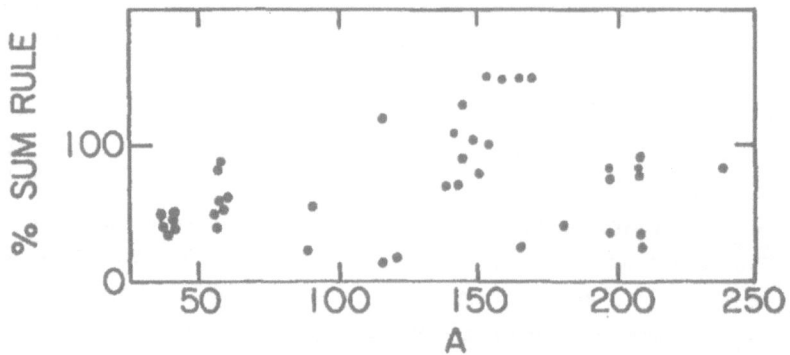

Fig. 51. Summary of the strengths of isoscalar E2 resonances
obtained from inelastic (X,X') and (e,e') experiments.

partly to the difficulty of obtaining definitive width measurements in the presence
of background and other multipoles but also to the difficulty of assigning widths
to the broad and structured distributions in the light nuclei.

The measured strengths of the E2 resonances are summarized in Fig. 51. Here
also there is a very large amount of scatter extending from about 10% to 150%.
These measurements are subject to the same difficulties as the width measurements.
As suggested in Fig. 46 many of the strength measurements may be too large because
of the failure to account properly for the presence of other multipoles. Clearly,
the determination of how much of the sum rule is exhausted by the E2 resonances
is still an open question.

V. The Isovector E2 Strength

The E2 strength discussed in the previous chapter is for the most part the isoscalar strength identified with quadrupole oscillations in which neutrons and
protons move in phase (see Fig. 1). In this section we shall examine the evidence
for isovector E2 strength identified with quadrupole oscillations in which protons
and neutrons move out of phase. In Table I we see that this $2\hbar\omega$ excitation is
pushed up to about $3\hbar\omega$ or about $130/A^{1/3}$ MeV in any typical shell model calculation [83]. The hydrodynamical model gives about $112/A^{1/3}$ MeV [83]. The very
limited experimental evidence presented below suggests a value of about $130/A^{1/3}$.
The isovector E2 sum rule that is analogous to the isoscalar sum rule Eq. (14)
is given by

$$\int (\sigma/E^2)\,dE = 0.255\,(A/4)\,<R^2>\ \mu b/MeV .\qquad (16)$$

A review of the evidence on the isovector E2 resonance has been recently given

[83]. In this section we shall summarize the evidence and give some examples.

V.1. Early evidence

As discussed in Chapter IV, evidence for E2 strength was obtained in (p,γ) and (γ,p) experiments as illustrated in Fig. 26. This strength was generally found to be spread throughout and above the region of the E1 resonance and was thought to be largely isovector in character. It now appears [90] that an appreciable amount of this strength could be of a direct or semi-direct character, which would contribute to both the isoscalar and isovector sum rules. Attempts to find a coherent isovector E2 resonance in the (p,γ_o) reaction will be discussed in Section V.3 below.

In the extensive (γ,n) measurements made throughout the nuclear table [104] significant strength is observed above the Lorentzian lineshape on the high side of the giant E1 resonance in several nuclei. This excess strength has been attributed variously to isospin splitting of the E1 resonance or to an E2 giant resonance. An example of this strength in ^{159}Tb is shown in Fig. 52. A critique of this type of evidence for an isovector E2 resonance is given in Ref. 104.

V.2. Electron scattering

Examples of the significant (e,e') measurements [76,107,120,121] that bear on the existence of an isovector E2 resonance are shown in Figs. 43, 53, and 58. In ^{208}Pb (Fig. 53 and Ref. 107) there appears to be definite structure of E2 character that would correspond to an isovector resonance at 22 MeV or $130/A^{1/3}$ MeV. The width of this resonance would be about 5 MeV. However, in ^{40}Ca (Fig. 43) there is no evidence for such a resonance. There may be E2 strength present but if so it spreads out and does not form a coherent resonant. Similar measurements on ^{90}Zr [120] indicate a situation rather similar to ^{208}Pb (see Fig. 58). Experiments from another laboratory [121] on ^{197}Au and ^{208}Pb find structure similar to that shown in Fig. 53.

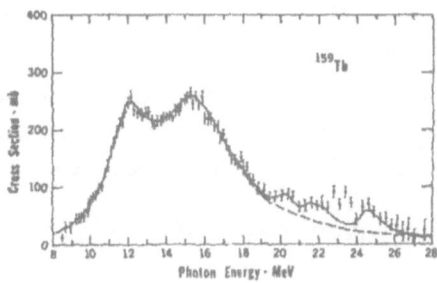

Fig. 52. The total photoneutron cross section for ^{159}Tb. Reference [104].

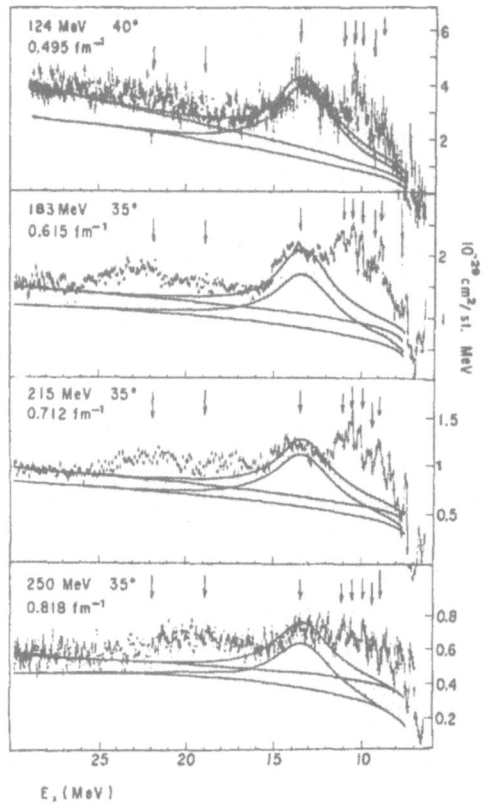

Fig. 53.   Inelastic electron scattering spectra from ^{208}Pb at various momentum transfers.   Reference [107].

From these examples we may conclude that (1) the evidence on the shape of the resonance in the heavy nuclei is fortuitous or (2) a coherent isovector resonance does exist in ^{208}Pb and perhaps in other heavy nuclei, but that in light nuclei such as ^{40}Ca the resonance is very broad and completely washed out.   Clearly, many more detailed (e,e') measurements are needed to establish the existence and universal nature of the isovector E2 resonance.

V.3. Proton capture measurements

The search for a coherent isovector E2 resonance has been carried out with the (p,γ) reaction in a series of measurements [83]. The technique used in this unpolarized work is to measure an "interference" parameter

$$a = \frac{W(55°) - W(125°)}{2A_o P_1(\cos 55°)} = a_1 - 0.68a_3 \sim \left|\frac{E2}{E1}\right| \cos\phi \tag{17}$$

which can then be used to obtain a pseudo E2 cross section

$$\sigma*(E2) = A_o a^2 \sim |E2|^2 \cos^2 \phi \qquad (18)$$

where the coefficients A_o, a_1 and a_3 are defined in Eq. (1).

Measurements of this quantity have been carried out over the expected region of the isovector resonance in ^{12}C, ^{16}O, ^{90}Zr, ^{206}Pb, and ^{207}Bi with interesting results [83]. Figure 54 shows the yield A_o, the asymmetry parameter a, and the quantity $A = aA_o$ for $^{208}Pb(p,\gamma)^{209}Bi$. The experiment actually sums over several gamma rays making transitions to excited states of ^{209}Bi and so is not strictly representative of giant strength built on the ground state. According to Eqs. (17,18) the pseudo E2 yield would be proportional to the product of the curves for a and A. We see that there is definite evidence for resonant structure at E_x = 23 MeV (E_p = 20 MeV) with a width of about 3.5 MeV. It is puzzling that this width is distinctly less than the width observed in electron scattering (Section V.2 above); nevertheless, the evidence from both experiments suggests that there is localized E2 strength at about $130/A^{1/3}$ MeV. The measurements in $^{205}Tl(p,\gamma)^{206}Pb$ give a somewhat similar result [83].

The picture in the light nuclei from ^{12}C to ^{90}Zr is quite different. Figure 55 shows the measurements in $^{89}Y(p,\gamma_o)^{90}Zr$ which are typical also of the lighter nuclei. In this case no resonant structure is observed; the pseudo cross section rises smoothly as the energy increases in a manner indicative of a direct capture process. The evidence here is somewhat in disagreement with that obtained in electron scattering [120] which tends to show a broad resonance in ^{90}Zr. However,

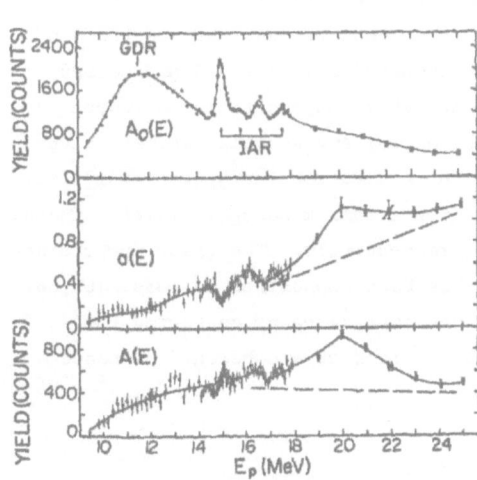

Fig. 54. Excitation curves for the reaction $^{208}Pb(p,\gamma_o\gamma_1\gamma_2)^{209}Bi$. The quantity a is given in Eq. (17) and $A = aA_o$. Reference [83].

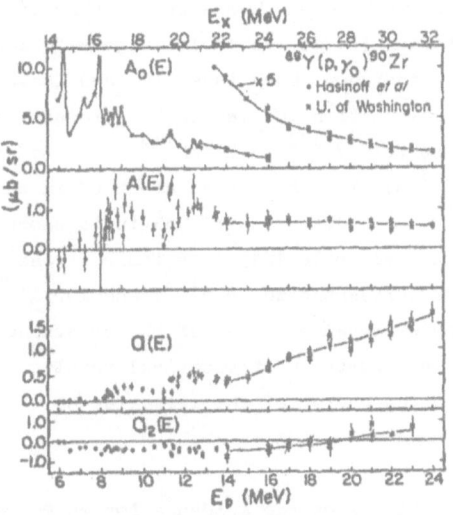

Fig. 55. Excitation curves for the reaction $^{89}Y(p,\gamma_o)^{90}Zr$. The quantity a is given in Eq. (17) and $A = aA_o$. Reference [83].

in evaluating all these (p,γ) measurements it is important to keep in mind the approximations inherent in obtaining the pseudo cross sections, especially the one which ignores variations in the phase factor $\cos^2\phi$ in Eq. (18).

V.4. Summary of evidence for an isovector E2 resonance

In both electron scattering and proton capture measurements there is evidence for a coherent isovector E2 resonance at about $130/A^{1/3}$ MeV in the heavy nuclei. However, in both types of experiments it is clear that further definitive measurements are needed. In the lighter nuclei, the evidence from both reactions suggests that this resonance is broadened to such an extent as to be indistinguishable from a direct or semi-direct mechanism. Thus, at the present time the isovector E2 resonance cannot be considered as firmly established as a universal coherent mode of excitation in nuclei.

VI. The E0 Strength

The location of the isoscalar electric monopole strength is one of the important problems of nuclear physics. As can be seen in Fig. 1 this oscillation corresponds to a compressional or breathing mode and thus relates directly to the compressibility of nuclear matter. Although estimates can be made of the compressibility and thus of the excitation energy of the monopole, in fact the compressibility is not well known and the experimental location of the monopole would provide information on this important property of nuclear matter.

There have now been several theoretical treatments of the electric monopole excitation [64-66,122-124]. Figure 56 reproduces the results of Ref. 122 for ^{208}Pb and shows both isoscalar and isovector $(T_>)$ excitations. We see that in this calculation the isoscalar strength varies anywhere from 15 to 22 MeV corresponding to $88/A^{1/3}$ to $130/A^{1/3}$ MeV. Reference 123 shows an excitation of about $80/A^{1/3}$ MeV falling off to about 60 in the light nuclei and another estimate gives $70/A^{1/3}$ MeV [66]. In Ref. 65 it is shown that the excitation energy is $\sqrt{2}\hbar\omega_o$ for a Suzuki monopole [64] (see Table I) and $\sqrt{6}\hbar\omega_o$ for a Zamick monopole [124]. These values correspond to $60/A^{1/3}$ and $100/A^{1/3}$ MeV, respectively. The great uncertainty in the expected position of the breathing mode has been emphasized in Refs. [65,66]. From this survey of theoretical results we see the importance of measuring the excitation energy of the isoscalar E0 resonance, if in fact a coherent resonance does exist.

VI.1. Status of the evidence for an E0 resonance

In our discussion of the E2 strength in Chapter IV we covered the attempts to identify giant isoscalar E0 strength in ^{40}Ca and ^{208}Pb in the region of $50/A^{1/3}$. In the case of ^{40}Ca this monopole strength is not established experimentally, while in ^{208}Pb high resolution studies [109,110] do not find E0 strength

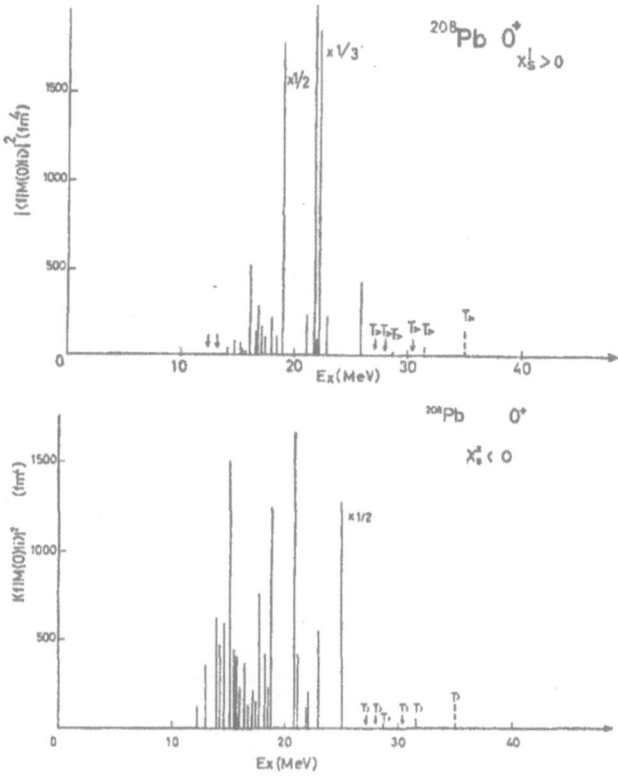

Fig. 56. Energies and transition strengths of calculated monopole states in ^{208}Pb for a positive and a negative
isoscalar coupling constant. Reference [122].

greater than about 13% of the sum rule in this region.

More recently E0 resonances have been claimed at 19.5 MeV ($67/A^{1/3}$ MeV) in
^{40}Ca, 17.5 MeV ($78/A^{1/3}$ MeV) in ^{90}Zr and 13.5 MeV ($80/A^{1/3}$ MeV) in ^{208}Pb in
dueteron scattering [125] and at about 17 MeV ($76/A^{1/3}$ MeV) in ^{90}Zr in electron
scattering [120].

The evidence from deuteron scattering on ^{40}Ca and ^{90}Zr is shown in Fig. 57.
The evidence for an E0 resonance is based on comparing the shape of the E2 resonance
in the (d,d') curves with the shape obtained from (α,α') curves (dashed line)
taken in a different laboratory [61]. The excess of (d,d') strength on the high
energy side of the (α,α') resonance is attributed to a monopole resonance with
an excitation energy of about $67/A^{1/3}$ MeV in ^{40}Ca and $80/A^{1/3}$ MeV in ^{90}Zr and
^{208}Pb. This argument rests on the fact that DWBA calculations predict a much larger
excitation of the monopole for (d,d') scattering than for (α,α') scattering.

The electron scattering result, shown in Fig. 58, is based on a decomposition
of the yield into E1, E2, and the proposed E0 resonance as shown in the figure.
Again, the shape of the E2 resonance is taken from the (α,α') work [61]. It is

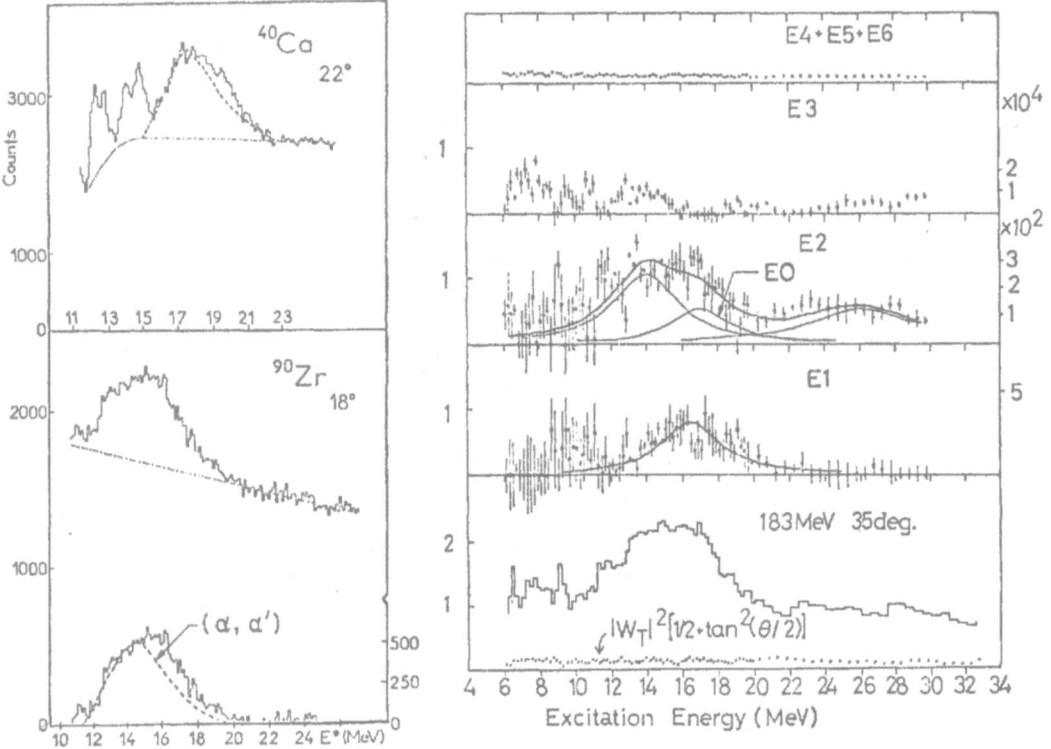

Fig. 57. The (d,d') spectra from ^{40}Ca and ^{90}Zr. The (α,α') resonance [61] is indicated.

Fig. 58. Decomposition of (e,e') curves into multipoles. The proposed E0 resonance is indicated [120].

noted that the shape and location of the E0 resonance agree with those obtained in the (d,d') experiment.

VI.2. Summary of evidence for an E0 resonance

The evidence shown above from (d,d') and (e,e') experiments for the existence of an E0 resonance rests primarily on the shape observed for E2 resonances compared to the shape observed in (α,α') experiments. It would of course be desirable to determine these shapes under controlled conditions in similar experimental arrangements. At present, the evidence for an E0 resonance is provocative but confirmation of these resonances must await further definitive observations.

VII. Other Multipole Resonances

In earlier chapters we have discussed the monopole, dipole, and electric quadrupole resonances. In this chapter we shall review the higher multipole resonances. By extending the modes shown in Fig. 1 we see that such resonances also correspond to simple vibrations of nuclear matter which could be expected to exist in nuclei.

In fact, giant excitations corresponding to several of these modes have been known for some time. A brief review of the earlier work is given in Ref. 5.

VII.1. The M2 and higher magnetic resonances

These resonances have been observed in electron scattering in the light nuclei. Figure 59 shows the electron response function measured for ^{12}C [126]. The major regions of strength are identified with the excitations in ^{12}C shown in Fig. 60. Although unresolved from other excitations, the giant M2 resonance at about 19 MeV constitutes a major part of the strength in the prominent peak in this region. A similar excitation of the giant M2 resonance has been observed in ^{16}O [127].

These giant M2 excitations can be satisfactorily explained in terms of the $p_{3/2}^{-1}d_{5/2}$ particle-hole configuration [126-128] which also accounts for the main component of the giant E1 resonance. In the dipole case, however, the strength is pushed up to 22.6 MeV.

The giant hexadecapole M4 excitation can also be identified in the electron scattering data in Fig. 59. To explain the dependence on momentum transfer of the 19-MeV peak definitely requires both an M2 and an M4 excitation [126]. This M4 excitation is also part of the $p_{3/2}^{-1}d_{5/2}$ configuration and the 4^- level is practically degenerate with the 2^- level as indicated in Fig. 60.

Giant magnetic strength has also been studied in ^{90}Zr [129]. Structure observed at 9 MeV at the backward angle of 155° can be explained as a superposition

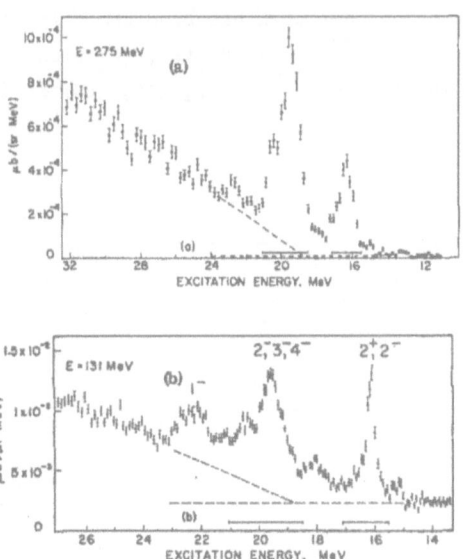

Fig. 59.   Inelastic electron cross sections at 135° from ^{12}C.
Reference [126].

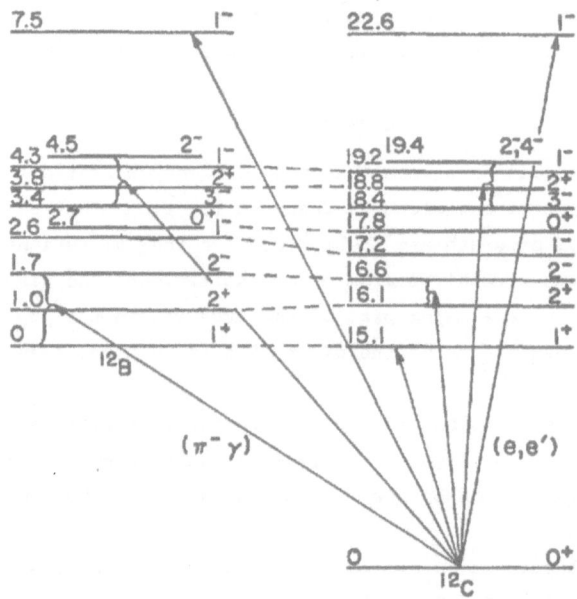

Fig. 60. Giant excitations in mass 12 observed in (e,e') and
(π^-,γ) reactions on ^{12}C.

of M2, M4, and M6 strength, whereas a similar superposition of the odd multipoles,
M1, M3, M5, and M7, fails to account for the observed yield.

Important evidence on the higher giant multipoles has been obtained by radiative pion capture. A review of this work has been recently given [130]. Figure 61
shows the yield measured in the $^{12}C(\pi^-,\gamma)$ capture reaction. By comparing this
curve with that obtained in electron scattering (Fig. 59) it is possible to identify

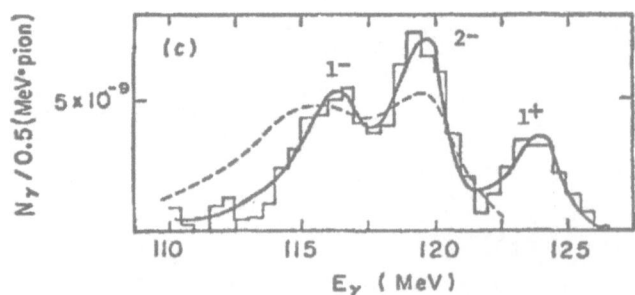

Fig. 61. Gamma-ray yield from π^- capture in ^{12}C. Levels are
identified in Fig. 60. Reference [130].

the giant analogue transitions to the 1^+(M1), 2^-(M2), and 1^-(E1) levels of ^{12}B. Further results from (π^-,γ) capture can be found in Reference [130].

In the heavy nuclei the most significant work has been on ^{208}Pb [55]. The results of this experiment are shown in Fig. 62. The peaks at 7.40 and 7.91 MeV are identified as components of the giant M2 resonance. The excitations at 6.93 and 7.22 MeV are also probably magnetic in character.

The results of a theoretical calculation [131] of the giant magnetic resonances in ^{208}Pb are shown in Fig. 63. We see that major M2 strength is predicted in the 7-8 MeV region. We note also the prediction of two levels for the M1 strength as discussed above in Section III.7 (see Table VI). No experimental evidence has yet been reported for the M3 and M4 resonances predicted by this calculation.

In general the evidence on M2 excitations is not extensive enough throughout the nuclear table to provide for a systematic description of its location and strength in terms of a sum rule. The evidence on the higher magnetic multipoles is even less well developed.

VII.2. The E3 and higher electric resonances

In our discussion of the other multipoles we have seen various instances in which E3 and higher electric multipoles have been invoked to explain the observed response functions. In ^{12}C(e,e'), excitation of the known 3^- level at 18.4 MeV (Fig. 60) has been included [126] in the fit to the momentum-transfer data of the

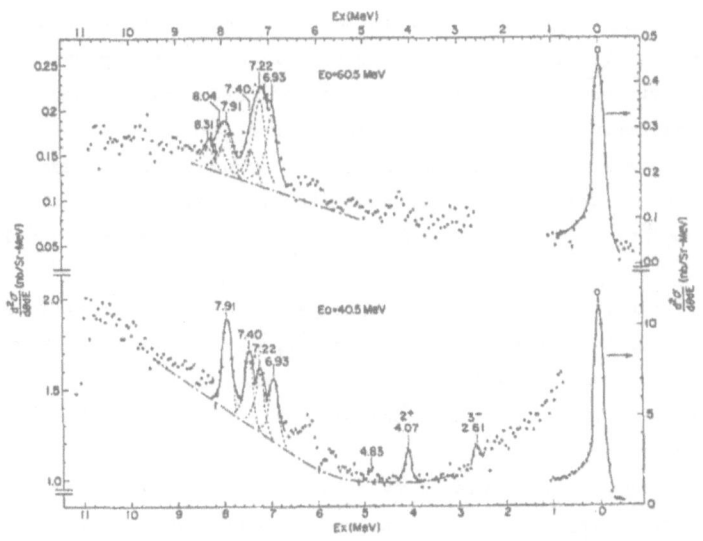

Fig. 62. Spectra of electrons scattered at 180° from ^{208}Pb. Reference [55].

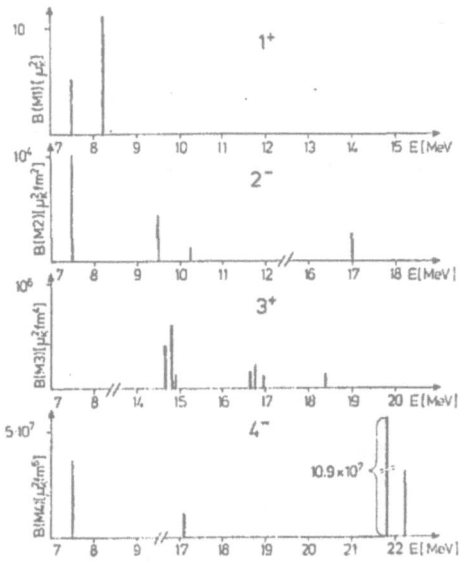

Fig. 63. Calculated results for giant magnetic resonances in 208Pb. Reference [131].

prominent structure observed at 19 MeV (Fig. 59). This excitation is also attributed to the $p_{3/2}^{-1}d_{5/2}$ configuration.

In the doorway-state model of inelastic excitation [81] evidence for higher multipoles has been found. In the analysis shown in Fig. 42 E3 strength appears in the region of 35 MeV in ^{16}O.

In the decomposition of the (e,e') yield from ^{40}Ca shown in Fig. 43 very broad (E_x = 10-22 MeV) E3 strength is invoked to explain the residual yield. A similar analysis [107] of the (e,e') results from ^{208}Pb, shown in Fig. 64, gives a broad and structured E3 response with a concentration of strength in 10 MeV region.

In the gross-structure analysis of (p,p') scattering from ^{208}Pb shown in Fig. 46 a large portion of the strength is attributed to an E4 resonance at 11 MeV.

There are now numerous calculations of the response functions of the higher electric multipoles. Typical results can be found in References [69,132].

It will be clear to the reader of this paper that much of it is in the nature of a progress report. The study of giant multipole resonances is still a vital and fast developing field. We can confidently look forward to many new and exciting discoveries and as a result to a deeper under standing of the nucleus and nuclear physics.

It is a pleasure to acknowledge my colleagues who have participated in the research at Stanford reported here. The experiments were carried out in collaboration with J. R. Calarco, G. A. Fisher, H. F. Glavish, G. King, E. Kuhlmann,

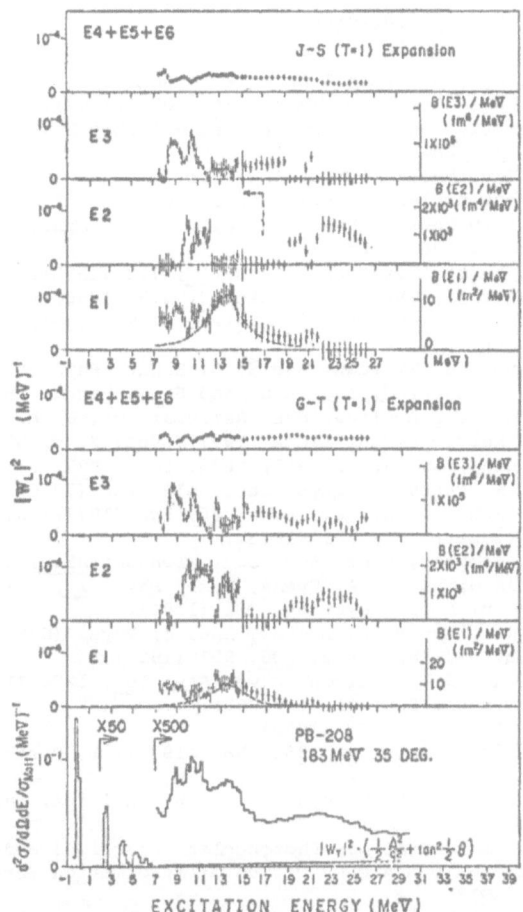

Fig. 64. Inelastic electron scattering at 183 MeV and 35° from ^{208}Pb. For T = 1
excitations: (top) Jensen-Steinwedel, (bottom) Goldhaber-Teller model.
For T = 0 excitations: Tassie model. Reference [107].

P. M. Kurjan, R. LaCanna, D. G. Mavis, and K. Wienhard. The calculations on the
El resonance of ^{16}O are due to H. F. Glavish and D. G. Mavis.

References

* Work supported in part by the National Science Foundation.

1   L. L. Foldy and J. D. Walecka, Nuovo Cimento 34, 1026 (1964).

2   H. Uberall, Electron Scattering from Complex Nuclei, Part B (Academic Press, N.Y. and London, 1971) pp. 664-5.

3   D. H. Wilkinson, Ann. Rev. Nucl. Sci. 9, 1 (1959).

4   E. G. Fuller et al., Photonuclear Reaction Data, NBS Special Publication 380 (1973).

5   S. S. Hanna, in Isospin in Nuclear Physics, ed. D. H. Wilkinson (North-Holland Publishing Co., Amsterdam, 1969), Ch. 12.

6   B. L. Berman and S. C. Fultz, Revs. Mod. Phys. 47, 713 (1975).

7   G. E. Brown and M. Bolsterli, Phys. Rev. Lett. 3, 472 (1959).

8   R. G. Allas, S. S. Hanna, L. Meyer-Schützmeister, R. E. Segel, P. P. Singh, and Z. Vager, Phys. Rev. Lett. 13, 187 (1964).

9   M. Hasinoff, G. A. Fisher, H. M. Kuan, and S. S. Hanna, Phys. Lett. 30B, 337 (1969).

10 W. J. O'Connell, G. L. Latshaw, J. L. Black, and S. S. Hanna, in Proc. Int. Conf. on Photonuclear Reactions and Applications, Asilomar, ed. B. L. Berman (U.S.A.E.C. Office of Information Services, Oak Ridge, 1973), Vol. 2, p. 939.

11 V. Gillet, M. A. Melkanoff, and J. Raynall, Nucl. Phys. A97, 631 (1967).

12 N. M. Kabachnik and V. N. Razuvaev, Phys. Lett. 61B, 420 (1976).

13 C. M. Shakin and W. L. Wang, Phys. Rev. Lett. 26, 902 (1971); W. L. Wang and C. M. Shakin, Phys. Rev. Lett. 30, 301 (1973).

14 J. P. Elliot and B. H. Flowers, Proc. Roy. Soc. (London) A242, 57 (1957).

15 G. E. Brown, L. Castillejo, and J. A. Evans, Nucl. Phys. 22, 1 (1961).

16 V. Gillet and N. Vinh-Mau, Nucl. Phys. 54, 321 (1964).

17 H. Feshbach, A. K. Kerman, and R. H. Lemmer, Ann. of Phys. (N.Y.) 41, 230 (1967).

18 E. Hayward and E. G. Fuller, Phys. Rev. 106, 991 (1957).

19 H. W. Kuehne, P. Axel, and D. C. Sutton, Phys. Rev. 163, 1278 (1967).

20 W. C. Barber, Ann. Rev. Nucl. Sci. 12, 1 (1962).

21 L. W. Fagg, Revs. Mod. Phys. 47, 683 (1975).

22 R. Pitthan and T. Walcher, Phys. Lett. 36B, 563 (1971); R. Pitthan, Z. Physik 260, 283 (1973).

23 C. D. Bowman, R. J. Baglan, B. L. Berman, and T. W. Phillips, Phys. Rev. Lett. 25, 1302 (1970).

24 H. E. Jackson, in Proc. Int. Conf. on Photonuclear Reactions and Applications, Asilomar, ed. B. L. Berman (U.S.A.E.C. Office of Information Services, Oak Ridge, 1973) Vol. 2, p. 817.

25 E. J. Winhold, B. H. Patrick, E. M. Bowey, and D. B. Gayther, in Proc. Int. Conf. on Photonuclear Reactions and Applications, Asilomar, ed. B. L. Berman (U.S.A.E.C. Office of Information Services, Oak Ridge, 1973), Vol. 1, p. 701.

26 L. M. Bollinger, in Proc. Int. Conf. on Photonuclear Reactions and Applications, Asilomar, ed. B. L. Berman (U.S.A.E.C. Office of Information Services, Oak Ridge, 1973), Vol. 2, p. 783.

27 E. Lipparini, S. Stringari, M. Traini, and R. Leonardi, Nuovo Cimento 31A, 207 (1976).

28 E. K. Warburton and J. Weneser, in Isospin in Nuclear Physics, ed. D. H. Wilkinson (North-Holland Publishing Col, Amsterdam, 1969), Ch. 5.

29 D. Kurath, Phys. Rev. 130, 1525 (1963), and private communication.

30 F. Ajzenberg-Selove and T. Lauritsen, A = 5-10, Nucl. Phys. A227, 1 (1974); F. Ajzenberg-Selove, A = 11-12, Ibid. A248, 1 (1975); A = 13-15, Ibid. A152, 1 (1970); A = 16-17, Ibid. A166, 1 (1971); A = 18-20, Ibid. A190, 1 (1972).

31 L. W. Fagg, R. A. Lindgren, W. L. Bendel, and E. C. Jones, Jr., preprint.

32 M. Harakeh, P. Paul, and K. A. Snover, Phys. Rev. C11, 998 (1975).

33 S. S. Hanna, in Proc. of the Sixth Masurian School in Nuclear Physics, Mikolajki, Nukleonika, Vol. 19, Nos. 7-8, 655-685 (1974).

34 P. M. Endt and C. Van der Leun, Nucl. Phys. A214, 1 (1973).

35 A. Richter, private communication.

36 J. Vernotte, S. Gales, M. Langevin, and J. M. Maison, Phys. Rev. C8, 178 (1973); A. Huck, G. J. Costa, G. Walter, M. M. Aleonard, J. Dalmas, P. Hubert, F. Leccia, P. Mennrath, J. Vernotte, M. Langevin, and J. M. Maison, Phys. Rev. C13, 1786 (1976).

37 R. A. Lindgren, W. L. Bendel, E. C. Jones, Jr., L. W. Fagg, X. K. Maruyama, J. W. Lightbody, and S. P. Fivozinsky, Phys. Rev. C, November, 1976.

38 F. R. Metzger, Nucl. Phys. A173, 141 (1971).

39 L. M. Bollinger and G. E. Thomas, Phys. Rev. C2, 1951 (1970); R. K. Smither and L. M. Bollinger, in Proc. Int. Symp. on Neutron Capture Gamma-Ray Spectroscopy, Studsvik (IAEA, 1969), p. 601.

40 G. A. Bartholomew, E. D. Earle, A. J. Ferguson, J. W. Knowles and M. A. Lone, in Advances in Nuclear Physics, eds. M. Baranger and E. Vogt (Plenum Press, 1973), Vol. 7, p. 229.

41 R. J. Holt and H. E. Jackson, Phys. Rev. C12, 56 (1975).

42 F. R. Buskirk et al., in Proc. Int. Conf. on Photonuclear Reactions and Applications, Asilomar, ed. B. L. Berman (U.S.A.E.C. Office of Information Services, Oak Ridge, 1973), Vol. 1, p. 703.

43 C. P. Swann, Phys. Rev. Lett. 32, 1449 (1974).

44 R. Del Vecchio, S. Freedman, G. T. Garvey and M. Oothoudt, Phys. Rev. Lett. 34, 1296 (1975).

45 R. J. Holt and H. E. Jackson, Phys. Rev. Lett. 36, 244 (1976), and private communication.

46 G. T. Garvey, private communication.

47 B. H. Wildenthal, private communication.

48 H. E. Gove, K. H. Purser, J. J. Schwartz, W. P. Alford, and D. Cline, Nucl. Phys. A116, 369 (1968).

49 M. J. A. De Voight and G. H. Wildenthal, Nucl. Phys. A206, 305 (1973), and references contained therein.

50 E. C. Halbert, J. B. McGrory, and B. H. Wildenthal, Phys. Rev. Lett. 20, 1112 (1968).

51 E. G. Adelberger, A. B. McDonald, C. L. Cocke, C. N. Davids, A. P. Shukla, H. B. Mak, and D. Ashery, Phys. Rev. C7, 889 (1973).

52 A. Arima and S. Yoshida, Nucl. Phys. A161, 492 (1971).

53 R. E. Marrs, E. G. Adelberger, K. A. Snover, and M. D. Cooper, Phys. Rev. Lett. 35, 202 (1975).

54 A. Wolf, R. Moreh, A. Nif, O. Shahal, and J. Tenebaum, Phys. Rev. C6, 2276 (1972).

55 R. A. Lindgren, W. L. Bendel, L. W. Fagg, and E. C. Jones, Jr., Phys. Rev. Lett. 35, 1423 (1975).

56 S. J. Freedman, C. A. Gagliardi, G. T. Garvey, M. A. Oothoudt, and B. Svetitsky, to be published.

57 P. Ring and J. Speth, Phys. Lett. 44B, 477 (1973).

58 M. B. Lewis and F. E. Bertrand, Nucl. Phys. A196, 337 (1972).

59 Progress Report, Nuclear Physics Laboratory, Stanford University, 1974; E. Kuhlmann, E. Ventura, J. R. Calarco, D. G. Mavis, and S. S. Hanna, Phys. Rev. C11, 1525 (1975).

60 S. S. Hanna, H. F. Glavish, R. Avida, J. R. Calarco, E. Kuhlmann, and R. LaCanna, Phys. Rev. Lett. 32, 114 (1974).

61 J. M. Moss, C. M. Rozsa, D. H. Youngblood, J. D. Bronson, and A. D. Bacher, Phys. Rev. Lett. 34, 748 (1975); D. H. Youngblood et al., Phys. Rev. C13, 994 (1976).

62 J. M. Blatt and V. F. Weisskopf, Theoretical Nuclear Physics (Wiley, New York, 1952), p. 656.

63 A. Bohr and B. R. Mottelson, Nuclear Structure, (W. A. Benjamin, N.Y., 1976), Vol. 2, p. 508.

64 T. Suzuki, Nucl. Phys. A217, 182 (1973).

65 E. Lipparini, G. Orlandini, and R. Leonardi, Phys. Rev. Lett. 36, 660 (1976); Phys. Lett. 64B, 21 (1976); and private communication.

66 J. Martorell, O. Bohigas, S. Fallieros, and A. M. Lane, Phys. Lett. 60B, 313 (1976) and private communication.

67 L. J. Tassie, private communication.

68 S. Krewald, J. Birkholz, A. Faessler, and J. Speth, Phys. Rev. Lett. 33, 1386 (1974); and private communication.

69 G. F. Bertsch and S. F. Tsai, Physics Reports, 18C, 125 (1975).

70 K. F. Liu and G. E. Brown, Nucl. Phys. A265, 385 (1976); K. F. Liu and N. V. Giai, to be published.

71 R. G. Allas, S. S. Hanna, L. Meyer-Schützmeister, and R. E. Segel, Nucl. Phys. 58, 122 (1964).
72 D. E. Frederick, R. J. J. Stewart, and R. C. Morrison, Phys. Rev. 186, 992 (1969).
73 L. Meyer-Schützmeister, Z. Vager, R. E. Segel, and P. P. Singh, Nucl. Phys. A108, 180 (1968).
74 R. B. Watson, D. Branford, J. L. Black, and W. J. Caelli, Nucl. Phys. A203, 209 (1973); G. S. Foote et al., Nucl. Phys. A263, 349 (1976).
75 K. A. Snover, E. G. Adelberger, and D. R. Brown, Phys. Rev. Lett. 32, 1061 (1974).
76 Y. Torizuka, K. Itoh, Y. M. Shin, K. Kawazoe, H. Matsugaki, and G. Takeda, Phys. Rev. C11, 1174 (1975).
77 N. Marty, M. Morlet, A. Willis, V. Comparat, and R. Frascaria, Nucl. Phys. A238, 93 (1975).
78 G. Perrin et al., in Proc. Int. Conf. on Nuclear Structure and Spectroscopy, Amsterdam, eds. H. P. Blok and A. E. L. Dieperink (Scholar's Press, Amsterdam, 1974), Vol. 1, p. 158; M. Buenerd et al., Phys. Rev. C, October 1976.
79 A. Moalem, W. Benenson, and G. M. Crawley, Phys. Rev. Lett. 31, 482 (1973).
80 C. C. Chang, F. E. Bertrand, and D. C. Kocher, Phys. Rev. Lett. 34, 221 (1975).
81 H. V. Geramb et al., Nucl. Phys. A199, 454 (1973); D. Lebrun et al., Nucl. Phys., to be published; and private communication.
82 R. E. Peschel, J. M. Long, H. D. Shay, and D. A. Bromley, Nucl. Phys. A232, 269 (1974).
83 P. Paul, in Proc. Symp. on Highly Excited States, Jülich, September 1975.
84 R. J. Philpott and P. P. Szydlik, Phys. Rev. 153, 1039 (1967).
85 T. Hoshino and A. Arima, Phys. Rev. Lett. 37, 266 (1976).
86 W. Knüpfer and M. G. Huber, Z. Physik A, to be published.
87 J. S. Dehesa, J. Speth, S. Krewald, and A. Faessler, to be published.
88 C. C. Yang, P. P. Singh, A. van der Woude, and A. G. Drentje, Phys. Rev. C13, 1376 (1976).
89 D. G. Mavis and H. F. Glavish, Bull. Am. Phys. Soc. 21, 636 (1976).
90 K. A. Snover, J. E. Bussoletti, K. Ebisawa, T. A. Trainor, and A. B. McDonald, Phys. Rev. Lett. 37, 273 (1976).
91 F. E. Bertrand, to be published in Ann. Rev. of Nucl. Sci.
92 A. Moalem, W. Benenson, and G. M. Crawley, Nucl. Phys. A236, 307 (1974).
93 K. T. Knöpfle, G. J. Wagner, H. Breuer, M. Rogge, and C. Mayer-Böricke, Phys. Rev. Lett. 35, 779 (1975).
94 M. N. Harakeh, A. R. Arends, M. J. A. de Voight, A. G. Drentje, S. Y. van der Werf and A. van der Woude, Nucl. Phys. A265, 189 (1976).
95 K. T. Knöpfle et al., private communication.
96 A. Hotta, K. Itoh, and T. Saito, Phys. Rev. Lett. 33, 790 (1974).
97 G. R. Hammerstein, H. McManus, A. Moalem, and T. T. S. Kuo, Phys. Lett. 49B, 235 (1974).
98 A. Moalem, W. Benenson, G. M. Crawley, and T. L. Khoo, Phys. Lett. 61B, 167 (1976).
99 T. Walcher, in Proc. Int. Conf. on Nuclear Physics, Munich, eds. J. de Boer and H. J. Mang (North-Holland Publishing Co., Amsterdam, 1973), p. 510.
100 G. R. Satchler, Nucl. Phys. A195, 1 (1972); Part. and Nucl. 5, 105 (1973).
101 J. F. Ziegler and G. A. Peterson, Phys. Rev. 165, 1337 (1968).
102 F. R. Buskirk et al., in Proc. Int. Conf. on Nuclear Structure and Spectroscopy, Amsterdam, eds. H. P. Blok and A. E. L. Dieperink (Scholar's Press, Amsterdam, 1974), Vol. 1, p. 205; R. Pitthan, private communication.
103 A. Veyssière, H. Beil, R. Bergère, P. Carlos, and A. Leprêtre, Nucl. Phys. A159, 561 (1970).
104 B. L. Berman and S. C. Fultz, Revs. Mod. Phys. 47, 713 (1975).
105 L. M. Young, thesis, University of Illinois. Data reproduced in Ref. 101.
106 A. Schwierczinski, R. Frey, A. Richter, E. Spamer, H. Theissen, O. Titze, T. Walcher, S. Krewald, and R. Rosenfelder, Phys. Rev. Lett. 35, 1244 (1975).
107 M. Nagao and Y. Torizuka, Phys. Rev. Lett. 30, 1068 (1973); M. Nagao Sasao and Y. Torizuka, preprint.
108 E. C. Halbert, J. B. McGrory, G. R. Satchler, and J. Speth, Nucl. Phys. A245, 189 (1975).
109 H. P. Morsch, P. Decowski, and W. Benenson, Phys. Rev. Lett. 37, 263 (1976).
110 H. K. Sherman, H. M. Ferdinande, K. H. Lokan, and C. K. Ross, Phys. Rev. Lett. 35, 1215 (1975).

111 R. Pitthan, F. R. Buskirk, and Th. Walcher, preprint.

112 E. G. Fuller and M. S. Weiss, Phys. Rev. 112, 560 (1958).

113 D. J. Horen, F. E. Bertrand, and M. B. Lewis, Phys. Rev. C9, 1607 (1974).

114 D. J. Horen, J. Arvieux, M. Buenard, J. Cole, G. Perrin, and P. Saintignon, Phys. Rev. C11, 1247 (1975).

115 T. Kishimoto, J. M. Moss, D. H. Youngblood, J. O. Bronson, C. M. Rozsa, D. R. Brown, and A. D. Bacher, Phys. Rev. Lett. 35, 552 (1975).

116 A. Schwierczinski, R. Frey, E. Spamer, H. Theissen, and Th. Walcher, Phys. Lett. 55B, 171 (1975).

117 G. L. Moore, F. R. Buskirk, E. B. Dally, J. N. Dyer, X. K. Maruyama, and R. Pitthan, Z. Naturforschung, to be published.

118 D. Zawischa and J. Speth, Phys. Rev. Lett. 36, 843 (1976).

119 G. Kyrchev, L. A. Malov, V. O. Nesterenko, and V. G. Soloviev, Phys. Lett., to be published.

120 S. Fukuda and Y. Torizuka, Phys. Rev. Lett. 29, 1109 (1972); and preprint.

121 R. Pitthan et al., Phys. Rev. Lett. 33, 849 (1974).

122 H. Sagawa, preprint.

123 S. Krewald, R. Rosenfelder, J. E. Galonska, and A. Faessler, preprint.

124 L. Zamick, Nucl. Phys. A232, 13 (1974).

125 N. Marty, M. Morlet, A. Willis, V. Comparet, and R. Frascaria, preprint.

126 T. W. Donnelly, J. D. Walecka, I. Sick, and E. B. Hughes, Phys. Rev. Lett. 21, 1196 (1968).

127 I. Sick, E. B. Hughes, T. W. Donnelly, J. D. Walecka, and G. E. Walker, Phys. Rev. Lett. 23, 1117 (1969).

128 N. Vinh-Mau and G. E. Brown, Nucl. Phys. 29, 89 (1962).

129 Y. Torizuka, in Proc. Colloque Franco-Japonais sur Spectroscopie Nucleaire et Reaction Nucleaire, Dogashima, Japan, September 1976.

130 H. W. Baer, K. M. Crowe, and P. Truöl, in Advances in Nuclear Physics, Vol. 9.

131 J. Speth, in Proc. Symposium on Nuclear Structure; Coexistence of Single Particle and Collective Types of Excitations, Balatonfüred, Hungary, September 1975.

132 A. Faessler, in Proc. XIII International Meeting on Nuclear Physics, Bormeo, Italy, January 1975.

Photon Scattering in the Energy Range 5-30 MeV

Evans Hayward
National Bureau of Standards
Washington, D. C. 20234

I. AN OVERVIEW

As a means of introducing the subject, the elastic scattering cross sections for two especially simple nuclei having rather low-level densities are shown in Figs. 1 and 2. These plots are largely artistic and are meant to illustrate the main phenomena taking place. Figure 1 shows the elastic scattering cross section for ^{12}C. The resonance fluorescence of the levels at 4.43, 12.73, and 15.1 MeV is indicated by the bars one MeV wide, the areas being proportional to their integral scattering cross sections. The scattering cross section in the continuum region has been calculated using the absorption cross section data of Ahrens et al. (75Ah3). This figure shows that the spectrum scattered by a carbon target is dominated by the resonance fluorescence of the 15.1 MeV level.

Figure 2 illustrates similar data for ^{208}Pb. The resonance fluorescence data came from the University of Illinois' theses of Laszewski (75La2) and Coope (75Co2). The giant resonance cross section has been calculated using the absorption cross section of Veyssière et al. (70Vel) and assuming that all the transitions are electric dipole.

A. Resonance Fluorescence

Below the particle emission threshold, photon scattering is the only process that can occur. The energy levels are discrete and well-defined, and the photon absorption cross section into such a level may be written as:

$$\sigma_a = 2\pi\lambda^2 \; \frac{2I+1}{2I_0+1} \; \frac{\gamma_0}{\Gamma} \; \frac{(E\Gamma)^2}{(E_0^2-E^2)^2 + (E\Gamma)^2} \; . \tag{1.1}$$

Here, I and I_0 refer to the spins of the excited and ground states, γ_0 is the ground-state radiation width, and Γ the total width of the level. Since Γ is the sum of all the partial widths associated with the decay to all the lower states,

$$\Gamma = \gamma_0 + \gamma_1 + \gamma_2 + - - - - \tag{1.2}$$

the states at higher excitation energy become progressively wider and wider. The width of the level then depends on the properties of all the states at lower excitation energy.

The integrated absorption cross section, on the other hand, depends only on the excitation energy and the ground-state radiation width since

$$\int \sigma_a(E)\,dE = \frac{\pi}{2} \; \sigma_a^0 \Gamma \; , \tag{1.3}$$

where σ_a^0 is the peak absorption cross section and,

$$\sigma_a^0 = 2\pi\lambda^2 \; \frac{2I+1}{2I_0+1} \; \frac{\gamma_0}{\Gamma} \; . \tag{1.4}$$

Combining (1.3) and (1.4), we obtain

$$\int \sigma_a(E)\, dE = (\pi \lambda)^2 \, \frac{2I+1}{2I_0+1} \, \gamma_0 \; . \tag{1.5}$$

The integrated scattering cross section is then

$$\int \sigma_s(E)\, dE = \frac{\gamma_0}{\Gamma} \int \sigma_a(E)\, dE = (\pi \lambda)^2 \, \frac{2I+1}{2I_0+1} \, \frac{\gamma_0^2}{\Gamma} \; . \tag{1.6}$$

This is the quantity measured when a target is irradiated with a spectrum that is broader than the level width, such as bremsstrahlung. The integral scattering cross sections plotted in Fig. 1 and tabulated in Table I were obtained from (1.6) using ground-state radiation widths given in (75Aj6).

In resonance fluorescence experiments there can be quite substantial target thickness corrections because the peak absorption cross sections are enormous and large compared to the electronic cross sections. The peak absorption cross section, defined by (1.4), can amount to hundreds of barns (see Table I). However, it is usually depressed by at least one of two important effects. The first of these is contained in the ratio γ_0/Γ; the peak absorption cross section in a level is inversely proportional to its total width. The second is thermal Doppler broadening, the effective broadening of the line as a result of the random motions of the nuclei in the target.

The absorption cross section is consequently expressed as a Doppler-broadened, Breit-Wigner line:

$$\sigma(x,t) = \sigma_a^0 \; \psi(x,t) \; , \tag{1.7}$$

where

$$x = 2(E-E_0)/\Gamma \text{ and } t = (\Delta/\Gamma)^2 \; . \tag{1.8}$$

The function

$$\psi(x,t) = \frac{1}{2(\pi t)^{\frac{1}{2}}} \int_{-\infty}^{\infty} \frac{e^{-(x-y)^2/4t}}{1+y^2} \; dy \; , \tag{1.9}$$

with $y = 2(E'-E_0)/\Gamma$ $\hspace{3cm}$ (1.10)

is available in tabular form (54Ro1). The Doppler width

$$\Delta = E_0 (2kT'/AMc^2)^{\frac{1}{2}} \; , \tag{1.11}$$

where k is the Boltzmann constant and T' is the effective temperature given by Lamb (39La1) in terms of the Debeye temperature. For a fully Doppler-broadened line, $t \gg 1$, the function $\psi(x,t)$ becomes a Gaussian:

$$\psi(x,t) = \frac{\pi^{\frac{1}{2}}\Gamma}{2\Delta} \; e^{(E-E_0)^2/\Delta^2} \; . \tag{1.12}$$

In the other extreme where t<<1, $\psi(x,t)$ is a Breit-Wigner line:

$$\psi(x,t) = \frac{\Gamma^2/4}{\Gamma^2/4 + (E-E_o)^2} \quad . \qquad (1.13)$$

These limiting forms are very useful for making estimates. Quantitative results, on the other hand, always require the use of the exact $\psi(x,t)$. Fortunately, a modern computer program is available (63Os2).

Resonance fluorescence studies have been made using as sources bremsstrahlung, the bremsstrahlung monochrometer (62Oc1, 63Ti1) and neutron capture γ-rays. The 1^+ states in the self-conjugate nuclei ^{12}C, ^{24}Mg, and ^{28}Si have traditionally been favorite candidates for resonance fluorescence studies. They have been studied using bremsstrahlung (57Ha1) and the brems-strahlung monochrometer (67Ku2). These techniques have now probably been superseded by inelastic electron scattering (75Fa8).

The odd-A nuclei in the s-d shell ^{19}F, ^{23}Na, ^{27}Al, ^{31}P, ^{35}Cl, and ^{39}K were studied (72Sh2) in an important experiment where 14 MeV bremsstrahlung was used in conjunction with a Ge(Li) detector. Many γ-rays were observed and ground-state radiation widths estimated. A comparison with inelastic electron scatter-ing data supports the idea that these γ-rays have an M1 character.

The results shown in Fig. 2 for ^{208}Pb are based on the University of Illinois' theses of Laszewski (75La2) and Coope (75Co2). In the first experiment, the bremsstrahlung monochro-meter was used in conjunction with a sodium iodide spectrometer to trace out the elastic scattering cross section with an energy resolution of ~ 100 keV. In the second, bremsstrahlung was scattered from the same target but the spectrum was analyzed using a Ge(Li) detector. A combination of these two results yields the energies of the states as well as their integral scattering cross sections. Assuming that there is only decay to the ground state, $\gamma_0 = \Gamma$, the ground-state radiation width can also be obtained (see Table II). Most of these levels had already been found by Khan and Knowles (72Kh1) who used a variable energy photon monochrometer using Compton-scattered, neutron-capture γ-rays. We presume all of these states to be 1^-. Those at 5.295 and 5.515 MeV have been shown to be 1^- in a study of the (d,pγ) reaction by Earle et al. (70Ea1). Del Vecchio et al. (75De5), have shown that the 4.843 MeV state is also 1^-.

Thermal neutron capture γ-rays have been used extensively to study photonuclear reactions, and especially in photon scattering experiments. As a source, neutron capture γ-rays offer extremely well-defined energies though one has to be satisfied with the choices made by nature. Nevertheless, they have been used in many very important experiments (73Ar15) both in the energy region where the energy levels are discrete and in the continuum. In the recent past this work has been carried on by Arad and Ben-David, Moreh, Jackson, and their collaborators. There are a remarkable number of chance coincidences between capture γ-ray energies and bound states in medium and heavy nuclei. Of course, the intensities vary over a large range.

The most famous example is the 7.277 MeV level in ^{208}Pb which is excited very strongly by a capture γ-ray in iron. This level is not illustrated in Fig. 2 because its ground-state radiation width is only 0.78 eV which corresponds to an integrated scattering cross section of only 0.17 MeV mb. Moreh and Friedman (68Mo1) have analyzed this radiation using a Compton polarimeter and found that the level at 7.277 MeV has positive parity. The only other state in ^{208}Pb having a definite 1^+ assignment is one above the (γ,n) threshold at 7.99 MeV. Holt and Jackson (76Ho1) have measured the polarization of the photoneutrons emitted by seven states below 8.5 MeV and found only this one with positive parity.

The elastic scattering cross sections below the particle threshold for ^{12}C and ^{208}Pb, illustrated in Figs. 1 and 2, are not typical. Most nuclei have high level densities making inelastic scattering more important and depressing the elastic scattering cross section. In other words, the average total absorption cross section is a smooth function of energy. In the energy region where the total cross section must go into scattering, the magnitude of the elastic scattering cross section depends on the properties of the low-lying levels through which deexcitation can occur.

In some nuclei, inelastic photon scattering leads to the production of isomeric states. The cross sections for their production have been obtained by measuring the radioactivities (60Bo2, 58Si1, 58Bo1, 56Mo3, 56Bo1) produced in ^{89}Y, ^{103}Rh,

^{107}Ag, ^{115}In, and ^{197}Au. Figure 3 shows a typical example (58Sil) of the cross section for the formation of the isomeric state in ^{89}Y as a function excitation energy. Below the particle emission threshold in these nuclei, inelastic scattering competes successfully with elastic scattering. Just above the (γ,n) threshold both the elastic and inelastic scattering cross sections drop as a result of competition with neutron emission. As the excitation energy is increased through the giant resonance, the isomer production cross section, as well as the elastic scattering cross section, go through a maximum.

B. Scattering in the Continuum

At excitation energies a few hundred kilovolts above the (γ,n) threshold the nuclear energy levels fade into the continuum. The nuclear scattering cross sections in the continuum can be calculated from the photonuclear absorption cross sections. The latter have been measured (75Ah3) directly for light elements and for heavy elements are equivalent to the neutron production cross sections. Since the coherent scattering cross section and the total absorption cross section can be expressed in terms of the same complex scattering amplitude, $R(E,\theta)$, it is possible to obtain the scattering cross section from the absorption cross section. The absorption cross section is related to the forward scattering amplitude by the so-called optical theorem,

$$\sigma_a(E) = 4\pi\lambda\,\text{Im}\,R(E,0). \qquad (1.14)$$

The coherent scattering cross section is

$$\frac{d\sigma}{d\Omega}(E,\Theta) = |R(E,\Theta)|^2 \quad . \tag{1.15}$$

The dispersion relation is the connection between the real and imaginary parts of the forward scattering amplitude,

$$\text{Re } R(E,0) = \frac{E}{2\pi^2\hbar c} \, P \int \frac{dE' \sigma_t(E')}{E'^2 - E^2} \quad . \tag{1.16}$$

If the total absorption cross section is known over a wide range of energies, then the scattering cross section can be obtained from it.

The nuclear scattering amplitude must be combined with three other scattering amplitudes. The Thomson scattering amplitude, D, is real and independent of energy

$$D = -\frac{Z^2 e^2}{AMc^2} \quad . \tag{1.17}$$

Rayleigh and Delbrück scattering are associated, respectively, with the photoelectric effect and pair production. Both are peaked very strongly forward. Rayleigh scattering can probably be safely neglected at the energies of interest here, but Delbrück scattering surely makes an important contribution at small angles.

The coherent scattering cross sections for ^{12}C and ^{208}Pb of Figs. 1 and 2 were calculated from the total absorption cross sections using (1.14) and (1.16) and assuming that the absorption is electric dipole.

The structures that appear in the scattering cross section are amplifications of those in the absorption cross section since the scattering cross section depends on the square of the scattering amplitude.

To avoid the necessity of making the dispersion integrals, photonuclear cross sections are often represented by one or more Lorentz lines. For example, if the electric dipole giant resonance can be described by a single Lorentz line, then the forward scattering amplitude is

$$R(E,0) = \frac{NZ\beta e^2}{AMc^2} E^2 \ \frac{E_o^2 - E^2 + i\Gamma E}{(E_o^2 - E^2)^2 + (\Gamma E)^2} \tag{1.18}$$

where E_o is the giant resonance energy and Γ its width. Then, using (1.14)

$$\sigma_t = \frac{4\pi e^2 \hbar}{Mc} \frac{NZ\beta}{A} \ \frac{E^2 \Gamma}{(E_o^2 - E^2)^2 + (E\Gamma)^2} \tag{1.19}$$

and

$$\int \sigma dE = \frac{2\pi^2 e^2 \hbar}{Mc} \frac{NZ\beta}{A} \ , \tag{1.20}$$

the nuclear dipole sum, where β is the factor by which the integrated absorption cross section exceeds a dipole sum. It is also sometimes useful to write the amplitude in terms of its maximum value, σ_o, and the width Γ:

$$R(E,0) = \frac{\sigma_o \Gamma}{4\pi\hbar c} E^2 \frac{E_o^2 - E^2 + i\Gamma E}{(E_o^2 - E^2)^2 + (\Gamma E)^2} . \qquad (1.21)$$

The splitting in two of the giant resonances of the deformed nuclei is one of the best-understood phenomena in photonuclear physics. These two resonances are associated with electric dipole charge oscillations along the one long ($\Delta K=0$) and two short axes ($\Delta K=\pm 1$) of the nuclear ellipsoid, and hence, the upper resonance has an integrated absorption cross section twice as big as the lower. The coherent scattering cross section results from the superposition of the two amplitudes. They interfere destructively in the energy region between the two resonances thus suppressing the coherent scattering associated with the low energy resonance. This is illustrated in Fig. 4 which shows the coherent scattering cross section calculated using the Saclay (73Vel) resonance parameters for ^{238}U. In this nucleus the two resonances in the absorption cross section are located at 10.96 and 14.04 MeV; the lower resonance appears only as a shoulder in the coherent scattering cross section.

The giant resonance splitting is associated with the fact that the nuclear polarizability is a tensor. This means that in addition to the coherent scattering just discussed, there is another important component in the scattering cross section which is incoherent with the incident radiation. If the forward scattering amplitudes associated with charge oscillations along the

one long and two short axes of the nuclear ellipsoid are labeled A and B and have the form of (1.18), then the coherent
scattering cross section is

$$\left.\frac{d\sigma}{d\Omega}\right)_0 = \left[\frac{A+2B}{3} + D\right]^2 \frac{1+\cos^2\theta}{2} . \qquad (1.22)$$

The incoherent, or Raman scattering cross section is

$$\left.\frac{d\sigma}{d\Omega}\right)_2 = \sum_{I_f}(I_0K_0 20 |I_fK_0)^2 |\tfrac{2}{3}(A-B)|^2 \frac{13+\cos^2\theta}{40} . \qquad (1.23)$$

This scattering cross section is also illustrated in Fig. 4.
In ^{238}U this represents the cross section for populating the 2^+
state of the ground-state rotational band, the only excited
state of that band that can be reached in dipole-dipole transitions. Since this state is only 45 keV above the ground state,
when the scattering cross section is measured using a NaI(Tℓ)
detector, the ground-state and first excited state cross sections
are summed and the result is just a smooth, uninteresting giant
resonance cross section.

The giant resonance scattering cross sections have been
studied using neutron capture γ-rays and Ge(Li) detectors by
Jackson, Moreh, and their collaborators (74Ja2, 75Ja1, 71Ha1,
74Ba6). They have spectroscopically separated these two components in the scattering and shown that, by and large, they obey
the simple, classical model outlined above. Fig. 5 shows the

Ge(Li) pulse height distribution produced by the 10.83 MeV photons scattered by ^{238}U as observed by Jackson and Wetzel (72Ja1).

Delbrück Scattering

One of the important results of these high resolution experiments was the observation of coherent scattering peaked strongly forward. For the first time Delbrück scattering was identified without ambiguity.

Delbrück scattering is the process in which an electron-positron pair is created and then annihilated in a static Coulomb field. It can be described by a complex scattering amplitude, the imaginary part being related to pair production and the real part to vacuum polarization. Some time ago, Ehlotzky and Sheppey (64Eh1) calculated these amplitudes, differential in polarization for selected energies below 20 MeV. In a recent paper, Papatzacos and Mork (75Pa9) have confirmed the imaginary parts of their amplitudes but found the real parts to be in error.

Meanwhile, Moreh and his collaborators (74Ka9, 74Ba6) have studied the coherent scattering cross sections for a number of targets in the energy range 7.9-11.4 MeV. Moreh recognized that the nuclear Thomson scattering and nuclear resonance scattering have the same angular momentum properties. Below the (γ,n) threshold the real part of the nuclear scattering amplitude and the Thomson scattering amplitude are opposite in sign so that it is possible to find an energy at which they very nearly cancel. Here the Delbrück scattering amplitude is the only one contributing to the coherent scattering cross section. For ^{181}Ta this energy is near 7.9 MeV. Figure 6 shows a comparison of the measured elastic scattering cross section for ^{181}Ta at this energy with the prediction of Papatzacos and Mork (75Pa9); the agreement is very satisfactory. The experimental cross sections for U and Th are systematically smaller than the prediction; whether this is a serious discrepancy remains to be seen.

A second experiment performed at only 2.75 MeV (75Sc15), where Rayleigh scattering cannot be neglected, has shown that the real and imaginary parts of all the amplitudes must be included. This is the one experiment where the four amplitudes, Rayleigh, Thomson, nuclear, and Delbrück, all contribute and the result is apparently understood.

II. ANGULAR DISTRIBUTIONS AND POLARIZATIONS

The foregoing has avoided any discussion of angular momentum considerations. Here we will summarize the angular distribution formulae for the scattering of both polarized and
unpolarized radiation. Photon scattering is, after all, just
a special case of the general angular correlation problem which
has already been treated extensively (59Gol).

In angular correlations the transition is regarded as a two-
step process through intermediate state I_k, each link being
described by its own factor, A_ν, so that

$$W(\theta) = \sum_\nu A_\nu(I_o I_k) \, A_\nu(I_k I_f) \, P_\nu(\cos\theta). \tag{2.1}$$

For photons the radiation parameters have been expressed as the
F-coefficients which are available in tabular form (68Apl). Then

$$A_\nu(I_o I_k) = F_\nu(L_1 L_1' I_o I_k)$$

$$\tag{2.2}$$

$$A_\nu(I_k I_f) = F_\nu(L_2 L_2' I_f I_k)$$

where the primes denote interfering radiations. The spin of
the excited state I_k is all-important. For $I_k = 0$ or $1/2$
the angular distribution is isotropic. It is also worth noting
that the complexity of the angular distribution is limited by
the condition that ν is less than or equal to the smaller of
$2L$ or $2I_k$. Electric or magnetic dipole excitation of a state

having spin 1 from a spin zero ground state has the familiar
dipole angular distribution:

$$1 + \cos^2\theta \qquad\qquad (2.3)$$

In general, as the spins of the states involved become larger,
the angular distribution becomes more isotropic.

For intermediate states of well-defined parity, only even
powers of ν appear in the sum. Odd powers occur when there are
overlapping levels of opposite parity. These can be important
in the photonuclear continuum wherever 1^+ or 2^+ states can lie
in the 1^- continuum.

It is not possible to determine the parity of a discrete
energy level by measuring the angular distribution of the resonance radiation. This can, however, be done by either using
plane-polarized incident radiation or by examining the polarization of the scattered radiation. The angular distribution
resulting from plane-polarized incident radiation is obtained
(55Sa1) by replacing $P_\nu(\cos\theta)$ of (2.1) by

$$P_\nu(\cos\theta) - \omega_{L'} f_\nu(L,L')p \; P_\nu^2(\cos\theta)\cos 2\phi \qquad\qquad (2.4)$$

where $\omega_{L'}$ is $\pm$ according to whether the transition is electric
or magnetic, p is the degree of polarization, $P_\nu^2(\cos\theta)$ is the
associated Legendre function and

$$f_\nu(L,L') = - \left[(\nu-2)!/(\nu+2)!\right]^{\frac{1}{2}} \begin{pmatrix} L & L' & \nu \\ 1 & 1 & -2 \end{pmatrix} \bigg/ \begin{pmatrix} L & L' & \nu \\ 1 & -1 & 0 \end{pmatrix} \qquad (2.5)$$

The angle ϕ is between the reaction plane and the electric vector. A few values of $P_\nu(\cos\Theta)$, $P_\nu^2(\cos\Theta)$, and $f_\nu(L,L')$ are given in Table III. In simple electric dipole transitions the electric vector is perpendicular to the plane of scattering; for magnetic dipole transitions, the electric vector lies in the plane of scattering.

As an illustration, consider the group of levels below the (γ,n) threshold in ^{208}Pb illustrated in Fig. 2. They are presumably 1^+ or 1^- and have been excited by bremsstrahlung. Their parities could be determined if they were excited by plane-polarized radiation and the scattered photons examined in the directions parallel and perpendicular to the electric vector.

Two sources of plane-polarized radiation come to mind. The first is bremsstrahlung itself. Here the bremsstrahlung converter must be made thin enough so that the multiple scattering does not destroy the polarization which can be as much as 20%. The second method is the Compton scattering of a laser beam by a high energy electron beam. In this interaction the high energy electron beam collides with the polarized laser beam and the highest energy photons are those backscattered against the electrons and are moving in the electron's initial direction. The highest energy photon has essentially complete polarization.

Alternatively, the polarization of the resonantly-scattered

photons can be studied. The parities of a few levels excited in resonance fluorescence experiments have been determined through the use of the Compton effect. Moreh and Rajewski (72Mo7) have given a very detailed description of the Compton polarimeter they have used to analyze the polarization of resonantly-scattered neutron capture γ-rays. They have shown that asymmetries as large as 20% can be obtained under favorable circumstances with the spin sequence $0 \to 1 \to 0$.

A second Compton polarimeter used by Swann (74Sw7) has been described by Litherland et al. (70Li3). Here, a planar Ge(Li) detector is used. It depends on the fact that the counting rate in the full energy peak is a maximum when the detector plane is perpendicular to the electric vector.

The Continuum

When we talk about scattering in the continuum, the angular distribution formulae given above are not so useful because, in general, we do not know the all-important spin of the intermediate state, I_k. In fact, several different spins may be contributing.

Fano (60Fa2) has presented a treatment that is much more appropriate to scattering from the continuum. Here, the important parameter is not the spin of the intermediate state, I_k, but the angular momentum transferred to the nucleus in the two-step scattering process, ν. In the usual electric dipole case, the nucleus can absorb 0, 1, or 2 units of angular momentum; the corresponding scattering cross sections have been labeled scalar, vector, and tensor.

The scalar scattering is the one that leaves the nucleus in its initial states of both energy and angular momentum; it is coherent with the incident wave. This is the scattering cross section that shares a complex scattering amplitude with the absorption cross section and can be obtained from the absorption cross section by means of the optical theorem and dispersion relation. The scalar scattering amplitude must, however, be combined with the other coherent scattering amplitudes - Rayleigh, Delbrück, and Thomson.

The vector ($\nu=1$) scattering amplitude is usually dismissed because in most models it depends on the difference between two nearly equal terms. The tensor scattering ($\nu=2$), otherwise known as Raman scattering, is a very important component in the scattering from deformed and vibrational nuclei, those having static as well as dynamic deformations.

Fano writes the electric dipole scattering cross section as the sum of three terms in which the scattering amplitudes, A_ν, and the geometrical parts, g_ν, are separate,

$$\frac{d\sigma}{d\Omega} = \frac{|A_\nu|^2}{2\nu+1} \, g_\nu(\Theta). \tag{2.6}$$

The angular distribution factors, $g_\nu(\Theta)$, for unpolarized incident radiation are

$$g_0 = \frac{1 + \cos^2\Theta}{6}$$

$$g_1 = \frac{2 + \sin^2\Theta}{4} \tag{2.7}$$

$$g_2 = \frac{13 + \cos^2\Theta}{12}$$

The scattering amplitude is

$$A_\nu = C_\nu \sum_k \langle I_f \| r \| I_k \rangle \left[\frac{1}{E_k - \hbar\omega - \frac{1}{2} i \Gamma_k} \right.$$

$$\left. + \frac{(-1)^\nu}{E_k + \hbar\omega' + \frac{1}{2} i \Gamma_k} \right] \langle I_k \| r \| I_o \rangle \begin{Bmatrix} I_o I_f & \nu \\ 1 & 1 & I_k \end{Bmatrix}$$

(2.8)

$$+ \sqrt{3} \, D \delta_{\nu o} \delta_{of}$$

and $C_\nu = (-1)^{I_o + I_f + \nu} (e/c)^2 (\omega'/\omega)^{\frac{1}{2}} \omega\omega' [(2\nu+1)/(2I_o+1)]^{\frac{1}{2}}$.

The summation is over all the intermediate states k that can be reached in electric dipole transitions and $\langle I_k \| r \| I_o \rangle$ is the reduced matrix element of the dipole operator connecting the states I_o and I_k. The symbols ω and ω' represent the frequencies of the incoming and outgoing photons and $\begin{Bmatrix} I_o I_f & \nu \\ 1 & 1 & I_k \end{Bmatrix}$ is the 6-j symbol that weights the different components that comprise the total scattering cross section. The delta functions insure that the Thomson scattering amplitude interferes only with the part of the nuclear scattering amplitude that is coherent with the incident beam. Note that the energy dependence of the vector ($\nu=1$) scattering amplitude differs slightly from that associated with the scalar ($\nu=0$) and tensor ($\nu=2$) scattering.

The scattering amplitude may also be expressed in terms of the oscillator strength

$$f_k = \frac{2ME_k}{\hbar^2} \; \frac{|\langle I_k \| r \| I_o \rangle|^2}{3(2I_o+1)} \tag{2.9}$$

$$A_\nu = C_\nu \sum_k (-1)^{I_o+1-I_k} \; 3(2I_o+1) \; \frac{\hbar^2}{2ME_k} \; f_k$$

$$\times \begin{Bmatrix} I_o & I_o & \nu \\ 1 & 1 & I_k \end{Bmatrix} \left[\frac{1}{E_k - \hbar\omega - \frac{1}{2}i\Gamma_k} + \frac{(-1)^\nu}{E_k + \hbar\omega + \frac{1}{2}i\Gamma_k} \right] \tag{2.10}$$

$$+ \sqrt{3} \; D\delta_{o\nu} \; .$$

In the continuum where the levels are overlapping the scattering depends then on the sum:

$$\sum_{I_k=I_o-1}^{I_o+1} (-1)^{I_o+1-I_k} \begin{Bmatrix} I_o & I_o & \nu \\ 1 & 1 & I_k \end{Bmatrix} f_k \; . \tag{2.11}$$

If we can assume that f_k is proportional to $2I_k+1$, then the only contribution to the scattering is the coherent ($\nu=0$) component. This results from the fact that

$$\sum_{I_k=I_o-1}^{I_o+1} (-1)^{I_o+1-I_k} (2I_k+1) \begin{Bmatrix} I_o & I_o & \nu \\ 1 & 1 & I_k \end{Bmatrix} = 0 \text{ for } \nu > 1. \tag{2.12}$$

This is the basis for the statement that the giant resonance scattering cross sections for heavy spherical nuclei have angular distributions typical of a classical oscillator: $1+\cos^2\theta$.

Arenhövel (67Ar1, 68Ar1) has extended Fano's treatment to include electric quadrupole scattering. Here, it is possible to transfer as many as four units of angular momentum to the nucleus. The more general angular distribution factor may be written as $g_\nu(LL'\theta)$ where L and L' indicate the multipolarities. For coherent scattering and unpolarized incident radiation these quantities are:

$$g_o(11\theta) = \frac{1 + \cos^2\theta}{6}$$

$$g_o(22\theta) = \frac{1 - 3\cos^2\theta + 4\cos^4\theta}{10} \qquad (2.13)$$

$$g_o(12\theta) = -\frac{\cos^3\theta}{\sqrt{15}}$$

Table IV contains all of the $g_\nu(LL'\theta)$ for electric dipole and quadrupole scattering for incident radiation plane-polarized perpendicular, $\perp$, and parallel, $\parallel$, to the plane of scattering. An average of the factors for the two polarizations will yield the result for unpolarized incident radiation, such as those given above.

In discussing nuclear photon scattering we need to distinguish rigid spheres and permanently deformed nuclei from those having important surface oscillations. The scattering cross section for a rigid sphere is just that given by (1.22)

with A and B equal. The coherent scattering cross section for
a deformed nucleus is given by (1.22) and the incoherent scat-
tering cross section by (1.23). The factor $(I_oK_o20|I_fK_o)^2$
determines the relative intensity of the radiation populating
those members of the ground-state rotational band that can be
reached in dipole-dipole transitions. For a spin zero nucleus,
such as ^{238}U, all of the Raman scattering is to the first 2^+
state. For a nucleus with odd spin (e.g., Ta or Ho) the ground
state, as well as the first two excited states of the ground-
state rotational band are populated. However, the total intensity is a constant because

$$\sum_{I_f} (I_oK_o20|I_fK_o)^2 = 1. \tag{2.14}$$

When measured with poor energy resolution, i.e., NaI(Tℓ)
detectors, the measured intensity is independent of the spin
of the ground state and as large as it can be for a classical
system.

As has already been pointed out, the nuclear Raman effect
has been studied by two groups using Ge(Li) detectors and
neutron capture γ-rays in the energy range 7.9 to 11.4 MeV.
These energies coincide with the low energy (ΔK=0) peak in the
giant resonance of the deformed nuclei. Figure 7 shows the
angular distributions for the coherent and incoherent scattering from ^{238}U at 10.83 MeV. The Raman scattering is almost
isotropic as it should be. The elastic scattering cross
section has the angular distribution approaching that of a

classical oscillator in the backward hemisphere but in the forward direction the scattering is dominated by Delbrück scattering.

The dynamic collective model (64Dal) takes into account the fact that the nuclear surface is soft and vibrating. This oscillation is coupled to the giant dipole resonance and has some consequences. For the deformed nuclei, it removes the degeneracy of the two upper resonances and provides some weaker satellite states. The deformed nucleus becomes dynamically triaxial. The oscillating spherical nucleus becomes dynamically deformed and its giant resonance acquires some subtle structures. The predicted giant resonance shapes are not, however, so unique that we can be certain that experimentally-observed cross sections are related to this model.

On the other hand, if the dipole oscillation is coupled to the quadrupole vibrations of the nuclear surface, then the giant resonance must decay through the low-lying vibrational states. These ideas are illustrated in Fig. 8 which compares the energy level diagrams for even-even, spherical and deformed nuclei. In the deformed nucleus, the scattering cross section would consist of the coherent scattering and the Raman ($\nu=2$) scattering to the 2^+ state of the ground-state rotational band. The dynamic collective model demands that the vibrational states that can be reached in electric dipole transitions are also populated. The transition to the K=2 γ-vibrational band head is particularly intense. Figure 9 shows the scattering cross sections for populating the ground and first two excited states of ^{166}Er (65Ar4, 67Ar1). Notice that the coherent scattering cross section is depressed in the energy region between the two resonances in the absorption cross section (they are at 12 and 16 MeV) and that the high energy side of the coherent scattering cross section is lifted by constructive interference with Thomson scattering. When measured with poor energy resolution, the Raman scattering to the first 2^+ state essentially fills in the valley in the coherent scattering cross section. The cross section for populating the γ-vibrational (K=2) band head is associated with absorption into the upper ($\Delta K=\pm 1$) resonance. For this reason, this component in the scattering has not been seen in the experiments (already mentioned) where the giant resonances of deformed nuclei were studied using neutron capture γ-rays. The scattering to both of these 2^+ states results from the transfer of two units of angular momentum to the nucleus ($\nu=2$) and has the corresponding radiation pattern.

The incoherent scattering makes even more important contributions to the scattering cross sections of spherical nuclei. This is illustrated in Fig. 10 which shows the coherent scattering cross section and those for populating the first two 2^+ states of the spherical vibrator ^{106}Pd (67Ar2). Here, the incoherent ($\nu=2$) components can be as much as 30% of the total scattering.

This incoherent ($\nu=2$) scattering, populating the low-lying vibrational states, is the true signature of the dynamic collective model. It is therefore important to measure the branching ratios to the various final states throughout the giant resonance. Since the giant resonance absorption cross section is a continuous function of energy, it is essential to use monochromatic photons. Possible sources are positron annihilation radiation, the bremsstrahlung monochrometer, the 17.6 MeV γ-rays from the ^{7}Li(p,γ) reaction, or Compton-scattered laser radiation. It is necessary that the energy spread in the source be smaller than the differences in the energies of the γ-rays being examined. This also requires high resolution spectroscopy for the emitted photons. None of these experiments has yet been done.

Another alternative is, of course, to analyze the scattered photons not according to their energy, but according to the angular momentum transfer, ν (68Ar2). This experiment has been done (75Ha3, 74Ha4) at the single energy of 15.1 MeV. The beam of 15.1 MeV plane-polarized photons was produced by irradiating a carbon target with 20 MeV bremsstrahlung. The 15.1 MeV resonance fluorescence radiation is then monochromatic and plane-polarized at 90°. This radiation was scattered a second time by fourteen natural targets in the atomic number range 48 to 92 and the scattered photons detected in large NaI(Tℓ) detectors. The poor resolution of the detectors al-lowed them to integrate over the ground state and the low-lying excited states populated in the scattering process.

Using (2.6) and the geometrical factors from Table IV the scattering cross sections at 90° parallel and perpendicular to the polarization vector are:

$$\frac{d\sigma^{\parallel}}{d\Omega} = \frac{|A_2|^2}{5}$$

and

$$\frac{d\sigma^{\perp}}{d\Omega} = \frac{|A_0|^2}{3} + \frac{7|A_2|^2}{30} \, .$$

The coherent ($\nu=0$) radiation cannot scatter along the polarization vector whereas the nearly isotropic ($\nu=2$) radiation has a small contribution in that direction. The latter is the signature of the coupling of the quadrupole oscillations of the nuclear surface with the dipole mode described in the dynamic collective model.

The experimental results are shown in Tables V and VI. No attempt was made to make the measurements absolute so that $d\sigma^{\parallel}/d\Omega)_F$ and $d\sigma^{\perp}/d\Omega)_F$ of Table V are the scattering cross sections in arbitrary units for the rather large solid angle used in the experiment. The symbol η_F stands for their ratio, and η is the ratio of these scattering cross sections after correction for the finite solid angle. The last column shows the results of the dynamic collective model; these have been supplied by Arenhövel.

Table VI is presented as a check on the experiment; it is a comparison of the result of this experiment with the total absorption cross sections measured at Saclay (68Be5, 70Vel, 73Vel, 74Lel, 71Cal). The quantity $|A_0|^2$ can be obtained from the polarized photon scattering experiment through the use of the relation

$$|A_0|^2 \sim d\sigma^\perp/d\Omega - 7/6 \, d\sigma^\parallel/d\Omega \; . \tag{2.16}$$

From the Saclay cross sections $|A_0|^2$ was obtained from the equation

$$|A_0|^2 = 3|R(E,0)|^2 = 3\left|\frac{A+2B}{3} + D\right|^2 \; . \tag{2.17}$$

The ratio of the $|A_0|^2$'s obtained from the two experiments is a constant, as it should be.

We take the result of this experiment to be positive evidence for the ideas behind the dynamic collective model. We are, of course, assuming that the 1^- continuum at 15.1 MeV is uncontaminated with states of opposite parity, 1^+ or 2^+. This survey has, necessarily, been limited to the single, sharply-defined energy of 15.1 MeV. We hope that some future experiment will be able to explore the whole giant resonance of some of these nuclei.

III. ELECTRIC QUADRUPOLE SCATTERING IN THE CONTINUUM

As an exercise, let us estimate the contribution of electric quadrupole scattering to the total coherent scattering cross section in a heavy nucleus, i.e., ^{208}Pb. In order to do so, we need to know the energies and strengths of the nuclear electric quadrupole oscillations.

Nuclear collective motions are, by definition, those in which many nucleons participate and the transition strengths associated with them are many times those connected with a single particle transition. For this discussion, we wish to distinguish those collective motions in which all the particles, neutrons, and protons move together from those in which the neutrons and protons move against each other. The former are called isoscalar or T=0 transitions, and the latter isovector, or T=1 transitions.

In nuclear physics we have two apparently opposing models, the independent particle model or shell model, and the hydrodynamic model - the unified model of Bohr and Mottelson brings these two together. We use the models interchangeably to describe the different features that may be portrayed best in each.

The electric dipole excitations are the best known and earliest recognized collective modes. They are made up of the coherent superposition of independent particle model transitions in which the particles move upward to the next higher shell of opposite parity $\Delta N=1$. This is an isovector excitation which would occur at $1\hbar\omega$. The isoscalar electric dipole mode is associated with the motion of the nuclear center of mass in the laboratory, nuclear Thomson scattering.

There are three kinds of electric quadrupole transitions. The most familiar are those associated with the low-lying 2^+ states and are connected with transitions within a major shell $\Delta N=0$. These represent only about 10% of the total isoscalar E2 strength. The rest is in transitions that do not change the parity but for which $\Delta N=2$ with an excitation of $2\hbar\omega$. The isovector E2 transitions, in which the neutrons and protons move against each other are also basically of the $2\hbar\omega$ type.

However, one of the important properties of the nuclear Hamiltonian is that it moves the isoscalar strength down and isovector strength up, producing an important separation. (See Table VII.) Bohr and Mottelson (75Bo10) give $60A^{-\frac{1}{3}}$ and $135A^{-\frac{1}{3}}$ as the energies of the isoscalar and isovector E2 giant resonances. Much more speculative and tentative are the monopole vibrations, isoscalar and isovector. According to Suzuki (73Su20), these would occur at $60A^{-\frac{1}{3}}$ and $178A^{-\frac{1}{3}}$. The isoscalar E0 and E2 oscillation should then overlap in energy.

The hydrodynamic model, which has been discussed by Danos and his colleague Greiner, really describes only the isovector modes. These resonate at the energies where $j_\lambda{}'(KR)$ are zero (74De12). These are given in Table VII under "Hydrodynamics".

In order to obtain the relative strengths of these excitations, we can use sum rules. The most important sum rule in photonuclear physics is THE electric dipole sum rule which is a conservation law that places a lower limit on the total integrated electric dipole absorption cross section. This sum rule is but one of a class of so-called energy weighted sum rules. "Energy weighted" because they result from taking the sum

$$\sum_{k} B(EL;0 \rightarrow k)(E_k - E_o)$$

where   (3.1)

$$B(EL;0 \rightarrow k) = \frac{|\langle k \| Q_L \| 0 \rangle|^2}{2I_o + 1} \quad .$$

The evaluation of the double commutator using only the kinetic energy term in the nuclear Hamiltonian yields for the general case (65Nal, 64La6)

$$\sum_{k} B(EL;0 \rightarrow k)(E_k - E_o) =$$

                                                                                                                  (3.2)

$$\frac{L(2L+1)^2}{4\pi} \quad \frac{\hbar^2}{2M} \quad Z \quad \frac{e^2}{\hbar c} \quad \langle r^{2L-2} \rangle .$$

The relationship between this energy weighted sum rule and the various moments of the integrated photon absorption cross section is (730cl)

$$\int \frac{\sigma dE}{E^{2L-2}} = \frac{(2\pi)^3(L+1)}{L[(2L+1)!!]^2} \quad \frac{1}{(\hbar c)^{2L-2}} \quad \sum_{k} B(EL;0 \rightarrow k)(E_k - E_o)$$ (3.3)

which yields (74Del2)

$$\int \frac{\sigma \, dE}{E^{2L-2}} = \pi^2 \left(\frac{e^2}{\hbar c}\right) \frac{L+1}{[(2L+1)!!]^2} \frac{\hbar^2}{M} Z \frac{\langle r^{2L-2}\rangle}{(\hbar c)^{2L-2}} . \qquad (3.4)$$

This sum includes both the T=0 and T=1 transitions since its derivation is based on the total operator:

$$Q_{LM} = e \sum_i \tfrac{1}{2}[1+\tau_3(i)] \, r_i^L \, Y_{LM}^*(\hat{r}_i) . \qquad (3.5)$$

The first term in the operator describes T=0, isoscalar, transitions and the second T=1, isovector, transitions. If there are no interferences between these two kinds of transitions, then these sum rules may be written separately as

$$\int \frac{\sigma \, dE}{E^{2L-2}} = \pi^2 \left(\frac{e^2}{\hbar c}\right) \frac{L+1}{[(2L-1)!!]^2} \frac{\hbar^2}{M} \frac{Z^2}{A} \frac{\langle r^{2L-2}\rangle}{(\hbar c)^{2L-2}} \qquad (3.6)$$

$$\text{for } T=0$$

$$\int \frac{\sigma \, dE}{E^{2L-2}} = \pi^2 \left(\frac{e^2}{\hbar c}\right) \frac{L+1}{[(2L-1)!!]^2} \frac{\hbar^2}{M} \frac{NZ}{A} \frac{\langle r^{2L-2}\rangle}{(\hbar c)^{2L-2}} . \qquad (3.7)$$

$$\text{for } T=1$$

For the electric dipole case, the isoscalar mode is of no interest since it does not disrupt the internal nuclear coordinates; it corresponds to the motion of the nucleus as a whole and is responsible for Thomson scattering. On the other hand, the nuclear electric dipole oscillations responsible for the giant resonance are known to be of the isovector type. The classical nuclear electric dipole sum rule is obtained by setting L=1 in the above equation.

$$\int \sigma \, dE = \frac{2\pi^2 e^2 \hbar}{Mc} \frac{NZ}{A} = 60 \frac{NZ}{A} \text{ MeV mb}. \tag{3.8}$$

For electric quadrupole transitions we have for T=0

$$\int \frac{\sigma \, dE}{E^2} = \frac{\pi^2}{3Mc^2} \frac{e^2}{\hbar c} \frac{Z^2}{A} \langle r^2 \rangle = \frac{\pi^2}{5Mc^2} \frac{e^2}{\hbar c} \frac{Z^2}{A} R^2. \tag{3.9}$$

This we recognize as the rule given by Gell-Mann and Telegdi for self-conjugate nuclei (53Ge1). For T=1

$$\int \frac{\sigma \, dE}{E^2} = \frac{\pi^2}{3Mc^2} \frac{e^2}{\hbar c} \frac{NZ}{A} \langle r^2 \rangle = \frac{\pi^2}{5Mc^2} \frac{e^2}{\hbar c} \frac{NZ}{A} R^2. \tag{3.10}$$

This result has been obtained and used by Ligensa and Greiner (67Li2). For L=3, these become

$$\int \frac{\sigma \, dE}{E^4} = \frac{4\pi^2}{225Mc^2} \frac{e^2}{\hbar c} \frac{Z^2}{A} \frac{\langle r^4 \rangle}{(\hbar c)^2}, \tag{3.11}$$

and

$$\int \frac{\sigma \, dE}{E^4} = \frac{4\pi^2}{225Mc^2} \frac{e^2}{\hbar c} \frac{NZ}{A} \frac{\langle r^4 \rangle}{(\hbar c)^2}. \tag{3.12}$$

The isoscalar sums represent much more reliable estimates than the isovector sums since the potential energy parts of the nuclear Hamiltonian are less likely to commute with the isovector part of the operator. So, even though the isoscalar and isovector sums depicted here are nearly the same size, we may expect the isovector sum to be somewhat larger.

A comparison of the magnitudes of the electric dipole and and electric quadrupole sums is now in order. Without doing too much violence, (3.9) and (3.10) may be written as

$$\int \sigma(E2,T=0)\,dE = \frac{\pi^2}{5Mc^2}\,\frac{e^2}{\hbar c}\,\frac{Z^2}{A}\,R^2 \sum_i f_i E_i^2 \tag{3.13}$$

and

$$\int \sigma(E2,T=1) = \frac{\pi^2}{5Mc^2}\,\frac{e^2}{\hbar c}\,\frac{NZ}{A}\,R^2 \sum_i f_i E_i^2 \tag{3.14}$$

where f_i is the fraction of the oscillator strength in a state at E_i, and where $\sum f_i = 1$. Assuming that all of the quadrupole strength is in a single state at E_T then

$$\frac{\int \sigma(E2,T=0)\,dE}{\int \sigma(E1,T=1)\,dE} = \frac{1}{10}\,\frac{Z}{N}\left(\frac{RE_0}{\hbar c}\right)^2$$

$$\tag{3.15}$$

$$\frac{\int \sigma(E2,T=1)\,dE}{\int \sigma(E1,T=1)\,dE} = \frac{1}{10}\left(\frac{RE_1}{\hbar c}\right)^2 .$$

For heavy elements, using the normalized hydrodynamic result
of Table VII,

$$RE_o = 1.2 \times 10^{-13} A^{\frac{1}{3}} \times 60 A^{-\frac{1}{3}} = 72 \ 10^{-13} \ \text{MeV cm}$$

$$(3.16)$$

$$RE_1 = 1.2 \times 10^{-13} A^{\frac{1}{3}} \times 128 A^{-\frac{1}{3}} = 154 \ 10^{-13} \ \text{MeV cm.}$$

For the ratio of the two isovector sums we then have

$$\frac{\int \sigma(E2,T=1) dE}{\int \sigma(E1,T=1) dE} = 0.06 \ , \tag{3.17}$$

independent of everything. Danos (52Da1), who solved this problem long ago, obtained 0.08; he informs me that the difference
is only that he used a slightly larger radius.

To calculate the scattering cross section for ^{208}Pb we
will take as the basic absorption cross section the Lorentz-
line fit to the Saclay data: $E_o = 13.42$ MeV, $\Gamma = 4.05$ MeV, and
$\sigma_o = 640$ mb; this is to be contrasted with the electric dipole
cross section of Fig. 2 calculated from the actual data. The
electric dipole forward scattering amplitude is then:

$$R(E1) = \frac{\sigma_o \Gamma}{4\pi\hbar c} \ E^2 \ \frac{E_o^2 - E^2 + i\Gamma E}{(E_o^2 - E^2)^2 + (\Gamma E)^2} \ . \tag{3.18}$$

The forward scattering amplitude associated with the isoscalar E2 strength would be

$$R(E2,T=0) = \frac{\sigma_o \Gamma}{4\pi\hbar c} \frac{Z}{N} \frac{1}{10} \left(\frac{R}{\hbar c}\right)^2 E^2$$

(3.19)

$$\times \sum_i E_i^2 \; f_i \; \frac{E_i^2-E^2 + i\Gamma_i E}{(E_i^2-E^2)^2 + (E\Gamma_i)^2}$$

where σ_o and Γ are now the parameters of the E1 resonance. We
assume that the bulk of this strength is distributed in three
lines (72Bu19) at 10.2, 10.6, and 11.2 MeV, each with a width
of 0.2 MeV and containing 30% of a sum.

The isovector quadrupole strength has been discussed by
Ligensa (66Li3) and has been observed for deformed nuclei in
three different laboratories (63Br1, 64Br1, 68Be5, and 73Hi6).
Figure 11 shows, as an example, a comparison of the neutron
production cross section of Tb from Saclay (68Be5) with the
Ligensa (66Li3) prediction. There should be five E2 states at
this excitation energy; the remarkable thing is that they are
experimentally observed to be only 0.6-0.7 MeV wide, one of
the great mysteries of photonuclear physics. As Ligensa
(67Li2) points out, the E2 giant resonance in a spherical nucleus, such as ^{208}Pb, should consist of a single state. The
forward scattering amplitude may be written as

$$R(E2,T=1) = \frac{\sigma_o \Gamma}{4\pi\hbar c} \left(\frac{RE_1}{\hbar c}\right)^2 \frac{E^2}{10} \frac{E_1^2-E^2 + i\Gamma_1}{(E_1^2-E^2)^2 + (E\Gamma_1)^2} .$$

(3.20)

Here, we locate it at 21.5 MeV and, using the deformed nucleus results as a guide, assume the width to be 1 MeV.

The coherent scattering cross section has three terms (67Ar1):

$$\frac{d\sigma(E1)}{d\Omega} = |A_o(E1)|^2 \left(\frac{1+\cos^2\theta}{6}\right) = |R(E1)+D|^2 \left(\frac{1+\cos^2\theta}{2}\right)$$

$$\frac{d\sigma(E2)}{d\Omega} = |A_o(E2)|^2 \left(\frac{1-3\cos^2\theta+\cos^4\theta}{10}\right) \qquad (3.21)$$

$$= |R(E2)|^2 \left(\frac{1-3\cos^2\theta+\cos^4\theta}{2}\right)$$

$$\frac{d\sigma(E1E2)}{d\Omega} = \frac{-2\cos^3\theta}{\sqrt{15}} R_e[A_o(E1) \; A_o^*(E2)]$$

$$= -2\cos^3\theta \; R_e[R(E1) \; R^*(E2)] \; .$$

The resulting scattering cross section at 90° is shown in Fig. 12. The scattering associated with the isoscalar electric quadrupole absorption is buried in the E1 giant resonance scattering, and hence could probably only be observed using polarized incident radiation. The isovector contribution is appreciable and may be observable. On the other hand, a recent measurement (74Sn5) of the $^{208}Pb(p,\gamma_o)^{209}Bi$ cross section suggests that this resonance is, in fact, 3.5 MeV wide which would make it much less prominent and very difficult to observe.

In conclusion, let us list the photon scattering experiments for which monochromatic plane-polarized photons would be useful:

1. Determination of the parity of the dipole states below the particle threshold in nuclei such as ^{208}Pb.

2. Study of the incoherent ($\nu=2$) scattering as a function of excitation energy in medium and heavy nuclei.

3. Measurement of the Delbrück scattering cross section.

4. Study of the electric quadrupole strength in nuclei.

ACKNOWLEDGMENT

The author wishes to thank P. Axel for providing the theses of Laszewski and Coope; W. R. Dodge for making the continuum calculations for Figs. 1 and 2; and T. C. Dunn for her care with the manuscript.

References

39La1   W. E. Lamb, Phys. Rev. 55, 190 (1939).

52Da1   M. Danos, Ann. Phys. 6, 18 (1952).

53Ge1   M. Gell-Mann and V. Telegdi, Phys. Rev. 91,
169 (1953).

54Ro1   M. E. Rose, W. Miranker, P. Leak, L. Rosenthal,
and J. K. Henrickson, Westinghouse Atomic Power
Division Report SR-506, 1954 (unpublished) Vols.
I and II.

55Sa1   G. R. Satchler, Proc. Phys. Soc. 68A, 1041 (1955).

56Bo1   O. V. Bogdankevich, L. E. Lazareva, and F. A.
Nikolaev, JETP 31, 405 (1956); Sov. Phys. JETP 4,
320 (1957).

56Fu1   E. G. Fuller and E. Hayward, Phys. Rev. 101,
692 (1956).

56Me3   Luise Meyer-Schützmeister and V. L. Telegdi,
Phys. Rev. 104, 185 (1956).

57Ha1   E. Hayward and E. G. Fuller, Phys. Rev. 106,
991 (1957).

58Bo1   O. V. Bogdankevich, B. S. Dolbilkin, L. E. Lazareva,
and F. A. Nikolaev, Comptes Rendus du Congres
International de Physique Nucleaire, Paris (1958),
Dunod, Paris (1959), p.697.

58Si1   E. Silva and J. Goldemberg, Phys. Rev. 110,
1102 (1958).

59Go2   L. J. B. Goldfarb, Nuclear Reactions I, edited by
         P. M. Endt and M. Demeur (North-Holland Publishing
         Company, Amsterdam, 1959).

59Me2   F. R. Metzger, Prog. Nucl. Phys. $\underline{7}$, 54 (1959).

60Bo2   O. V. Bogdankevich, L. E. Lazareva, and A. M. Moiseev,
         JETP $\underline{39}$, 1224 (1960); Sov. Phys. JETP $\underline{12}$, 853 (1961).

60Fa2   U. Fano, NBS Technical Note 83; U.S. Government
         Printing Office, Washington, D.C. 20402, 1960.

63Br1   R. L. Bramblett, J. T. Caldwell, G. F. Auchampaugh,
         and S. C. Fultz, Phys. Rev. $\underline{129}$, 2723 (1963).

63Os2   D. M. O'Shea and H. C. Thacher, Am. Nucl. Soc. $\underline{6}$,
         36 (1963).

64Br1   R. L. Bramblett, J. T. Caldwell, R. R. Harvey, and
         S. C. Fultz, Phys. Rev. $\underline{133}$, B868 (1964).

64Da1   M. Danos and W. Greiner, Phys. Rev. $\underline{134}$, B284 (1964).

64Eh1   F. Ehlotzky and G. C. Sheppey, Nuovo Cimento $\underline{33}$,
         1185 (1964).

64La6   A. M. Lane, Nuclear Theory, W. A. Benjamin, Inc.,
         New York, 1964.

65Ar4   H. Arenhövel and W. Greiner, Phys. Lett. $\underline{18}$,
         136 (1965).

65Na1   O. Nathan and S. G. Nilsson, Alpha-, Beta-, and
         Gamma-Ray Spectroscopy, Vol. I, edited by K. Siegbahn,
         North-Holland Publishing Company, Amsterdam, 1965).

66Li3   R. Ligensa, W. Greiner, and M. Danos, Phys. Lett. $\underline{16}$,
         B535 (1966).

67Ar1   H. Arenhövel, M. Danos, and W. Greiner, Phys.
         Rev. 157, 1109 (1967).

67Ar2   H. Arenhövel and H. J. Weber, Nucl. Phys. A91,
         145 (1967).

67Ku2   H. W. Kuehne, P. Axel, D. C. Sutton, Phys. Rev.
         163, 1278 (1967).

67Li2   R. Ligensa and W. Greiner, Nucl. Phys. A92, 673 (1967).

68Ap1   H. Appel in Landolt-Börnstein, Vol. 3, Numerical
         Tables for Angular Correlation Computations in α-,
         β-, and γ-Spectroscopy: 3f-, 6f-, 9f-symbols, F- and
         Γ-Coefficients, edited by H. Schopper (Springer-Verlag,
         New York, 1968).

68Ar1   H. Arenhövel and W. Greiner, Prog. Nucl. Phys. 10,
         167 (1968).

68Ar2   H. Arenhövel and E. Hayward, Phys. Rev. 165, 1170
         (1968).

68Be5   R. Bergère, H. Beil, A. Veyssière, Nucl. Phys.
         A121, 463 (1968).

68Mo1   R. Moreh and M. Friedman, Phys. Lett. 26B, 579 (1968);
         31B, 642 (1968).

70Ea1   E. D. Earle, A. J. Ferguson, G. Van Middelkoop,
         G. A. Bartholomew, and I. Bergqvist, Phys. Lett. 32B,
         471 (1970).

70Li3   A. E. Litherland, G. T. Ewan, and S. T. Lam,
         Can. J. Phys. 48, 2320 (1970).

70Ve1   A. Veyssière, H. Beil, R. Bergère, P. Carlos, and
         A. Leprêtre, Nucl. Phys. A159, 561 (1970).

71Cal   P. Carlos, H. Beil, R. Bergère, A. Leprêtre, and
         A. Veyssière, Nucl. Phys. A172, 437 (1971).

71Hal   M. Hass, R. Moreh, and D. Salzman, Phys. Lett. 36B,
         68 (1971).

72Bul9   F. R. Buskirk, H. D. Graf, R. Pitthan, H. Theissen,
         O. Titze, and Th. Walcher, Phys. Lett. 42B, 194
         (1972).

72Jal   H. E. Jackson, K. J. Wetzel, Phys. Rev. Lett. 28,
         513 (1972).

72Khl   A. M. Khan and J. W. Knowles, Nucl. Phys. A179,
         333 (1972).

72Mol   R. Moreh and J. Rajewski, Nucl. Instrum. Methods
         98, 13 (1972).

72Sh2   N. Shikazono and Y. Kawarasaki, Nucl. Phys. A188,
         461 (1972).

73Arl5   B. Arad and G. Ben-David, Rev. Mod. Phys. 45, 230
         (1973).

73Hi6   R. S. Hicks and B. M. Spicer, Aust. J. Phys. 26,
         585 (1973).

73Ocl   J. S. O'Connell, Proceedings of the International
         Conference on Photonuclear Reactions and Applications,
         Pacific Grove, California, 1973, edited by B. L.
         Berman, CONF 730301 (Lawrence Livermore Laboratory,
         Livermore, Calif., 1973).

73Su20   T. Suzuki, Nucl. Phys. A217, 182 (1973).

73Vel   A. Veyssière, H. Beil, R. Bergère, P. Carlos,
         A. Leprêtre, and K. Kernbach, Nucl. Phys. A199, 45 (1973).

74Ba6   T. Bar-Noy and R. Moreh, Nucl. Phys. A229, 417 (1974).

74De12   A. deShalit and H. Feshbach, Theoretical Nuclear
         Physics, Vol. I: Nuclear Structure, John Wiley and
         Sons, Inc., New York, 1974).

74Ja2   H. E. Jackson, G. E. Thomas, and K. J. Wetzel,
         Phys. Rev. C9, 1153 (1974).

74Ka9   S. Kahane and R. Moreh, Phys. Rev. C9, 2384 (1974).

74Le1   A. Leprêtre, H. Beil, R. Bergère, P. Carlos,
         A. De Miniac, A. Veyssière, and K. Kernbach,
         Nucl. Phys. A219, 39 (1974).

74Sn5   K. A. Snover, K. Ebisawa, D. R. Brown, and P. Paul,
         Phys. Rev. Lett. 32, 317 (1974).

75Ah3   J. Ahrens, H. Borchert, K. H. Czock, H. B. Epper,
         H. Gimm, H. Gundrum, M. Kröning, P. Riehn, G. Sita
         Ram, A. Zieger, and B. Ziegler, Nucl. Phys. A251,
         479 (1975).

75Aj6   F. Ajzenberg-Selove, Nucl. Phys. A248, 1 (1975).

75Bo10   A. Bohr and B. R. Mottelson, Nuclear Structure, Vol. II,
         W. A. Benjamin, Inc., Reading, Mass. (1975).

75Co2   D. F. Coope, Thesis, University of Illinois, 1975.

75De5   R. Del Vecchio, S. Freedman, G. T. Garvey, and
         M. Oothoudt, Phys. Rev. Lett. 34, 1296 (1975).

75Ja1   H. E. Jackson, G. E. Thomas, and K. J. Wetzel,
         Phys. Rev. C11, 1664 (1975).

75La2   R. M. Laszewski, Thesis, University of Illinois, 1975.

75Pa9   P. Papatzacos and K. Mork, Phys. Rev. $\underline{D12}$, 206 (1975).

75Sc15   M. Schumacher, I. Borchert, F. Smend, and P. Rullhusen, Phys. Lett. $\underline{59B}$, 134 (1975).

76Ho1   R. J. Holt and H. E. Jackson, Phys. Rev. Lett. $\underline{36}$, 244 (1976).

Table I

Resonance Parameters for Levels in ^{12}C

E_o	I^+	γ_o/Γ	σ_a^o (b)	γ_o (eV)	$\int \sigma_s dE$ (eVb)
4.442	2^+	1	624	10.1×10^{-3}	9.9
12.73	1^+	0.85	38.76	0.35	21.3
15.1	1^+	0.92	29.78	37	1730

Table II

Resonance Parameters for Levels in ^{208}Pb

E	γ_o (eV)	$\int \sigma_s dE = 3\pi^2\lambda^2\gamma_o$ (*) (MeVmb)
7.335	48.8±2.9	10.5
7.085	9.2±0.6	2.13
7.065	19.2±1.3	4.47
6.723	16.0±1.6	4.11
5.515	24.4±2.2	9.31
5.295	7.8±1.1	2.89
4.843	7.6±1.1	3.76

*Assuming that $\Gamma=\gamma_o$.

Table III

Some Special Functions for L= 1,2

$P_0(\cos\theta) = 1$

$P_1(\cos\theta) = \cos\theta$

$P_2(\cos\theta) = 1/2(3\cos^2\theta-1)$

$P_3(\cos\theta) = 1/2(5\cos^3\theta-3\cos\theta)$

$P_4(\cos\theta) = 1/8(35\cos^4\theta-30\cos^2\theta+3)$

$P_2^2(\cos\theta) = 3\sin^2\theta$

$P_3^2(\cos\theta) = 15\sin^2\theta\cos\theta$

$P_4^2(\cos\theta) = 15/2(\sin^2\theta)(7\cos^2\theta-1)$

$f_2(11) = -1/2$

$f_2(12) = -1/6$

$f_2(22) = 1/2$

$f_3(12) = -1/6$

$f_3(22) = 0$

$f_4(22) = -1/12$

Table IV

Angular Distribution Factors, $g(LL'\theta)$, for $L=1,2$

$g_0^\perp(11\theta) = 1/3$

$g_0^\parallel(11\theta) = 1/3\cos^2\theta$

$g_1^\perp(11\theta) = 1/2$

$g_1^\parallel(11\theta) = 1/2(2-\cos^2\theta)$

$g_2^\perp(11\theta) = 7/6$

$g_2^\parallel(11\theta) = 1+1/6\cos^2\theta$

$g_0^\perp(22\theta) = 1/5\cos^2\theta$

$g_0^\parallel(22\theta) = 1/5(1-4\cos^2\theta\sin^2\theta)$

$g_1^\perp(22\theta) = 4/9(2-\cos^2\theta)$

$g_1^\parallel(22\theta) = 4/9(1+16\cos^2\theta\sin^2\theta)$

$g_2^\perp(22\theta) = 1/14(6+\cos^2\theta)$

$g_2^\parallel(22\theta) = 1/14(7-16\cos^2\theta\sin^2\theta)$

$g_3^\perp(22\theta) = 1/5(3-\cos^4\theta+\sin^4\theta)$

$g_3^\parallel(22\theta) = 1/5(3-\cos^4\theta-\sin^4\theta)$

$g_4^\perp(22\theta) = 2/35(10+4\cos^2\theta)$

$g_4^\parallel(22\theta) = 2/35(14-\cos^2\theta+\cos^4\theta)$

$g_0^\perp(12\theta) = [\cos\theta(\sin^2\theta-\cos^2\theta)]/\sqrt{15}$

$g_0^\parallel(12\theta) = -\cos\theta/\sqrt{15}$

$g_1^\perp(12\theta) = -\cos\theta(1+4\sin^2\theta)/2\sqrt{5}$

$g_1^\parallel(12\theta) = -\cos\theta/2\sqrt{5}$

$g_2^\perp(12\theta) = -\cos\theta(1+4\cos^2\theta)/2\sqrt{21}$

$g_2^\parallel(12\theta) = -5\cos\theta/2\sqrt{21}$

$g_0^\perp(11\pi/2) = 1/3$

$g_0^\parallel(11\pi/2) = 0$

$g_1^\perp(11\pi/2) = 1/2$

$g_1^\parallel(11\pi/2) = 1$

$g_2^\perp(11\pi/2) = 7/6$

$g_2^\parallel(11\pi/2) = 1$

$g_0^\perp(22\pi/2) = 0$

$g_0^\parallel(22\pi/2) = 1/5$

$g_1^\perp(22\pi/2) = 8/9$

$g_1^\parallel(22\pi/2) = 4/9$

$g_2^\perp(22\pi/2) = 3/7$

$g_2^\parallel(22\pi/2) = 1/2$

$g_3^\perp(22\pi/2) = 4/5$

$g_3^\parallel(22\pi/2) = 2/5$

$g_4^\perp(22\pi/2) = 4/7$

$g_4^\parallel(22\pi/2) = 4/5$

$g_0^\perp(12\pi/2) = 0$

$g_0^\parallel(12\pi/2) = 0$

$g_1^\perp(12\pi/2) = 0$

$g_1^\parallel(12\pi/2) = 0$

$g_2^\perp(12\pi/2) = 0$

$g_2^\parallel(12\pi/2) = 0$

Table V

Results

Target	$d\sigma^{\parallel}/d\Omega_F$ (Arbitrary	$d\sigma^{\perp}/d\Omega_F$ Units)	η_F	η	η(DCM)
Cd	0.042±0.028	0.39±0.05	0.11±0.07	0.09±0.07	0.19
In	0.026±0.020	0.54±0.04	0.05±0.04	0.03±0.04	0.19
Sn	0.084±0.036	0.65±0.06	0.13±0.06	0.11±0.06	0.07
Sb	0.14 ±0.030	0.77±0.05	0.18±0.05	0.16±0.05	
Nd	0.14 ±0.07	1.03±0.10	0.14±0.07	0.12±0.07	
Ta	0.24 ±0.10	1.47±0.14	0.16±0.07	0.14±0.07	0.20
W	0.52 ±0.10	1.66±0.12	0.31±0.07	0.29±0.07	0.20
Pt	0.23 ±0.08	1.94±0.13	0.12±0.04	0.10±0.04	0.08
Au	0.39 ±0.11	2.08±0.15	0.19±0.06	0.17±0.06	0.07
Hg	0.33 ±0.09	2.16±0.15	0.15±0.04	0.13±0.04	0.03
Pb	0.19 ±0.14	2.42±0.19	0.08±0.06	0.06±0.06	0
Bi	0.10 ±0.15	2.65±0.26	0.04±0.06	0.02±0.06	0
Th	0.31 ±0.12	2.26±0.19	0.14±0.05	0.12±0.05	0.07
U	0.21 ±0.11	2.38±0.19	0.09±0.05	0.07±0.05	0.08

Table VI

Comparison with Saclay Data

Target	$|A_o|^2$ This experiment (Arbitrary units)	$|A_o|^2$ Saclay (mb)	Ratio
Cd	0.337±0.058	0.508	0.663±0.114
In	0.507±0.046	0.591	0.859±0.078
Sn	0.550±0.072	0.822	0.669±0.096
Sb	0.590±0.061	0.794	0.743±0.077
Nd	0.837±0.100	1.170	0.715±0.086
Ta	1.19 ±0.18	1.88	0.633±0.096
W	1.05 ±0.17	2.05	0.512±0.083
Pt	1.67 ±0.16	2.70	0.619±0.059
Au	1.62 ±0.20	2.92	0.555±0.068
Hg	2.16 ±0.20	3.29	0.540±0.060
Pb	2.20 ±0.27	3.43	0.641±0.078
Bi	2.53 ±0.31	3.43	0.737±0.090
Th	1.89 ±0.22	2.73	0.692±0.080
U	2.13 ±0.22	2.83	0.754±0.077
			0.656±0.021

Table VII

Energy of Collective Nuclear Excitations

L	T	Oscillator	Suzuki[*]	Hydrodynamics
1	0	Thomson Scattering		
1	1	$1\hbar\omega \rightarrow 2\hbar\omega$		$80A^{-\frac{1}{3}}$
2	0	$0\hbar\omega \rightarrow 0\hbar\omega$		
2	0	$2\hbar\omega \rightarrow 1.5\hbar\omega$	$60A^{-\frac{1}{3}}$	
2	1	$2\hbar\omega \rightarrow 3\hbar\omega$	$135A^{-\frac{1}{3}}$	$128A^{-\frac{1}{3}}$
0	0	$2\hbar\omega \rightarrow 1.5\hbar\omega$	$60A^{-\frac{1}{3}}$·	
0	1	$2\hbar\omega \rightarrow 4\hbar\omega$	$178A^{-\frac{1}{3}}$	$173A^{-\frac{1}{3}}$

[*]See Ref. 73Su20.

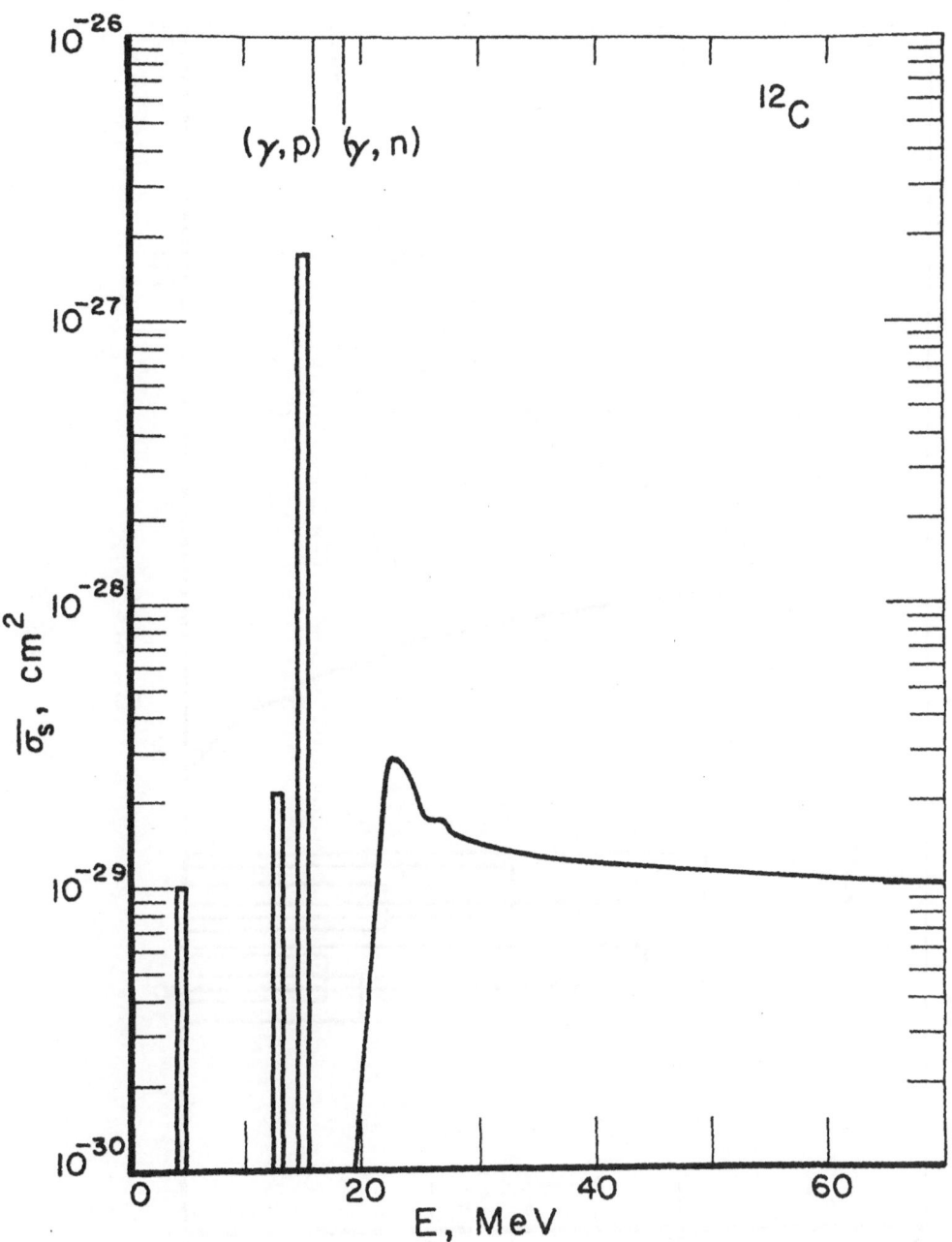

Fig. 1

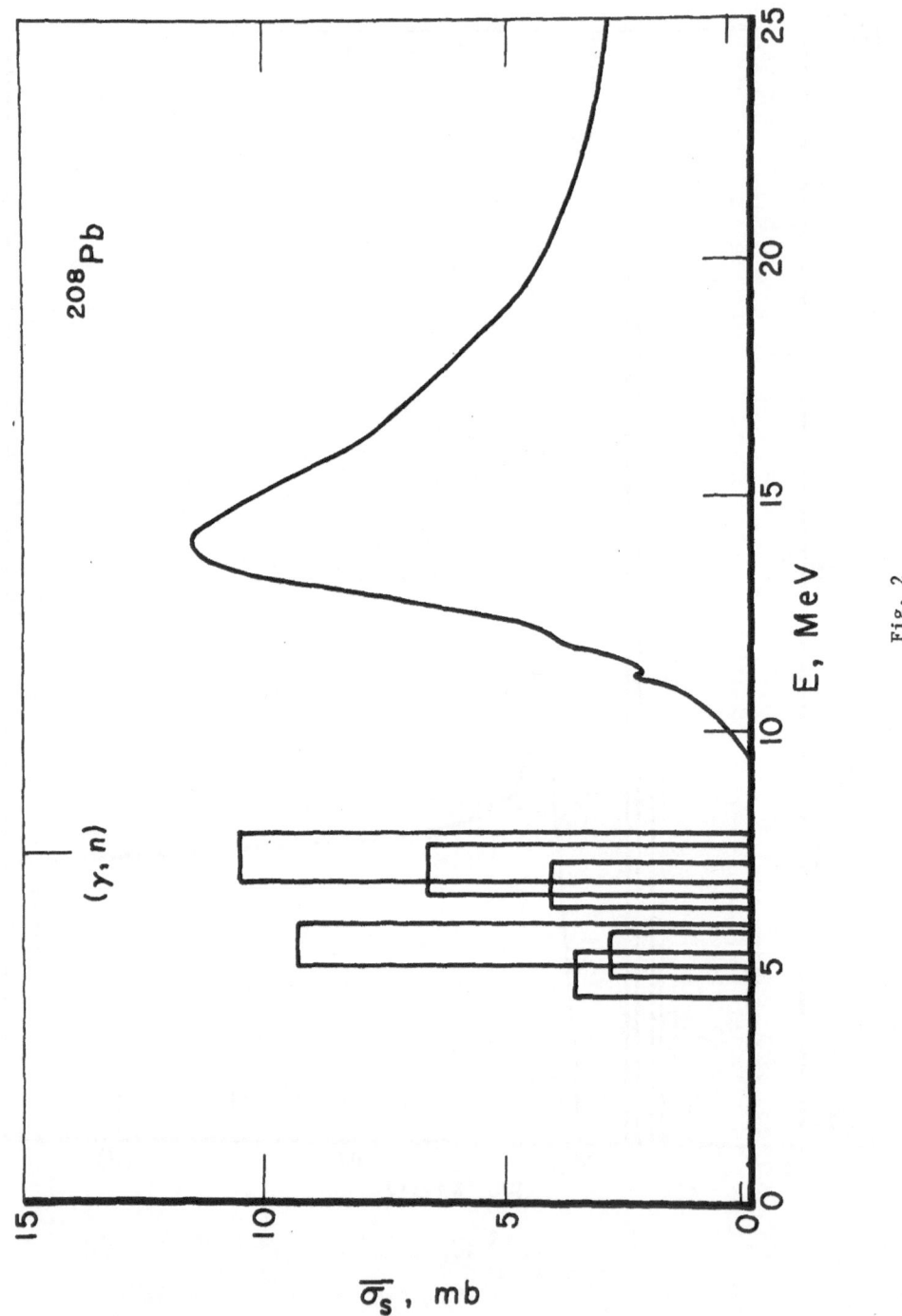

Fig. 2

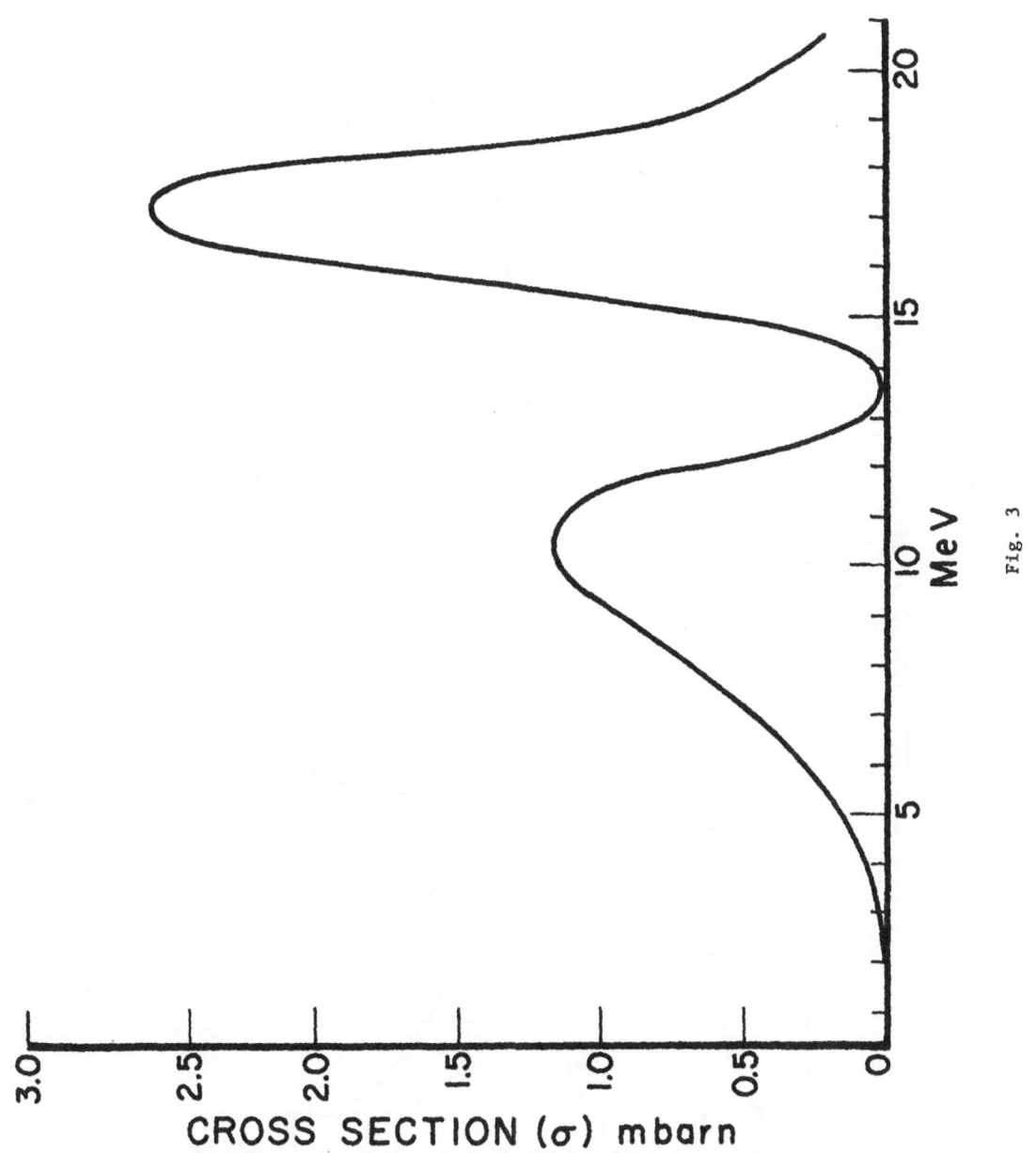

Fig. 3

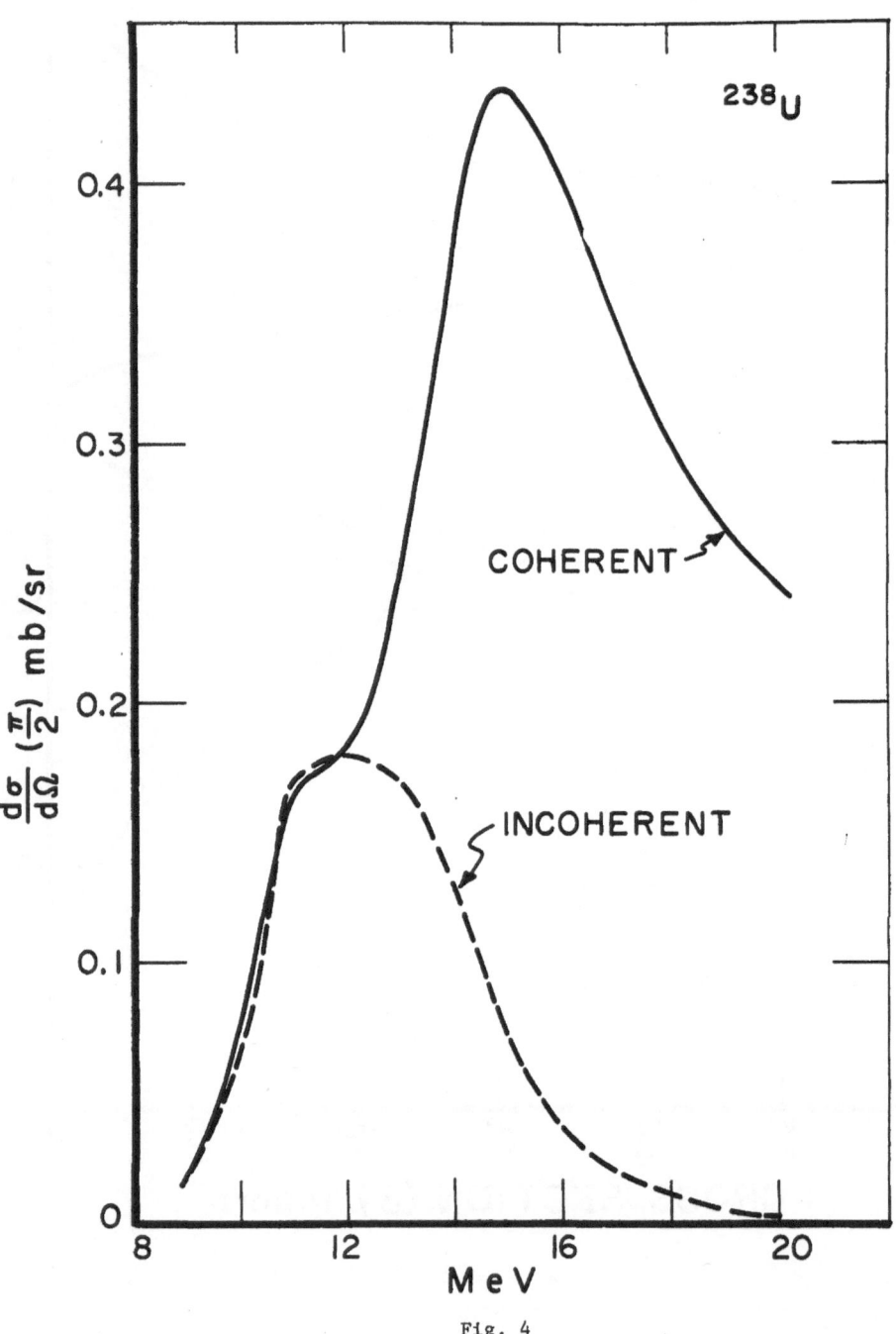

Fig. 4

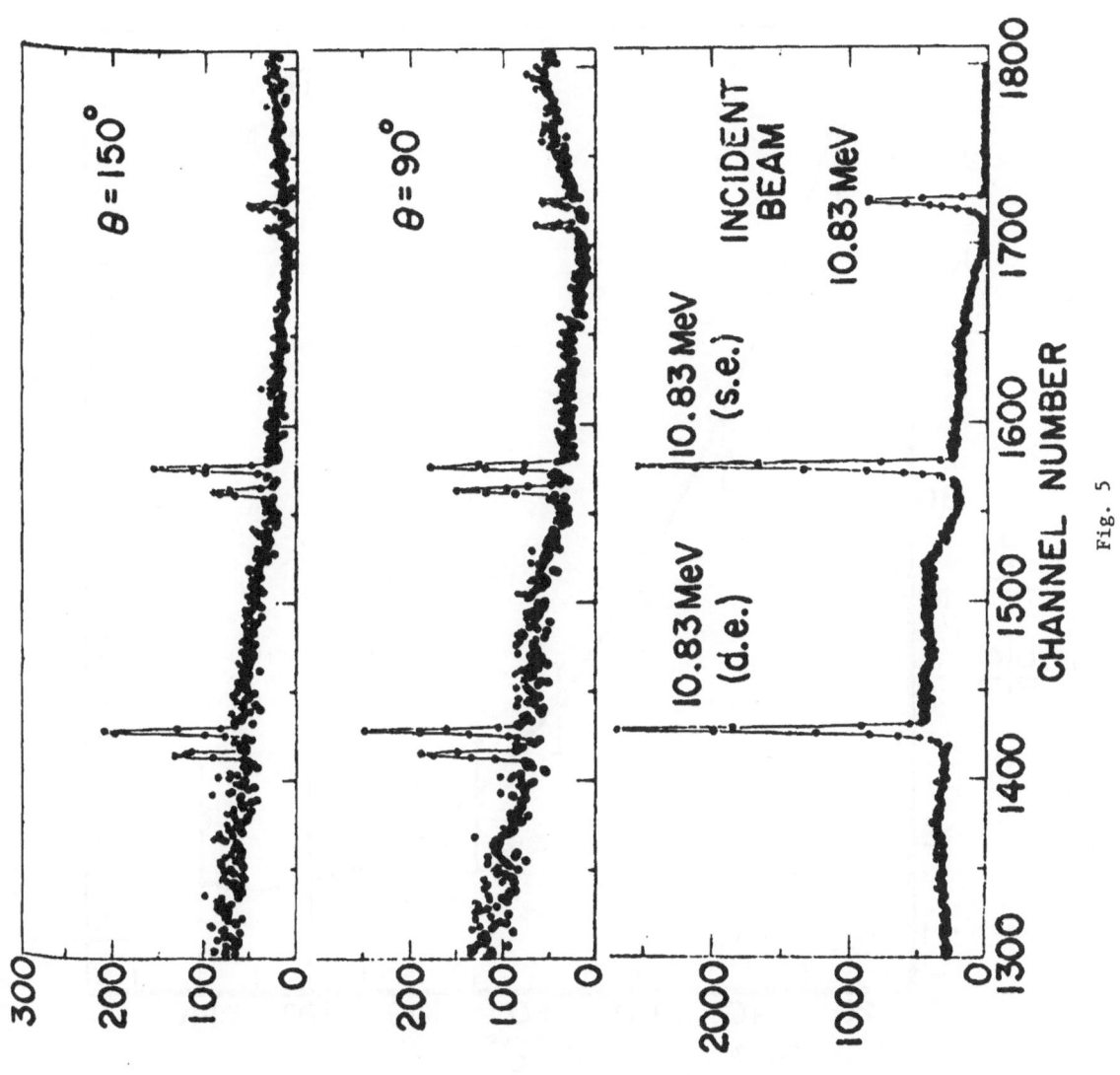

COUNTS/CHANNEL

Fig. 5

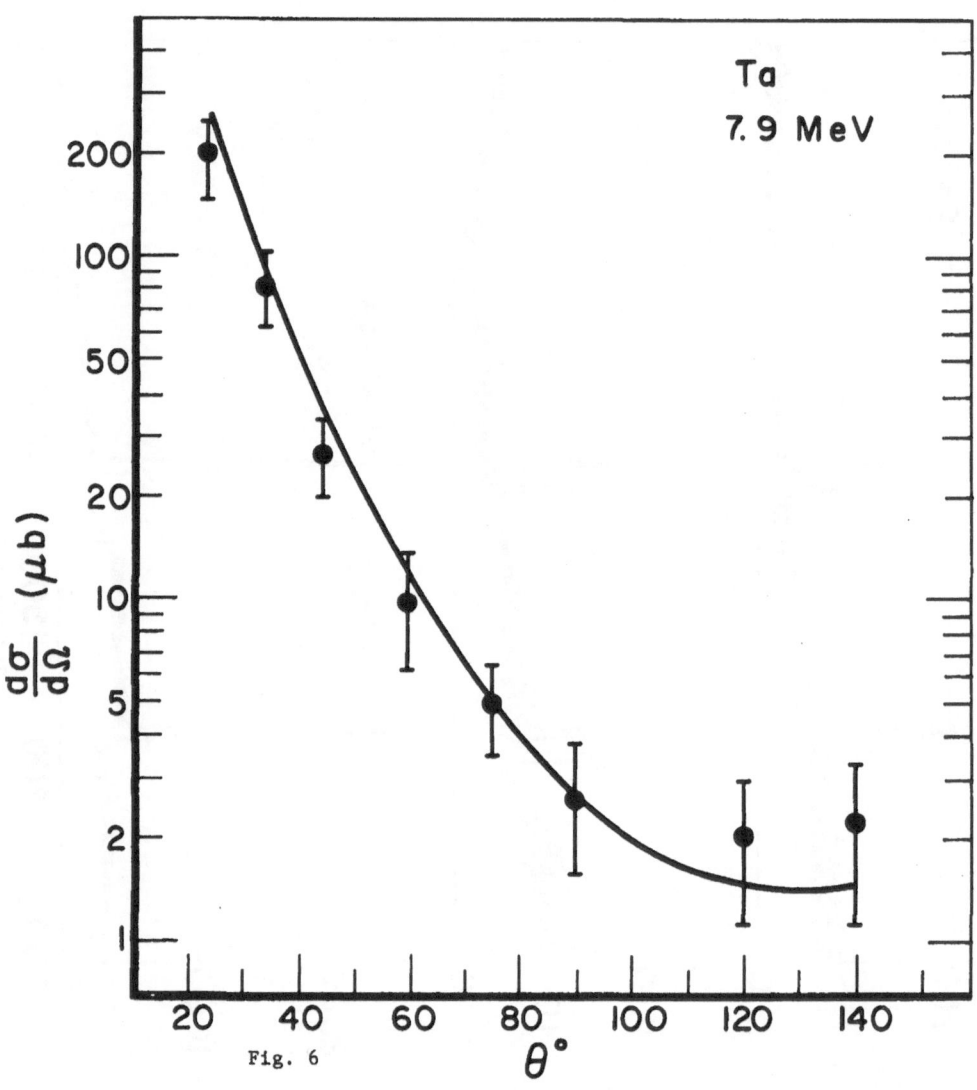

Fig. 6

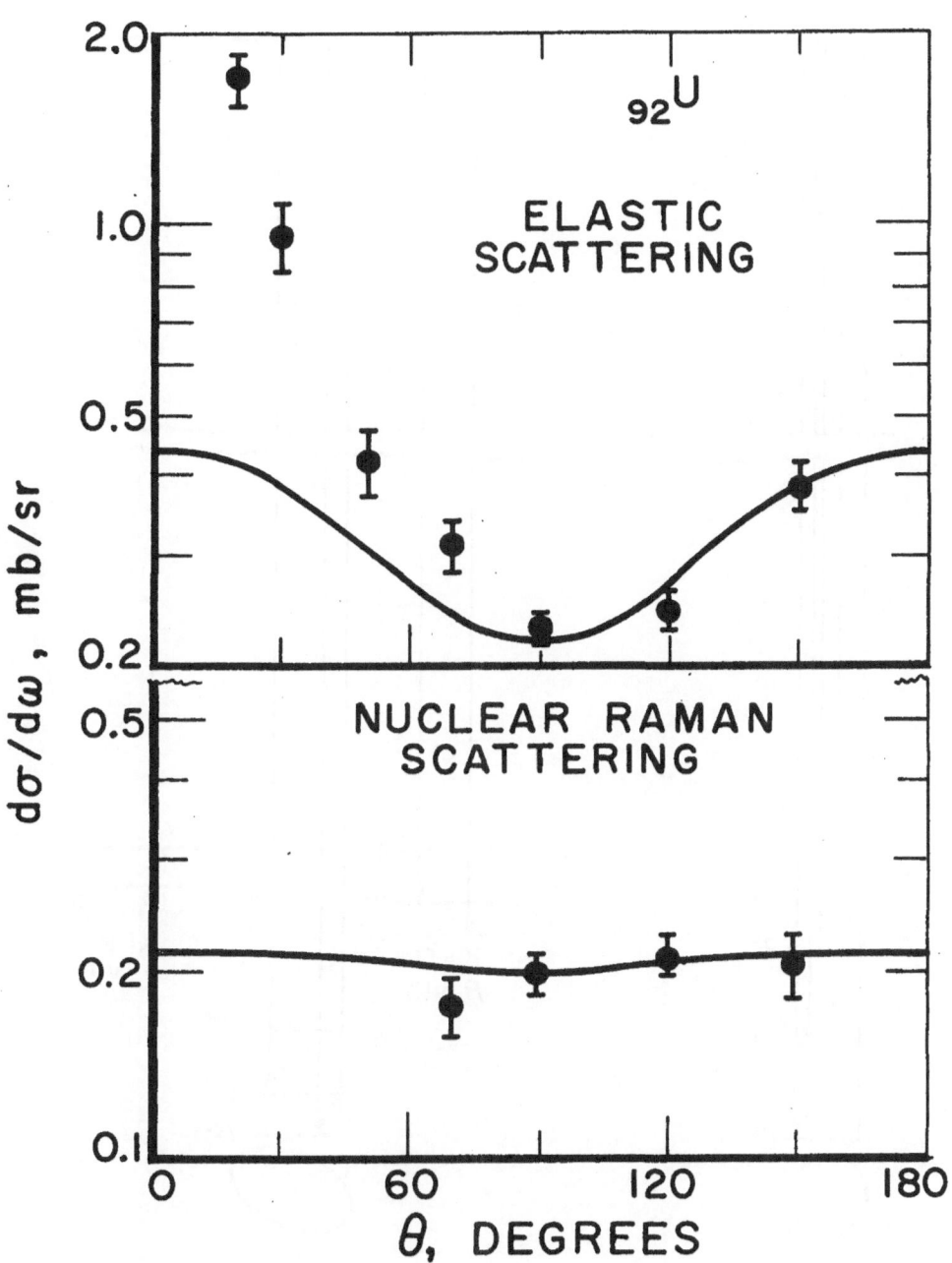

Fig. 7

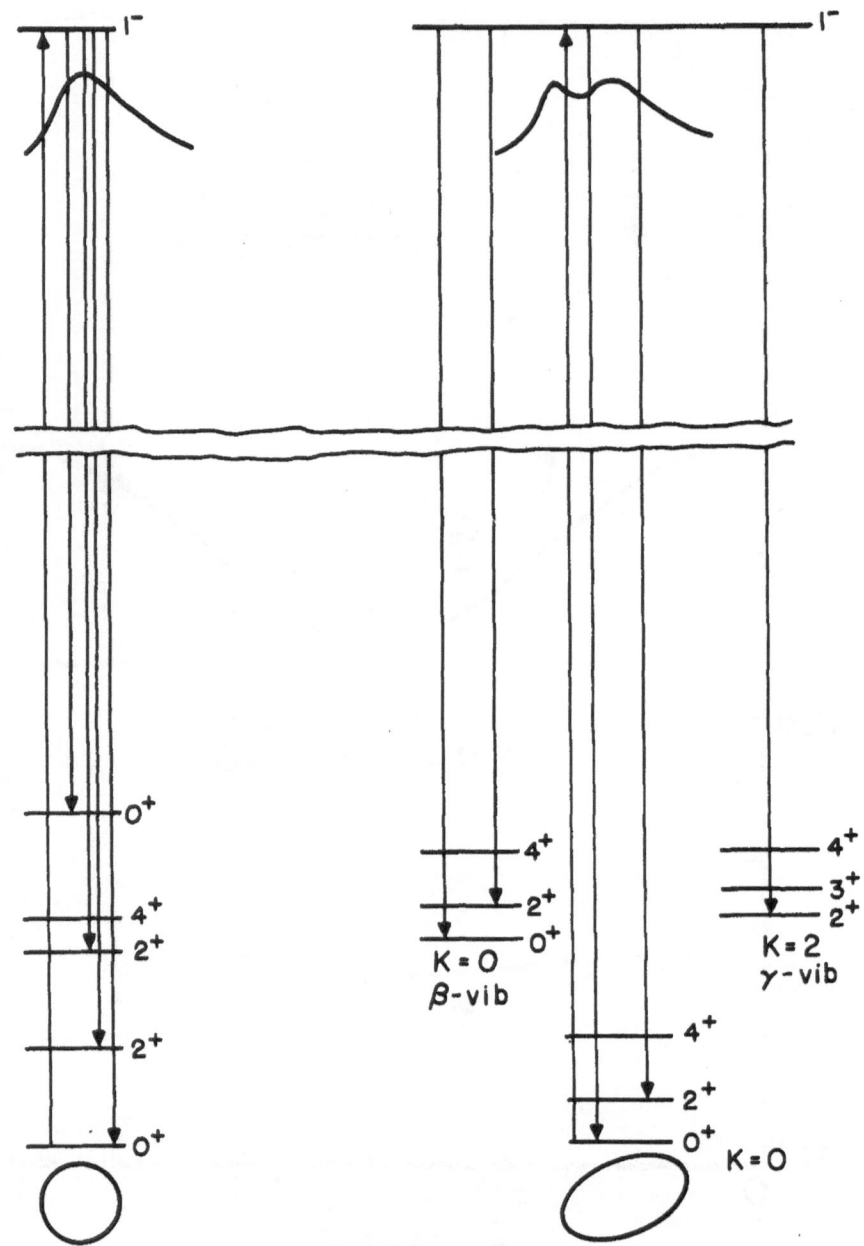

Fig. 8

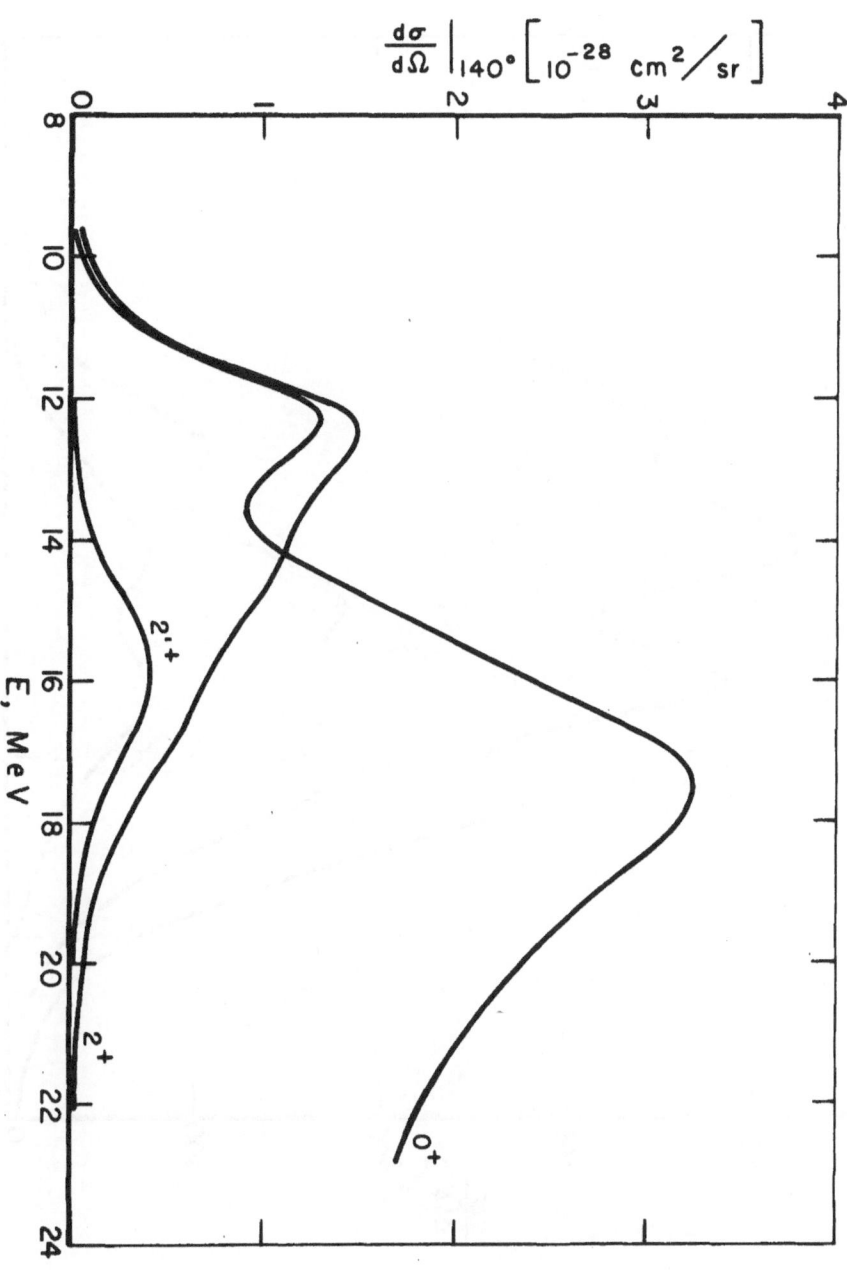

Fig. 9

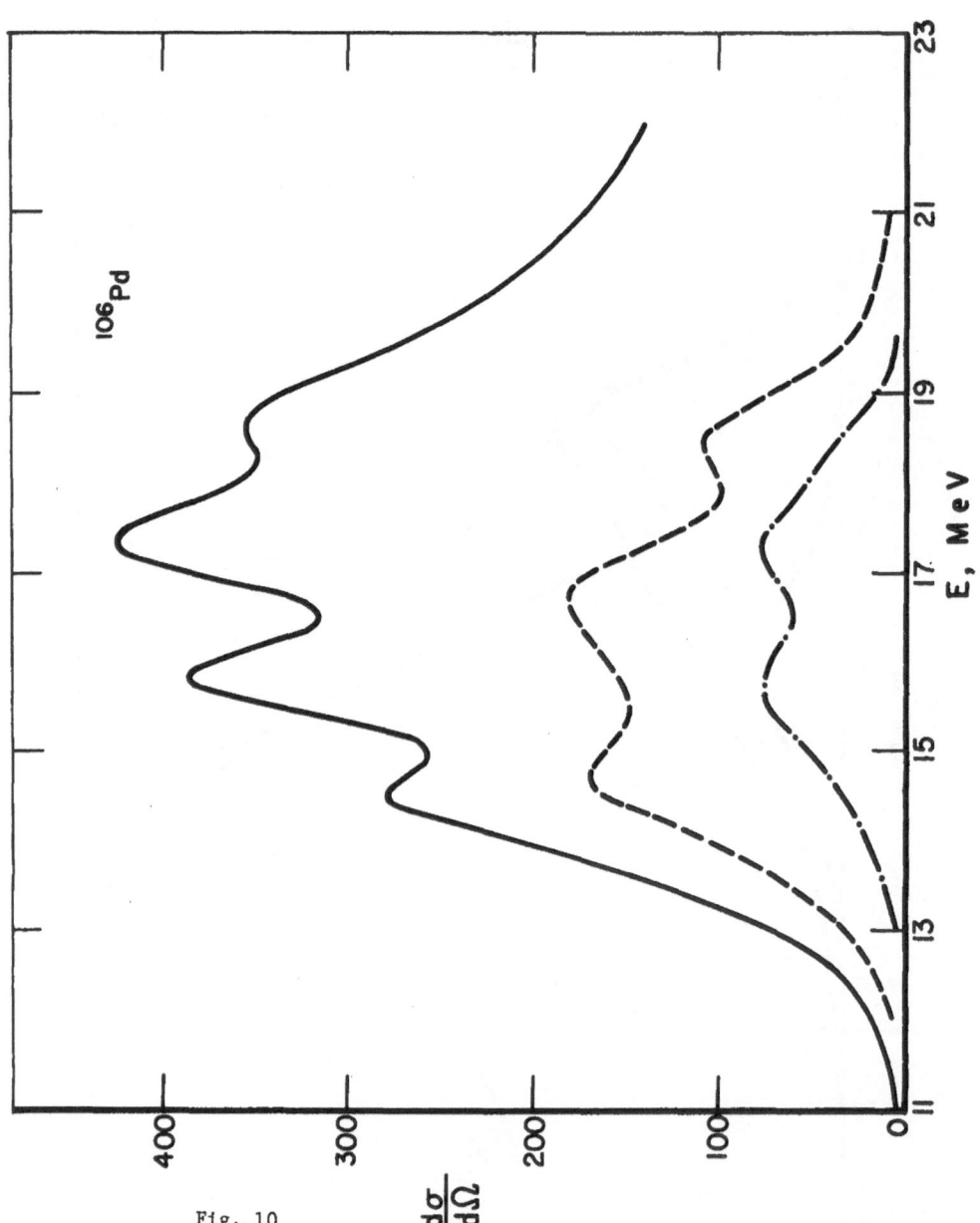

Fig. 10

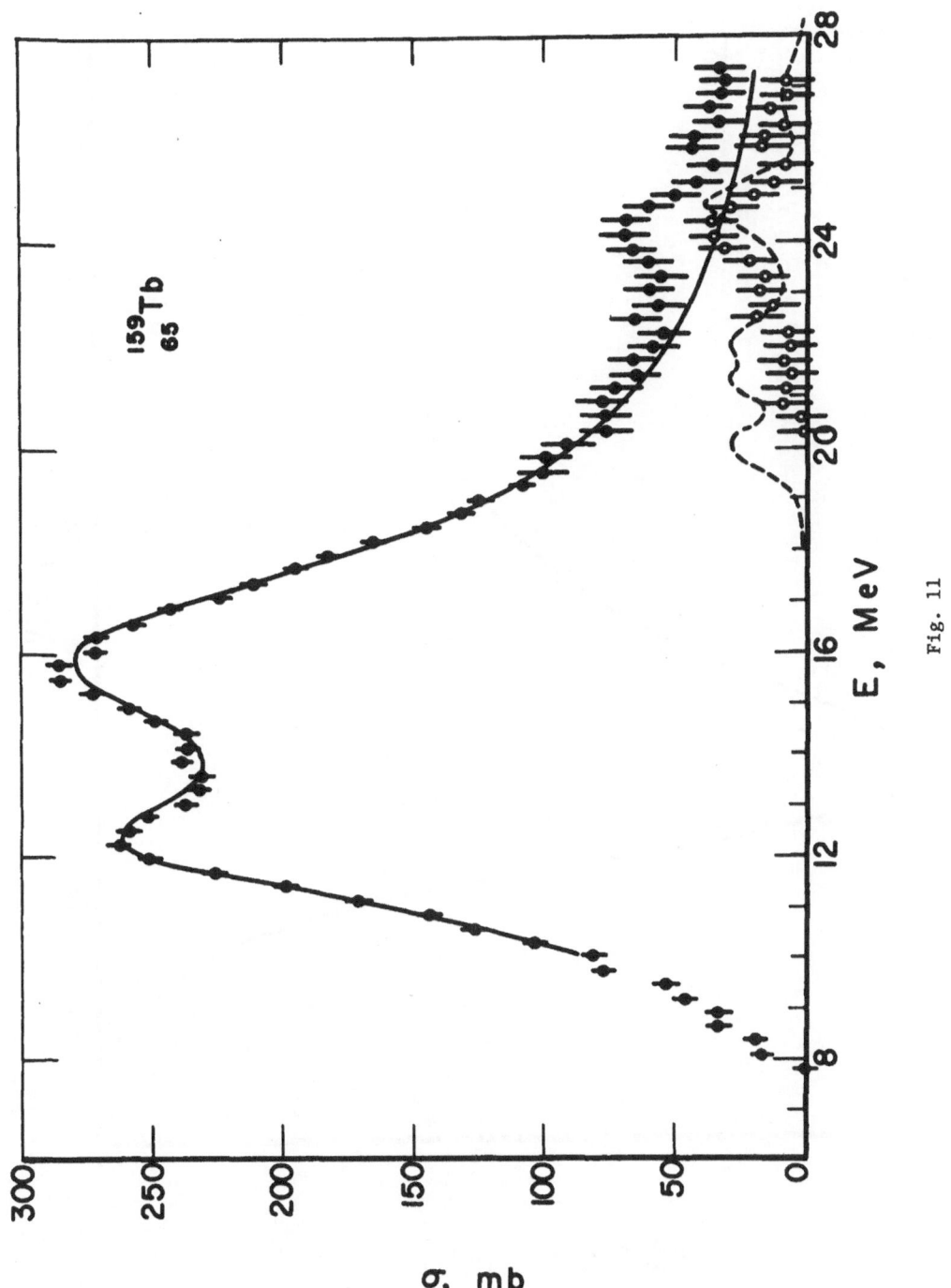

$^{159}_{65}$Tb

E, MeV

σ, mb

Fig. 11

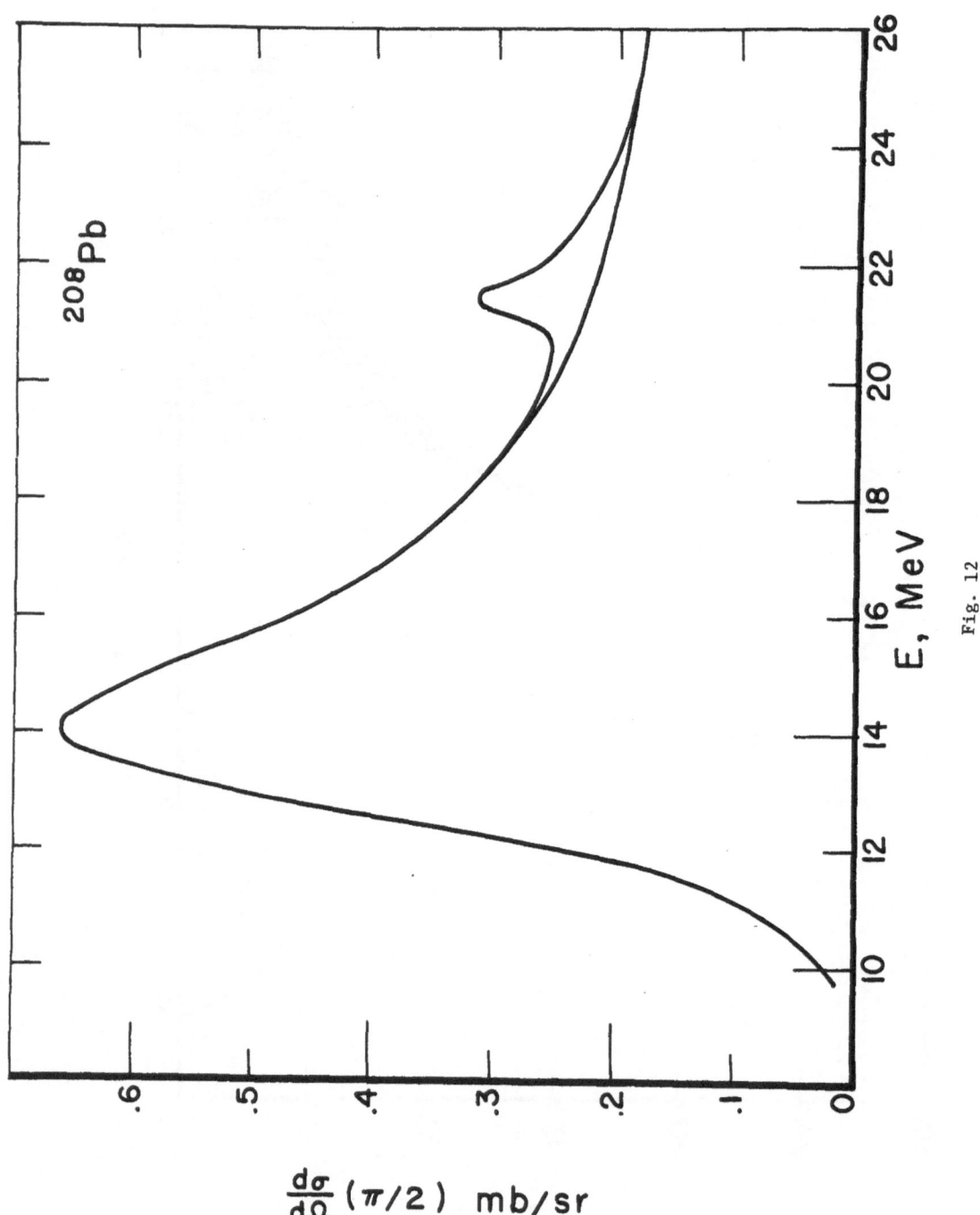

Fig. 12

Figure Captions

1. The average elastic scattering cross section for ^{12}C. The
    scattering cross sections for the 2^+ state at 4.44 MeV
    as well as the 1^+ states at 12.73 and 15.1 MeV are rep-
    resented as histograms 1 MeV wide with areas proportion-
    al to their integrated scattering cross sections. The
    scattering from the continuum has been calculated from
    the total absorption cross section of Ahrens et al.
    (75Ah3) using the optical theorem and the dispersion rela-
    tion and assuming that only electric dipole transitions
    participate. The (γ,n) and (γ,p) thresholds are indicated
    by vertical lines at the top of the figure. Note that the
    scattering from ^{12}C is dominated by the resonance fluores-
    cence of the 15.1 MeV level.

2. The average elastic scattering cross section for ^{208}Pb.
    The resonance fluorescence data came from the University
    of Illinois theses of Laszewski (75La2) and Coope (75Co2).
    Each line has been represented as a histogram 1 MeV wide
    and with a height proportional to the integrated scatter-
    ing cross sections; in fact, the states at 7.065 and
    7.085 MeV are shown as unresolved. In addition, it has
    been assumed that the ground-state radiation width is the
    total level width. The continuum scattering cross section
    has been calculated from the Saclay neutron production
    cross section (70Vel) assuming that all the transitions

are electric dipole. The vertical line at the top of the
figure indicates the energy of the (γ,n) threshold.

3. The cross section for the production of the isomer of ^{89}Y,
   a special kind of inelastic scattering. These data were
   taken from (58Sil).

4. The coherent and incoherent scattering cross section for
   ^{238}U at 90°. These were obtained using (1.22) and (1.23)
   and the resonance parameters given in (73Vel).

5. The Ge(Li) pulse height distribution produced by 10.83 MeV
   photons scattered by ^{238}U as observed by Jackson and
   Wetzel (72Jal). The incident spectrum of pulses shows the
   full-energy peak at 10.83 MeV as well as the single- and
   double-escape peaks. The scattered spectrum reflects this
   distribution but now each peak has a partner resulting
   from transitions to the 2^{+} state 45 keV above the ground
   state.

6. A comparison of the measured (74Ka9) elastic scattering
   cross section of ^{181}Ta with the predicted Delbrück scattering cross section (75Pa9) at 7.9 MeV. At this energy, the
   nuclear Thomson scattering and nuclear resonance scattering
   amplitudes cancel almost exactly. The only remaining component in the scattering cross section is Delbrück scattering.

7. The measured (72Ja1) angular distributions for the
   elastic and Raman scattering from ^{238}U. The elastic scattering cross section obeys the $1+\cos^2\theta$ rule for simple
   dipole scattering only in the backward hemisphere. In the
   forward direction the elastic scattering cross section is
   dominated by Delbrück scattering. The Raman scattering,
   on the other hand, is almost isotropic varying as
   $13+\cos^2\theta$.

8. A comparison of the level schemes for even-even spherical
   and deformed nuclei. The possible electric dipole de-
   excitations of the giant resonance are indicated.

9. The predicted scattering cross sections (65Ar4) for ^{166}Er
   showing the different energy dependence for the coherent
   scattering cross section, the Raman scattering to the 2^+
   state of the ground-state, rotational band, and the scattering that populates the 2^+ state of the γ-vibrational
   (K=2) band.

10. The predicted (67Ar1) scattering cross sections for ^{106}Pd
    showing the coherent scattering cross section and those
    for populating the 1st and 2nd 2^+ states.

11. The neutron production cross section (68Be5) of ^{159}Tb.
    The smooth curve is the sum of two Lorentz lines that fit
    the data below 18 MeV. The open circles are the difference

between the data on the smooth curve above 20 MeV. This difference is compared with the predicted (66Li1) E2 cross section.

12. A calculated ^{208}Pb coherent scattering cross section at 90° including electric dipole and electric quadrupole contributions. The resonance parameters that fit the ^{208}Pb Saclay (70Vel) data were used for the electric dipole part. The isoscalar E2 contribution was assumed to be made up of three resonances each 0.2 MeV wide and each containing 0.3 of an isoscalar E2 sum. They were located at 10.2, 10.6, and 11.2 MeV. The total isovector E2 strength was placed in a resonance at 21.5 MeV and 1 MeV in width.

RUB TP II / 151
May 1976

Mechanisms of photonuclear reactions at intermediate energies (40-140 MeV)

H. Hebach

Institut für Theoretische Physik
Ruhr-Universität Bochum

I. Introduction

It is the purpose of these lectures to discuss the
dynamical aspects of photonuclear reactions at energies above
the giant resonance region and below the pion threshold. In the
last years there has been considerable progress in this field,
mainly on the experimental side. Due to an impressive development in experimental techniques a great deal of new information on various photonuclear cross sections has become available.
Total nuclear photon absorption cross sections on several light
nuclei have been measured for energies up to or even beyond the
meson threshold. Partial cross sections like (γ,p), (γ,n) and
(γ,pn) have also been studied extensively. Energy spectra of
fast nucleons emitted from a large number of nuclei have been

Institut für Theoretische Physik, Ruhr-Universität Bochum
Universitätsstraße 150, PO Box 10 21 48, D 4630 Bochum 1

investigated in great detail. For several light nuclei we know the angular distributions of nucleons emitted in (γ,p) and (γ,n) reactions from definite shells in the target.

The increasing experimental information allows quantitative checks and improvements of models describing the photon absorption mechanism for energies above the giant resonances. In the past a variety of photonuclear data has been analysed in terms of the quasideuteron model. Partial success has been reported mainly in the interpretation of the energy spectra of the outgoing nucleons. Cross sections of processes like (γ,p), (γ,n) and (γ,pn) have been calculated in the frame of the shell model with and without nucleon-nucleon correlations. These calculations have been only partially successful in explaining the data and the answers given by different authors sometimes have been contradictory.

In section II we give a short review of these earlier theoretical attempts. We start with a discussion of total photonuclear absorption cross sections. Some recent calculations of the dipole sum rule for various light nuclei are reported here. Further, we give a survey of experimental results obtained for (γ,N) reactions in light nuclei and of the interpreation of these data by the models mentioned above.

In section III we outline a model for photonuclear reactions developed by the author in collaboration with M. Gari. This treatment starts from a shell model description and describes nucleon-nucleon correlations by means of closure to the giant resonance states. The importance of the gauge contributions to the transition matrix for higher photon energies below the pion threshold is emphasized. Thereby, the success of the quasideuteron model is explained. We present numerical results for the reactions (γ,p) and (γ,n) on ^{4}He, ^{12}C and ^{16}O, for the capture reaction (p,γ) on ^{3}H and for the reaction (γ,pn) on ^{16}O. The results are in good agreement with the data.

II. Nuclear photoeffect above the giant resonance region

II.1. Total nuclear photoabsorption

One of the most interesting quantities in photonuclear physics above the giant resonance region is the total absorption cross section σ_T (E_γ) for photons of energy E_γ. From the measurements of Ziegler and his group /1/ at the Mainz linear accelerator we know σ_T for several light nuclei (Li, Be, C, O, Al, Si and Ca) from E_γ = 10 MeV up to photon energies beyond the meson production threshold. Above the giant resonance the cross sections fall off smoothly towards higher energies. For instance, the total cross section for Carbon is about 17 mb in the maximum of the giant resonance near 22 MeV. At higher energies the cross section decreases from about 4 mb at 40 MeV to approximately 1 mb at 100 MeV. An important result is that the total photoabsorption cross sections $\sigma_T(E_\gamma)$, integrated from 10 MeV up to meson threshold (140 MeV) exceed the classical dipole sum by a factor of 1.4 to 2:

$$\sum_o (10, \hat{E}) = \int_{10 \, MeV}^{\hat{E} \, MeV} \sigma_T (E_\gamma) \, dE_\gamma \xrightarrow[(\hat{E} = 140 \, MeV)]{} (1.4-2) \sum_{cl} , \quad (II.1)$$

$$\sum_{cl} = 60 \frac{NZ}{A} [MeV \, mb]. \quad (II.2)$$

This large value of Σ_0 has initiated new calculations /2-6/ of the electric dipole sum rule

$$\Sigma^{E1} = \int_0^\infty \sigma^{E1}(E_\gamma)\, dE_\gamma = 60 \frac{NZ}{A}(1+\varkappa) \; [MeVmb] \qquad (II.3)$$

where the enhancement $\varkappa$ is the ground state expectation value

$$\varkappa = \frac{A}{NZ} \frac{M}{\hbar^2} \langle \psi_0 | [D_z, [V, D_z]] | \psi_0 \rangle . \qquad (II.4)$$

V is the nucleon-nucleon potential and D_z is the electric dipole operator

$$D_z = \sum_{\alpha=1}^{A} \frac{1}{2} \tau^z(\alpha)\, Z_\alpha . \qquad (II.5)$$

In table I we list the values of $\varkappa$ calculated recently /6/ for a number of nuclei and for three potentials: Hamada-Johnston (HJ), Reid soft core (RSC) and super-soft-core (SSC). In these calculations the nuclear ground-state wave functions $|\psi_0\rangle$

Potential	^2H	^4He	^{12}C	^{16}O	^{28}Si	^{32}S	^{40}Ca
HJ	0.27	0.49	0.50	0.65	0.57	0.60	0.68
RSC	0.24	0.50	0.49	0.62	0.53	0.55	0.60
SSC	0.18	0.38	0.37	0.44	0.41	0.42	0.48

Table I   Values of $\varkappa$ for several nuclei /6/.

include two-body correlations obtained from the Bethe-Goldstone formalism. In figure 1 we compare the results for Σ^{E1}/Σ_{cl} derived in ref. /6/ with experimental data. Black vertical bars show the dependence of this quantity on the NN-potential of use (see table I). The rectangles on top of the black bars indicate

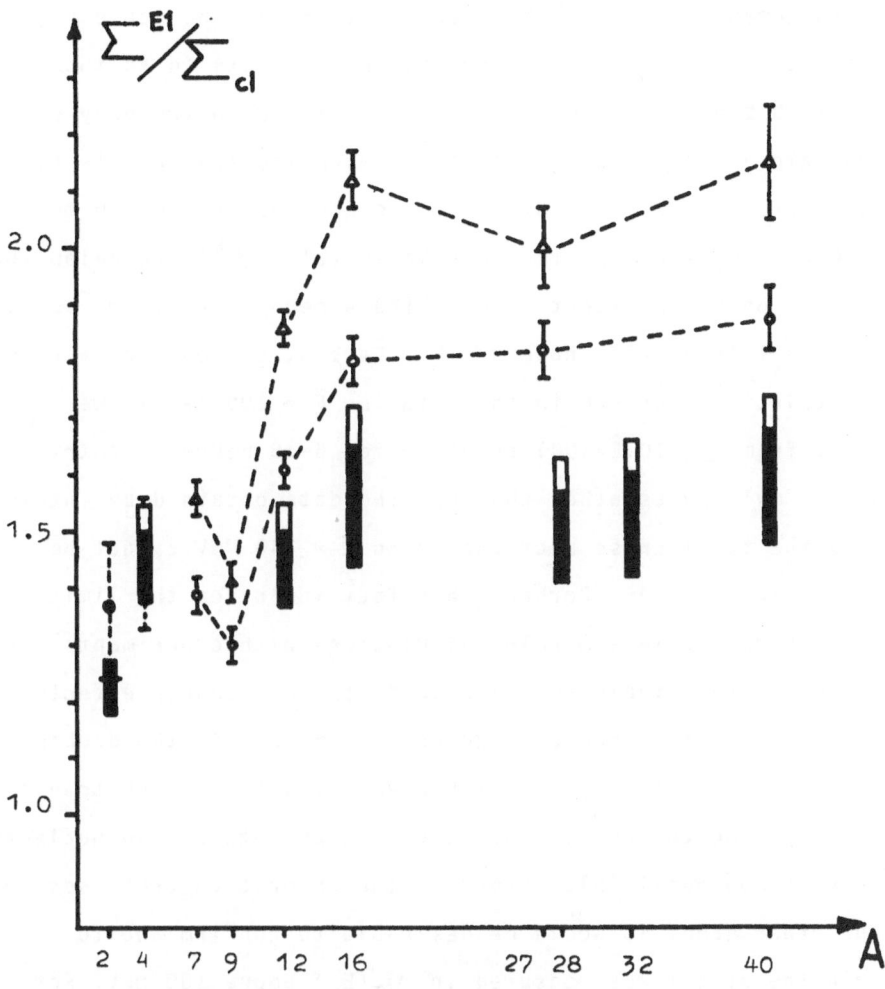

Figure 1

Calculated values of $\sum^{E1}$ (ref. /6/) in units of the classical sum $\sum_{cl}$. The meaning of the vertical bars is explained in the text. For comparison the integrated cross sections $\sum_0 (10, \hat{E})$ from ref. /1/ are shown for two values of $\hat{E}$: ($\frac{I}{Q}$) $\hat{E}$ = 100 MeV , ($\frac{I}{Q}$) $\hat{E}$ = 140 MeV. Deuteron data are shown according to ref. /7/. Helium data are taken from ref. /8/.

the enhancement when three-body correlations in the nuclear ground state are taken into account. This amounts to an increase of about 10 % of the values of $\varkappa$ calculated with two-body correlations alone /5/. Figure 1 shows the experimental results for $\Sigma_0(10,\hat{E})$ for integration of σ_T up to $\hat{E} = 100$ MeV and up to $\hat{E} = 140$ MeV. We see that the calculated vales Σ^{E1} are below the numbers of both experimental sets. The A dependence seems to be reproduced fairly well, however. For Carbon, Oxygen and Calcium the calculations agree with the data for $\hat{E} = 100$ MeV if we subtract from $\Sigma_0(10,\hat{E}=100)$ an estimated 5-10 percent contribution from multipoles other than E1. The data obtained by integrating the total cross sections up to $\hat{E} = 140$ MeV cannot be explained in this way. Perhaps this fact indicates the limits for comparing dipole sum rule calculations with experiments. For instance, in the dipole sum rule we describe exchange effects by the use of static nucleon-nucleon potentials in the double commutator of eq. (II.4). At higher energies the energy transfer from the photon to the pion being exchanged between two nucleons may have a non-negligible effect on the absorption cross sections. On the other hand, it would be desirable to confirm and to explain the structures measured in $\sigma_T(E_\gamma)$ above 100 MeV. For Oxygen the values of σ_T vary between zero and four mb, for the range from 100 to 140 MeV. Therefore, at present the only possible conclusion is that calculations of the dipole sum rule give a reliable lower limit for total photoabsorption cross sections integrated up to 100 MeV.

In ref. /1/ the measured total cross sections $\sigma_T(E_\gamma)$ are interpreted in terms of Levinger's quasideuteron model /9/ which connects the total photonuclear cross sections with the cross section σ_D for the photodisintegration of the deuteron:

$$\sigma_T(E_\gamma) = L \frac{NZ}{A} \sigma_D(E_\gamma). \qquad\qquad (II.6)$$

It is found that for the elements Li, Be, C and O this formula explains the energy dependence and the absolute magnitude of the cross sections in the energy region from 40 to 100 MeV, if L is chosen equal to 8. Therefore it is concluded that the dominant absorption mechanism above 40 MeV is the absorption of photons by correlated neutron-proton pairs. Indeed this idea is supported by the following facts:

(i)   From the Mainz data and from (γ,p) and (γ,n) cross sections on ^{12}C and ^{16}O the ratio of the total absorption cross section and the cross section for single nucleon emission can be estimated to be equal to 4 near 60 MeV and equal to about 20 near 100 MeV. Presumably, a large fraction of the total cross section is made up by the (γ,pn) reaction , at least at the higher energies around 100 MeV.

(ii)   Neutron-proton pairs from the $^{12}C(\gamma,np)$ reaction have been observed by Gorbunov and Taran /10/. The $^{12}C(\gamma,np)$ and $^{12}C(\gamma,p)$ cross sections have been found to be comparable between 40 MeV and 100 MeV photon energy, and the (γ,np) cross section to be a factor of about ten greater than the (γ,p) cross section for energies 100 MeV$< E_\gamma <$ 170 MeV.

There is a correspondence between the quasideuteron model and the commutator $[V,D_z]$ which appears within the double commutator in eq. (II.4). $[V,D_z]$ represents the contribution from

charge exchange parts in the nuclear potential V to electric
dipole transitions. Matrix elements of $[V, D_z]$ are different
from zero only for neutron-proton states as can be seen from the
relations

$$[V, D_z]|pn\rangle \sim |np\rangle, \quad [V, D_z]|pp\rangle = 0, \quad [V, D_z]|nn\rangle = 0. \quad (II.7)$$

In section III we shall discuss the importance of this exchange
(or "gauge") contribution in photonuclear reactions like (γ,p),
(γ,n) and (γ,pn).

II.2. (γ,N) reactions: experiments and theoretical models

We consider (γ,n) and (γ,p) reactions on light nuclei
which have been investigated rather extensively in the last few
years. Since later in this paper numerical results will be
discussed for the target nuclei ^{4}He, ^{12}C and ^{16}O, we mention
the measurements reported in references /11-25/. (γ,N) reactions
have been studied also for ^{3}He, ^{6,7}Li and ^{9}Be. From the ^{12}C(γ,N)
and ^{16}O(γ,N) data, for instance, one obtains the following
results:

(i)   In the energy spectra of the emitted protons or neutrons
        there is a peak at the high energy side corresponding to
        the removal of a nucleon from the p-shell. The residual
        nucleus is left in its ground state or in low-lying
        excited states. The peak is followed by a tail for lower
        nucleon energies.

(ii)   The angular distributions of the protons and neutrons
        (with the residual nucleus left with zero or small excita-
        tion) show a forward asymmetry which increases slightly
        with increasing photon energy.

(iii) The total cross sections (this means here: integrated over
      angles) are of comparable magnitude for (γ,p) and (γ,n).

In the past years many attempts have been made to explain these
features of (γ,N) reactions. The motivation for experiments and
calculations had been to obtain information on high momentum
components in the nuclear wavefunctions. If the nucleon is
emitted from the nucleons with maximum kinetic energy, momentum
components of about 300 - 500 $\frac{MeV}{c}$ are needed during the absorption
act, for the energy range considered here. This is because little
momentum is carried by the incident photon. Essentially two kinds
of model calculations have been started: (i) calculations in the
frame of an independent particle model with and without nucleon-
nucleon correlations, (ii) calculations based on the quasideuteron
model. In the following we discuss a few of these attempts and
their results.

II.2.A. <u>Independent particle model for (γ,N)</u>

Early descriptions /26, 27/ of the photoemission of nucleons
from definite shell model orbits have assumed a single particle
mechanism. In this picture the interaction of the photon with
the nucleus is described by a one-body operator. The photonucleon
undergoes a transition from a bound shell model state (e.g.,
calculated in a harmonic oscillator or Woods-Saxon well) to a
final state in the continuum (e.g., optical model solution or
plane wave). The cross sections obtained in various papers differ
by several orders of magnitude depending on the wave functions
used in the initial and final states. For instance, the (γ,p)

cross section may be reproduced in some cases when a plane wave approximation is chosen for the outgoing protons. Such a calculations, however, fails completely in describing (γ,n) data. Furthermore, if the plane wave is replaced by a scattering solution in a complex optical model potential, the (γ,p) cross section goes down by one or two orders of magnitude /24, 28/. The same conclusion is reached when the required orthogonality of the initial and final wave functions is observed /29/, i.e., if one calculates the single particle bound and scattering states in the same shell model potential.

II.2.B. Shell model including nucleon-nucleon correlations

In other papers the shell model (SM) treatment has been modified by taking into account short-range nucleon-nucleon correlations which have been supposed to remedy the deficiency of high-momentum components in pure shell-model wavefunctions. Usually a Jastrow ansatz (or modifications thereof) has been chosen:

$$\psi(1,\dots,A) = \phi_{SM}(1,\dots,A) \prod_{i<j}^{A} f(r_{ij}),$$
(II.8)

where the correlation factor $f(r_{ij})$ suppresses the relative wavefunction of a nucleon pair at short distances. The resulting (γ,N) cross sections obtained by several authors again differed widely from each other. Shklyarevskij /30/ obtained an enhancement of pure SM results of about one order of magnitude. He could explain $^{12}C(\gamma,p)$ data given in ref. /17/. Weise and Huber /31, 28, 24/ parametrize the correlation factor in the form

$$f(r) = 1 - \int dq' \, w(q') \, j_0(q'r),$$

$$w(q') = \delta(q-q'), \quad \text{or:} \quad w(q') \sim \frac{1}{\Delta q} \exp\left\{-\frac{(q'-q)^2}{\Delta q^2}\right\},$$

(II.9)

corresponding to an exchange of momenta $\hbar q'$ between two nucleons. The calculations carried out for (γ, N) reactions on ^{6,7}Li, ^{12}C and ^{16}O show that the SM results are increased by one order of magnitude or more for $\hbar q \approx 300 \pm 50 \frac{MeV}{c}$. In addition, (γ, p) angular distributions in ^{6}Li and ^{12}C can be explained, at least for certain photon energies.

Fink et al. /29/ use a Jastrow factor which, for relative s-states, equals the ground-state defect wavefunction obtained from the solution of the Bethe-Goldstone equation with a hard-core potential. Below 100 MeV photon energy the SM results are not altered essentially in this way. Malecki and Picchi /32/ also find that short range correlations are of little influence below 100 MeV. Only for energies above 100 MeV the SM cross sections for ^{16}O(γ, p) are enhanced considerably.

Nucleon-nucleon correlations in an extended shell model frame have also been considered in two papers by Brown /33/ and Fujii and Sugimoto /34/. Their main ideas can be sketched as follows. The nuclear Hamiltonian is written in the form

$$H = H_0 + v$$

(II.10)

where H_0 is a shell model Hamiltonian with eigenstates $|\phi\rangle$ and v is an effective two-body interaction. The final state $|\psi_f\rangle$ of the nuclear system is written as

$$|\psi_f\rangle = |\phi_f\rangle + \frac{1}{E_f - H - i\varepsilon} \, v \, |\phi_f\rangle. \qquad (\mathrm{II.11})$$

For a (γ, N) reaction on a closed shell nucleus (in ref. /34/ the $^{16}O(\gamma, N)$ reactions are considered) $|\phi_f\rangle$ is a one particle-one hole state. With the approximation $|\psi_i\rangle \approx |\phi_i\rangle$, the transition matrix for electric transitions is

$$M_{fi}^{(EL)} \approx \langle \phi_f | \mathcal{E}_\lambda^{(L)} | \phi_i \rangle + \langle \phi_f | v \frac{1}{E_f - H + i\varepsilon} \mathcal{E}_\lambda^{(L)} | \phi_i \rangle \qquad (\mathrm{II.12})$$

with

$$\mathcal{E}_\lambda^{(L)} \sim r^L Y_\lambda^{(L)}(\hat{r}). \qquad (\mathrm{II.13})$$

The second term of eq. (II.12) is treated by using closure to the giant resonance states. The operator $\mathcal{E}_\lambda^{(L)}$, acting on $|\phi_i\rangle$, predominantly excites intermediate states in the neighborhood of the giant resonances. Therefore, one may write to a good approximation

$$M_{fi}^{(EL)} \approx \langle \phi_f | \mathcal{E}_\lambda^{(L)} | \phi_i \rangle + \frac{1}{E_f - E_L^{res}} \langle \phi_f | v \mathcal{E}_\lambda^{(L)} | \phi_i \rangle, \qquad (\mathrm{II.14})$$

where

$$E_L^{res} = E_L - i \frac{\Gamma_L}{2} \qquad (\mathrm{II.15})$$

is given by the position and width of the resonances. The first term in eq. (II.14) is the "direct" (or shell-model) transition, while the second term is called the "semi-direct" transition because the nuclear system goes through the dipole, quadrupole etc. resonance states. In ref. /33/ a schematic model calculation shows that the semi-direct contribution can enhance the cross sections by a factor of about ten in the region of the giant resonance, and that its effects persist to relatively high energies.

In ref. /34/ a δ-force is used for the residual interaction, and ground state correlations are considered in first order perturbation theory. For the reaction $^{16}O(\gamma,n)$ the calculation gives a forward asymmetry of the angular distribution for photon energies above 40 MeV. The theoretical angular distributions of the reactions $^{16}O(\gamma,p)$ and $^{16}O(\gamma,n)$ are presented in the form of an integral $\int (d\sigma/d\Omega)(1/E_\gamma)\,dE_\gamma$ for an interval 26.6 MeV $\leqslant E_\gamma \leqslant$ 170 MeV. This quantity shows a small forward asymmetry for both (γ,p) and (γ,n) reactions and its absolute magnitude is very similar in both cases. The results indicate that this model gives a possible frame for the description of (γ,N) data above the giant resonance region.

II.2.C. Calculations based on the quasideuteron model (QDM)

Earlier in this paper we mentioned the interpretation of total photonuclear absorption cross sections /1/ in terms of Levinger's quasideuteron model /9/. In many other papers reporting photoemission data in the energy range from 50 to 150 MeV the quasideuteron model or modified versions thereof /35/ have also been applied with considerable success (see the detailed review article given by Costa /36/). We list a few of these investigations:

Matthews et al. /37/ compare photoproton spectra from ^{6}Li for 102 MeV bremsstrahlung with the predictions of the QDM. For various angles of the outgoing protons the shape of the spectra is very well explained, except at the highest proton energies.

Costa et al. /38/ also find that the low-energy tail of photoprotons from ^{9}Be for 50 MeV quasi-monochromatic photons can be explained in the QDM, while the yield of the highest energy protons (emitted from the p-shell) is not accounted for in this model.

Miller et al. /16/ note that the results of a QDM calculation do not agree with their data for the angular distributions of neutrons emitted from the p-shell of ^{12}C. A better fit is obtained with the quasi-alpha model developed by Mamasakhlisov and Jibuti /39/. In contrast to these findings Schier and Schoch /14/ report good agreement of QDM calculations with the measured energy dependence and angular distribution of the reaction ^{12}C(γ,n) and ^{16}O(γ,n), for neutrons leaving the residual nuclei in the ground state or low-excited states.

In a series of investigations the QDM has been extended to include secondary interactions following the initial photo-absorption act. Secondary interactions have been described in the intranuclear-cascade model /40-42/. Each of the particles knocked from its position is traced through the nucleus until it collides with another particle. In this way a cascade may be generated. Monte-Carlo methods are used and each collision is described by free particle-particle cross sections. In several papers, for instance in refs. /43-45/, the QD plus intranuclear-cascade model has been applied to the interpretation of photo-nucleon spectra above 40 MeV photon energy, and good agreement has been achieved for absolute cross sections, energy spectra shape and mass dependence.

In the next section we shall describe calculations for the reactions (γ,p), (γ,n) and (γ,pn) based on a model developed by a Bochum group /46-51/. The different aspects of the photo-nuclear reactions mechanism encountered in this section will be discussed separately. We start from a shell model description and introduce correlations by using closure to the giant resonance states. The transition matrix is decomposed into contributions from different physical processes: (i) shell model contribution, (ii) nucleon-nucleon correlations in the initial and final states, (iii) gauge (or exchange) contributions which arise from the direct coupling of the photon to the correlations between a neutron-pair pair. In the case of (γ,N) these contributions are studied as a function of the photon energy. It turns out that for $E_\gamma \gtrsim 60$ MeV the gauge contributions are the dominant parts of the (γ,N) transition matrix. Moreover, the (γ,pn) cross section calculated from the gauge terms is larger than the (γ,N) cross section by a factor of ten, for $E_\gamma \approx 100$ MeV. This gives an explanation of the success of the quasideuteron model in this energy range.

III. Transition matrix for photonuclear processes

III.1. General considerations

We have to evaluate transition matrix elements

$$M_{fi} = \left\langle \psi_f \left| -\frac{1}{c}\sqrt{\frac{2\pi\hbar c}{k_\gamma}} \int \vec{j}(\vec{r}) \cdot \vec{\varepsilon}_\lambda e^{i\vec{k}_\gamma \cdot \vec{r}} d\tau \right| \psi_i \right\rangle \qquad (III.1)$$

for absorption of a photon of momentum $\hbar\vec{k}_\gamma$ while the nuclear system makes a transition from the ground state $|\psi_i\rangle$ to a state $|\psi_f\rangle$ with a certain number of nucleons emitted from the nucleus. The cross section is given by

$$\delta_{fi} = \frac{2\pi}{\hbar c} |M_{fi}|^2 \rho_f(E_f), \qquad (III.2)$$

where $\rho_f(E_f)$ is the density of final states. $|\psi_i\rangle$ and $|\psi_f\rangle$ are eigenstates of the nuclear Hamiltonian $H = T + V$. We assume that the nuclear potential V is a sum of two-body interactions. The nuclear current density $\vec{j}(\vec{r})$ in eq. (III.1) satisfies the continuity equation

$$\vec{\nabla} \cdot \vec{j} + \frac{i}{\hbar}[H, \rho] = 0, \qquad (III.3)$$

where ρ is the charge density operator. The charges are assumed to be located at the positions of the nucleons:

$$\rho(\vec{r}) = \sum_{\alpha=1}^{A} e_\alpha \, \delta(\vec{r} - \vec{r}_\alpha), \qquad (III.4)$$

e_α being the charge of nucleon α. If the two-body potentials contain exchange parts, i.e., terms with an isospin factor $\vec{\tau}(\alpha) \cdot \vec{\tau}(\beta)$, the nuclear current density consists of one-body and two-body parts. The one-body current density (convection current plus spin current),

$$\vec{j}_{[1]} = \sum_{\alpha=1}^{A} \frac{e_\alpha}{2M} \left\{ \vec{p}_\alpha \, \delta(\vec{r}-\vec{r}_\alpha) + \delta(\vec{r}-\vec{r}_\alpha)\vec{p}_\alpha \right\}$$

$$+ \sum_{\alpha=1}^{A} \frac{e\hbar}{2M} \mu_\alpha \, \vec{\nabla}_r \times \vec{\sigma}_\alpha \, \delta(\vec{r}-\vec{r}_\alpha), \tag{III.5}$$

satisfies the equation

$$\vec{\nabla} \cdot \vec{j}_{[1]} + \frac{i}{\hbar} \, [T, \rho] = 0, \tag{III.6}$$

and the two-body current density $\vec{j}_{[2]}$ (exchange current) is a solution of the equation

$$\vec{\nabla} \cdot \vec{j}_{[2]} + \frac{i}{\hbar} \, [V, \rho] = 0. \tag{III.7}$$

The total current density $\vec{j} = \vec{j}_{[1]} + \vec{j}_{[2]}$ fulfils the continuity equation (III.3). Hereby the gauge invariance of the electromagnetic interaction is ensured. The current densities $\vec{j}_{[1]}$ and $\vec{j}_{[2]}$ are visualized in figure 2.

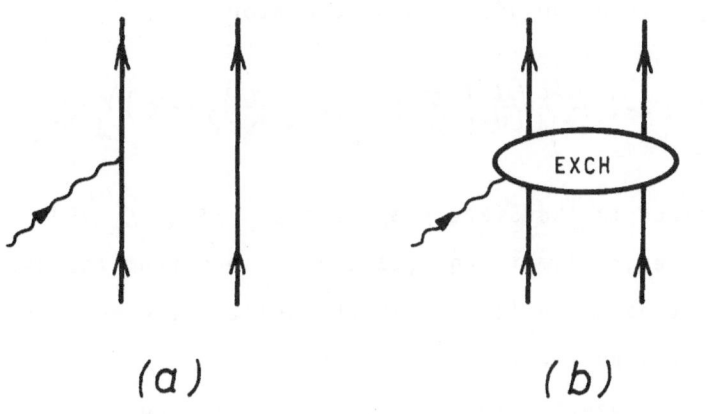

(a)   (b)

Figure 2   Contributions from (a) convection current,
(b) exchange current, to the electromagnetic
transition operator

For the evaluation of the transition matrix eq. (III.1) a multipole expansion of the photon field turns out to be a useful concept. Our calculation shows that even at photon energies around 100 MeV the contributions from various multipoles decrease rapidly with increasing multipole order. The expansion reads

$$\vec{\xi}_\lambda \, e^{i\vec{k}_\gamma \cdot \vec{r}} = \sum_{L=1}^{\infty} \left(\frac{2\pi(2L+1)}{L(L+1)} \right)^{\frac{1}{2}} \left\{ -i^{L+1} \frac{1}{k_\gamma} \vec{\nabla} \left[\left(1 + r\frac{d}{dr}\right) j_L(k_\gamma r) Y_\lambda^{(L)}(\hat{r}) \right] \right.$$

$$- i^{L+1} k_\gamma \, \vec{r} \, j_L(k_\gamma r) Y_\lambda^{(L)}(\hat{r}) \qquad\qquad (III.8)$$

$$\left. - \lambda \, i^L j_L(k_\gamma r) \left[\vec{L} \, Y_\lambda^{(L)}(\hat{r}) \right] \right\}.$$

The first term of this expansion, giving rise to electric multipole transitions (EL), is the most important one in a description of the qualitative features of the nuclear photoeffect. It gives the following contribution to the transition matrix eq. (III.1)

$$M_{fi}^{(EL)} = \langle \psi_f | [H, Q_\lambda^{(L)}] | \psi_i \rangle, \qquad\qquad (III.9)$$

where $Q_\lambda^{(L)}$ is a sum of one-body operators:

$$Q_\lambda^{(L)} = \sum_{\alpha=1}^{A} Q_\lambda^{(L)}(\alpha) = -i^L \frac{2\pi\hbar c}{\sqrt{E_\gamma}} \sqrt{\frac{2L+1}{L(L+1)}} \sum_\alpha e_\alpha \frac{1}{k_\gamma} \left(1 + r_\alpha \frac{d}{dr_\alpha}\right) j_L(k_\gamma r_\alpha) Y_\lambda^{(L)}(\hat{r}_\alpha), \qquad (III.10)$$

which reduce to the Siegert-operators $\sim r_\alpha^L \, Y_\lambda^{(L)}(\hat{r}_\alpha)$ in the long-wavelength limit. Eq. (III.9) follows from the use of the continuity equation (III.3) which replaces the nuclear current density $\vec{j}(\vec{r})$ by the charge density operator $\rho(\vec{r})$. This is the content of Siegert's theorem /52/. Usually, eq. (III.9) is written in the form

$$M_{fi}^{(EL)} = E_\gamma \langle \psi_f | Q_\lambda^{(L)} | \psi_i \rangle. \qquad (III.11)$$

Using this expression we do not have to worry about mesonic corrections. However, in eq. (III.11) exchange currents are taken into account already as can be seen from the equivalent expression (III.9) which may be written as

$$M_{fi}^{(EL)} = \langle \psi_f | [T, Q_\lambda^{(L)}] | \psi_i \rangle + \langle \psi_f | [V, Q_\lambda^{(L)}] | \psi_i \rangle. \qquad (III.12)$$

Here, the commutator $[T, Q_\lambda^{(L)}]$ originates from eq. (III.6) for the one-body current density $\vec{J}_{[1]}$ whereas the commutator $[V, Q_\lambda^{(L)}]$ follows from eq. (III.7) for the two-body current density $\vec{J}_{[2]}$. Eq. (III.12) is illustrated by figure 3. The first term of

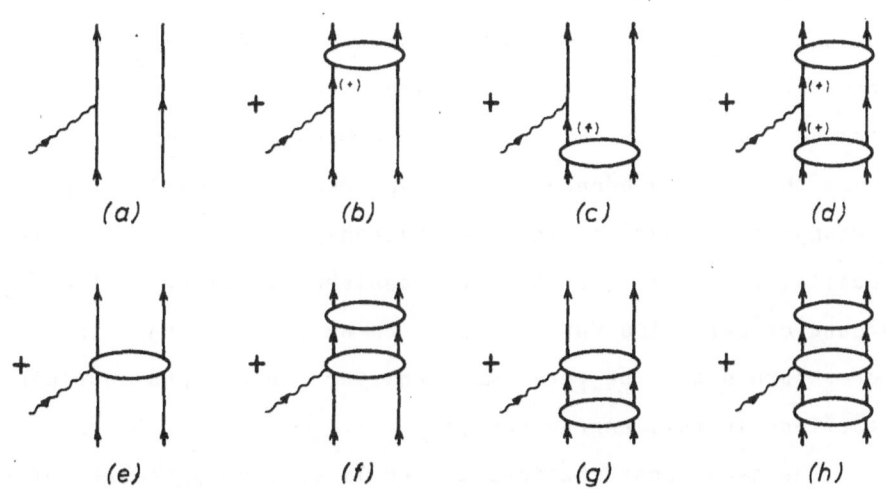

(a)   (b)   (c)   (d)

(e)   (f)   (g)   (h)

Figure 3   Contributions to the transition matrix eq.
(III.12), in the case of a two-nucleon system.
The blobs indicate nucleon-nucleon correlations.
In diagrams (a-d) the photon interacts with a
single nucleon in a positive energy state
(denoted by a(+)-sign). Diagrams (e-h)
illustrate the gauge contributions.

eq. (III.12) is represented by diagrams 3 (a-d) where the photon interacts with a single nucleon. The second term of eq. (III.12) can be interpreted by diagrams 3 (e-h). In these processes (gauge contributions) the photon field is coupled directly to the nucleon-nucleon correlations.

For the discussion of specific photoemission processes like (γ,p), (γ,n) or (γ,pn) it is important to note the isospin dependence of the commutator

$$\left[V(1,2), Q_\lambda^{(L)} \right] \sim \left(\vec{\tau}(1) \times \vec{\tau}(2) \right)^3. \tag{III.13}$$

This operator acts on two-nucleon states as follows

$$\left[\; \right] |pn\rangle \approx |np\rangle, \quad \left[\; \right] |pp\rangle = 0, \quad \left[\; \right] |nn\rangle = 0. \tag{III.14}$$

Therefore, the exchange terms contribute to the reaction (γ,pn) but not to the reactions (γ,pp) and (γ,nn), at least in the first order process of diagram 3e. The reactions (γ,p) and (γ,n) receive equal contributions from the exchange terms. Eq. (III.14) gives the correspondence to the quasideuteron model as the exchange contributions involve neutron-proton pairs only. The qualitative success of the quasi-deuteron model even at energies around or below 100 MeV is explained by the fact that the gauge terms turn out to be very important for certain photonuclear reactions in this energy region.

The gauge contributions are built up from processes of the type shown in figure 4 where we consider the exchange of charged π-mesons. The photon is coupled to a nucleon in a negative energy state (pair term, diagram 4(a)) and to exchanged mesons (meson exchange term diagram 4(b)). We note that diagrams (4a) plus (4b), in the non-relativistic limit, just give the commutator $\left[V_{\pi}\,,\,Q_{\lambda}^{(L)}\right]$ for the electric interaction, where V_{π} is the static one-pion exchange potential.

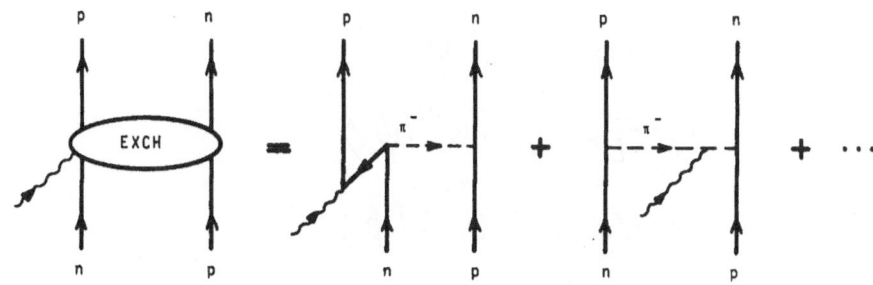

Figure 4   Exchange contributions to the interaction of a photon with the neutron-proton system

III.2. Model assumptions

In the following we summarize our method /51/ for the treatment of the transition matrix eq. (III.11). As described earlier in this paper (see eqs. (II.10-15)) where we have discussed the work of Brown /33/ and Fujii and Sugimoto /34/ the transition matrix can be written as

$$
\begin{aligned}
M_{fi}^{(EL)} \cong\; & E_{\gamma} \langle \phi_f | Q_{\lambda}^{(L)} | \phi_i \rangle \\[2mm]
& + E_{\gamma} \langle \phi_f | \frac{\upsilon}{E_f - H + i\varepsilon} Q_{\lambda}^{(L)} | \phi_i \rangle \\[2mm]
& + E_{\gamma} \langle \phi_f | Q_{\lambda}^{(L)} \frac{\upsilon}{E_i - H_o} | \phi_i \rangle,
\end{aligned}
\tag{III.15}
$$

where $|\phi_i\rangle$ and $|\phi_f\rangle$ are eigenstates of a shell model Hamiltonian $H_0 = T + U$ and v is an effective two-body potential. In addition to eq. (II.14) a first order perturbation of the ground state $|\phi_i\rangle$ has been added in eq. (III.15). Using closure to the giant resonance states we obtain (putting $E_i = 0$ and $E_f = E_\gamma$):

$$M_{fi}^{(EL)} \cong E_\gamma \langle \phi_f | Q_\lambda^{(L)} | \phi_i \rangle$$
$$+ \frac{E_\gamma}{E_\gamma - E_L^{res}} \langle \phi_f | v \, Q_\lambda^{(L)} | \phi_i \rangle \qquad (III.16)$$
$$- \frac{E_\gamma}{E_\gamma + E_L^{res}} \langle \phi_f | Q_\lambda^{(L)} v | \phi_i \rangle,$$

where E_L^{res} is given by the positions and the widths of the giant resonances in the electric multipole L, according to eq. (II.15). The coupling to the resonance modes should be an appropriate method to describe initial and final state correlations in photonuclear transitions above 40 MeV photon energy. Since an exact calculation of the correlations, at least in the final states, is not yet possible, we are going to use the physical information obtained from experiments. This information is provided by the existence of giant resonances. At least in the dipole case the parameters of E_L^{res} are well defined. In a preliminary treatment /49, 50/ of (γ,N) reactions on ^{16}O the closure argument has not been used. Instead, an explicit integration has been carried out in eq. (III.15) over intermediate particle-hole states. The numerical results have not been very different from those obtained with the closure assumption which simplifies the calculation very much. This supports our confidence that closure to L = 2 and L = 3 giant resonance states is justified although their parameters are not as well established as in the dipole case.

A few comments should be added concerning the corrections
to electric transitions arising from the second part in the
multipole expansion eq. (III.8) and concerning the magnetic
transitions due to the third part of eq. (III.8). Roughly, one
should expect 5-10 % contributions from these terms to the total
cross sections in the energy range from 40 to 100 MeV photon
energy. For instance, in the calculations by Partovi /53/ the
Siegert terms give about 90 % of the total photodisintegration
cross section of the deuteron at E_γ = 80 MeV. Fabian, Arenhövel
and Miller /54/ consider one-pion exchange contributions to the
d(γ,p)n total cross section and find them to be in the order of
1 % for photon energies 5 MeV < E_γ < 140 MeV. Contributions from
isobar admixtures are shown to be much more important at energies
above 80 MeV.

We have estimated the magnetic contributions to $^{16}O(\gamma,N)$
cross sections to be in the order of a few percent in the energy
range of interest. Calculations have been done for shell model
transitions involving the single-particle current (III.5) and, as
far as pion exchange currents are concerned, the Sachs term has
been considered. Each of these contributions does not exceed one
percent of the total cross sections.

Therefore, the problem of interpreting experimental (γ,N)
data reduces to a reliable treatment of the transition matrix eq.
(III.11) which we expect to explain the qualitative features of the
cross sections,i.e., the correct order of magnitude, the energy
dependence and the angular distributions. We shall see that this
can be achieved by using the transition matrix in the approximation given by eq. (III.16). In the numerical calculations we
change eq. (III.16) slightly insofar as in the first term, which

is the shell model contribution, we use instead of the Siegert operator $E_\gamma Q_\lambda^{(L)}$ the full electric part of the interaction between the photon and the convection current. This operator is called $\Omega_\lambda^{(L)}$. The reason for doing this is that the shell model matrix-element depends rather sensitively on the long-range behaviour of the operator.

Magnetic transitions are not included in the numerical results presented below. From the above discussion we expect them to be within the limits of our model calculation. In the case of (γ,N) reactions, for instance, it is our aim here to make up the discrepancy of one order of magnitude between the experimental total cross sections and the results obtained in pure shell model calculations.

Therefore, we use the transition matrix in the form

$$
\begin{aligned}
M_{fi}^{(EL)} = &\langle \phi_f | \Omega_\lambda^{(L)} | \phi_i \rangle &\text{(SM)}\\[1em]
&+ \langle \phi_f | [v, Q_\lambda^{(L)}] | \phi_i \rangle &\text{(EXCH)}\\[1em]
&+ \left(-1 + \frac{E_\gamma}{E_\gamma - E_L^{res}}\right) \langle \phi_f | v\, Q_\lambda^{(L)} | \phi_i \rangle &\text{(KORR}(f)) &\qquad \text{(III.17)}\\[1em]
&+ \left(1 - \frac{E_\gamma}{E_\gamma + E_L^{res}}\right) \langle \phi_f | Q_\lambda^{(L)} v | \phi_i \rangle. &\text{(KORR}(i))
\end{aligned}
$$

The first term in this expression is the shell model contribution (SM). The second and third terms in eq. (III.16) have been decomposed giving the second, third and fourth term in eq. (III.17).

The second term in eq. (III.17) shows the gauge terms (or exchange terms (EXCH)) explicitly, corresponding to diagram figure (3e). The third and fourth terms can be interpreted as the contributions from final state correlations (KORR(f)) and from initial state correlations (KORR(i)) according to diagrams figure (3b) and (3c), respectively. A much better understanding of photo-reactions can be achieved by a separate discussion of the different processes which altogether make up the transition matrix. We note that the contributions from the initial and final state correlations tend to become small with increasing photon energy E_γ due to the factors in front of the matrix elements. The gauge contributions do not show such an explicit energy dependence. Therefore, in reactions like (γ, pn), (γ, p) and (γ, n), the exchange contributions are expected to be dominant for energies near 100 MeV and up to the pion threshold.

We have studied numerically (γ, p) and (γ, n) reactions on the target nuclei ^{4}He, ^{12}C and ^{16}O. For these nuclei with closed shells (or subshells, in the case of ^{12}C) the different parts of the transition matrix eq. (III.17) can be represented by the diagrams in figure 5. Diagram (5a) gives the shell model contribution (SM) describing the "direct" transition of a nucleon (neutron or proton) from a single particle bound state $|k\rangle$ into a continuum state $|a\rangle$.

Diagram (5b) shows the exchange contribution (EXCH). The corresponding matrix element gives equal contributions to (γ, p) and to (γ, n) reactions, apart from a sign. This symmetry property explains why (γ, p) and (γ, n) cross sections are of comparable magnitude.

The final state correlations (KORR(f)) are represented by diagrams (5c) and (5d). The operator $Q_\lambda^{(L)}$ produces one-particle

one-hole intermediate states. As the latter are expected to
be concentrated near characteristic transition energies ($\approx E_L^{res}$)
this illustrates again the assumption of closure to the multipole
resonances. Diagrams (5e) and (5f) show the contributions from
the initial state correlations (KORR(i)). While the energy of the
configuration (k,a) is $E_f = E_\gamma$, the intermediate states should be
concentrated near $E_\gamma + E_L^{res}$ as an additional particle-hole pair
(d,ν) connected by the operator $Q_\lambda^{(L)}$ is present there.

In our calculation we have used a shell model potential of Woods-
Saxon type

$$U(r) = - U_0 \frac{1}{1 + exp\left(\frac{r-r_0}{a}\right)} . \qquad (III.18)$$

For a definite nucleus the same potential U has been used for
protons and neutrons. The Coulomb energy has been neglected. The
parameters U_0, r_0 and a (listed in table II) have been chosen to
give (i) a binding energy of 20.7 MeV for an s-shell nucleon in
^{4}He, (ii) a binding energy of a p-shell nucleon of 17 MeV in ^{16}O
and of 17.3 MeV in ^{12}C, these values being an average of the
experimental neutron and proton separation energies.

We note that the continuum state $|a\rangle$ of the outgoing nucleon
in a (γ,N) reaction has been taken as an eigenstate of the same
shell model Hamiltonian $H_0 = T + U$ which defines also the bound
single particle states $|k\rangle$ etc. In this way we ensure orthogona-
lity of our initial and final states.

The residual interaction has been chosen as

$$v(r) = - V_0 \frac{e^{-\mu r}}{\mu r} \left[a_0 + a_\sigma \left(\vec{\sigma}(1)\cdot\vec{\sigma}(2)\right) + a_\tau \left(\vec{\tau}(1)\cdot\vec{\tau}(2)\right) \right.$$
$$\left. + a_{\sigma\tau} \left(\vec{\sigma}(1)\cdot\vec{\sigma}(2)\right)\left(\vec{\tau}(1)\cdot\vec{\tau}(2)\right) \right]. \qquad (III.19)$$

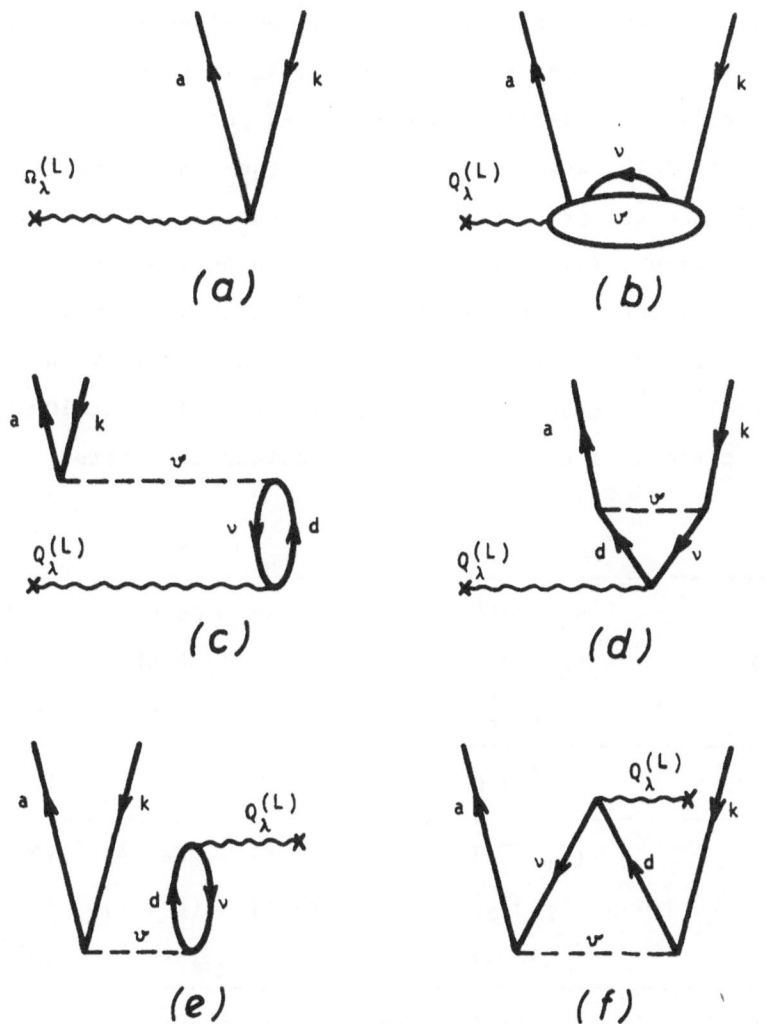

Figure 5   Contributions to the transition matrix eq.
(III.17) in the case of (γ,p) and (γ,n) reactions:
(a) shell model transition (SM), (b) exchange
contribution (EXCH), (c+d) final state correlations (KORR(f)), (e+f) initial state correlations
(KORR(i)).

Our results to be shown below have been obtained for the case of
a Rosenfeld mixture ($a_0 = a_\sigma = -0.0025$, $a_\tau = -0.1025$,
$a_{\sigma\tau} = -0.2325$). The parameter μ has been put equal to
$\sqrt{(0.71)^2 - (k_\gamma/2)^2}$, assuming that in the one-pion exchange
(figure 4) the energy of the photon is transferred equally to
the two nucleons. This should give a rough estimate of the effect
of the energy transfer on the range of the nucleon-nucleon potential. Such a correction gives a slight enhancement of the total
cross sections at higher photon energies (about 5% at 100 MeV).

The parameters used in the calculations are listed in
table II.

		^{4}He	^{12}C	^{16}O
	U_0 [MeV]	64.8	66.6	58.5
U(r):	r_0 [fm]	1.67	2.52	2.77
	a [fm]	0.4	0.5	0.5
v(r):	V_0 [MeV]	95	60	55
	$(E_1; \Gamma_1)$	(22; 4)	(22; 4)	(22; 4)
(E_L, Γ_L) [MeV]:	$(E_2; \Gamma_2)$	(33; 8)	(26; 8)	(26; 8)
	$(E_3; \Gamma_3)$	——	(48; 10)	(48; 10)

Table II: Input data for shell model potential U(r),
nucleon-nucleon potential v(r), and multipole
resonances.

In the calculations for (γ,p) and (γ,n) reactions on ^{4}He electric
dipole and quadrupole transitions have been considered. For the
target nuclei ^{12}C and ^{16}O electric multipoles up to L = 3 have
been taken into account (only in the shell model contributions
we proceed to L = 4). Effective kinematical charges for protons

and neutrons have been used in the form

$$e_{eff}^{L}(p) = e \frac{1}{A^L}\left[(A-1)^L + (-1)^L(Z-1)\right], \qquad e_{eff}^{L}(n) = eZ\left(-\frac{1}{A}\right)^L. \qquad (III.20)$$

III.3. Results for (γ,N) reactions

In figure 6 we show as a typical example the angular distribution for the reaction $^{16}O(\gamma,p)^{15}N_{g.s.}$ for a photon energy
E_γ = 82 MeV. The residual nucleus is left in its ground state.
The experimental data shown in the figure have been obtained by
the Glasgow group /25/. The curves I-IV exhibit the contributions
of the different pieces to the transition matrix eq. (III.17).
Curve I is the angular distribution obtained from the full
transition matrix. Curve II results if in the transition matrix
only the shell model term (SM) plus the exchange (or qauge) contribution (EXCH) are taken into account. In curve III we show
the angular distribution calculated from the shell model plus
correlation (SM+KORR) contributions. Since the initial state
correlations (KORR(i)) are rather small compared to the final
state correlations (KORR(f)) we do not present these contributions separately. Curve IV is the angular distribution obtained
in the pure shell model.

We see that the shell model (curve IV) and the combined
contributions from shell model plus correlation terms (curve III)
give cross sections which are much too small. The most important
part of the transition matrix at these higher energies obviously
is the exchange contribution. This piece and the shell model term
together (curve II) nearly explain the experimental (γ,p) data.

Adding the correlations does not change the results very much
(curve I).

It should be mentioned that due to the interference of the
various parts in the transition matrix the contributions from
different multipoles decrease rapidly with increasing multipole
order. Table III shows the multipole decomposition of the total
cross sections corresponding to curves I-IV in figure 6.

	SM (curve IV)	SM+KORR (curve III)	SM+EXCH (curve II)	total (curve I)
E1	1	36	40	65
E2	55	31	44	26
E3	38	25	15	8
E4	6	8	1	1

Table III   Contributions (in %) of different electric
             multipoles to the total cross section of the
             reaction $^{16}O(\gamma,p)^{15}N_{g.s.}$ for 82 MeV photon
             energy.

In the pure shell model (SM) cross section E2 and E3 dominate
strongly at this energy (82 MeV). The dipole contributions are
enhanced when either the correlations or the exchange contributions are added to the shell model term. However, only in the
total transition matrix (last column in table III) the dipole
contribution is seen to be the dominant one and good convergence
with respect to increasing multipole order is achieved.

It should be noted that the shape of the angular distributions essentially is given by the multipoles taken into consideration. A variation of the input data (range and depth of the
potentials, parameters of the multipole resonances) does not
alter the differential cross sections very much.

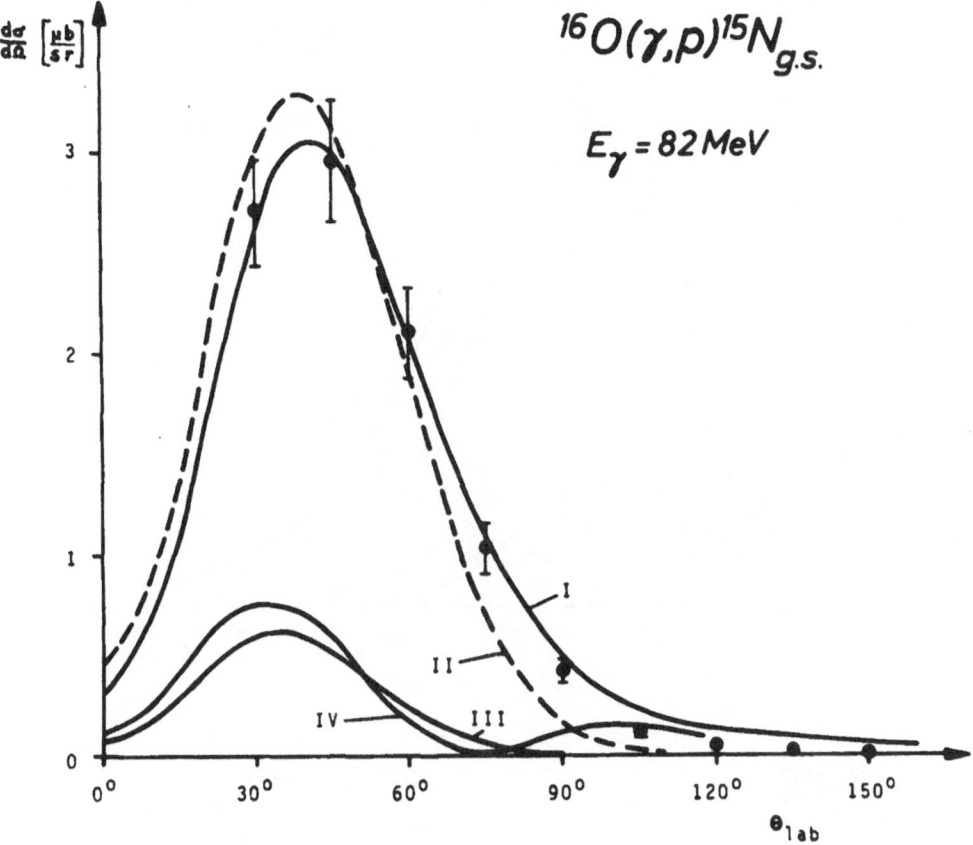

Figure 6. Angular distribution of the reaction $^{16}O(\gamma,p)^{15}N_{g.s.}$ for the photon energy E_γ = 82 MeV: I - total transitions matrix eq. (III.17), II- shell model plus exchange contributions (SM + EXCH), III - shell model plus initial and final state correlations (SM+KORR), IV - shell model (SM). The experimental values are taken from ref. /25/.

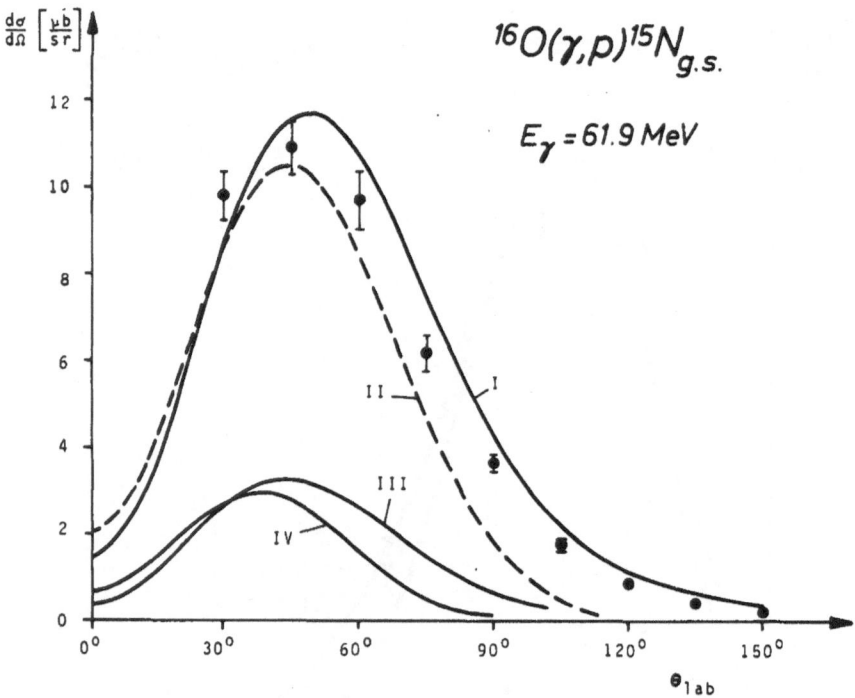

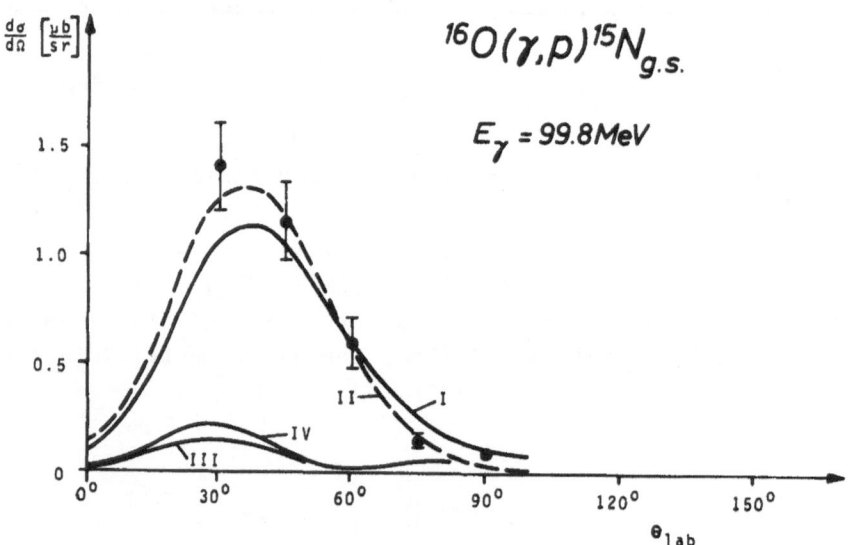

Figure 7   Same as figure 6, for the photon energies 61.9 and 99.8 MeV

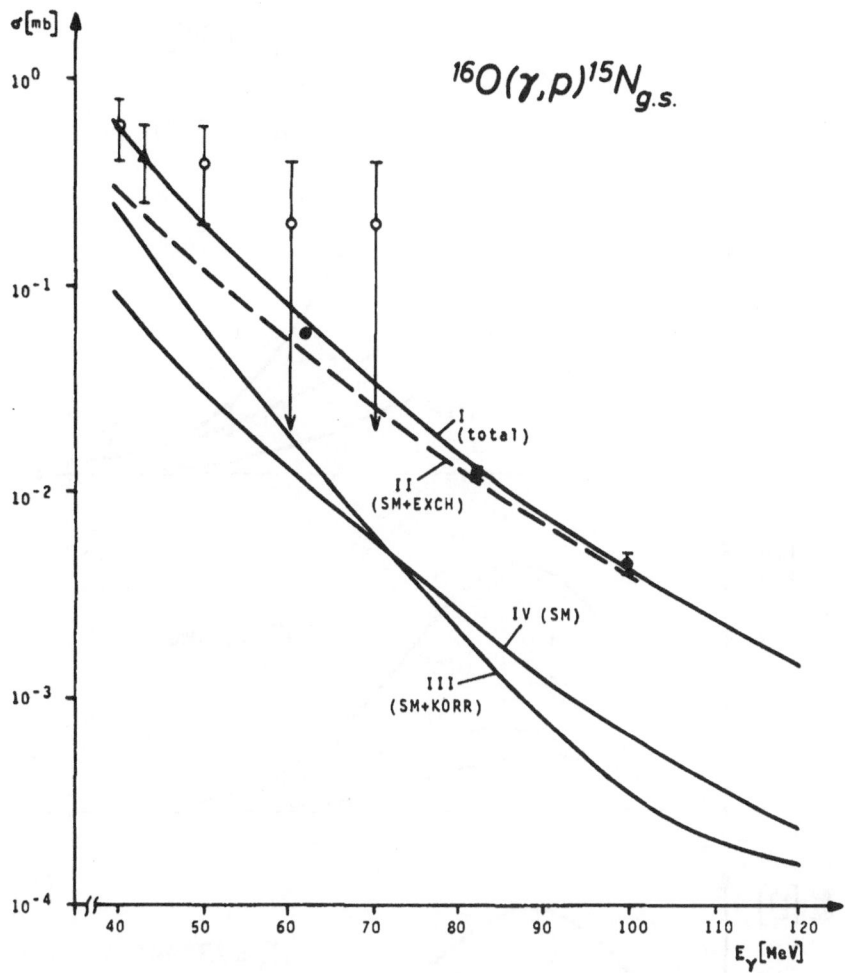

Figure 8   Total cross section of the reaction $^{16}O(\gamma,p)^{15}N_{g.s.}$
as a function of the photon energy. Curves I - IV
display the different contributions to the transition matrix in the same way as defined in figure 6.
Experiments are taken from ref. /25/ ( ),
ref. /20/ ( ) and ref. /21/ ( ).

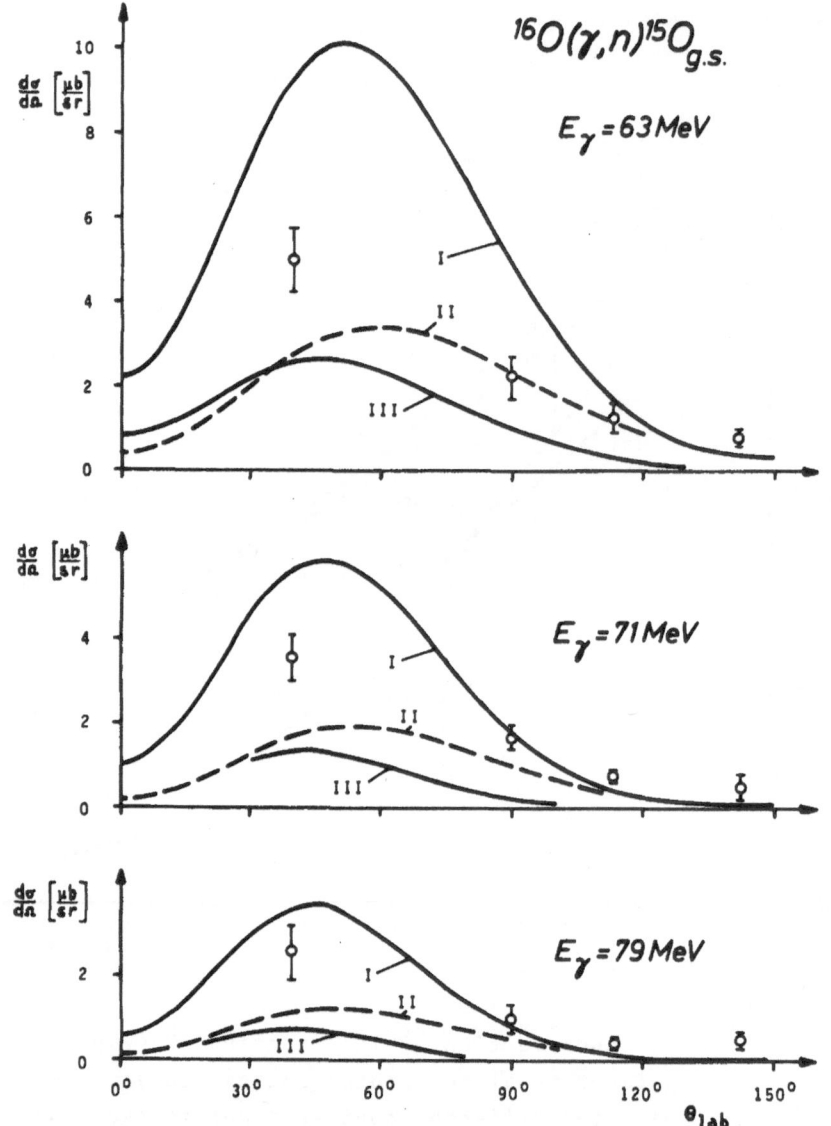

Figure 9   Angular distributions of the reaction $^{16}O(\gamma,n)^{15}N_{g.s.}$ for three photon energies. I - total transition matrix, II- shell model plus exchange contributions, III - shell model plus initial and final state correlations. Experiments are taken from refs. /14, 15/.

Figure 7 shows the angular distributions of the reactions $^{16}O(\gamma,p)^{15}N_{g.s.}$ for 61.9 and 99.8 MeV photon energy. The remarks made above (for 82 MeV) concerning the dominance of the gauge contributions apply here as well. The energy dependence and the forward shift of the maximum of the angular distribution with increasing energy is reproduced. Curve I (full transition matrix) and curve II (shell model plus gauge contributions) are close to one another for higher energies ($\approx$ 100 MeV) which shows that contributions from initial and final state correlations are rather unimportant here. This can also be seen from figure 8 where we have plotted the total cross section of $^{16}O(\gamma,p)^{15}N_{g.s.}$ as a function of the photon energy. At energies $E_\gamma \geq 60$ MeV the shell model plus exchange contributions (curve II) are close to the experimental values. At these energies the pure shell model (curve IV) and the shell model plus correlations (curve III) give cross sections which are one order of magnitude smaller than the experimental values. The correlations are important only at lower energies. Curve I, calculated from the total transition matrix, is in good agreement with the data in the range 40 MeV$< E_\gamma <$ 100 MeV.

Figure 9 shows the angular distributions of the reaction $^{16}O(\gamma,n)^{15}O_{g.s.}$ for the photon energies $E_\gamma =$ 63, 71 and 79 MeV in comparison with the data obtained by Schier and Schoch /14,15/. The meaning of curves I-III is the same as before. The pure shell model contribution is not shown as it is very small due to the small effective charges of the neutron for the multipoles L = 2 and L = 3. One notes that the exchange contributions are not as dominant as in the (γ,p) reaction on ^{16}O. At least for small forward angles $\Theta< 40^o$ the correlations are equally important.

Therefore the (γ,n) cross section depends more sensitively upon details of the correlations than does the (γ,p) cross section. However, in view of the fact that the pure shell model results are about two orders of magnitude too small, we may say that our model calculation is in fair agreement with the present data. This is true especially for the photon energy E_γ = 79 MeV where we expect the correlations to be less important than at the lower energies.

Our results obtained for the reaction $^{12}C(\gamma,p)^{11}B$ are presented in figures 10 and 11. Figure 10 gives the angular distributions for various photon energies including 60, 80 and 100 MeV. For these energies the differential cross sections between 30^0 and 150^0 have been measured by the Glasgow group /24/. The data in figure 10 show the sum of the experimental cross sections for the transitions into the ground state and the 2.1 MeV state of ^{11}B. For comparison with our calculations we have assumed that these two states are the main components of a $(1p_{3/2})^{-1}$ configuration. The qualitative discussion concerning the relative contributions of the shell model, the gauge terms and the correlations is very similar to that given above in the case of the reaction $^{16}O(\gamma,p)^{15}N$. In figure 10 we show for each energy the curves I (total transition matrix) and curves II (shell model plus gauge contributions). The shaded areas between these curves indicate the effect of the initial and final state correlations. Clearly their relative contributions to the cross sections become smaller with increasing energy. Figure 10 shows that the calculated angular shape as well as the energy dependence of the cross sections are in good agreement with the

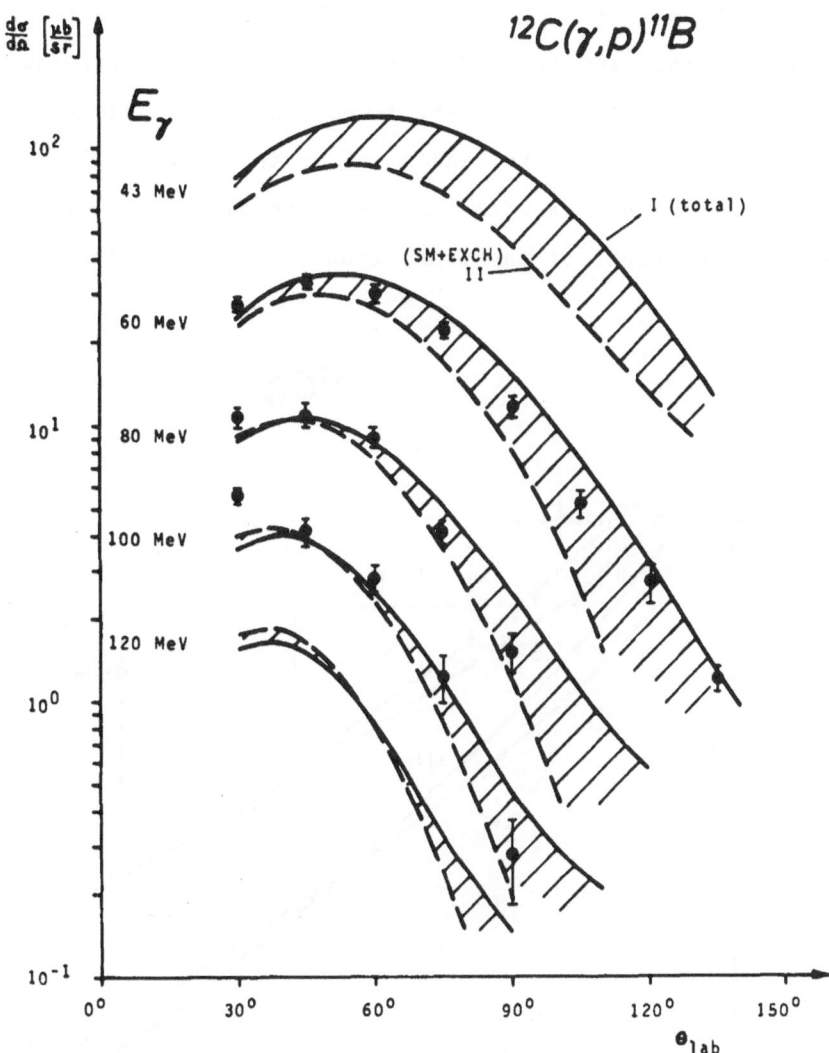

Figure 10   Angular distributions of the reaction
$^{12}C(\gamma,p)^{11}B$ for various photon energies.
The experimental values are taken from
ref. /24/.

experimental data. The calculated total cross sections for $^{12}C(\gamma,p)^{11}B$ presented in figure 11 also follow the experimental trend. Again we see that the shell model with and without the correlations in the initial and final states (curves III and IV) cannot explain the data. The gauge contributions are the most important components of the transition matrix at photon energies above 50 MeV.

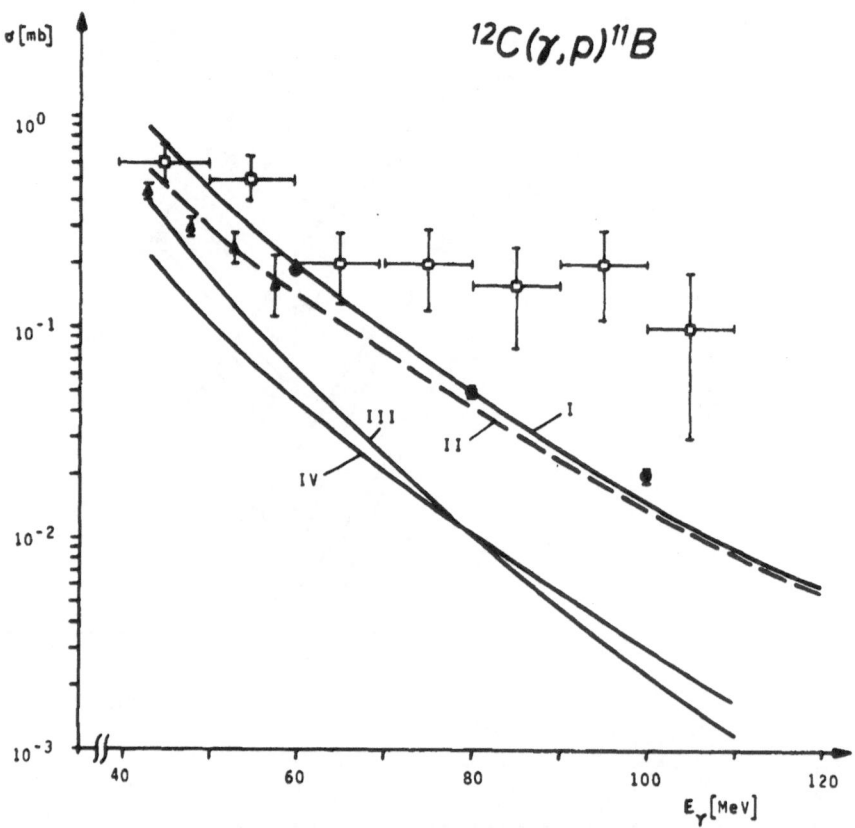

Figure 11 Total cross sections of the reaction $^{12}C(\gamma,p)^{11}B$ as a function of the photon energy: I - total transition matrix, II - shell model plus exchange contributions, III - shell model plus initial and final state correlations, IV - pure shell model. The experimental values are taken from ref. /24/ ( ), ref. /17/ ( ) and ref. /18/ ( ).

Angular distributions of the reaction $^{12}C(\gamma,n)^{11}C$ are shown in figures 12 and 13. The neutron is assumed to be ejected from a $1p_{3/2}$ orbit in ^{12}C. The corresponding transition probability

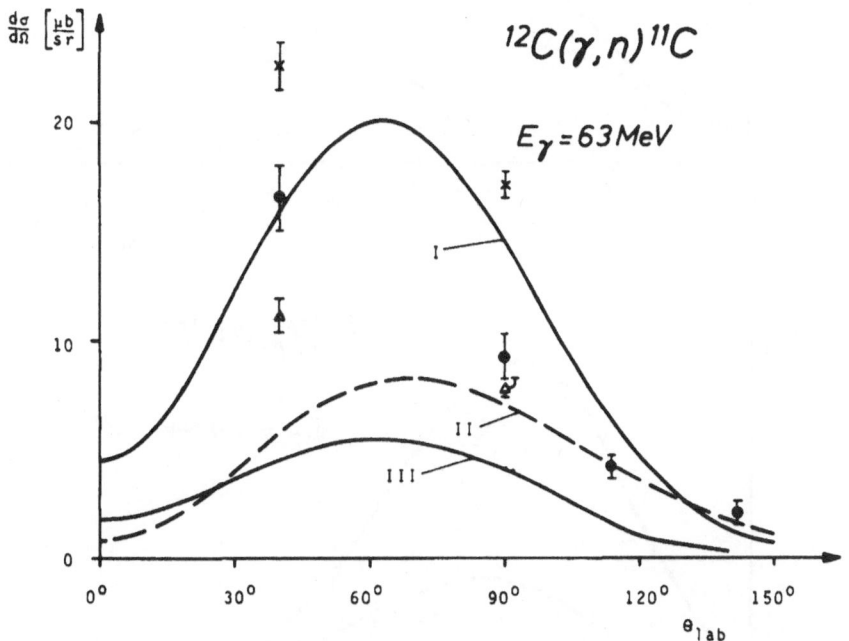

Figure 12   Angular distribution of the reaction $^{12}C(\gamma,n)^{11}C$ for E_γ = 63 MeV: I - total transition matrix, II - shell model plus exchange contributions, III - shell model plus initial and final state correlations. The experimental values are from refs. /14, 15/.

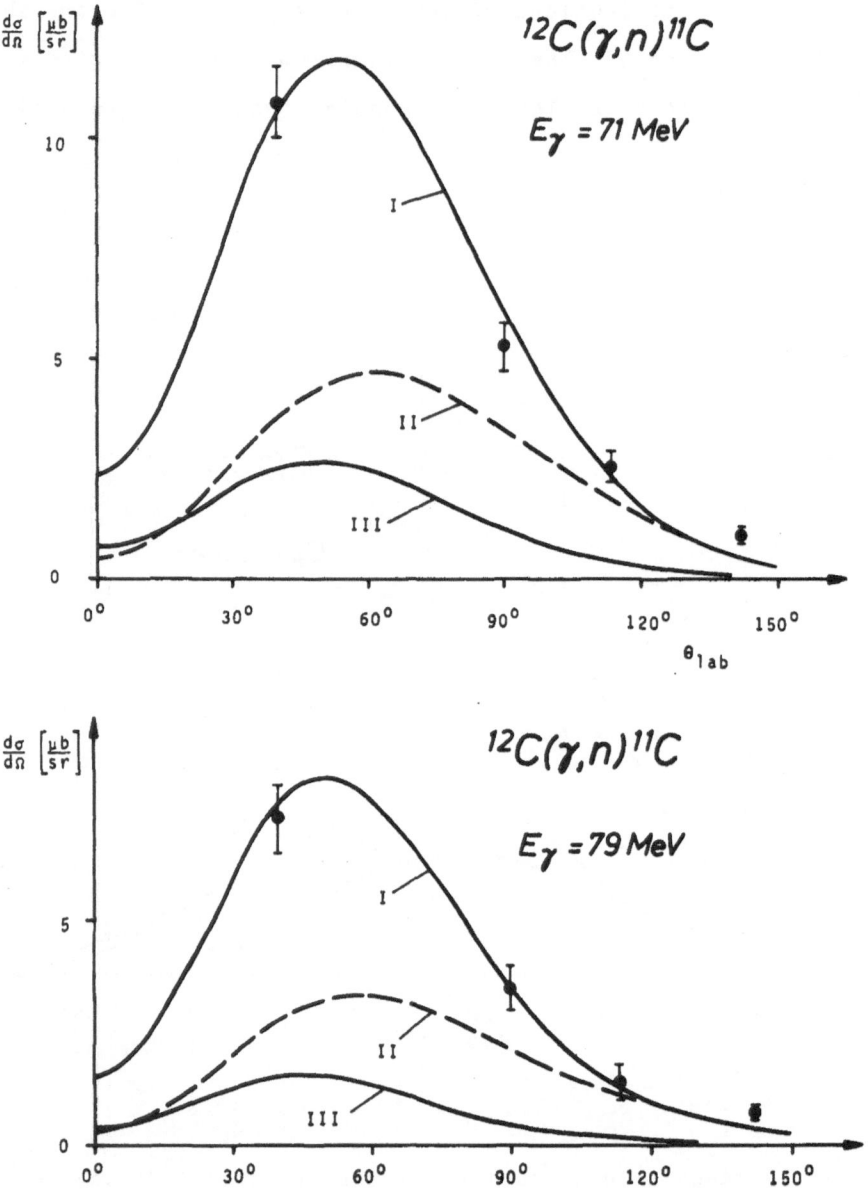

Figure 13   The same as figure 12, for E_γ = 71 and 79 MeV.

may be shared among the transitions into the ground state and
several low-lying excited states in the residual nucleus ^{11}C.
Full circles in figures 12 and 13 denote measurements /14, 15/
which presumably include transitions to the 2.0 MeV level in ^{11}C,
in addition to ground state transitions. We see that these
values are in good agreement with our calculated results
(curve I). For E_γ = 63 MeV, the calculated cross sections are
larger than the measured ground state transitions (denoted by
triangles at angles 40° and 90°) and are smaller than the measured cross sections which include levels up to 4.79 MeV in ^{11}C
(denoted by crosses). The qualitative discussion concerning the
contributions from different pieces (shell model, correlations
and exchange parts) is identical with that given above in connection with the reaction ^{16}O$(\gamma,n)^{15}$O.

In figures 14 and 15 we show the total cross sections for the
reactions ^{4}He$(\gamma,p)^{3}$H and ^{4}He$(\gamma,n)^{3}$He. The curves I-IV have the
same meaning as in the preceding discussion. The results obtai-
ned with the full transition matrix (curves I) explain the
energy dependence of the cross sections fairly well. The exchange parts again are the most important contributions to the
transition matrix for photon energies above 60 MeV. In ref. /47/
it has been demonstrated that also the angular distributions for
(γ,p) and (γ,n) reactions on ^{4}He can be explained in the frame
of the model calculation presented here. It should be noted that
Noguchi and Prats /55/ have treated the (γ,N) reactions on ^{4}He
for 65 MeV $\leq E_\gamma \leq$ 170 MeV by applying a quasideuteron mechanism.
They have shown that the quasideuteron contribution is much
larger than the contribution from the "direct" mechanism, for

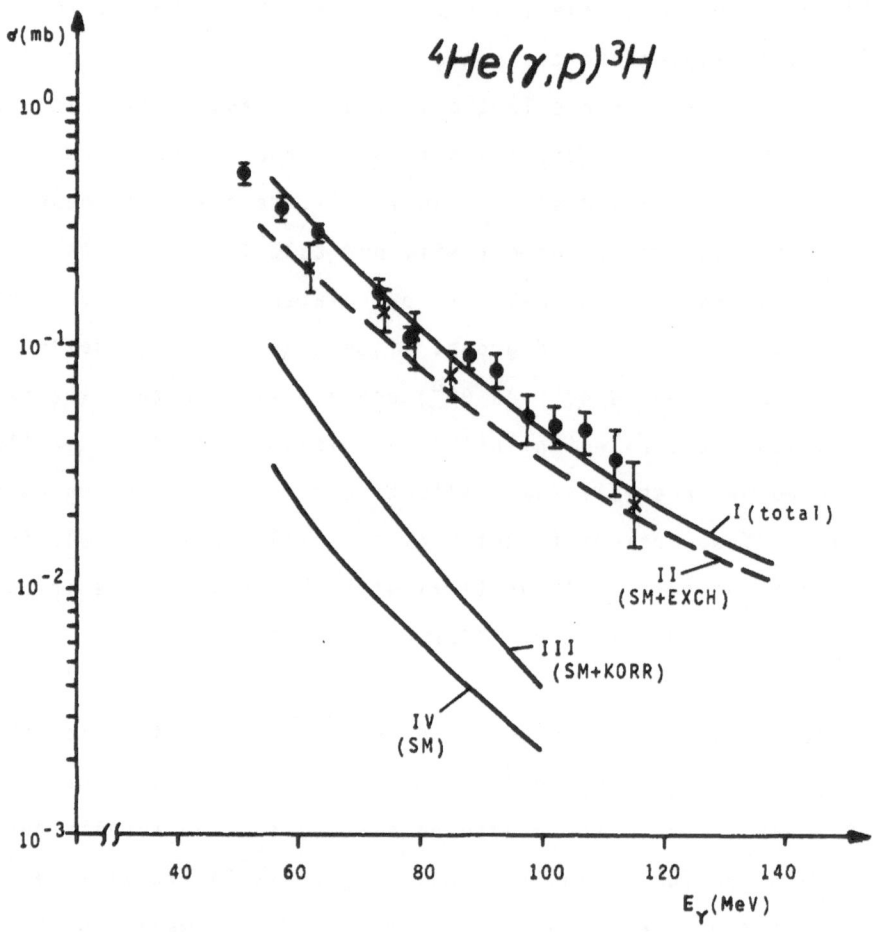

Figure 14   Total cross section of the reaction $^4He(\gamma,p)^3H$ as a function of the photon energy. The experimental values are taken from ref. /11/ ( ) and from ref. /12/ ( ).

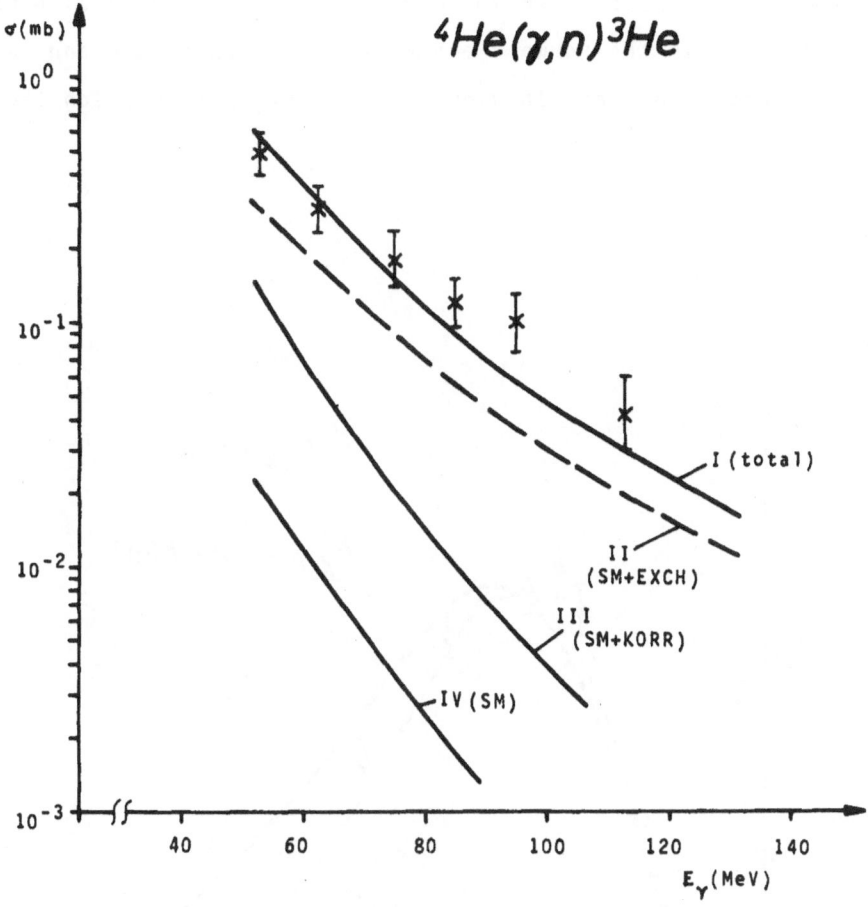

Figure 15 Total cross section of the reaction ^{4}He$(\gamma,n)^3$He as a function of the photon energy. The experimental values are taken from ref. /13/.

energies around and above 100 MeV. These authors have also been able to explain the forward asymmetry for both proton and neutron angular distributions. In view of our earlier discussion there is

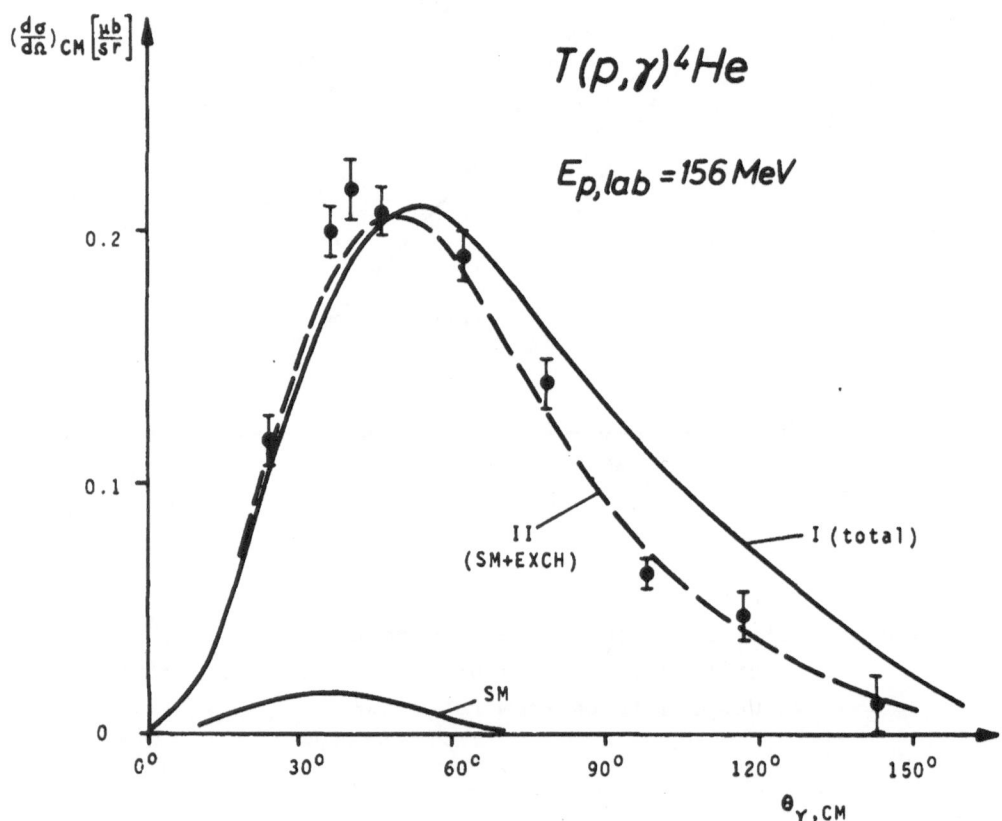

Figure 16   Angular distribution of the reaction T(p, γ)^{4}He for 156 MeV protons: I - full transition matrix, II - shell model plus gauge contributions, SM - pure shell model contribution. The experimental data are given in ref. /56/.

an obvious correspondence between the method of ref. /55/ and our treatment: the "quasideuteron aspect" is represented by the gauge terms in our terminology.

Figure 16 shows the angular distribution of the photon in the capture reaction $^3H(p,\gamma)^4He$ for an energy of 156 MeV of the incident proton beam. This corresponds to a photon energy of 137 MeV for the inverse reaction $^4He(\gamma,p)^3H$. The calculations have been carried out with the same parameters which give the (γ,N) results on ^4He presented in figures 14 and 15. Again one notes that the shell model cross section (denoted by SM here) is much too small. However, the shell model plus exchange contributions (curve II) give an almost perfect explanation of the data. Curve I is calculated from the full transition matrix. However, since the correlations possibly are not very well described by our closure argument at this high energy, curve II is perhaps more reliable   for comparison with experiments. From this example we expect the presented model to explain capture reactions (N,γ) in heavier nuclei as well.

III.4. The (γ,pn) reaction on ^{16}O

As an example for two-nucleon photoemission we have considered the reaction (γ,pn) on ^{16}O. In the absence of detailed experimental information in the energy region below 140 MeV we have calculated the integrated cross sections $\delta(E_\gamma)$ only. Moreover, we have taken into account only the exchange contributions to the transition matrix neglecting initial and final state correlations. The shell model contribution vanishes due to orthogonality of the initial and final states. I.e., in the transition matrix eq. (III.17) we consider the second term (EXCH) only. Since the initial and final state correlations are not expected to be very important at higher energies ($\gtrsim$ 60 MeV) we are certain to obtain a reliable prediction for the order of magnitude of the (γ,pn) cross section.

In figure 17 the result of our calculation of the reaction $^{16}O(\gamma,pn)^{14}N$ is given by curve B. Electric multipoles L = 1 and L = 2 have been considered. Furthermore, we have summed over the three possibilities for the two nucleons to be ejected (i) both from the p-shell, (ii) both from the s-shell and (iii) one nucleon from the p-shell, one nucleon from the s-shell. A few details are summarized in table IV. We note that emission from

E_γ [MeV]	$(1p)^2$	(1p1s)	$(1s)^2$	E1
60	81	17	2	96.5
80	70	21	9	93.9
100	66	23	11	90.7
120	61	24	15	87.3

Table IV   Decomposition of the $^{16}O(\gamma,pn)^{14}N$ total cross sections into contributions from different shells. The last column shows the dipole part (in %) in the total cross sections.

the p-shell for both nucleons dominates clearly the other possibilities. The dipole contribution is seen to make up more than 90 % of the (γ,pn) cross section up to 100 MeV photon energy. Since for these higher energies the (γ,pn) reaction probably constitutes a large fraction of the total nuclear photoabsorption we may conclude that the dipole dominance holds for the latter as well.

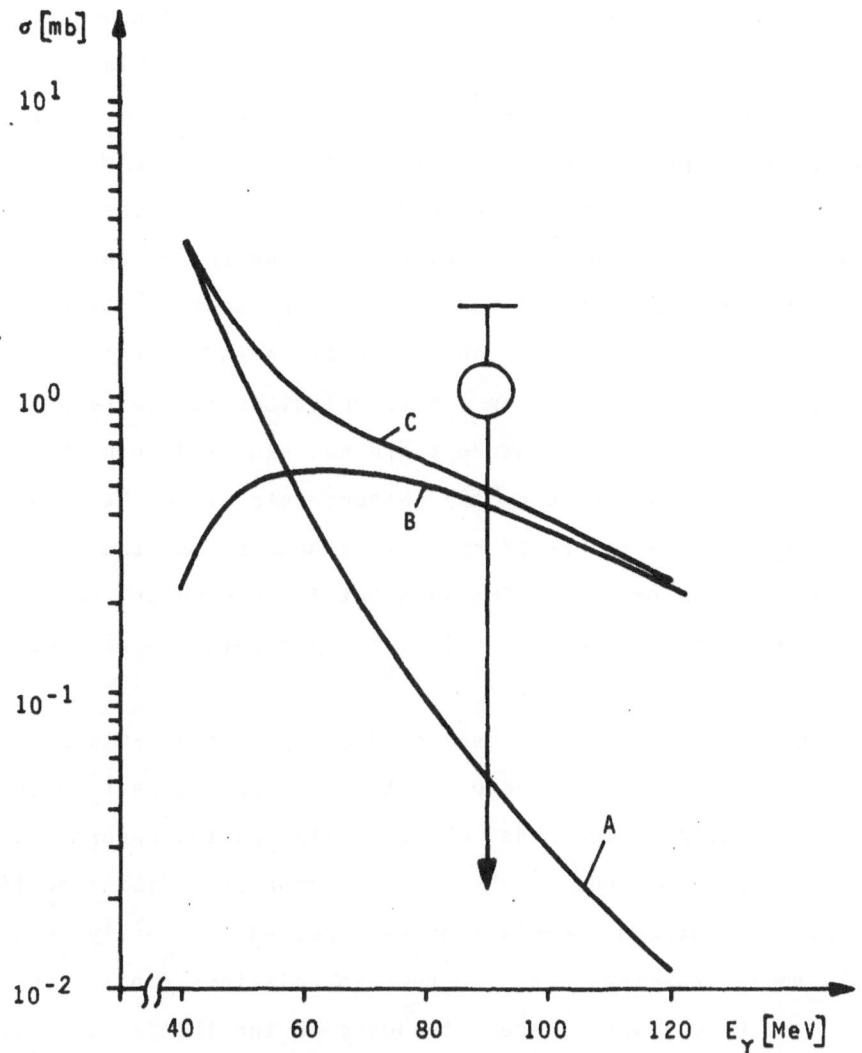

Figure 17   Comparison of the total cross section of the reaction
$^{16}O(\gamma,pn)^{14}N$ (curve B) and the cross section for
single nucleon emission from the p-shell of ^{16}O
(curve A), as a function of photon energy. Curve C
shows the sum of both calculated cross sections.
The arrow indicates the measurements of ref. /1/.

Curve A shows the sum of the calculated total cross sections of the reactions (γ,p) and (γ,n) on ^{16}O for emission of the nucleon from the 1p shell. We see that, according to our model, the (γ,pn) cross section is larger than the single nucleon emission cross sections above 60 MeV. Near 100 MeV the ratio is about equal to ten. Curve C, which shows the sum of the cross sections given in curves A and B, should present a reliable lower limit for the total nuclear photoabsorption in the energy range shown, as contributions from the correlations to the (γ,pn) reaction and from other channels are not expected to be very large. This is supported by the measurements of the Mainz group /1/. The arrow in figure 17 shows that near 90 MeV the total photoabsorption cross section is about 1 mb. However, due to the large statistical errors no definite statements can be made at present.

(γ,pn) reactions have been studied by several authors /57-60/ in the model of independent pairs. In contrast to the original quasideuteron model which relates photon absorption from a complex nucleus to the free deuteron photodisintegration, the nucleon-nucleon correlations are treated explicitly in a shell model framework. Jastrow type correlations (mostly in the form used by Dabrowski) are introduced in the initial states. In the final states the center-of-mass motion of the outgoing neutron-proton pair is taken to be a plane wave and two-nucleon scattering solutions with empirical phase parameters are used to describe the relative motion of the pair. The cross sections depend sensitively upon details of the final state wavefunctions but the initial state correlations are rather unimportant below the meson threshold.

Our calculation for the (γ,pn) reaction differs from that reported in refs. /57-60/ insofar as we consider explicitly the contribution of the gauge terms to the transition matrix, i.e., we calculate matrixelements of the commutator $\left[v,Q_\lambda^{(L)}\right]$ between neutron-proton states. The initial as well as the final state wavefunctions are eigenstates of the shell model Hamiltonian. Our description is preliminary in the sense that initial and final state correlations are not yet included. However, our preceding discussion shows that from the gauge contributions alone we obtain (γ,pn) cross sections in the correct order of magnitude.

IV. Summary and conclusive remarks

In the preceding section we have presented a model for photonuclear reactions in the energy range above the giant resonance and below the pion threshold. The transition matrix has been decomposed into three pieces: (i) the shell model contribution; (ii) the contributions from initial and final state correlations; these are processes in which the photon interacts with a nucleon in a positive energy state after and before a nucleon-nucleon interaction takes place; (iii) the terms necessary to achieve gauge invariance of the description; these terms involve processes in which the photon interacts with nucleons in negative energy states and with mesons mediating the nucleon-nucleon correlations.

At present detailed comparison of the model predictions with experiments is possible for (γ,p) and (γ,n) reactions on light nuclei. The calculations show that long-standing discrepancies between experimental data and earlier model treatments can be solved:

a) The order of magnitude and the energy dependence of total
   (γ,p) and (γ,n) cross sections on ^{4}He, ^{12}C and ^{16}O is explained satisfactorily.

b) Even more important is the agreement of the calculated angular
   distributions with experimental data. The model explains the
   observed forward asymmetry of both (γ,p) and (γ,n) differential cross sections.

We have shown that for photon energies above 60 MeV the
gauge terms give the main contribution to the transition matrix.
The correlations in the usual sense tend to become small with
increasing energy. For energies greater than 60 MeV the shell
model plus exchange pieces nearly explain the experimental (γ,p)
cross sections on ^{12}C and ^{16}O. Introducing the correlations
gives only minor changes. In the (γ,n) reactions the exchange
contributions also dominate but the effect of the correlations
is larger than in the (γ,p) reactions. The cross sections calculated from the shell model term alone are far below the measurements. The contributions from the correlations, mainly from
those in the final state, are largest in the energy range from
40 to 60 MeV where the coupling to the giant resonance is still
appreciable.

The dominance of the gauge contributions explains why a
phenomenological treatment via the quasideuteron model has been
partially successful in the interpretation of photonuclear data
above 40 MeV photon energy. The gauge terms describe the interaction of the photon with a neutron-proton pair. Thereby (γ,p)
and (γ,n) transition matrix elements are effected symmetrically
which explains the observed similarity of the cross sections.
For the (γ,pn) reactions the exchange contributions are even

more important. We have seen that, near 100 MeV, the (γ,pn) cross sections for ^{16}O are greater than the (γ,N) cross sections by one order of magnitude.

In conclusion we remark that the presented model gives a satisfactory description of the (γ,N) data. A more detailed calculation of (γ,pn) cross sections will become meaningful when more experimental information in the energy range considered here will be available. Calculations with more refined forms of the single-particle potential and the two-body interaction (including tensor parts) could be carried out as further steps of the theoretical development, and magnetic transitions should be included. However, such efforts seem to be useful only in connection with an improved treatment of the nuclear many-body problem. For instance, the interactions of the outgoing nucleons with the residual nucleus have to be properly understood before definite conclusions concerning the parameters of one-body and two-body potentials are possible.

References

/1/ J. Ahrens, H. Borchert, K.H. Czock, H.B. Eppler, H. Gimm, H. Gundrum, M. Kröning, P. Riehn, G. Sita Ram, A. Zieger and B. Ziegler; Proc. Intern. Conf. on Photonuclear Reactions and Applications, Asilomar, 1973; Nucl. Phys. A251 (1975) 479

/2/ A. Arima, G.E. Brown, H. Hyuga and M. Ichimura, Nucl. Phys. A205 (1973) 27

/3/ W.T. Weng, T.T.S. Kuo and G.E. Brown, Phys. Letters 46B (1973) 329

/4/   M. Fink, M. Gari and H. Hebach, Phys. Letters 49B (1974) 20

/5/   M. Fink, M. Gari, H. Hebach and J.G. Zabolitzky, Phys. Letters 51B (1974) 320

/6/   K. Grassau, Diplomarbeit, Bochum 1976

/7/   J.G. Lucas and M.L. Rustgi, Nucl. Phys. A112 (1968) 503

/8/   W.E. Meyerhof and S. Fiarman, Proc. Asilomar Conf. (1973) 385

/9/   J.S. Levinger, Phys. Rev. 84 (1951) 43

/10/   G. Taran and A.N. Gorbunov, Sov. Phys. JETP 19 (1964) 1010

/11/   Y.M. Arkatov, P.I. Vatset, V.I. Volshchuk, V.V. Kirichenko, I.M. Prokhorets and A.F. Khodyachikh, Sov. J. Nucl. Phys. 12 (1971) 123

/12/   A.N. Gorbunov, JETP Letters 8 (1968) 88

/13/   A.N. Gorbunov, Phys. Letters 27B (1968) 436

/14/   H. Schier and B. Schoch, Nucl. Phys. A229 (1974) 93

/15/   H. Schier and B. Schoch, Il Nuovo Cimento Lett. 12 (1975) 334

/16/   H.G. Miller, W. Buss and J.A. Rawlins, Nucl. Phys. A163 (1971) 637

/17/   S. Penner and J. Leiss, Phys. Rev. 114 (1959) 1101

/18/   G.G. Taran and A.N. Gorbunov, Sov. J. Nucl. Phys. 6 (1968) 816

/19/   B.C. Cook, J.E.E. Gaglin, J.N. Bradford and J.E. Griffin, Phys. Rev. 143 (1966) 712, 724

/20/   A.N. Gorbunov and V.A. Osipova, JETP (Sov. Phys.) 16 (1963) 27

/21/   V.P. Denisov, A.P. Komar and L.A. Kulchitsky, Nucl. Phys. A113 (1968) 289

/22/ G. Manuzia, G. Ricco, M. Sanzone and L. Ferrero, Nucl. Phys. A133 (1969) 225

/23/ E. Mancini, G. Ricco, M. Sanzone, S. Costa and L. Ferrero, Il Nuovo Cimento 15A (1973) 705

/24/ D.J.S. Findlay, S.N. Gardiner, J.L. Matthews and R.O. Owens, J. Phys. A7 (1974) L157

/25/ R.O. Owens; private communication
D.J.S. Findlay, thesis, Glasgow University (1975)

/26/ G.M. Shklyarevskii, JETP (Sov. Phys.) 9 (1959) 1057

/27/ S. Fujii, Progr. Theor. Phys. 29 (1963) 374

/28/ W. Weise, Phys. Lett. 38B (1972) 301

/29/ M. Fink, H. Hebach and H. Kümmel, Nucl. Phys. A186 (1972) 353

/30/ G.M. Shklyarevskii, JETP (Sov. Phys.) 14 (1962) 324

/31/ W. Weise and M.G. Huber, Nucl. Phys. A162 (1971) 330

/32/ A. Malecki and P. Picchi, Nuovo Cim. Lett. 8 (1973) 16

/33/ G.E. Brown, Nucl. Phys. 57 (1964) 339

/34/ S. Fujii and O. Sugimoto, Nucl. Phys. 56 (1964) 73

/35/ K. Gottfried, Nucl. Phys. 5 (1958) 557

/36/ S. Costa, Proc. Asilomar Conf. (1973) 1319

/37/ J.L. Matthews, W. Bertozzi, S. Kowalski, C.P. Sargent and W. Turchinetz, Nucl. Phys. A112 (1968) 654

/38/ S. Costa, L. Ferrero, L. Pasqualini and M. Sanzone, Lettere Nuovo Cim. 1 (1971) 448

/39/ V.I. Mamasakhlisov and R.I. Jibuti, Soviet Phys. JETP 14 (1962) 1066

/40/ R. Serber, Phys. Rev. 72 (1947) 1114

/41/ N. Metropolis et al., Phys. Rev. 110 (1958) 185

/42/ H.W. Bertini, Phys. Rev. 131 (1963) 1801

/43/ N.N. Kaushal et al., Phys. Rev. **175** (1968) 1330

/44/ T.A. Gabriel + R.G. Alsmiller, Phys. Rev. **182** (1969) 1035

/45/ S. Costa, L. Ferrero, L. Pasqualini and E. Mancini, Lettere Nuovo Cimento **2** (1971) 665

/46/ M. Gari and H. Hebach, Phys. Letters **49B** (1974) 29

/47/ M. Gari and H. Hebach, Phys. Rev. **C10** (1974) 1629

/48/ M. Gari, Mesonic Effects in Nuclear Structure (Bibliographisches Institut Mannheim/Wien/Zürich, 1975)

/49/ H. Hebach, Interaction Studies in Nuclei (Proc. Intern. Symposium, Max-Planck-Institute for Chemistry, Nuclear Physics Division, Mainz 1975, North-Holland Publ. Co.)

/50/ R. Blamauer, Diplomarbeit, Bochum, 1974

/51/ H. Hebach, A. Wortberg and M. Gari, Nucl. Phys. (in print)

/52/ A.J.F. Siegert, Phys. Rev. **52** (1937) 787

/53/ F. Partovi, Annals of Phys. **27** (1964) 79

/54/ H. Arenhövel, W. Fabian and H.G. Miller, Phys. Letters **52B** (1974) 303

/55/ C. Noguchi and F. Prats, Phys. Rev. Letters **33** (1974) 1168

/56/ J.P. Didelez, H. Langevin-Joliot, N. Bijedić and Z. Marić, Il Nuovo Cimento **67A** (1970) 388

/57/ T. Kopaleishvili and R.I. Jibuti, Nucl. Phys. **44** (1963) 34

/58/ A. Reitan, Nucl. Phys. **64** (1965) 113

/59/ E. Østgaard, Nucl. Phys. **64** (1965) 289

/60/ E. Bramanis, Nucl. Phys. **A175** (1971) 17; **A193** (1972) 323

Note added: Shortly before the opening of the School the author received the preprint of a paper by M. Marangoni, P.L. Ottaviani and A.M. Saruis. In this paper the reactions $^{16}O(\gamma,N_o)$ are studied from the particle threshold up to 100 MeV in the frame of the Tamm-Dancoff approximation. Interference of E1 and E2 amplitudes leads to a forward asymmetry of both (γ,p) and (γ,n) angular distributions for photon energies above 45 MeV.

REAL AND VIRTUAL PHOTONS

B.BOSCO

Istituto di Fisica dell'Università - Firenze

Istituto Nazionale di Fisica Nucleare -
Sezione di Firenze.

1) Introduction and definitions.

It is well known from the classical electromagnetic theory that introducing the electromagnetic tensor $F_{\mu\nu}$ in terms of the quadripotential A_μ defined as[1]:

$$F_{\mu\nu} = \frac{\partial}{\partial x^\mu} A_\nu - \frac{\partial}{\partial x^\nu} A_\mu \qquad (1)$$

the Maxwell equations are equivalent to the following equations:

$$\frac{\partial}{\partial x^\sigma} \frac{\partial}{\partial x_\sigma} A_\mu = \Box A_\mu = 0 \qquad (2)$$

$$\chi = \frac{\partial}{\partial x_\mu} A_\mu = 0 \qquad (3)$$

The equation (3) is treated as a subsidiary condition and one recognizes the usual relation known as Lorentz condition. Let me recall that the condition (3) is not yet sufficient to define uniquely the potential A_μ given by equation (1) since one can always perform the so called gauge trans= formation

$$A_\mu \longrightarrow A'_\mu + \frac{\partial \varphi}{\partial x^\mu} \qquad (4)$$

which leaves $F_{\mu\nu}$ and therefore the electromagnetic field unaltered, and the equation (2) will also remain unaffected provided the φ satisfies the equation:

$$\frac{\partial}{\partial x^\mu} \frac{\partial}{\partial x_\mu} \varphi = 0 \qquad (5)$$

Let us now develop the radiation potential in a four
dimensional Fourier integral

$$A_\mu(x) = \frac{1}{(2\pi)^2} \int b_\mu(k)\, e^{ikx}\, d^4k \qquad (6)$$

(remember that $k \cdot x = k^\mu x_\mu = g_{\mu\nu} k^\mu x^\nu = k^0 x^0 - k\,x$)
Of course one has:

$$b_\mu(k) = \frac{1}{(2\pi)^2} \int A_\mu(x)\, e^{-ik \cdot x}\, d^4x \qquad (7)$$

The equation for the radiation potential will become
an algebraic equation for the amplitude in the Fourier
space. Using equation (6), equation (2) can be rewritten as:

$$k^2 b_\mu(k) = 0 \qquad (8)$$

Hence $b_\mu(k)$ can be different from zero only if $k^2 = 0$.
The solution of (8) shall therefore have the form:

$$b_\mu(k) = \delta(k^2)\, c_\mu(\vec{k}) \qquad (9)$$

Remembering the well known property of the δ :

$$\delta(k^2) = \delta(k_0^2 - k^2) = \frac{1}{2\omega} \left[\delta(k^0 - \omega) + \delta(k^0 + \omega) \right] \qquad (10)$$

with $\omega = |\vec{k}|$

The expression (6) can be written after the integration
on k_0 has been carried out:

$$A_\mu(x) = \frac{1}{(2\pi)^2} \int \frac{1}{2\omega} \left\{ e^{i(\omega x_0 - \vec{k} \cdot \vec{x})} + e^{i(\omega x_0 + \vec{k} \cdot \vec{x})} \right\}$$

$$\cdot c_\mu(\vec{k})\, d^3k \qquad (11)$$

By changing $\vec{k}$ in $-\vec{k}$ in the second term and
by introducing the quantity:

$$a_\mu(\vec{k}) = \frac{1}{\sqrt{4\pi\omega}} \, c_\mu(\vec{k}) \qquad (12)$$

$A_\mu(x)$ can be cast in the form:

$$A_\mu(x) = \frac{1}{(2\pi)^{3/2}} \int \frac{d^3k}{2\omega} \left[a_\mu(\vec{k}) \, e^{ik\cdot x} + a_\mu^*(\vec{k}) \, e^{-ik\cdot x} \right] \qquad (13)$$

if A_μ has to be real.

A very useful gauge when no source is present is the 'so called' Coulomb or transverse gauge. In this case, as it is known, we can put $A_o = 0$, and therefore one has:

$$\vec{\nabla} \cdot \vec{A} = 0 \qquad (14)$$

Then it follows:

$$\vec{E} = - \frac{1}{c} \frac{\partial \vec{A}}{\partial t} \qquad \vec{B} = \vec{\nabla} \times \vec{A} \qquad (15)$$

From these relations and from equation (13) the transversality of the radiation field follows.

When the electromagnetic field is quantized the a_μ and a_μ^* in equation (13) will become destruction and creation operators for the photon, but what is most relevant for us is that equation (8) and the relative solution (9) must be satisfied. We shall call a photon for which such a condition is valid, a real photon. We can notice by the way that the condition $k^2 = 0$ is just the equivalent, for a massless particle, of what is usually called 'on mass shell' condition for a particle of mass m and momentum p whose relation is expressed by the Einstein formula $p^2 = m^2$.

2) The Fermi idea and the Weizsacker-Williams method.

What nowadays is going under the name of Weizsacker-Williams method is nothing else that a brilliant idea of Fermi put forward in 1924[2] and applied by himself at that time to some atomic processes with great success.
Let us first outline the idea, which is very simple, and then we shall enter in its theoretical details.
A charge q gives rise in its own system of reference to an elctric field, which when viewed by a relative moving system is equivalent to an elctromagnetic field due to the transformation law of the electromagnetic tensor.
Any system which is experiencing the near-by passing of a charged particle is therefore experiencing the presence of such an electromagnetic field which we are now going to treat in detail.
Let us consider two relative systems of coordinates K and K', one of which (K') is moving in the direction x_3 with relative velocity v. The systems are such that for t = 0   K = K'. See fig. 1.

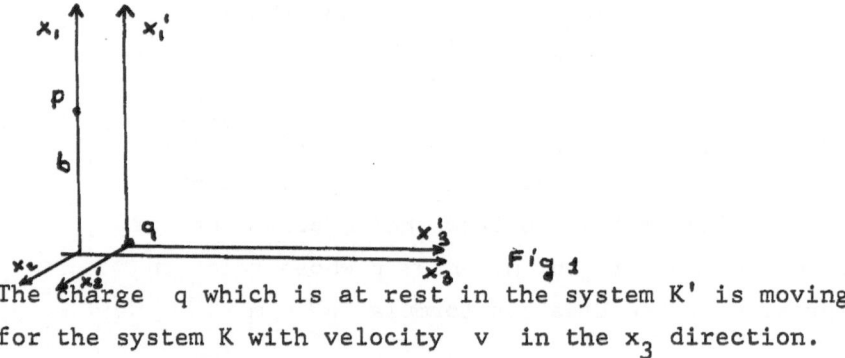

Fig 1

The charge q which is at rest in the system K' is moving for the system K with velocity v in the x_3 direction.

Let a point P in the K system be located at distance $x_1 = b$ on the x_1 axis. When observed from q the coordinate of P will be:

$$P = (x^{1'} = b, \ x^{2'} = 0, \ x^{3'} = -vt')$$

and the distance r' between q and P will be given by:

$$r' = (b^2 + v^2 t'^2)^{1/2}$$

The elctric field generated by q in P can be written:

$$\vec{E'} = -\frac{q}{r'^2} \ \text{vers} \ \vec{r'} \tag{16}$$

and the three components along the axis are:

$$E_1' = -\frac{qb}{r'^2} \tag{17a}$$

$$E_2' = 0 \tag{17b}$$

$$E_3' = -\frac{qvt'}{r'^2} \tag{17c}$$

while the three components of the H field are zero.
Now we wish to compute the components of $\vec{E}$ and $\vec{H}$ in the K system which can be obtained by the well known relation between the tensors:

$$F_{\mu\nu} = a_\mu{}^\sigma \ a_\nu{}^\beta \ F'_{\sigma\beta} \tag{18}$$

where $a_\mu{}^\sigma$ are the coefficients of the Lorentz transformation[3].

After some calculations one gets from (18)

$$E_1(t) = -\frac{\gamma \ b \ q}{(b^2 + \gamma^2 v^2 t^2)^{3/2}} \tag{18a}$$

$$H_1 = 0 \tag{18d}$$

$$E_2(t) = 0 \tag{18b}$$

$$H_2 = \beta E_1 \tag{18e}$$

$$E_3(t) = -\frac{q \ \gamma \ v \ t}{(b^2 + \gamma^2 v^2 t^2)^{3/2}} \tag{18c}$$

$$H_3 = 0 \tag{18f}$$

where $\gamma = \frac{1}{\sqrt{1-\beta^2}}$ and $\beta = \frac{v}{c}$

For our purposes the Fourier transforms of the previous components of the fields are important. These, by intro= ducing the variable $x = \dfrac{\gamma v t}{b}$, can be written as:

$$E_1(\omega) = \frac{1}{\sqrt{2\pi}} \frac{q}{bv} \int_{-\infty}^{+\infty} \frac{e^{i\frac{\omega b}{\gamma v} x}}{(1 + x^2)^{3/2}} \, dx \qquad (19a)$$

$$E_3(\omega) = - \frac{1}{\sqrt{2\pi}} \frac{q}{bv\gamma} \int_{-\infty}^{+\infty} \frac{e^{i\frac{\omega b}{\gamma v} x} x}{(1 + x^2)^{3/2}} \, dx \qquad (19b)$$

$$H_2(\omega) = \beta E_1(\omega) \qquad (19c)$$

all other components being zero.

Now the integrals on the r.h.s. of equations (19a) and (19b) are the integral representations of the so called Kelvin function or modified Bessel function of the third kind[4]:

$$\int_{-\infty}^{+\infty} \frac{e^{i\frac{\omega b}{\gamma v} x}}{(1 + x^2)^{3/2}} \, dx = 2 \frac{\omega}{\gamma} \frac{b}{v} K_1(-\frac{\omega}{\gamma} \frac{b}{v}) \qquad (20a)$$

$$\int_{-\infty}^{+\infty} \frac{e^{i\frac{\omega b}{\gamma v} x} x}{(1 + x^2)^{3/2}} \, dx = 2i \frac{\omega}{\gamma v} b K_0(-\frac{\omega}{\gamma v} b) \qquad (20b)$$

The K_ν are connected to the more familiar Henkel functions of first kind

$$K_\nu(x) = \frac{\pi}{2} i^{\nu+1} H_\nu^{(1)}(ix) \qquad (21)$$

For our further purposes the asymptotic behaviours are relevant:

$$\text{for } x \ll 1 : \quad K_\nu(x) \longrightarrow \begin{cases} -(\log \frac{x}{2} + 0.5772 + \ldots) & \text{for } \nu = 0 \\ \frac{\Gamma(\nu)}{2} (\frac{2}{x})^\nu & \text{for } \nu = 0 \end{cases}$$

for $x \gg 1$: $K_\nu(x) \longrightarrow \sqrt{\dfrac{\pi}{2x}} \, e^{-x} \left(1 + 0\left(\tfrac{1}{x}\right) \right)$

Using the equations (20), the expressions for $E_1(\omega)$ and $E_3(\omega)$ can now be written as:

$$E_1(\omega) = \frac{q}{bv} \left(\frac{2}{\pi}\right)^{1/2} \left(\frac{\omega b}{\gamma v} K_1\left(\frac{\omega b}{\gamma v}\right) \right) \qquad (22a)$$

$$E_3(\omega) = - \frac{i \cdot q}{b \gamma v} \left(\frac{2}{\pi}\right)^{1/2} \left(\frac{\omega b}{\gamma v} K_0\left(\frac{\omega b}{\gamma v}\right) \right) \qquad (22b)$$

After these mathematical peripheralia and before entering into the physics of the radiation generated by our moving particles, let me recall some results strictly valid for electromagnetic radiations made up by real photons. At any given time t the radiation intensity is given by the Poynting vector at that time:

$$\vec{S}(t) = \frac{c}{4\pi} \, \vec{E}(t) \times \vec{H}(t) \qquad (23)$$

For a free wave $\left|\vec{E}(t)\right| = \left|\vec{H}(t)\right|$, and the two fields are furhtermore orthogonal. Therefore:

$$\left|\vec{S}(t)\right| = \frac{c}{4\pi} \left|E(t)\right|^2 \qquad (24)$$

The total energy which radiates for unit surface from the time $-\infty$ to $+\infty$ will therefore be:

$$W = \int_{-\infty}^{+\infty} \left|\vec{S}(t)\right| \, dt = \frac{c}{4\pi} \int_{-\infty}^{+\infty} \left|\vec{E}(t)\right|^2 \, dt = \frac{c}{4\pi} \int_{-\infty}^{+\infty} \left|\vec{E}(\omega)\right|^2 \, d\omega \qquad (25)$$

where use has been made of the well known Parseval theorem:

$$\int_{-\infty}^{+\infty} \left|E(t)\right|^2 \, dt = \int_{-\infty}^{+\infty} \left|E(\omega)\right|^2 \, d\omega$$

Now if E(t) is real, since the sign of the frequence has no physical meaning, one can write:

$$W = \frac{c}{2\pi} \int_0^\infty |E(\omega)|^2 \, d\omega \qquad (26)$$

Let us finally return to the radiation generated by our moving charge q.

At this stage it is better to look back at the expressions (18) and draw the fig. 2 where the blob is the targed object.

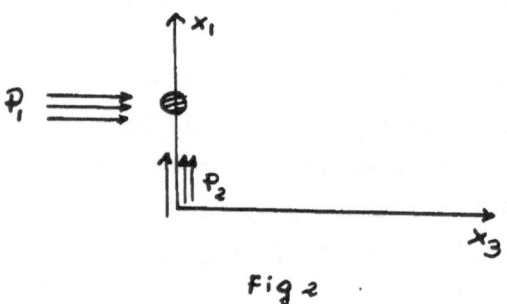

Fig 2

For a charge q moving at an extremely relativistic speed $\gamma \approx 1$, we can assume by (18e) that $H_2 = E_1$, and we can conclude that the E_1 and H_2 will give rise to a pulse P_1, equivalent to an electromagnetic plane wave moving in the x_3 direction with the property of similar waves. Then we are left with the component $E_3(t)$ with no $H_2(t)$ companion component of equal intensity to form a plane wave in the x_1 direction. We can however add arbitrarily such a component, provided that a posteriori the resulting pulse P_2 is in general negligible compared to P_1.
Let us remark that the condition for which such a physical situation is realized will be the condition for the physical validity of our approximation.
Since the two pulses are not interfering, we can write by use of (26) for the total intensity:

$$I(\omega) = I_1(\omega) + I_2(\omega) \qquad (27)$$

where

$$I_1(\omega) = \frac{c}{2\pi} |E_1(\omega)|^2 \tag{28a}$$

$$I_2(\omega) = \frac{c}{2\pi} |E_3(\omega)|^2 \tag{28b}$$

Introducing the new variable $\xi = \frac{\omega b}{\gamma v}$ we can write, using (20a) and (20b):

$$I(\omega,b) = I_1 + I_2 = \frac{1}{\pi^2} \frac{q^2 c}{v^2} \frac{1}{b^2} \left\{ \xi^2 K_1^2(\xi) + \frac{1}{\gamma^2} \xi^2 K_0^2(\xi) \right\} \tag{29}$$

where we have put in evidence the existing functional dependence from the impact parameter b.

The formula (29) justifies already qualitatively our assumption that the pulse P_2 given by the second term in the curl bracket will be negligible for a relativistic particle, due to the factor $\frac{1}{\gamma^2}$.

The total frequency which is radiated with frequency ω, for all the values of the impact parameter b, namely $\mathcal{J}(\omega)$, will be given by:

$$\mathcal{J}(\omega) = 2\pi \int_{b_{min}}^{\infty} b \, db \, I(\omega,b) = \frac{2}{\pi} \frac{q^2 c}{v^2} \int_{\xi_{min}}^{\infty} \left\{ \xi K_1^2(\xi) + \frac{1}{\gamma^2} \xi K_0^2(\xi) \right\} d\xi \tag{30}$$

The integral appearing on the r.h.s. of equation (30) can be computed analytically, and this is done in appendix A. The result is:

$$\mathcal{J}(\omega) = \frac{2}{\pi} \frac{q^2 c}{v^2} \left\{ \xi_{min} K_1(\xi_{min}) K_0(\xi_{min}) \right.$$

$$\left. + \frac{1}{2} \frac{v^2}{c^2} \xi_{min}^2 \left[K_0^2(\xi_{min}) - K_1^2(\xi_{min}) \right] \right\} \tag{31}$$

We would like to spend a few words on b_{min} or, what is equivalent, on ξ_{min}. b_{min} represents the shortest distance at which the charged particle is effective on the sys-tem,

and this is clearly depending on the process under consi=
deration, and it is therefore in some way a free parameter
in our theory.

The physical consideration of the process will dictate
the order of magnitude of this parameter.

It is now an easy matter to obtain from equation (31)
the spectrum of equivalent photons generated by a moving
charge. In fact considering that the radiated energy between
ω and $\omega + d\omega$ is given by $J(\omega)\, d\omega$, and that on
the other side a photon with frequency ω is carrying
an energy $\hbar\omega$ we can write:

$$J(\omega)\, d\omega = \hbar\omega \; N(\hbar\omega)\, d(\hbar\omega) \qquad\qquad (32)$$

from which the function $N(\hbar\omega)$, namely the number of
equivalent photons of frequency ω , can be derived
using equation (31).

We shall not write explicitly this function $N(\hbar\omega) = N(p_i, \omega)$,
which can be immediately derived by using equation (31)
in the l.h.s. of equation (32). We would like to call
instead the attention to the fact that two simple limits
exist for this function, namely for the limiting cases
$\xi_{min} \ll 1$ and $\xi_{min} \gg 1$, limits which can immediately
be derived using the asymptotic expressions of $K_\nu(z)$
which have been previously quoted.

Clearly the equivalent spectrum of the photons produced
by a charged particle is quite different from the one of
the real photons. The most striking difference is that
they have a longitudinal component which is lacking in
the case of the real ones. One should therefore expect
that as probing objects they may be more useful in
giving more information about the target.

However the method we have just studied, as useful as
it may be for the evaluation of the cross-section, may
sometimes be right only for the order of magnitude.
A more accurate treatment is the one based on the Müller
potential, which we are now going to describe.

3) The Müller potential.

The Müller potential is a potential which plays
the role of the four vector potential in a transition
in which a spin 1/2 particle is involved. For people
who are familiar with Feynmanology it is rather trivial
to write it down immediately. Let me instead follow
the more classical way and use the correspondence principle.

In other words let us suppose that we have an
electron with four momentum $p_\mu^{(1)}$, which undergoing a
transition will jump to the four momentum $p_\mu^{(2)}$.
We would like to determine up to the first order the
equivalent electromagnetic potential due to this transition.

We shall use the standard γ matrices representation,
in which the γ^k are antihermitian, i.e. $(\gamma^k)^+ = - \gamma^k$
for k=1,2,3 , and γ^o is hermitian.

It is well known from the study of the Dirac
equation that it is convenient to use, together with
the spinor ψ , the adjoint spinor $\overline{\psi} = \psi^+ \gamma^o$.
Then one shows that the fundamental equation of the theory
is the following equation:

$$\frac{\partial j^\mu}{\partial x^\mu} = 0 \qquad\qquad (33)$$

with j^μ defined by:

$$j^{\mu} = e c \overline{\psi} \gamma^{\mu} \psi \qquad (34)$$

Equation (33) can be interpreted either as the equation of continuity which conserves the norm or as the equation of the current conservation.

Let us recall that in the free Dirac particle theory the structure of the ψ (p,x) is factorized in the following way:

$$\psi (p,x) = u(p) \, e^{-ip \cdot x} \qquad (35)$$

Let us now go back to our problem.
The correspondence principle tells us how to construct a current when we have, instead of a stationary state, a transition from the continuum $p_{\mu}^{(1)}$ to a continuum $p_{\mu}^{(2)}$; our current must be:

$$j^{\mu} = e \, \bar{u}(p_2) \, \gamma^{\mu} u(p_1) \, e^{i(p_2 - p_1) \cdot x} \qquad (36)$$

This current will in turn generate an electromagnetic field ϕ^{μ} which satisfies the usual equations:

$$\Box \phi^{\mu} = - 4\pi \, j^{\mu} \qquad (37)$$

The inhomogeneous solution of equation (37), when j^{μ} is given by equation (36), is

$$\phi^{\mu} = 4\pi \, e \, \frac{\bar{u}(p2) \, \gamma^{\mu} u(p1)}{(p_2 - p_1)^2} \, e^{i(p_2 - p_1) \cdot x} \qquad (38)$$

and this is the Müller potential.

Everyone knowing the elements of Feynman-graphology will recognize at once in the expression (38) (but the exponential factor) the following graph:

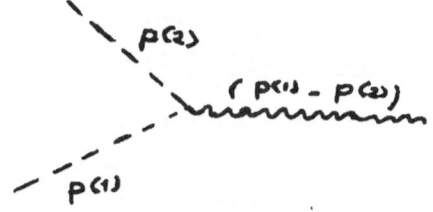

It is clear that $\dfrac{1}{(p^{(1)} - p^{(2)})^2}$ behaves like a propagator

of a photon. However this photon has properties quite different from the properties of the real one.

In fact if we call $k = p^{(1)} - p^{(2)}$ we get immediately

$k^2 = (p^{(1)} - p^{(2)})^2 = p^{(1)2} + p^{(2)2} - 2p(1) \cdot p(2) =$

$= 2m^2 - 2 \, k_o^{(1)} k_o^{(2)} + 2 \, \vec{p}^{(1)} \, \vec{p}^{(2)}$.

Therefore $k^2 = 2 \, (m^2 - k_o^{(1)} k_o^{(2)} + \left| \vec{p}^{(1)} \right| \left| \vec{p}^{(2)} \right| \cos \alpha)$

where α is the scattering angle.

An important property of (38) is that it satisfies the Lorentz condition:

$$\frac{\partial \phi^\mu}{\partial x^\mu} = 0 \tag{39}$$

In fact $\dfrac{\partial \phi^\mu}{\partial x^\mu} = i4\pi e \, \dfrac{\bar{u}(p_2) \, \gamma^\mu \, u(p_1)}{(p_2 - p_1)^2} \, (p^{(2)} - p^{(1)})_\mu \, e^{i(p_2 - p_1) \cdot x}$

Now the r.h.s. is zero due to the Dirac equation. This follows from the two equations

$(\gamma^\mu p_\mu - mc) \, u(p) = 0$ and $\bar{u}(p) \, (\gamma^\mu p_\mu - mc) = 0$

which allow us to write:

$\bar{u}(p_2) \, \gamma^\mu \, (p_\mu^{(2)} - p_\mu^{(1)}) \, u(p_1) =$

$= \bar{u}(p_2) \, \gamma^\mu \, p_\mu^{(2)} \, u(p_1) - \bar{u}(p_2) \, \gamma^\mu \, p_\mu^{(1)} \, u(p_1) = 0$

4) <u>The application of the Müller potential to the study of</u>
<u>the connection between photon and electron induced phenomena</u>.

In this section we shall outline how the Müller
potential can be used in establishing a connection between
the phenomena induced by electrons and those induced
by photons.

We shall first derive some general formulas.
Let us suppose that our potential will interact with a
current $J_\mu(x)$, from which we only require at the moment
to be conserved:

$$\frac{\partial J^\mu(x)}{\partial x^\mu} = 0 \tag{40}$$

Then the interaction hamiltonian will be given by:

$$H_{int} = \int \phi^\mu(x)\, J_\mu(x)\, d^3x \tag{41}$$

If the current is furthermore invariant under space-time
translations we can easily perform such an integral.
One has to observe that we are interested indeed not
in H_{int} itself but in the matrix elements of the
transition between the nuclear states $|\alpha\rangle$ and $|\alpha'\rangle$, which
we shall suppose to be eigenstates of the four momentum
operator P^μ with eigenvalues Q_1 and Q_2 respectively.

The fact that $J_\mu(x)$ is invariant under translations
can be expressed in one of the following ways:

$$i\, \frac{\partial}{\partial x^\nu}\, J^\mu(x) = \left[J^\mu(x),\, P_\nu \right] \tag{42}$$

or $\quad J^\mu(x) = e^{i\, P \cdot x}\, J(0)\, e^{-i\, P \cdot x}$

since the P_μ are the generators of the infinitesimal translations.

Now we shall make use of the fact that $|\alpha\rangle$ and $|\alpha'\rangle$ are eigenstates of P_μ and will write:

$$\langle \alpha'| M | \alpha \rangle = \langle \alpha'| \int_{-\infty}^{+\infty} dt\, H_{int} | \alpha \rangle$$

$$= \int d^4x\, 4\pi e\, \frac{\bar{u}(p2)\, \gamma^\mu\, u(p1)}{(p^{(2)} - p^{(1)})^2} \langle \alpha'| J_\mu(0)|\alpha\rangle\, e^{i(P_2+Q_2-P_1-Q_1)\cdot x}$$

$$= (2\pi)^4\, 4\pi e\, \frac{\bar{u}(p2)\, \gamma^\mu\, u(p1)}{(p^{(2)} - p^{(1)})^2} \langle \alpha'| J_\mu(0)|\alpha\rangle\, \delta^4(P_2+Q_2-P_1-Q_1) \qquad (43)$$

The δ has just the role of imposing the conservation of the energy and of the momentum.

The probability of the transition rate from which the cross-section can be computed is given by:

$$W = \left| \langle \alpha'| M | \alpha \rangle \right|^2 =$$

$$= (2\pi)^8\, (4\pi e)^2 \left| \frac{\bar{u}(p2)\, \gamma^\mu\, u(p1)}{(p^{(2)} - p^{(1)})^2} \langle \alpha'| J_\mu(0)|\alpha\rangle \right|^2 \left[\delta^4(P_2+Q_2-P_1-Q_1) \right]^2$$

$$(44)$$

and as it stands it is manifestly infinite due to the square of the δ.

This is physically due to the fact that in this way we are computing all the transitions for the whole space and the whole time from $-\infty$ to $+\infty$.

We may write identically:

$$\left| \delta^4(P_2+Q_2-P_1-Q_1) \right|^2 = \delta^4(0)\, \delta^4(P_2+Q_2-P_1-Q_1)$$

Now δ^4 stands for $\delta \cdot \delta^3$, the first being a function of the time component and the second of the space components.

Let us first consider the one relative to the time component. We can write:

$$\delta(0) = \lim_{T\to\infty} \lim_{\alpha\to 0} \frac{1}{2\pi} \int_{-\frac{T}{2}}^{\frac{T}{2}} e^{i\alpha t}\, dt = \lim_{T\to\infty} \frac{T}{2\pi}$$

having inverted the operations of the limit and of the integral with one of the "classical unorthodox" operations. In similar way one can deal with δ^3 referring to the space components. In particular one can think of δ^3 in terms of three components of a cube of volume L^3. Then for each side, let us call the particular one L_j, one can write:

$$\delta(0) = \lim_{L_j\to\infty} \lim_{\alpha\to 0} \frac{1}{2\pi} \int_{-\frac{L_j}{2}}^{\frac{L_j}{2}} e^{i\alpha x_j}\, dx_j = \lim_{L_j\to\infty} \frac{L_j}{2\pi}$$

so that $\delta^3(0) = \lim_{V\to\infty} \dfrac{V}{(2\pi)^3}$

From this discussion it follows that the pro-bability of transition rate per unit time and unit volume will be:

$$W = (2\pi)^4 (4\pi e)^2 \left| \bar{u}(p_2)\gamma^\mu u(p_1) \langle \alpha' | J_\mu(0) | \alpha \rangle \right|^2 \cdot$$

$$\delta^4(p_2+Q_2-p_1-Q_1) \tag{45}$$

This is as far as we can go using only general arguments. It is clear from equation (45) that our matrix element will be factorized in one factor depending only from the Müller potential (and therefore only from the electron part), and in a second one depending only from the matrix element $\langle \alpha' | J_\mu(0) | \alpha \rangle$ of the current J^μ.

We can still proceed a little further, using the conservation laws for ϕ_μ and J_μ expressed by equations (39) and (40), in order to rewrite equation (45) in terms of the space part only, both for the Müller potential and for the current J_μ .

In fact in the momentum space, with $k_\mu = p_\mu^{(2)} - p_\mu^{(1)}$, these two equations can be rewritten as:

$$k_o A_o - \vec{k}\cdot\vec{A} = 0 \tag{39'}$$

$$k_o J_o - \vec{k}\cdot\vec{J} = 0 \tag{40'}$$

and the matrix elements of the transition can be cast in the form:

$$\langle \alpha'|M|\alpha\rangle = \vec{A'}\cdot\langle\alpha'|\vec{J}|\alpha\rangle \tag{46}$$

where

$$\vec{A'} = -\frac{4}{(p^{(2)}-p^{(1)})^2}\frac{\pi}{}\frac{e}{}\left[\bar{u}(p_2)\,\vec{\gamma}\,u(p_1) - \vec{k}\,\frac{\vec{k}}{k_o^2}\cdot\bar{u}(p_2)\,\vec{\gamma}\,u(p_1)\right] \tag{47}$$

The cross-section for electron induced phenomena can then be written as:

$$\frac{d^2\sigma_{el}}{d\Omega_N\,d\Omega_e} = \frac{2\pi}{v_e} N_{ij}\,J_{ij}\,\rho_e\,\rho_N \tag{48}$$

where, as it follows from (45):

$$N_{ij} = \frac{1}{2}\sum_{\text{final electron spin}}\sum_{\text{initial electron spin}} A_i^*\,A_j \tag{49}$$

v_e = initial electron velocity

$$J_{ij} = S\,\langle\alpha'|J_i|\alpha\rangle^*\langle\alpha'|J_j|\alpha\rangle \tag{50}$$

and S means sum over the possible spin states other than the electron ones.

ρ_e and ρ_N are the densities of final states for the elctrons and for the nuclear products respectively.

A formula similar to (48) is valid for the photo= disintegration by real photons:

$$\frac{d\sigma_{ph}}{d\Omega_N} = \frac{2\pi}{k} \, \varepsilon_i \; \varepsilon_j \; J_{ij} \; \rho_N \tag{51}$$

This time k has to be the momentum of the real photon and the tensor N_{ij} is now simply constructed with the polarization vector of the photon.

We shall not reproduce here the expression of N_{ij} which has been given many times[5].

We shall instead call your attention to the fact that the k of equation (51) and the momentum tranfer which appears in N_{ij} of equation (49) must be such that the relative energy available to the final system is the same.

Finally we would like to discuss the tensor J_{ij}. It is almost obvious that all the physics of any process is contained in it, or in other words in the current which gives rise to the transition from the state $|\alpha\rangle$ to the state $|\alpha'\rangle$.

The case in which this current is simply given by the ordinary multipoles has been discussed in great detail a long time ago[6].

If one would like to introduce exchange currents etc., he needs a great care, and a full and complete analysis is in our opinion still lacking.

We hope to have been able in these lectures to outline the general methods by which such an analysis should be carried out.

Appendix A.

We would like to compute in this appendix the integral appearing in equation (30).

The integral is:

$$I = \int_{\xi_{min}}^{\infty} \left\{ \xi \, K_1^2(\xi) + \frac{1}{\gamma^2} \, \xi \, K_0^2(\xi) \right\} d\xi \qquad (A1)$$

Let us put:

$$I_1 = \int K_1^2(\xi) \, \xi \, d\xi \qquad (A2)$$

$$I_0 = \int K_0^2(\xi) \, \xi \, d\xi \qquad (A3)$$

The following recursion formulas are valid[7]:

$$K_{-\nu}(z) = K_\nu(z) \qquad (A4)$$

$$K_{\nu-1}(z) - K_{\nu+1}(z) = -\frac{2\nu}{z} K_\nu(z) \qquad (A5)$$

$$K_{\nu-1}(z) + K_{\nu+1}(z) = -2 K_\nu'(z) \qquad (A6)$$

From (A6) and (A4) it follows at once:

$$K_1(z) = -K_0'(z) \qquad (A7)$$

Therefore by an integration by parts one gets:

$$
\begin{aligned}
I_1 &= \int K_1^2(\xi) \, \xi \, d\xi = -\int K_1(\xi) \, K_0'(\xi) \, \xi \, d\xi \\
&= -K_1(\xi) \, K_0(\xi) + \int \frac{d}{d\xi}(K_1(\xi)\xi) \, K_0(\xi) \, d\xi \\
&= -K_1(\xi) \, K_0(\xi) + \int K_1(\xi) \, K_0(\xi) \, d\xi + \int K_1'(\xi) K_0(\xi) \xi d\xi \\
&= -K_1(\xi) K_0(\xi) - \frac{1}{2} K_0^{\;2}(\xi) - \frac{1}{2}\int \left[K_0(\xi) + K_2(\xi) \right] K_0(\xi) \xi \, d\xi
\end{aligned}
$$

$$(A8)$$

where (A7) has been used to evaluate the first integral on the r.h.s. and (A6) to eliminate $K_1'(\xi)$ from the last integral. We can therefore write:

$$I_1 = - K_1(\xi)\, K_0(\xi)\,\xi - \tfrac{1}{2} K_0^2(\xi) - \tfrac{1}{2} I_0 - \tfrac{1}{2} \int K_2(\xi)\, K_0(\xi)\,\xi\, d\xi \tag{A9}$$

Now (A5) allows us to write:

$$\int K_2(\xi)\, K_0(\xi)\,\xi\, d\xi = \int \left[K_0(\xi) + \tfrac{2}{\xi} K_1(\xi) \right] \xi\, K_0(\xi)\, d\xi$$

$$= I_0 + 2 \int K_1(\xi)\, K_0(\xi)\, d\xi = I_0 - K_0^2(\xi)$$

and by subsituting it in (A9) one gets:

$$I_1 = - K_1(\xi)\, K_0(\xi) - I_0 \tag{A10}$$

We must now evaluate I_0 .

$$I_0 = \int \xi\, K_0^2(\xi)\, d\xi = \tfrac{1}{2} \xi^2\, K_0^2(\xi) - \int \xi^2\, K_0(\xi)\, K_0'(\xi)\, d\xi$$

$$= \tfrac{1}{2} \xi^2\, K_0^2(\xi) - B \tag{A11}$$

Let us consider B:

$$B = \int \xi^2\, K_0(\xi)\, K_0'(\xi)\, d\xi = - \int \xi^2\, K_0(\xi)\, K_1(\xi)\, d\xi \tag{A12}$$

by use of (A7).
Now using (A5):

$$B = - \int \xi^2\, K_1(\xi) \left[K_2(\xi) - \tfrac{2}{\xi} K_1(\xi) \right] d\xi$$

$$= - \int \xi^2\, K_1(\xi)\, K_2(\xi)\, d\xi + 2 \int \xi\, K_1^2(\xi)\, d\xi \tag{A13}$$

and using (A6) to eliminate $K_2(\xi)$ in the first integral:

$$B = \int \xi^2\, K_1(\xi) \left[K_0(\xi) + 2 K_1'(\xi) \right] d\xi + 2 \int \xi\, K_1^2(\xi)\, d\xi$$

$$= \int \xi^2\, K_0(\xi)\, K_1(\xi)\, d\xi + 2 \int \xi^2\, K_1(\xi)\, K_1'(\xi)\, d\xi + 2 \int \xi\, K_1^2(\xi)\, d\xi$$

Therefore:

$$B = - \int \xi^2 \, K_0(\xi) \, K_0'(\xi) \, d\xi + \xi^2 \, K_1^2(\xi) - 2 \int \xi \, K_1^2(\xi) \, d\xi + 2 \int \xi \, K_1^2(\xi) \, d\xi$$

(A14)

having used (A7) in the first integral and performed partial integration in the second.

Therefore:

$$B = - B + \xi^2 \, K_1^2(\xi)$$

(A15)

From the last equation it follows:

$$B = \frac{1}{2} \xi^2 \, K_1^2(\xi)$$

(A16)

By substituting this last expression in (A11) one gets:

$$I_0 = \frac{1}{2} \xi^2 \, K_0^2(\xi) - \frac{1}{2} \xi^2 \, K_1^2(\xi)$$

(A17)

By direct substitution of (A8) and (A17) in (A1) one gets the formula (31) of the text.

6) See reference (5) and J.M.Eisenberg, Phys.Rev. <u>132</u>, 2243(1963).

7) See reference (4).

Footnotes and references.

(+) Work supported in part by Nato Grant 553.

1) The electromagentic tensor satisfies the homogeneous
   Maxwell equations

$$\frac{\partial f_{\mu\nu}}{\partial x^\sigma} + \frac{\partial f_{\nu\sigma}}{\partial x^\mu} + \frac{\partial f_{\sigma\mu}}{\partial x^\nu} = 0$$

$$\frac{\partial f_{\mu\nu}}{\partial x_\mu} = 0$$

Remember that with the metric we are using ($g_{oo} = 1$, $g_{kk} = -1$), the components of the electromagnetic tensor $F_{\mu\nu}$ as defined by us and the components of $\vec{E}$ and $\vec{H}$ are connected by the following relations:

$F_{ok} = E_k$ $\qquad$ $F_{il} = -H_k$ $\qquad$ provided that (ilk) are in the cyclic order (1,2,3).

2) E.Fermi, Z.Physik $\underline{29}$, 315(1924) and Nuovo Cimento $\underline{2}$, 143(1925).

3) Due to our metric all the $a_\mu{}^\sigma$ are real, and they are:

$$a_\mu{}^\sigma = \begin{pmatrix} \gamma & \gamma\beta & 0 & 0 \\ \gamma\beta & \gamma & 0 & 0 \\ 0 & 0 & 1 & 0 \\ 0 & 0 & 0 & 1 \end{pmatrix}$$

4) W.Magnus und F.Oberhettinger, Formeln und Sätze für die speziellen Funktionen der Mathematischen Physik, Springer Verlag, Berlin 1948, p.28.

5) See e.g. R.H.Dalitz and D.R.Yennie, Phys.Rev. $\underline{105}$, 1598(1957); B.Bosco and P.Quarati, Nuovo Cimento $\underline{33}$, 527(1964); B.Bosco, in: Proceedings of International Conference on Electromagnetic Interaction at Intermediate and Low Energy, Dubna, February 7-15, 1967.

SUM RULES IN PHOTONUCLEAR PHYSICS

W. Weise

Institute for Theoretical Physics
University of Erlangen

and

Department of Physics
University of Regensburg, Germany

I. Introduction and Motivation

The investigation of sum rules for photon scattering is a
method to describe the integral properties of the response
of a many-body system to an electromagnetic field. Let us
begin with an example from atomic physics, where we know
that the response of a system of many electrons to photons
is governed by electric dipole phenomena. This situation
is summarized by saying that the total integrated cross
section for (photon scattering from the atomic electrons
exhausts the Thomas-Reiche-Kuhn (TRK) sum rule,

$$\int_0^\infty d\omega \ \sigma(\omega) = 2\pi^2 \frac{e}{m} Q \ , \tag{1}$$

where $Q = Ze$ is the total charge of Z electrons with mass m.
In other words: integrating the response over all energies
is just counting the electron charges. (The limit $\omega \to \infty$ here
means $k = \omega/c \gg R^{-1}$, where R measures atomic dimensions,
but is still small compared to nuclear dimensions).

Now, an atom is a comparatively simple system. Electrons
do not have an intrinsic structure, that is, charges are
strictly localized at the positions of the (pointlike)
electrons. For such a system, Eq.(1) will always be valid,
independent of how the electrons interact with each other
in detail.

In a nucleus, the situation is quite different. Nucleons
are not at all structureless; they can be excited into well
defined states (e.g. the 3.3 resonance, or Δ isobar). They
are dressed by clouds of mesons. Pairs of nucleons interact

via the exchange of mesons which, if they are charged, contribute to the total current of the system. Clearly, charges are generally not localized at the positions of the individual nucleons. These explicit mesonic degrees of freedom are commonly referred to as exchange currents, as opposed to the usual convection currents which are proportional to $\vec{p}/M$, where $\vec{p}$ is the nucleon momentum and M its mass.

One of the basic questions is now to what extent these exchange currents take part actively in the nuclear electromagnetic response. If there are more degrees of freedom than just the nucleonic convection currents, then the integrated total photon cross section should reveal more information than a mere counting of charges. In fact, in the following chapter, we shall see that this is indeed the case. Experimentally, one observes a clear enhancement over the TRK sum. The interpretation of this enhancement in terms of exchange current phenomena, and its relation to properties of the instrinsic structure of nucleons, is a challenging problem.

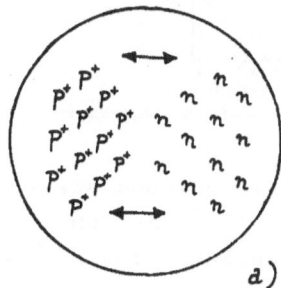 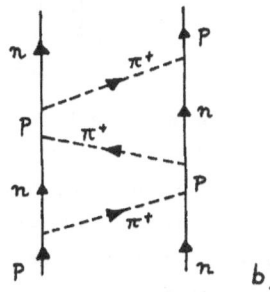

Figure 1: Illustration of dipole oscillation mechanisms in nuclei:
a) Giant dipole mode; protons vibrate collectively against neutrons.
b) Charge oscillation due to meson exchange currents.

We illustrate and close this introductory discussion by showing, in Fig.1, the two basic types of charge oscillation mechanisms present in a nucleus. The first one is the collective giant dipole mode, where all protons and neutrons are coherently involved. The frequency of this mode is inversely proportional to the nuclear radius, $\omega_{dip} \simeq 80 \, A^{-1/3}$ MeV. The second one shows a charge oscillation carried by meson exchange currents, which takes place on a smaller scale (of the order of the pion Compton wavelength) and contributes to photon absorption at frequencies larger than ω_{dip}.

II. <u>Survey of Experimental Data</u>

The structure of a free nucleon, as seen by a photon, is illustrated most directly in terms of the total cross section for γ p $\rightarrow$ hadrons (Figure 2).

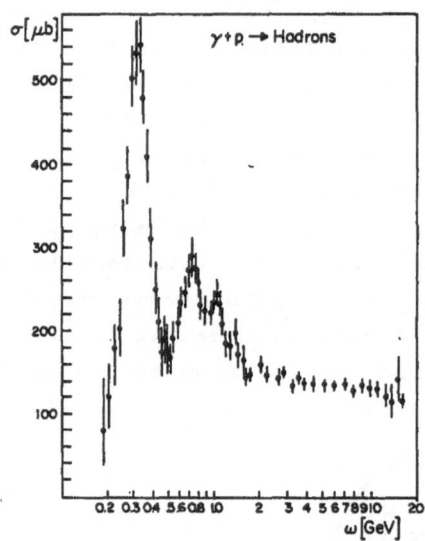

Fig.2 :

Total cross section for the photoproduction of any hadrons from protons as a function of photon energy (see Ref. [2]).

Below a photon energy $\omega = m_\pi = 140$ MeV, nucleons are transparent to photons. Above threshold, photoproduction of pions takes place, and a number of resonances indicates a sequence of excited states of the nucleon. The cross section becomes smooth above 2 GeV, and there is no indication of a drop-off at very high energy.

Total photon cross sections for nuclei have been measured by the Mainz group [1] . The various integrated cross sections are represented as

$$\Sigma_n (E) = \int_o^E d\omega \; \omega^n \sigma(\omega) \qquad (2)$$

and listed in Table 1. We note that Σ_{-2} measures essentially the nuclear polarisability, and Σ_{-1} is proportional to the nuclear mean square radius.

	$\Sigma_0\,(m_\pi)$		$\Sigma_{-1}\,(m_\pi)$		$\Sigma_{-2}\,(m_\pi)$	
	(mb · MeV) ± (%)		(mb) ± (%)		(mb/MeV) ± (%)	
^{6}Li	161	1.9	4.79	1.0	0.197	1.1
^{9}Be	189	2.1	5.33	1.5	0.194	2.5
^{12}C	334	2.2	9.18	1.2	0.316	1.7
^{16}O	509	2.5	15.10	1.3	0.585	1.6
^{27}Al	807	3.9	26.3	1.7	1.11	1.8
^{40}Ca	1290	4.6	46.8	1.7	2.23	1.2

Table 1: Experimental values for various moments $\Sigma_n(m_\pi)$ of the total photon-nucleus cross section, integrated up to the pion production threshold (from Ref. [1])

Figure 3 shows the integrated cross sections $\Sigma_0(E)$ for $E = 35$ MeV and $E = m_\pi$. They are expressed in units of the classical TRK sum,

$$S = \frac{2\pi^2 e^2}{M} \simeq 60\ MeV \cdot mb. \tag{3}$$

Clearly, the giant resonance region already exhausts one classical dipole sum. The measured enhancement κ in

$$\Sigma_0(m_\pi) = (1 + \kappa)\,S\,\frac{NZ}{A} \tag{4}$$

is of the order of 100 % for nuclei beyond A = 16.

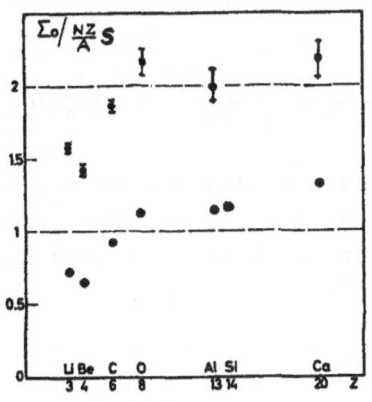

Figure 3: Integrated cross sections
$$\Sigma_0(E)$$
for E = 35 MeV (lower points) and for E = m_π (upper data) (from Ref. [1]).

Thus, if we would like to interpret κ as a measure of exchange current phenomena, then their importance is clearly enormous. We note that, some years ago, a generally accepted value for κ was 0.4, estimated on the basis of an early version of the Gell-Mann - Goldberger - Thirring (GGT) sum [3] rule, which we shall discuss in some detail later.

We shall now proceed to give a review of the different approaches to understand the size of κ.

III. Nuclear Compton Scattering and the Derivation of Sum Rules

Our basic quantity of interest is the amplitude for forward Compton scattering from nuclei, $F(\omega)$, shown pictorially in Figure 4, up to second order in the photon field.

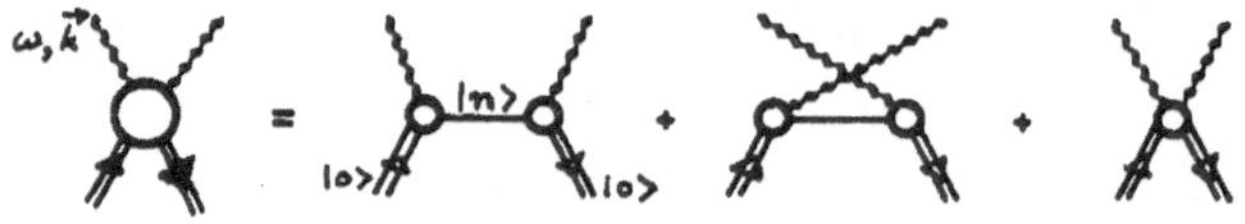

Figure 4: Amplitude for nuclear Compton scattering. First two terms involve all the excited states of the system. Third term is the gauge contribution.

The first two terms involve intermediate excitations of the system. The third (gauge, or seagull) term carries a contact interaction and is purely real. Following well known developments, we have

$$F(\omega) = \sum_n \frac{|\langle o|\mathcal{E}_\mu J^\mu(o)|n,\vec{k}\rangle|^2}{E_n - E_o - \omega - i\delta} + \sum_n \frac{|\langle o|\mathcal{E}_\mu J^\mu(o)|n,-\vec{k}\rangle|^2}{E_n - E_o + \omega + i\delta} + seagull. \quad (5)$$

Here J^μ is the total (nucleonic and mesonic) electromagnetic current of the system. The sum goes over all possible excited states $|n\rangle$, including mesonic degrees of freedom, and E_n are the corresponding energies. In the ground state $|0\rangle$, the nucleus is assumed to be at rest. The photon polarisation vector is denoted by ε_μ, and we shall usually use the transverse gauge, $\varepsilon_\mu = (0, \vec{\varepsilon})$. (For more details, see Refs.[4,5]).

The low energy limit of $F(\omega)$, up to second order in the frequency ω, is determined by the Thomson scattering (which survives as $\omega \to 0$) and by the electric polarisability α_E and the susceptibility χ_M [6,7] :

$$F(\omega) \simeq -\frac{(Ze)^2}{AM} + \left[\frac{1}{3}\frac{Ze^2}{AM}\langle r^2\rangle + \alpha_E\right]\omega^2 + \chi_M \omega^2 + O(\omega^4) \qquad (6)$$

where
$$\chi_M = -\frac{1}{6}\frac{Ze^2}{M}\langle r^2\rangle + \frac{e^2}{6AM}\langle \vec{D}^2\rangle + \chi_{para}$$

(diamagnetic terms)

(Here $\vec{D}$ is the usual dipole operator). The paramagnetic susceptibility χ_{para} is a quantity of special interest, since this is probably the place where mesonic effects might show up most clearly.

The amplitude $F(\omega)$ satisfies some general conditions, like causality, crossing symmetry and unitarity, which are the basis for the construction of dispersion relations. In particular, unitarity gives the optical theorem,

$$\sigma(\omega) = \frac{4\pi}{\omega} \mathcal{I}m\, F(\omega). \qquad (7)$$

From Eq.(5), we have

$$\sigma(\omega) = \frac{4\pi^2}{\omega} \sum_n |\langle 0|\vec{\varepsilon}\cdot\vec{J}|n,\vec{k}\rangle|^2 \delta(E_n - E_0 - \omega). \qquad (8)$$

1) Dipole Sum Rules

Let us now make contact with the familiar photonuclear dipole sum rule. In the long wavelength limit, electric dipole transitions dominate, and we have

$$\vec{J} = e\,[H, \vec{D}], \tag{9}$$

where $\vec{D}$ is the dipole operator, which, for a system of point nucleons, is given by

$$\vec{D} = \sum_{i=1}^{A} t_3(i)\,\vec{r}_i \tag{10}$$

(t_3 is the z-component of isospin, $\vec{r}_i$ is the position of the i-th nucleon), and H is the nuclear Hamiltonian. From there it is simple to show that

$$\Sigma_o^D = \int_o^\infty d\omega\,\sigma_D(\omega) = 2\pi^2 e^2\,\langle o|[D_z,[H,D_z]]|o\rangle \tag{11}$$

$$= \frac{NZ}{A}\,S\,(1 + \kappa_D),$$

$$\kappa_D = \frac{A}{NZ}\,M\,\langle o|[D_z,[V,D_z]]|o\rangle, \tag{12}$$

where $V = \frac{1}{2}\sum_{ij} V_{ij}$ in terms of the nucleon-nucleon potential V_{ij}. We realize in Eq.(12) that $\kappa_D \neq 0$ only if V is velocity dependent and/or contains terms like $t_+(i)t_-(j)$ which do not commute with the dipole operator.

In deriving the TRK sum rule $\Sigma_o^D(\infty)$, we have obviously introduced rather drastic assumptions. In Eq.(9), the complicated dynamical structure of currents has been reduced, in terms of the dipole operator $\vec{D}$, to a statement about (effective) charges localized at the positions of the nucleons. This is more or less just the essence of Siegert's theorem. Mesonic degrees of freedom are appearing only through the exchange of charged mesons in the (usually static) nucleon-nucleon potential V. Thus there is no a priori reason to expect that κ_D should coincide with the measured enhancement κ. Nevertheless, for theoretical reasons, it is interesting to find out what κ_D would be, on the basis of nuclear models.

Further sum rules that can be derived in the dipole approximation are

$$\sum_{-1}^{\mathcal{D}} = \int_0^\infty \frac{d\omega}{\omega} \; \sigma_{\mathcal{D}}(\omega) = 4\pi^2 e^2 \langle 0 | \mathcal{D}_z^2 | 0 \rangle , \qquad (13)$$

and

$$\sum_{-2}^{\mathcal{D}} = \int_0^\infty \frac{d\omega}{\omega^2} \; \sigma_{\mathcal{D}}(\omega) = 4\pi^2 e^2 \sum_n \frac{|\langle 0 | \mathcal{D}_z | n \rangle|^2}{E_n - E_0} = 2\pi^2 \alpha_\varepsilon . \qquad (14)$$

Both the measured $\sum_{-1}$ and $\sum_{-2}$ are quite well understood in terms of Eqs.(13-14) using conventional nuclear models, because of the dominance of the giant dipole resonance. We shall not go into details, but refer to Ref.[8] for a discussion. Finally, we note that the paramagnetic susceptibility, in the dipole approximation, is given by

$$\chi_{para} = 2 \sum_n \frac{|\langle 0 | \vec{m} | n \rangle|^2}{E_n - E_0} ,$$

where $\vec{m}$ is the magnetic dipole moment operator. We mention that the isovector part of $\vec{m}$ carries the same quantum numbers as a p-wave pion. Consequently, the intermediate states $|n\rangle$ will, for example, directly involve the virtual excitation of the 3.3 resonance inside a nucleus. Thus, from the point of view of mesonic degrees of freedom, the paramagnetic susceptibility is a particularly interesting quantity to look at.

2) The GGT Sum Rule

Starting from the properties of unitarity, crossing symmetry and causality, a dispersion relation can be written for any forward Compton amplitude $F(\omega)$:

$$Re \; F(\omega) = F(0) + \frac{\omega^2}{2\pi^2} P \int_0^\infty d\omega' \frac{\sigma(\omega')}{\omega'^2 - \omega^2} . \qquad (15)$$

Using this relation for the nuclear amplitude, $F_{\gamma A}(\omega)$, and subtracting the same relation for A times the Compton amplitude for the elementary nucleon,

$$F_{\gamma N}(\omega) = \frac{Z}{A} F_{\gamma p}(\omega) + \frac{N}{A} F_{\gamma n}(\omega) ,$$

we obtain the GGT sum rule

$$\Sigma_o(m_\pi) = \int_o^{m_\pi} d\omega \; \sigma_{\gamma A}(\omega) = \frac{NZ}{A} S(1+\kappa), \qquad (16)$$

$$\kappa = \frac{A}{NZ} S^{-1} \int_{m_\pi}^\infty d\omega \left[A \sigma_{\gamma N}(\omega) - \sigma_{\gamma A}(\omega) \right]. \qquad (16a)$$

The strong assumption that went into (16) is that the limit

$$\lim_{\omega \to \infty} \left[F_{\gamma A}(\omega) - A F_{\gamma N}(\omega) \right] = 0 \qquad (17)$$

is approached rapidly enough. This means that the nuclear electromagnetic response at asymptotic energies behaves as if the nucleus were a collection of A free nucleons. This assumption is wrong, as we shall discuss later. For the moment, let us first understand the physical picture behind Eq.(16). Basically, the GGT sum rule relates the enhancement κ of Σ_o below pion threshold to physical meson photoproduction for $\omega \geqslant m_\pi$. Such a relation becomes intuitively clear, since the direct coupling of a photon to exchange currents below pion threshold can be interpreted as virtual meson photoproduction processes inside the nucleus. On the other hand, virtual and real pion photoproduction are connected to each other, because the dynamical mechanism is basically the same, by the analyticity properties of the Compton amplitude as a function of frequency ω, which have gone decisively into the derivation of Eq.(16).

The following limiting situation is worth noting: suppose that meson photoproduction does not take place. From our previous discussion, this would also imply that exchange currents cannot be realised. Let us take $m_\pi \to \infty$ to simulate such a situation. Then, in fact,

$$\Sigma_o(\infty) = \int_o^\infty d\omega \; \sigma_{\gamma A}(\omega) = \frac{NZ}{A} S, \qquad (18)$$

i.e. the classical TRK sum holds, without ever mentioning the dipole approximation. The reason is that higher multipoles cancel against retardation effects, as has been proven in various different ways in the literature. It is not correct, however, to conclude that this is still valid if exchange currents are present (see the discussion in Ref.[9]); there is no obvious way to relate the $\kappa_{\mathcal{D}}$ of Eq.(12) to the κ of Eq.(16a).

IV. Survey of Dipole Sum Rule Calculations

Let us now review the calculations that have been performed on the dipole sum enhancement

$$\kappa_{\mathcal{D}} = \frac{A}{NZ} M \langle o | [\mathcal{D}_{\mathcal{z}}, [V, \mathcal{D}_{\mathcal{z}}]] | o \rangle . \tag{19}$$

Given the nuclear ground state wave function, the size of $\kappa_{\mathcal{D}}$ then depends on the precise form of the nucleon-nucleon potential. To be more specific, if this potential is local, then the non-vanishing contribution will come from those parts of the nucleon-nucleon interaction which are proportional to $\vec{\tau}_{1} \cdot \vec{\tau}_{2}$. Such pieces involve the exchange of charged bosons, like the pion and the ρ meson.

First simple estimates of $\kappa_{\mathcal{D}}$ have been carried out in Refs.[10, 11]. For example, Arima et al.[10] used the one-pion exchange (OPE) potential for V and a simple scheme to construct the nuclear ground state wave function. They arrived at $\kappa_{\mathcal{D}} \sim 1.4$ and observed that most of this came from the tensor part of OPE. Furthermore, they found that $\kappa_{\mathcal{D}}$ is quite insensitive to the behaviour of nucleon-nucleon dynamics at very short distances ($r < 1$ fm). These features can be readily understood, since the radial structure of the operator in Eq.(19) is essentially $r^2 V(r)$, and consequently cuts off short distances, but enhances intermediate ranges, which are governed by the tensor force. We mention this because these features are qualitatively common to all such investigations.

The most elaborate calculations of this kind have been performed by Fink et al. [12] , and recently by Hebach et al. [13] . They combine Brueckner-Hartree-Fock and Bethe-Goldstone equations in order to generate their ground state wave functions and then study κ_D for different nucleon-nucleon potentials. Most recent results are listed in Table 2.

	RSC	HJ
^{2}H	0.24	0.27
^{4}He	0.50	0.49
^{12}C	0.49	0.50
^{16}O	0.62	0.65
^{28}Si	0.53	0.57
^{32}S	0.55	0.60
^{40}Ca	0.60	0.68

Table 2: Values of the dipole enhancement κ_D , as calculated by Grassau and Hebach [13] , for different nucleon-nucleon potentials: Reid Soft Core (RSC) and Hamada-Johnston (HJ).

These results are remarkable in two respects. First, κ_D is smaller than the measured κ by roughly a factor of 2 for most nuclei. Second, κ_D is almost independent of nuclear mass number (with the exception of the deuteron, of course), which also disagrees with the trends in Fig. 3. Further improvements of the calculations do not help. For example, three-particle correlations, treated according to Ref. [14], would lead to an increase of κ_D by not more than 10 % [13].

Recalling the strong assumptions that went into the derivation of the dipole sum rule, the situation as such is not so much of a surprise. The fact that the action of mesonic currents has been reduced to the static properties of the exchange parts in the nucleon-nucleon force is a drastic simplification of meson dynamics in nuclei. It is clear that,

at photon energies of the order of 100 MeV, not far away
from pion production threshold, the virtual pions can
eventually carry a large amount of energy, which brings
them closer to their mass shell and therefore effectively increases the range of the one-pion exchange interaction. This aspect (among others) is completely missed
in the long-wavelength limit and Siegert's theorem.

In essence, we cannot claim to understand, on a
microscopic basis, the measured value of κ . Part of it
is certainly coming from static exchange forces, but a
large fraction (and, in particular, the A-dependence of κ)
still calls for an explanation.

V. Dispersion Relation Sum Rule Considerations

A brief look at experimental data in the high energy region
reveals that the assumption Eq.(17), on which the GGT
dispersion relation sum rule has been primarily based,
cannot be true. Instead of $\sigma_{\gamma A} = A \, \sigma_{\gamma N}$ at very high energy,
the data show a behaviour

$$\sigma_{\gamma A}(\omega) = A_{eff}(\omega) \, \sigma_{\gamma N}(\omega) , \tag{20}$$

with
$$A_{eff} \simeq A^{0.9}, \tag{21}$$

almost independent of energy, in the region between 5 GeV
and 20 GeV, as shown in Figure 5.

This phenomenon, the socalled shadowing effect, is well
understood in terms of the primarily hadronic nature of
photon interactions at very high energies. In fact, at
energies larger than about 5 GeV. the photon reveals its
hadronic components [15] . The photon decays into virtual
hadronic intermediate states. At such high energies, these
hadronic states live long enough so that they have plenty of
time to interact strongly with the nucleons in the nucleus.

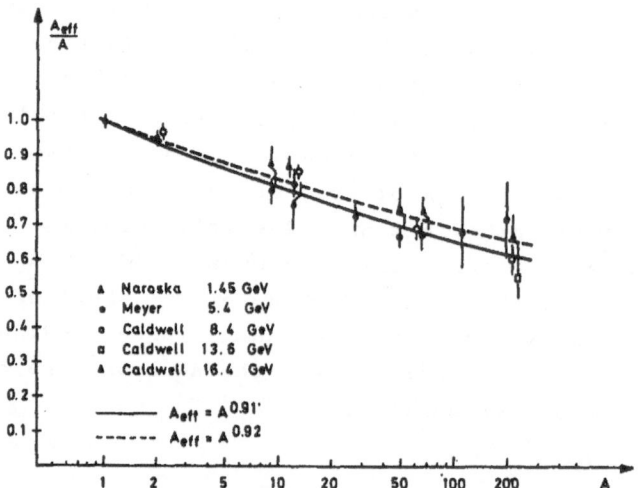

Figure 5: Effective nucleon number A_{eff}, as measured in various photonuclear experiments at high energies (for references see Ref. [15]).

The most important intermediate hadrons are the vector mesons, and among those, the ρ meson dominates. Photon-nucleus scattering between 5 GeV and 20 GeV can then be viewed as vector meson scattering the transition of the photon into vector mesons. Due to the strength of the hadronic interaction, the incoming photon beam then scatters essentially from surface nucleons, so that the effective number of nucleons taking part in the scattering process, A_{eff}, is smaller than A.

Clearly, this hadronic shadowing effects has to be incorporated into the dispersion relation sum rule. This need a more complicated subtraction procedure, from which we obtain the following generalized form of the GGT sum rule [16] :

$$\Sigma_o(m_\pi) = \frac{NZ}{A} S (1 + \kappa) \tag{22}$$

$$\kappa = \frac{A}{NZ} \left\{ \left(\frac{A_{eff}(\infty)}{A} - 1 \right) Z + \frac{R+I}{S} \right\} \tag{23}$$

$$R = 2\pi^2 \left[A_{eff}(\infty) \, \text{Re} \, F_{\gamma N}(\infty) - \text{Re} \, F_{\gamma A}(\infty) \right] \tag{24}$$

$$I = \int_{m_\pi}^{\infty} d\omega \left[A_{eff}(\infty) - A_{eff}(\omega) \right] \sigma_{\gamma N}(\omega) . \tag{25}$$

(This is a simplified version, assuming that R and I in Eqs.(24-25) remain finite. If this is not the case, a more refined form has to be used, as explained in detail in Ref. [16]). Let us now postulate the simplest possible behaviour of the asymptotic scattering amplitude, i.e. assume that "infinity" is actually reached at about 20 GeV. This need not be the case, of course, and different possibilities are discussed in Ref. [15]. To evaluate the quantity I of Eq.(25), we can then use the experimental $A_{eff}(\omega)$ between 2 GeV and 20 GeV, together with the measured $\sigma_{\gamma N}$. Data between $\omega = m_\pi$ and $\omega = 2$ GeV do not yet exist for nuclei, but we can parametrize the shadowing behaviour in the region of nucleon resonances by introducing an average shadowing exponent α :

$$A^\alpha = \frac{\int_{m_\pi}^{2 GeV} d\omega \, \sigma_{\gamma A}(\omega)}{\int_{m_\pi}^{2 GeV} d\omega \, \sigma_{\gamma N}(\omega)} , \tag{26}$$

a measurable quantity in principle. The quantity R of Eq.(24) can be estimated rather reliably using high energy multiple scattering models, with the known amplitude $F_{\gamma N}$ as an input. Once this is done, the resonance shadowing exponent α is determined by comparison with the measured values of $\Sigma_o(m_\pi)$. An example of such a consistency analysis is presented in Fig.6 and yields $\alpha \simeq 0.8$.

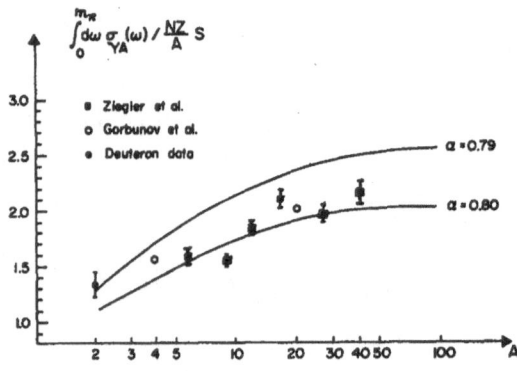

Figure 6:
Dispersion relation
analysis of the sum
rule, assuming that
the asymptotic domain
is reached at about
20 GeV. The quantity
is the average shadowing
exponent in the region
of nucleon resonances.
(taken from Ref. [16])

Clearly, the band of possible values for the parameter α
is very narrow, once the asymptotic limit for the photon-
nucleus amplitude is specified. Conversely, given α as
an experimental input, we would be in a position to put
strong constraints on the asymptotic behaviour of photon
cross sections.

Two features are remarkable in this investigation,
despite of the simplicity of asymptotic assumptions.
First, the proposed α is rather close to the average
shadowing behaviour observed in nuclear pion scattering
in the same region of energy. Second, the integral
consistency conditions implied by the dispersion relation
automatically yield the correct dependence of $\Sigma_o(m_\pi)$ on
nuclear mass number.

This, however, exhausts the possibilities of a
sum rule analysis based on dispersion relations. It
emphasises the strong consistency bounds between low and
high energy phenomena in a way such that the relations
between meson exchange currents at low energy and meson
photoproduction processes at high energies become evident.
But otherwise, we cannot learn more about the properties
of mesonic degrees of freedom, on a microscopic scale.

Nevertheless, it is certainly desirable to have
experimental photonuclear data covering the region of the
nucleon resonances, not only because the generalized GGT
sum rule demands such data as an input, but also in order
to study the collaboration between nuclear and nucleonic
degrees of freedom in their response to the photon field.

VI. <u>Summary</u>

In the course of these notes, we have emphasised that the enhancement κ of the integrated total photon-nucleus cross section $\Sigma_o(m_\pi)$ over the classical dipole sum contains integral information about explicit mesonic degrees of freedom in nuclei. The "classical" modes of excitation, like the giant dipole resonance, exhaust roughly one dipole sum, and an equal amount is left to explain, for all but the lightest nuclei, in terms of exchange current phenomena.

Two complementary approaches have been discussed. The dispersion relation sum rule of the GGT type draws global connections of analyticity between exchange currents and physical meson photoproduction, but is incapable of a microscopic description of the underlying mechanisms. The dipole sum rule calculations, on the other hand, are missing all but the low energy (static) aspects of the exchange current phenomenon. There is obviously an important link still missing. We are not yet able to understand, on a microscopic basis, the nature of exchange currents in the transition region just below and around the meson production threshold.

References

[1] J. Ahrens, H. Borchert, K.H. Czock, H.B. Eppler,
    H. Gimm, H, Gundrum, M. Krönig, P. Riehn, G. Sita Ram,
    A. Zieger and B. Ziegler, Nucl. Phys. A 251 (1975) 479

[2] D. Lüke and F. Söding, Springer Tracts in Mod. Phys.
    59 (1971) 39

[3] M. Gell-Mann, M.L. Goldberger and W. Thirring,
    Phys. Rev. 95 (1954) 1612

[4] J.L. Friar, Ann. of Phys. 96 (1976) 158

[5] P. Christillin and M. Rosa-Clot, Nuovo Cim. 28 A (1975) 29

[6] T.E.O. Ericson, in: Interaction Studies in Nuclei (Mainz)
    North-Holland (1975), and ref. therein

[7] J.L. Friar, Phys. Rev. Lett. 36 (1976) 510

[8] A.M. Lane and A.Z. Mekjian, Phys. Rev. C 8 (1973) 1981

[9] W. Weise, in: Interaction Studies in Nuclei (Mainz),
    North-Holland (1975), p. 679

[10] A. Arima, G.E. Brown, H, Hyuga and M. Ichimura,
     Nucl. Phys. A 205 (1973) 27

[11] W.T. Weng, T.T.S. Kuo and G.E. Brown, Phys. Lett.
     46B (1973) 329

[12] M. Fink, M. Gari and H. Hebach, Phys. Lett. 49B (1974) 2o

[13] H. Hebach, private communication

[14] M. Fink, M. Gari, H. Hebach and J.G. Zabolitzky,
     Phys. Lett. 51B (1974) 320

[15] W. Weise, Phys. Reports 13C (1974) 53

[16] W. Weise, Phys. Rev. Lett. 31 (1973) 773

Renzo Leonardi

Istituto di Fisica dell'Università and INFN, Bologna, Italy.
and
Facoltà di Scienze, Libera Università, Trento, Italy.

I. INTRODUCTION

The dipole operator is one among a variety of excitations carring a unity of isospin. The subject of these lectures is mainly concerned with studying isospin effects which are manifest in nuclear dipole excitation. In particular we discuss the way in which the centroid energies of the various fragments of the dipole excitation split among the various available isospin channels. As we shall see, the magnitude of the energy spacing between these fragments yields informations about the isobaric spin dependence of the effective nuclear forces and on the neutron and proton distribution in nuclei. Three types of transitions are under consideration in our case: those with $\Delta T_3 = 0$, leading to states in the same nucleus, and the charge transfer reactions with $\Delta T_3 = \pm 1$ leading to states in neighbouring nuclei. (See Fig. 1)

To study the variety of phenomena associated with the coupling of the isospin of the (isovector) dipole mode to that of the target we have to introduce an appropria te formalism and some definitions. This will be done in sec. II. In sec. III we will focus on the practical evaluations of quantities as isospin splitting energy of the giant resonance, differences on the neutron and proton radii etc. In sec. IV we para metrize the theory with isoscalar, isovector and isotensor effective interactions clea ring up the origins and the meaning of these interactions. In sec. V it is shown that isovector and isotensor energies emerges naturally from a schematic model approach to the problem.

II. FORMALISM

We adopt the following definitions:

Dipole operator $\qquad D_a = \sum_i z_i \, t_{ia}$

Bremsstrahlung dipole cross section:

$$\sigma_{-1} = \int \frac{\sigma^{(E)}}{E} \, dE = \frac{4\pi^2}{137} \sum_n |\langle n \mid D_3 \mid o \rangle|^2 =$$

$$= \frac{4\pi^2}{137} \langle o \mid D_3 \, D_3 \mid o \rangle \qquad\qquad (1)$$

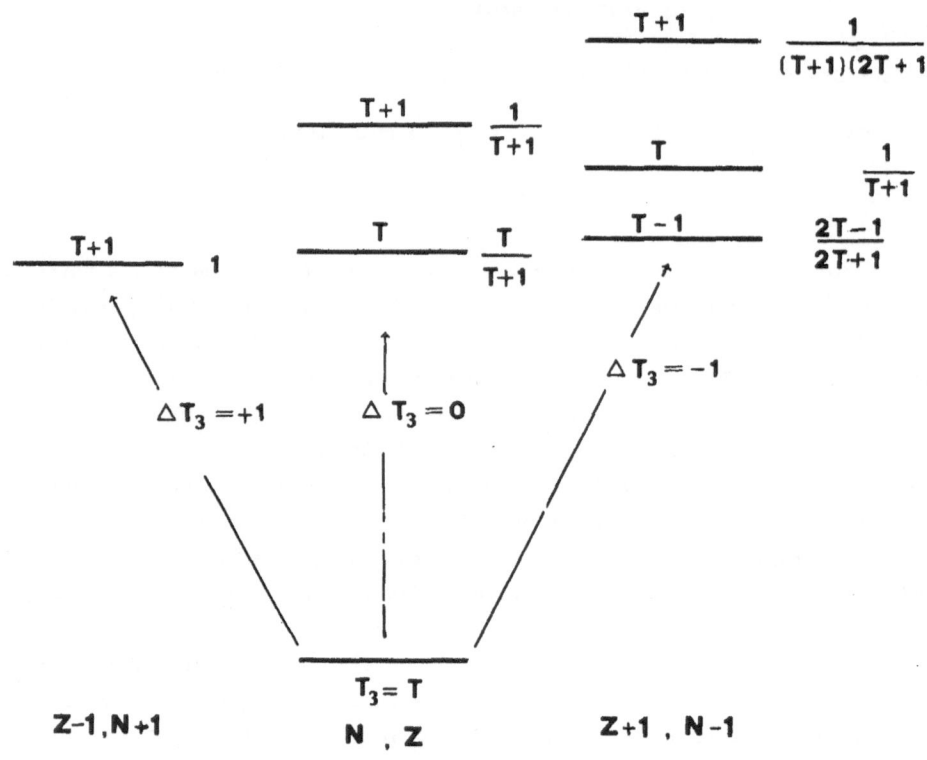

Fig. 1: The various isospin fragments exicited by a dipole excitation operator acting on a target with T_3 = T. There are also reported the various isogeometrical strenths.

Integrated dipole phto cross-section:

$$\sigma_o = \int \sigma(E)\,dE = \frac{4\pi^2}{137} \sum_n E_n \left|\langle n \mid D_3 \mid o\rangle\right|^2 = \frac{4\pi^2}{137} \langle o \mid D_3 \, H \, D_3 \mid o\rangle \quad (2)$$

Where H is the target hamiltonian and $H \mid n\rangle = E_n \mid n\rangle$ and E_o has been choosen conventionally zero.

Giant resonance:
$$E_{gr} = \frac{\sigma_o}{\sigma_{-1}} = \frac{\langle o \mid D_3 \, H \, D_3 \mid o\rangle}{\langle o \mid D_3 \, D_3 \mid o\rangle} \quad (3)$$

We develop now an appropriate formalism for taking into account the isospin quantum numbers.

Let be $\mid 0\rangle = \mid T, T_3\rangle$ and $\mid n\rangle = \mid n, T' \, T'_3\rangle$

Then the Projector operator on $\mid T' \, T'_3\rangle$ can be defined:

$$P_{T'T'_3} = \sum_n \mid n, T'T'_3\rangle \langle n, T', T'_3 \mid$$

and since D_3 is (the third component of) an isovector, then

$$D_3 \mid T,T_3 \rangle = \sum_{\substack{T'=T+1,T,T-1 \\ T'_3 = T_3}} P_{T'T'_3} \ D_3 \mid T \ T_3 \rangle$$

so that we can define cross-sections in a definite isospin channel:

$$\sigma_{-1}(T',T_3) = \frac{4\pi^2}{137} \ \langle T \ T_3 \mid D_3 \ P_{T',T_3} \ D_3 \mid T \ T_3 \rangle$$

$$\sigma_0(T',T_3) = \frac{4\pi^2}{137} \ \langle T \ T_3 \mid D_3 \ P_{T'} \ H \ P_{T'} \ D_3 \mid T \ T_3 \rangle$$

$$T' = T + 1, \ T, \ T - 1$$

Obviously

$$\sigma_{-1,0} = \sum_{T'} \ \sigma_{-1,0}(T', T_3)$$

One can eliminate the dependence on T_3 introducing the reduced cross-sections:[*]

$$\sigma_{-1,0}(T',T_3) = (2T'+1) \begin{pmatrix} T' & T & 1 \\ T_3 & -T_3 & 0 \end{pmatrix}^2 \sigma_{-1,0}^{T'} \qquad (4)$$

in particular for $T_3 = T$ nuclei

$$\sigma_{-1,0} = \frac{1}{T+1} \ \sigma_{-1,0}^{T+1} + \frac{T}{T+1} \ \sigma_{-1,0}^{T} \qquad (5)$$

The giant resonance energy in a defined isospin channel can now be defined as:

$$E_{T'} = \frac{\sigma_0(T',T_3)}{\sigma_{-1}(T',T_3)} = \frac{\sigma_0^{T'}}{\sigma_{-1}^{T'}} = \frac{\langle 0 \mid D_a \ P_{T'} \ H \ P_{T'} \ D_a \mid 0 \rangle}{\langle 0 \mid D_a \ P_{T'} \ D_a \mid 0 \rangle} \qquad (6)$$

The isospin geometry can be considered from another point of view.

We have just seen that

---

[*] Different conventions can be used in this respect: the one we use avoid as far as possible cumbersome factors all through the lectures.

$$\langle 0 | D_a D_b | 0 \rangle = \sum_{T'} \langle 0 | D_a P_{T'} D_b | 0 \rangle$$

so that we introduced the reduced cross-sections $\sigma_{-1}^{T'}$ in the channel T' ($T-1$, T, $T+1$). From another hand the following Identity holds:

$$2 D_a D_b = \frac{1}{3} \ \text{trace} \ \{D_a , D_b\} + [D_a , D_b] + \{D_a , D_b\} - $$

$$ - \frac{1}{3} \ \text{trace} \ \{D_a , D_b\} $$

These three pieces are the isoscalar, isovector and isotensor parts of $D_a D_b$ so that one is amened to define isoscalar, isovector and isotensor reduced cross-sections[x] in the following way:

$$\sigma_{-1}^{s} = \frac{4\pi^2}{137} \ \frac{1}{3} \ \sum_{a} \ \langle TT_3 | \ _a D_a D_a | TT_3 \rangle$$

$$\sigma_{-1}^{v} = \frac{4\pi^2}{137} \ \frac{1}{2T_3} \ \langle TT_3 | [D^- , D^+] | TT_3 \rangle \qquad (7)$$

(we use $[\tau^+, \tau^-] = -2\tau_3$ and $\frac{1}{2} \vec{\tau} = \vec{t}$)

$$\sigma_{-1}^{t} = \frac{4\pi^2}{137} \ \frac{1}{2 (3T_3^2 - T(T+1))} \ \langle TT_3 | 3 D_3 D_3 - \sum_{a} D_a D_a | TT_3 \rangle \ .$$

for targets $| TT_3 \rangle$ with $T_3 = T$
one has

$$\sigma_{-1} = \sigma_{-1}^{s} + \frac{2}{3} \ T(2T-1) \ \sigma_{-1}^{t} \ .$$

Similarly one can separate out

$$(D_a H D_b) = (D_a H D_b) \ \text{isoscalar} \ +$$

$$(D_a H D_b) \ \text{isovector} \ +$$

$$(D_a H D_b) \ \text{isotensor}$$

---

[x] Once more different conventions can be used: the one we use here has the only merit to be in line with the one used in our original articles.

and define reduced integrated cross-sections:

$$\sigma_0^s = \frac{4\pi^2}{137} \quad \frac{1}{6} \quad \langle TT_3 | \sum_a \left[[D_a , H] , D_a \right] | TT_3 \rangle$$

$$\sigma_0^v = \frac{4\pi^2}{137} \quad \frac{1}{4T_3} \quad \langle TT_3 | \left\{ [D^- , H] , D^+ \right\} - \left\{ [D^+ , H] , D^- \right\} | TT_3 \rangle \tag{8}$$

$$\sigma_0^t = \frac{4\pi^2}{137} \quad \frac{1}{4 \, (3T_3^2 - T(T+1))} \cdot$$

$$\cdot \langle TT_3 | 3 \left[[D_3 , H] , D_3 \right] - \sum_a \left[[D_a , H] , D_a \right] | TT_3 \rangle$$

These two ways of making the isospin geometry are connected by a proper iso-geometrical transformation involving 6-j coefficients:
one easily proves the follwoing:

$$\begin{pmatrix} \sigma_{-1,0}^{T-1} \\[2em] \sigma_{-1,0}^{T} \\[2em] \sigma_{-1,0}^{T+1} \end{pmatrix} = \begin{pmatrix} 1 & T+1 & -\frac{(2T+3)(T+1)}{3} \\[2em] 1 & 1 & \frac{(2T-1)(2T+3)}{3} \\[2em] 1 & -T & -\frac{T(2T-1)}{3} \end{pmatrix} \begin{pmatrix} \sigma_{-1,0}^{s} \\[2em] \sigma_{-1,0}^{v} \\[2em] \sigma_{-1,0}^{t} \end{pmatrix} \tag{9}$$

if one define $\quad x = T'(T' + 1) - T(T + 1) - 2 \quad (T' = T + 1, T, T - 1)$ then the second column of the matrix is

$$- \frac{x}{2} \quad \text{and the third column is} \quad \frac{4}{3} T (T + 1) - \frac{1}{2} x (x + 1).$$

In this way the dipole cross-section for the different isospin channels T' can be expressed through the six quantities $\quad \sigma_{-1,0}^{s,v,t}$ The quantities characterize comple_tely the isospin structure of the giant resonance, namely the relative intensity of the $T + 1$ fragment and its location in energy.
The relative intensity of the $T + 1$ fragment can be expressed as

$$\frac{\sigma_{-1} (T+1)}{\sigma_{-1}} \qquad \text{and the energy splitting as}$$

$$\Delta E^+ = \frac{\sigma_0^{T+1}}{\sigma_{-1}^{T+1}} - \frac{\sigma_0^T}{\sigma_{-1}^T} = E_{T+1} - E_T.$$

Explicitly we obtain:

$$\Delta E^+ = \frac{(T+1) \, E_{gr} \left(\dfrac{\sigma_{-1}^v}{\sigma_{-1}} \, K - \dfrac{\sigma_0^v}{\sigma_{-1}} \, K' \right)}{\left(1 + \dfrac{\sigma_{-1}^v}{\sigma_{-1}} \, K \right) \left(1 - T \, \dfrac{\sigma_{-1}^v}{\sigma_{-1}} \, K \right)} \qquad (10)$$

and

$$\frac{\sigma_{-1} (T+1)}{\sigma_{-1}} = \frac{1}{T+1} \left(1 - T \, \frac{\sigma_{-1}^v}{\sigma_{-1}} \, K \right) \qquad (11)$$

where

$$K = 1 + \gamma \qquad \gamma = \frac{(2T-1) \, \sigma_{-1}^t}{\sigma_{-1}^v} \qquad (12)$$

$$K' = 1 + \gamma' \qquad \gamma' = \frac{(2T-1) \, \sigma_0^t}{\sigma_0^v} \qquad (13)$$

In practice, since σ_0 and σ_{-1} can be considered known from experiments we have to calculate only four quantities; say $\sigma_{-1,0}^{v,t}$; furthermore, in the case of $T = \frac{1}{2}$ targets the situation is even more simple: in fact in this case the $T - 1$ channel does not exist $\left(\sigma_{-1,0}^{T-1} \equiv 0 \right)$ and as a geometrical counterpart $\sigma_{-1,0}^t \equiv 0$. In fact $\langle \frac{1}{2} |$ tensor of rank $2 | \frac{1}{2} \rangle \equiv 0$. As a consequence in the case of $T = \frac{1}{2}$ targets one has to calculate only two quantities namely σ_{-1}^v and σ_0^v.

III. NUMERICAL RESULTS.

Let first analyze the case of $T = \frac{1}{2}$ nuclei.

Working out $\sigma_{-1}{}^v$ one can obtain simply

$$\sigma_{-1}{}^v = \frac{\pi^2}{137} \quad \frac{1}{T_3} \quad \langle TT_3 | \sum_{i} z_i{}^2 \, \tau_{3i} \, | TT_3 \rangle$$

$$= \frac{N\langle r^2{}_n\rangle - Z\langle r^2{}_p\rangle}{2T} \cdot \frac{2\pi^2}{3} \quad \frac{1}{137} \simeq 0.048 \, \frac{3}{5} \, r_0{}^2 \, A^{2/3} \, \text{fm}^2$$

$$(14)$$

The calculation of $\sigma_0{}^v$ is more complicate and for the details of the calculation we refer back to Leonardi and Lipparini (Phys. Rev. C11, 2079, 1975). One can prove that

$$\frac{\sigma_0{}^v}{\sigma_0} \simeq \frac{2}{A} \quad . \quad \text{Combining this result with } \sigma_{-1} \simeq 0.035 A^{4/3} \text{ fm}^2$$

so that

$$\frac{\sigma_{-1}{}^v}{\sigma_{-1}} = 1.2 A^{-2/3}$$

One can easily calculate $\dfrac{\sigma_{-1}\,(3/2)}{\sigma_{-1}}$ and ΔE.

It is customary to write ΔE in the form

$$\Delta E = U \, \frac{(T + 1)}{A} \qquad :$$

from our previous formulae we obtain for U (in MeV) 24, 21, 35, 41, 50 for $A = 11, 13, 15, 25, 41$ respectively. It can be said that for light $T = \frac{1}{2}$ nuclei the symmetry energy U tend to fall to lower values in respect to the so called 60 MeV law. (This empirical law predict $U \simeq 60$ MeV). This behaviour has been recently nicely confirmed by experiments. (P. Paul and collaborators, Phys. Rev. C 12, 1410 (75) and C 11, 1008 (75)).

Let us now study the $T > \frac{1}{2}$ nuclei. Working our expression $\sigma_{-1}{}^t$ for $T = |T_3|$ nuclei we obtain:

$$\sigma_{-1}{}^t = \frac{2\pi^2}{137} \quad \frac{1}{3} \quad \frac{\left\langle \sum_{ij} (\vec{r}_i \cdot \vec{r}_j) \left(3\,(\tau_{3i}\,\tau_{3j}) - (\vec{\tau}_i \cdot \vec{\tau}_j) \right) \right\rangle}{4T\,(2T - 1)} \quad .$$

The operator $3\,(\tau_{3i} \cdot \tau_{3j}) - (\vec{\tau}_i \cdot \vec{\tau}_j)$ is twice the projection operator over neutron-excess-neutron-excess pairs[*] so that we can write

$$\sigma_{-1}^t = \frac{2\pi^2}{137} \quad \frac{1}{3} \quad \frac{\left\langle \sum_{ij} (\vec{r}_i \cdot \vec{r}_j) \right\rangle}{2T(2T-1)} \qquad i,j \text{ run over neutron excess only.}$$

$$= \frac{2\pi^{-2}}{137} \quad \frac{1}{3} \quad \left\langle \vec{r}_{ne} \cdot \vec{r}_{ne} \right\rangle \qquad (15)$$

where $\left\langle \vec{r}_{ne} \cdot \vec{r}_{ne} \right\rangle$ is the mean correlation $\left\langle \vec{r}_i \cdot \vec{r}_j \right\rangle$ averaged over the $2T(2T-1)$ neutron-excess pairs. It is interesting to compare σ_{-1}^t with σ_{-1}.

To do this comparison let us write σ_{-1} in the following way:

$$\sigma_{-1} = \frac{4\pi^2}{137} \left\langle D_3 D_3 \right\rangle = \frac{\pi^2}{137} \quad \frac{1}{3} \quad \left\langle \sum_{ij} \tau_{3i} \tau_{3j} (\vec{r}_i \cdot \vec{r}_j) \right\rangle$$

using now the fact that $\sum_i \vec{r}_i = 0$ (center of mass condition) we can write

$$\sigma_{-1} = \frac{\pi^2}{137} \quad \frac{1}{3} \left\langle \sum_{ij}' (\vec{r}_i \cdot \vec{r}_j)(\tau_{3i} \tau_{3j} - 1) \right\rangle$$

The operator $1 - \tau_{3i} \tau_{3j}$ is twice the projection operator over the neutron-proton pairs so that

$$\sigma_{-1} = -\frac{2\pi^2}{137} \quad \frac{1}{3} \quad \left\langle \sum_{ij} (\vec{r}_i \cdot \vec{r}_j) \right\rangle_{(i,j \text{ neutron-proton pairs})}.$$ Intro-

ducing now the mean correlation $\left\langle \vec{r}_n \cdot \vec{r}_p \right\rangle$ over the NZ n-p pairs we have finally:

$$\sigma_{-1} = -\frac{2\pi^2}{137} \quad \frac{2NZ}{3} \left\langle \vec{r}_n \cdot \vec{r}_p \right\rangle \qquad (16)$$

and

$$\frac{\sigma_{-1}^t}{\sigma_{-1}} = -\frac{1}{2NZ} \quad \frac{\left\langle \vec{r}_{ne} \cdot \vec{r}_{ne} \right\rangle}{\left\langle \vec{r}_p \cdot \vec{r}_n \right\rangle} \qquad (17)$$

We estimate $\left\langle \vec{r}_{ne} \cdot \vec{r}_{ne} \right\rangle$ and $\left\langle \vec{r}_p \cdot \vec{r}_n \right\rangle$ within a shell model. One has to remember at this purpose that the main part of the previous correlations are center of mass correla-tions so that shell-model can be used to obtain the order of magnitude of those corre-lations.

Let be $\vec{R}$ the c.m. coordinate in the (shell model) laboratory frame; and r'_i the laboratory coordinates.

---

[*] The isotensor term involves at least a pair of excess-neutrons, or equivalently T must be $> 1/2$.

Then $\vec{r}_i = \vec{r}'_i - \vec{R}$ and

$$\left(\langle \vec{r}'_{ne} \cdot \vec{r}'_{ne} \rangle - \langle \vec{R}^2 \rangle \right) = \langle (\vec{r}_{ne} \cdot \vec{r}_{ne}) \rangle$$

$$\left(\langle \vec{r}'_n \cdot \vec{r}'_p \rangle - \langle \vec{R}^2 \rangle \right) = \langle (\vec{r}_n \cdot \vec{r}_p) \rangle$$

It turns out that

$$\langle \vec{r}'_n \cdot \vec{r}'_p \rangle \text{ shell model } = 0$$

$$\langle \vec{r}'_{ne} \cdot \vec{r}'_{ne} \rangle \text{ shell model } = 0$$

(for more details see Leonardi Phys. Rev. C 14, 389 (1976))

so that
$$\frac{\sigma_{-1}^t}{\sigma_{-1}} = - \frac{1}{2NZ} \tag{18}$$

within a shell model. An experimental measure of $\langle \vec{r}_n \cdot \vec{r}_p \rangle$ is given by σ_{-1}. (See formula 16).

We know from experiments that $\sigma_{-1} \simeq 0.3A^{4/3}$ mb whereas (σ_{-1}) shell model =

$$= \frac{4\pi^2}{3} NZ \langle R^2 \rangle = \frac{1}{A} \frac{3}{2} \frac{\hbar}{M\omega} \simeq .36A^{4/3} \text{ mb}$$

so that $\langle \vec{r}_n \cdot \vec{r}_p \rangle$ shell model has the correct order of magnitude: on the light of these facts we expect that $\langle \vec{r}_{ne} \cdot \vec{r}_{ne} \rangle$ shell model has the correct order of magnitude too, so that the shell model estimate of $\sigma_{-1}^t / \sigma_{-1}$ should be reasonable.

We evaluate now σ_0^t compared to σ_0. (for details see Phys. Rev. C 14, 389 (1976).
Let us remember that

$$\sigma_0 = \frac{4\pi^2}{137} \frac{1}{2} \langle 0 | [D_3, [H, D_3]] | 0 \rangle$$

$$= \frac{4\pi^2}{137} \left(\frac{NZ}{2MA} + \text{exchange} \right)$$

The first term is model independent and corresponds to the commutators involving the Kinetic energy part and the isospin independent part of the potential and lead to the famous Thomas-Reiche-Khun result; the second term arises from the exchange part of the potential and numerically it turns out positive and of the order of magnitude of half the first term.

The quantity $\sigma_0 t$ can be calculated working out the commutators indicated in the previous section, formulae 8. The result can be putted in the form:

$$\sigma_0{}^t = - \frac{4\pi^2}{137} \left(\frac{1}{4MA} + \chi \frac{exchange}{2NZ} \right) \qquad (19)$$

Where the first term arises from commutators involving the kinetic energy and the second depends on the potential used; a part from the factor χ , the amount of "exchange" is exactly proportional to the exchange entering in σ_0 . It can be shown that χ ranges from 1 (for an Hamada-Jonston potential) to - 2 (for a Rosenfeld mixture) so that one finally obtains

$$\frac{\sigma_0{}^t}{\sigma_0} \lesssim \frac{1}{2NZ} \qquad (20)$$

With these results one can study the isospin splitting ΔE^+ ($= E_{T+1} - E_T$) for $T \gtrsim 1$ nuclei.

If one writes $\Delta E^+ = \frac{T+1}{A} U$ it turns out that U is 50 ÷ 70 MeV for various nuclei with $50 \leq A \leq 100$. For $A > 100, U$ tend to increase drastically with the previous inputs for $\sigma_{-1}{}^t$ and $\sigma_0{}^t$ unless one use $\langle r^2{}_n \rangle$ slightly smaller than $\langle r^2{}_p \rangle$. To obtain $U \lesssim 70$ MeV one has to use $\langle r^2{}_n \rangle^{1/2} - \langle r^2{}_p \rangle^{1/2} \simeq - 0.1 \div - 0.2$ fm for large T nuclei. (For details see for example R. Leonardi, in Proceedings of the Senday Conference 1972 pag· 443, (Editors Skoda and Ui)). This quite surprising result merits some furthermore comments. In fact it is interesting to point out that the difference $\langle r^2{}_n \rangle - \langle r^2{}_p \rangle$ can be studied within our isospin sum rules in a quite direct way? In fact, properly combining our three quantities σ_{-1} , $\sigma_{-1}{}^v$, and $\sigma_{-1}{}^t$, the following relation can be obtained:

$$\delta r = \frac{(1 - (T + 1) x) \sigma_{-1}}{\frac{4\pi^2}{3} \frac{1}{137} K \frac{N}{2} \langle r^2{}_p \rangle^{1/2}} - \frac{T}{N} \langle r^2{}_p \rangle^{1/2} \qquad (21)$$

where

$$x = \frac{\sigma_{-1}{}^{(T+1)}}{\sigma_{-1}} = \frac{1}{T+1} \frac{\sigma_{-1}{}^{T+1}}{\sigma_{-1}}$$

$$\delta r = \langle r^2{}_n \rangle^{1/2} - \langle r^2{}_p \rangle^{1/2}$$

and the approximation $\langle r^2{}_n \rangle + \langle r^2{}_p \rangle \simeq 2 \langle r^2{}_p \rangle$ has been used .

Furthermore $K = 1 + \gamma$ and $\gamma = - \frac{(2T-1) \sigma_{-1}{}^t}{\sigma_{-1}{}^v}$.

Since $\dfrac{\sigma_{-1}^{t}}{\sigma_{-1}}$ has been proved to be of the order of $-\dfrac{1}{2NZ}$ and since

$\dfrac{\sigma_{-1}}{\sigma_{-1}v}$ is of order of $\dfrac{2}{3} A^{2/3}$, one easily obtains the order of magnitude of γ :

$$\gamma \simeq + \frac{(2T-1) A^{2/3}}{3NZ} \lesssim 0.05 \qquad (22)$$

In rel. (21) the first and the second terms can be evaluated once x (i.e. the fraction of the T-upper giant resonance), σ_{-1} and $\langle r^2_p \rangle$ are known. In particular, since $x \geqslant 0$ one can write

$$\delta r \lesssim \frac{\sigma_{-1}}{0.097 \frac{KN}{2} \langle r^2_p \rangle^{1/2}} - \frac{T}{N} \langle r^2_p \rangle^{1/2} \qquad (23)$$

This upper limit is particularly interesting in the case of lead: in this case $\sigma_{-1} \simeq 270$ mb, $N = 126$, $\langle r^2_p \rangle \simeq 5.5$ fm: using $K = 0.95$ then

$$\delta r \lesssim - 0.15 \text{ fm} \quad \text{i.e.} \quad \langle r^2_n \rangle < \langle r^2_p \rangle \ !$$

(For further discussion see R. Leonardi, Phys. Letters 43B, 455 (1973)).

IV ISOSCALAR, ISOVECTOR, AND ISOTENSOR ENERGIES

In the previous sections we have seen that the isospin geometry involved in the isovector dipole excitation, from one hand suggest the introduction of reduced cross-sections in a definite isospin channel $\sigma_{-1,0}^{T'}$ from the other hand suggest to paramatrize the therory with isoscalar, isovector, and isotensor reduced cross-sections $\sigma_{-1,0}^{s,v,t}$. The first set of quantities $\sigma_{-1,0}^{T'}$ is connected to the second through a geometrical transformation involving $6-j$ coefficients (see ref. (9)). We have also introduced the quantities $E_{T'}$ (giant resonances in a definite isospin channel (see ref. 6)). In this section we will prove that as a geometrical counterpart of the quantities E_{T+1}, E_T, E_{T-1} it exists a set of quantities U_s, U_v, U_t which are in some sense "isoscalar, isovector and isotensor energies" connected to $E_{T'}$ by the same transformation connecting $\sigma^{T'}$ to $\sigma^{s,v,t}$. Let us remember that our "isovector dipole excitation" is a process in which a unity of isospin is adsorbed by a target. Let us distinguish between the "in" representation as the more convenient representation describing the system before the absorbtion of the isospin quantum, and the "out" representation as the more convenient for the system after the adsorbtion of the isospin quantum. Furthermore let be H_{in} the hamiltonian governing the evolution of the states before the adsorbtion of the quantum of isospin (in-states) and H_{out} the hamiltonian governing the states after the adsorbtion (out-states). It is obvious that formally [x]

$$H_{in} = H \text{ isospin field} + H \text{ target} + U \text{ interaction between target and field}$$

and

$$H_{out} = H_{target}$$

so that

$$\langle in| \ H_{in} \ |in\rangle = \langle in|H_{field}|in\rangle + \langle in| \ U \ |in\rangle +$$

$$\langle in| \ H_{target} \ |in\rangle \qquad (24)$$

let us now call $\langle in| \ H_{field} \ |in\rangle = \hbar\omega$ (it corresponds to the energy of the quantum of isospin) and let us assume as zero energy the energy of the target in its ground state, so that

$$\langle in| H_{target} |in\rangle = 0 .$$

The energy conservation imposes

$$\langle in| \ H_{in} \ |in\rangle = \langle out| \ H_{out} \ |out\rangle$$

and if the total isospin (quantum + target) is a good quantum number then the previous relation must hold in each isospin channel available, i.e.

$$\langle in| \ P_{T'} \ H_{in} \ P_{T'} \ |in\rangle = \langle out| \ P_{T'} \ H_{out} \ P_{T'} \ |out\rangle$$

where $T' = T+1, T, T-1$. (Remember that $\vec{T}' = \vec{T} \otimes \vec{\tau}$, $\vec{\tau}$ being a 3x3 matrix corresponding to the unity of isospin adsorbed and T the isospin of the target in its ground state).

Let us now indicate $P_{T'} \ |in\rangle$ as $|T', in\rangle$ and $P_{T'} \ |out\rangle$ as $|T', out\rangle$.

The "out" states are built simply by acting with the dipole operator D_3 on the target ground state $|TT_3\rangle$ and properly normalizing:

$$|out\rangle = \frac{D_3 \ |TT_3\rangle}{\langle TT_3 | D_3 \ D_3 \ |TT_3\rangle^{1/2}}$$

and projecting now on T'

$$|out\rangle = \frac{\sum P_{T'} \ D_3|TT_3\rangle}{\langle TT_3 | D_3 \ D_3 \ |TT_3\rangle^{1/2}} =$$

---

[x] It may be of some help the analogy with a nucleus in an electromagnetic field, adsorbing a unity say of energy and angular momentum (i.e. a photon).

$$= \sum_{T'} \left(\frac{\langle D_3 \, P_{T'} \, D_3 \rangle}{\langle D_3 \, D_3 \rangle} \right)^{1/2} \frac{P_{T'} \, D_3 \mid TT_3 \rangle}{(\langle TT_3 \mid D_3 \, P_{T'} \, D_3 \mid TT_3 \rangle)^{1/2}}$$

$$= \sum_{T'} \left(\frac{\langle D_3 \, P_{T'} \, D_3 \rangle}{\langle D_3 \, D_3 \rangle} \right)^{1/2} \mid \text{out, } T' \rangle$$

where

$$\mid \text{out, } T' \rangle = \frac{P_{T'} \, D_3 \mid TT_3 \rangle}{\langle TT_3 \mid D_3 \, P_{T'} \, D_3 \mid TT_3 \rangle^{1/2}}$$

and the coefficients $\dfrac{\langle D_3 \, P_{T'} \, D_3 \rangle}{\langle D_3 \, D_3 \rangle}$ gives the amount of $\mid \text{out, } T' \rangle$ states in the $\mid \text{out} \rangle$ state.

As a consequence

$$\langle \text{out, } T' \mid H_{\text{out}} \mid \text{out, } T' \rangle = \frac{\langle TT_3 \mid D_3 \, P_{T'} \, H \, P_{T'} \, D_3 \mid TT_3 \rangle}{\langle TT_3 \mid D_3 \, P_{T'} \, D_3 \mid TT_3 \rangle}$$

where H is the Hamiltonian of the target; the previous ratio is exactly what in our section II rel. 6 has been defined $E_{T'}$, i.e. the dipole giant resonance in the channel T'.

So finally we have obtained

$$\hbar\omega + \langle T', \text{in} \mid U \mid T', \text{out} \rangle = E_{T'} . \qquad (25)$$

Let us now study the isospin properties of U. Three isospin spaces are relevant in our analysis: the isospin space of the target $\vec{T}$, the isospin space of the isovector field $\vec{\tau}$ and the isospin space $\vec{T}'$ obtained by coupling $\vec{\tau}$ and $\vec{T}$, $T' = \vec{\tau} \otimes \vec{T}$. Since $\vec{\tau}^2 = \tau(\tau+1) = 2$ then $T' = T+1$, T, T-1. Note that isospin invariance demands only that T' is a good quantum number whereas neither $\vec{T}$ nor $\vec{\tau}$ must be conserved separately. (Analogy with $\vec{S}, \vec{L}, \vec{J}$ and the rotational invariance).

With this in mind let us come back to U.

a) U must be a scalar (SU_2 invariant) in the full isospin space $(\vec{T} + \vec{\tau})$

b) It must, at most, depend on the isospin of the target $\vec{T}$ and on the isospin $\vec{\tau}$ of the quantum .

So to construct U we must construct all the independent isoscalars from τ and T. Tensorial algebra tell us that three scalar S_λ can be built from τ and T .

$$S_\lambda = \left[\left[\vec{\tau} \otimes \vec{\tau} \right]^\lambda \otimes \left[\vec{T} \otimes \vec{T} \right] \right]^0$$

Where $\lambda = 0,1,2$ since $\left[\tau \otimes \tau \right]^{\lambda > 2} = 0 . (\tau^2 = 2 \, !)$

More explicitly one obtains:

$$S_0 \propto \text{Identity}, \quad S_1 \propto (\vec{\tau}.\vec{T}),$$

$$S_2 \propto -2 (\vec{\tau}.\vec{T})^2 - (\vec{\tau}.\vec{T}) + \frac{4}{3} T (T+1)$$

and U is a linear combination of S_λ :

$$= S_0 \, U_s + S_1 \, U_v + S_2 \, U_t \qquad (26)$$

Note that S_0 behaves like an isoscalar even under separate rotations in the space $\vec{\tau}$ and $\vec{T}$ whereas S_1 behaves like a vector under separate rotations and finally S_2 behaves like an isotensor under separate rotations. For this reason it is natural to call the coefficients of $S_{0,1,2}$ the isoscalar (U_s), isovector (U_v) and iso-tensor (U_t) potentials and finally

$$E_{T'} = \hbar\omega + U_s \langle T' | S_0 | T' \rangle +$$

$$U_v \langle T' | S_1 | T' \rangle + U_t \langle T' | S_2 | T' \rangle \qquad (27)$$

and the isospin properties of U have been separated out from the dynamics which is now entirely contained in the parameters U_s, U_v, U_t ; When the following convenctions are adopted:

$$S_0 = I , \quad S_1 = - (\vec{\tau}.\vec{T}) \quad S_2 = -2 (\vec{\tau}.\vec{T})^2 - (\vec{\tau}.\vec{T}) + \frac{4}{3} T (T+1)$$

then one has

$$\begin{pmatrix} E_{T+1} \\ \\ E_T \\ \\ E_{T-1} \end{pmatrix} = \begin{pmatrix} 1 & -T & -\dfrac{T(2T-1)}{3} \\ \\ 1 & +1 & +\dfrac{(2T-1)(2T+3)}{3} \\ \\ 1 & T+1 & -\dfrac{(T+1)(2T+3)}{3} \end{pmatrix} \begin{pmatrix} U_s + \hbar\omega \\ \\ U_v \\ \\ U_t \end{pmatrix} \qquad (28)$$

In particular

$$E_{T+1} - E_T = \Delta E^+ = (T+1) \left[-U_v - U_t (2T-1) \right] \qquad (29)$$

$$E_T - E_{T-1} = \Delta E^- = T \left[-U_v + U_t (2T+3) \right] \qquad (30)$$

Inverting the system one has

$$
\begin{pmatrix} \hbar\omega + U_s \\ \\ U_v \\ \\ U_t \end{pmatrix} = \begin{pmatrix} \dfrac{1}{3}\ \dfrac{2T+3}{2T+1} & \dfrac{1}{3} & \dfrac{1}{3}\ \dfrac{2T-1}{2T+1} \\[2ex] \dfrac{2T+3}{2(2T+1)(T+1)} & \dfrac{1}{2T(2T+1)} & \dfrac{2T-1}{2T(2T+1)} \\[2ex] -\dfrac{1}{2(2T+1)(T+1)} & \dfrac{1}{2T(T+1)} & -\dfrac{1}{2T(2T+1)} \end{pmatrix} \begin{pmatrix} E_{T+1} \\ \\ E_T \\ \\ E_{T-1} \end{pmatrix} \tag{31}
$$

For large T nuclei one has

$$
\hbar\omega + U_s \simeq \frac{1}{3} \ (E_{T+1} + E_T + E_{T-1})
$$

$$
U_v \simeq \frac{1}{2T} \ (E_{T-1} - E_{T+1})
$$

$$
U_t \simeq - \frac{1}{4T^2} \ \Big[(E_{T+1} - E_T) - (E_T - E_{T-1}) \Big]
$$

Incidentally we note that if one introduce

$$
E_{+-} = E_{T+1} \ , \quad E_{00} = \frac{1}{T+1} E_{T+1} + \frac{T}{T+1} E_T
$$

$$
E_{-+} = \frac{1}{(2T+1)(T+1)} E_{T+1} + \frac{1}{T+1} E_T + \frac{2T-1}{2T+1} E_{T-1}
$$

then one has exactly (for any T)

$$
U_s + \hbar\omega = \frac{1}{3} \ (E_{+-} + E_{00} + E_{-+})
$$
$$
U_v = \frac{1}{2T} \ (E_{-+} - E_{+-}) \tag{32}
$$
$$
U_t = \frac{1}{2T(2T-1)} \ \Big[(E_{00} - E_{+-}) - (E_{-+} - E_{00}) \Big]
$$

An interesting consequence can be drawn from this parametrization:
The term usually qualified as symmetry energy U entering in the formula of the isospin splitting energy

$$
\Delta E^+ = \frac{T+1}{A} \ U
$$

is in reality a sum of an isovector + isotensor energy $(\Delta E^+ = (T+1)\left[- U_v - U_t (2T-1)\right])$

It is obviously an important goal at this point to study $\dfrac{U_t (2T-1)}{U_v}$.

In the following sections we will discuss a theory predicting U_s, U_v and U_t.

V. ISOTENSOR ENERGIES AND SCHEMATIC MODEL

In the schematic model the energy of the giant dipole resonance is explained assuming that the nucleon-nucleon interaction behaves like a dipole-dipole interaction so that the energy shift of the giant resonance turns out proportional to the total strength of the dipole operator. If one want to take into account the isospin quantum numbers then one has to remember that the interacting nucleons of the target can adsorb a unity of isospin (τ) from the external isospin dipole field so that if i and j label two interacting nucleons; then $\vec{t}_i + \vec{t}_j$ is no longer conserved and SU_2 invariance requires simply that the effective isospin interaction is proportional to $(\vec{t}_i \cdot \vec{\tau}) \cdot (\vec{t}_j \cdot \vec{\tau})$ rather than simply to $(\vec{t}_i \cdot \vec{t}_j)$. With this in mind, the Dipole-Dipole Hamiltonian takes the form

$$H_D^{nm} = g \sum_i z_i z_j \tau_{im} \tau_m \tau_{jm} \tau_n \tag{33}$$

where g is the strength of the Dipole-Dipole interaction and $\vec{t}_i = \dfrac{1}{2} \vec{\tau}_i$. H^{nm} couples the space $\vec{T}$ and the space $\vec{\tau}$ thereby shifting the T' states and removing the degeneracy in respect to $T' = T + 1, T, T - 1$.

In the following we study quantitatively the amount of these shifts. The isospin part of the wave function of the system (isospin dipole field + target) is simply $|\tau \tau_3 \rangle | TT_3 \rangle$. This in turn can be decomposed into the

$| T'T'_3 (\tau T) \rangle$ basis as

$$|\tau \tau_3 \rangle | TT_3 \rangle = \sum_{T'} P_{T'} |\tau \tau_3 \rangle | TT_3 \rangle$$

where

$$P_{T'} = | T'T'_3 (\tau T) \rangle \langle T'T'_3 (\tau T)|$$

and

$$T' = T+1, T, T-1.$$

Let us call

$$\langle T'T'_3 (\tau T) | H_D^{33} | T'T'_3 (\tau T) \rangle = \Delta E_{T'}$$

since H^{33} is nothing but $g \frac{1}{4} D_3 D_3$

then we obtain the simple result (see sect. II)

$$\Delta E_{T'} = g \frac{137}{2\pi^2} \ 3 \ \sigma_{-1}^{T'} \tag{34}$$

and since we have seen in the previous sections that $\sigma_{-1}^{T'}$ can be expressed in terms of $\sigma_{-1}^{s,v,t}$ through relation (9) then we obtain that within a schematic model approach U_s, U_v, U_t are simply $g \frac{137 \times 3}{2\pi^2} \sigma_{-1}^{s,v,t}$.

The coupling constant can be fixed by the condition

$$E_{gr} = g \frac{137 \times 3}{2\pi^2} \sigma_{-1} + \hbar\omega$$

where

$$E_{gr} \simeq 80 \text{ MeV } A^{-\frac{1}{3}} , \quad \hbar\omega \simeq 40 \text{ MeV } A^{-\frac{1}{3}}$$

$$\sigma_{-1} \simeq 0.25 \ A^{4/3} \text{ mb}.$$

Using now our previous estimates for $\dfrac{\sigma_{-1}^v}{\sigma_{-1}}$ and

$\dfrac{\sigma_{-1}^t}{\sigma_{-1}}$ (rel. 14 and 18) we obtain

$$U_v = g \frac{137 \times 3}{2\pi^2} \ \sigma_{-1}^v \simeq + \frac{50}{A} \text{ MeV} \tag{35}$$

$$U_t = g \frac{137 \times 3}{2\pi^2} \ \sigma_{-1}^t \simeq - \frac{80}{A^{7/3}} \text{ MeV} \tag{36}$$

Let us remember that $\Delta E_{T'}$ are the collective shifts in a pure schematic model: to obtain the total centroid energy $E_{T'}$ one must add to $\Delta E_{T'}$ the unperturbed single particle energies ($\hbar\omega$) and the energy corrections which split neutron states from proton states. This last energy is referred as the Lane symmetry energy and derives from the familiar Lane potential $(\simeq \frac{100}{A} \text{ MeV})$

One finally obtains

$$E_{T+1} - E_T = (T+1) \left(\frac{100}{A} - U_v - (2T-1) \ U_t \right)$$

$$= {}^{(T+1)} \left(\frac{100}{A} \quad - \quad g \, \frac{137 \times 3}{2\pi^2} \, \sigma_{-1}{}^v \quad - \quad g \, \frac{137 \times 3}{2\pi^2} \, (2T-1) \, \sigma_{-1}{}^t \right)$$

$$\simeq {}^{(T+1)} \left(\frac{100}{A} \quad - \quad \frac{50}{A} \quad + \quad (2T-1) \, \frac{80}{A^{7/3}} \right) \text{MeV}$$

$$\simeq {}^{(T+1)} \left(\frac{50}{A} \quad + \quad (2T-1) \, \frac{80}{A^{7/3}} \right) \text{MeV} \tag{37}$$

We can conclude that from a schematic model approach an "isotensor" energy however *it* *emerges quite naturally;* turns out to be rather small since it is proportional to $\sigma_{-1}{}^t$.

The numerical results of the schematic model however cannot be taken too seriously. In fact we can prove easily that a small value of $\sigma_{-1}{}^t$ should correspond to a large value of U_t . (For details we refer to R. Leonardi, Phys. Rev. 14, 385 (1976)). In fact let us remember that from (31)

$$U_t \simeq - \frac{1}{4T^2} \left[(E_{t+1} - E_T) - (E_T - E_{T-1}) \right]$$

and

$$E_{T+1} - E_T = \Delta E^+ = \frac{\sigma_o{}^{T+1}}{T+1 \atop -1} - \frac{\sigma_o{}^T}{T \atop -1} = f^+ \left(E_{gr}, \sigma_{\theta,-1}{}^v, \sigma_{0,-1}{}^t \right)$$

and

$$E_T - E_{T-1} = \Delta E^- = \frac{\sigma_o{}^T}{\sigma{}^T_{-1}} - \frac{\sigma_o{}^{T-1}}{\sigma{}^{T-1}_{-1}} = f^- \left(E_{gr}, \sigma_{0,-1}{}^v, \sigma_{0,-1}{}^t \right)$$

Where f^+ and f^- are two functions of the quantities $E_{gr}, \sigma_{0,1}{}^v, \sigma_{0,-1}{}^t$: one can combine $\Delta E^{\ddagger}$ and ΔE^- to extract U_t in function of $E_{gr}, \sigma_{0,1}{}^v$ and $\sigma_{0,-1}{}^t$.

one obtains

$$- 4T^2 U_t \, \propto \, \frac{(2T-1) \, \Delta E^+}{T+1} \quad + \quad \frac{(2T-1)}{\sigma_{-1}{}^v} \, \sigma_{-1}{}^t \tag{38}$$

This relation, apart some factors of order unity omitted for semplicity, is an exact relation since it follows from the direct définition of U_t . This relation tell us that when $\sigma_{-1} t$ is small then $- 4T^2 U_t$ is of the order of $\triangle E^+$ itself, in contradiction with the finding of the schematic model.

In a more ambitious approach one should extract the isotensor isospin effects from a microscopic calculation. In the following we shall briefly comment on this possibility. Let us focus on the particle-hole excitation carrying the unity of isospin adsorbed from the dipole electromagnetic field. This pair ($\tau = 1$) couples with the excess neutrons T, to bring about the three isospin states $T' = T + 1$, T, T - 1. The total energy of the T' states is the sum of the umperturbed energy of the pair, the energy of the rest of the target and the interaction V between the target and the pair. One can prove that, when this interaction V is treated perturbatively, then, isotensor energies emerge from second order effects.

Let be t_h and t_p the isospin of the particle-hole pair and $t_{\alpha,\beta}$ the isospin of the remaining nucleons; then the following can be proved: isotensor effects emerge from second order corrections and can be expressed:

a) through an effective two-body p-h interaction, depending however on the direction of the isospin of the excess neutrons:

$$\text{isotensor} \propto K_{p-h} \left[(\vec{t}_p \cdot \vec{T})(\vec{t}_h \cdot \vec{T}) - \frac{1}{3} T (T+1)(\vec{t}_p \cdot \vec{t}_h) \right]$$

b) through an effective interaction affecting the nucleons of the target, depending however on the direction of the isospin $\vec{\tau}$ of the p-h pair

$$\text{isotensor} \propto \sum_{\alpha\beta} K_{\alpha\beta} \; (\vec{t}_\alpha \cdot \vec{\tau})(\vec{t}_\beta \cdot \vec{\tau}) - \frac{1}{3} \tau (\tau + 1)(\vec{t}_\alpha \cdot \vec{t}_\beta)$$

c) or finally through an effective interaction depending on the mutual direction of $\vec{\tau}$ and $\vec{T}$;

$$\text{isotensor} \propto U_t \; (\frac{1}{2} (\vec{\tau} \cdot \vec{T})^2 + \frac{1}{4} (\vec{\tau} \cdot \vec{T}) - \frac{1}{3} T (T+1))$$

Furthermore, assuming a zero range approximation for the interaction $V(r_p - r_\alpha) = G_{\alpha p} \delta (r_p - r_\alpha)$ and $V(r_h - r_\alpha) = G_{\alpha h} \delta (r_h - r_\alpha)$ one has the following radial shapes for $K_{\alpha\beta}$, K_{ph} and U_t :

$$K_{h-p} \propto \frac{2 \, G_{\alpha h} \, G_{\alpha p}}{E} \; \rho_{en} \; (\vec{r}_p, \vec{r}_h)$$

$$K_{\alpha\beta} \propto \frac{2 \, G_{\alpha h} \, G_{\beta \beta}}{E} \; \rho \; (\vec{r}_\alpha, \vec{r}_\beta)$$

$$U_t \propto \frac{1}{E} \; G_{\alpha h} \; G_{\beta p} \int \rho_{en} \; (\vec{r}, \vec{r}') \; \rho \; (\vec{r}, \vec{r}') \; d\vec{r} d\vec{r}'$$

Where ρ_{en} is the density ditribution function of the neutron excess pair, calculated however in the point r_p and r_h ; ρ is the density distribution function of the target pairs and $\frac{1}{E}$ is an appropriate mean energy denominator arising from the treatment of the microscopic interaction to second order.

ELECTROMAGNETIC AND HADRONIC INTERACTIONS WITH

THE FEW-BODY SYSTEMS AT INTERMEDIATE ENERGIES

C. Ciofi degli Atti

Physics Laboratory, Istituto Superiore di Sanità
and
Istituto Nazionale di Fisica Nucleare, Sezione Sanità
Rome, Italy

Abstract. In this series of lectures a review will be presented on e-
lectromagnetic and hadronic interactions at intermediate energies with
the few-body systems, with the aim of establishing whether realistic
conventional two-body forces and the non-relativistic many-body Schroedinger equation can account for the experimental data and the extent
to which information on non conventional nuclear physics might be obtained. To this end the lectures are divided in the following Chapters:
1. Introduction and motivation. The Gordian knot of Nuclear Physics;
2. The two-body system and its interaction with electrons, photons and
hadrons; 3. The three-body system and its electromagnetic interactions;
4. The four-body system and heavier nuclei; 5. Summary and conclusions.

1. Introduction and motivation. The Gordian knot of Nuclear Physics.

The "fundamental problem of Nuclear Physics", i.e. the calculation of nuclear properties in terms of the basic Nucleon-Nucleon (NN)
interaction, is at present limited to the following approach:

i) solve the non relativistic equation (m is the nucleon mass)

$$\left\{ \sum_{i} \frac{P_i^2}{2m} + \sum_{i<j} \upsilon(ij) \right\} \psi(\vec{r}_1 \cdots \vec{r}_A) = E \, \psi(\vec{r}_1 \cdots \vec{r}_A) \qquad (1.1)$$

with the "realistic" interaction $v(ij)$ taken from two-body bound
and scattering data (the correct solution of eqn (1.1) has the form

$$\Psi(\vec{r}_1 \cdots \vec{r}_A) = e^{i\vec{K}\vec{R}} \quad \Psi_{int}(\vec{\xi}_1 \cdots \vec{\xi}_{A-1}) \tag{1.2}$$

where $\vec{K}$ is the linear momentum of the nucleus, $\vec{R} = \sum_1^A \vec{r}_i / A$ and $\vec{\xi}_i$ is an intrinsic coordinate;

ii) use Ψ obtained from eqn (1.1) to calculate various bound and scattering properties of nucleon systems; assume that the discrepancies with the experimental data are due to the failure of the non relativistic approach (eqn (1.1)) and attempt to remove them by introducing new effects (e.g. baryonic resonances, many-body forces, mesonic exchange effects,etc.), in order to obtain information on "non conventional" nuclear physics.

However, even this restricted approach is faced with two severe difficulties: i) eqn (1.1) cannot be solved exactly, except for A = 2, and ii) different models for the realistic two-body interactions exist. For these reasons when some disagreement with the experimental data is found, it is very difficult to understand whether it is due:

a) to the failure of eqn (1.1) and, consequently, to the necessity of new effects not contained in it;

b) to a wrong model of the two-body interaction;

c) to the approximations introduced in order to solve eqn (1.1).

This very complicated situation has been recently called the Gordian knot of Nuclear Physics[1]. It is clear that the possibility to untie the Gordian knot strongly depends upon our ability to reduce the number of approximations introduced to obtain the solution of eqn (1.1), since only in this case is the removal of the discrepancies with the experimental data by means of model non conventional effects free from ambiguities.

In this series of lectures a review will be presented on electromagnetic and hadronic interactions with nuclei with the aim of establishing whether conventional two-body forces and the non relativistic Schroedinger equation can account for the existing experimental data, and the extent to which information on non conventional effects can be obtained. To this end our attention will be focused on:

1) the two-body system, for which eqn (1.1) can be solved exactly. It will be shown that the interaction of photons, electrons and protons with the deuteron (d) can be accounted for to a large extent by conventional nuclear physics, so that points a) and b) considered above might be useful in removing only some slight discrepancies with the experimental data available;

2) the three-body systems, for which eqn (1.1) can be very accurately solved. It will be shown that unlike the case of the two-body system, the solution of eqn (1.1) with conventional two-body forces leads to strong disagreements with the experimental data concerning both the static (e.g. binding energy) and dynamic (e.g. electron form factors, photodisintegration cross sections, etc.) properties. It appears, therefore, that for the three-body systems point a) and b) should be considered with particular attention;

3) the many-body system (A > 3); it will be emphasized, in this case, that the ambiguities arising from the approximate solution of eqn (1.1) may have considerable effect on various scattering processes, so that quantitative information concerning the importance of points a) and b) is very difficult to obtain.

2. The two-body system and its interaction with electrons, photons and protons.

2.1. Realistic interactions and the two-body system

The wave function of deuteron is

$$\Psi_d(\vec{r}) = \frac{u(r)}{r} \mathcal{Y}_{1M}^{01}\left(\frac{\vec{r}}{r}\right) + \frac{w(r)}{r} \mathcal{Y}_{1M}^{21}\left(\frac{\vec{r}}{r}\right) \qquad (2.1)$$

where $\vec{r}$ is the distance between the two nucleons and

$$\mathcal{Y}_{JM}^{LS}\left(\frac{\vec{r}}{r}\right) = \sum_{M_L M_S} \langle LM_L SM_S | JM \rangle Y_{LM_L}\left(\frac{\vec{r}}{r}\right) \chi_{SM_S} \qquad (2.2)$$

The radial functions are normalized in the usual way

$$U(o) = W(o) = O \qquad \int (U^2 + W^2) \, dr = 1 \qquad (2.3)$$

	Tabakin	HJ	RSC	dTS	Experiment
E_2(MeV)	-2.25	-2.269	-2.225	-2.224	-2.2245
Q (F^2)	0.104	0.285	0.279	0.262	0.2875
p_D (%)	3.2	6.97	6.47	4.43	3-9

Table 2.1 - Deuteron bound state properties calculated with various two-body interactions (Ref. 2). E_2 is the binding energy, Q the quadrupole moment and p_D the D wave probability (eqn (2.4)). HJ -Hamada-Jhonstone; RSC - Reid Soft Core; dTS - de Tourreil and Sprung (Ref. 2).

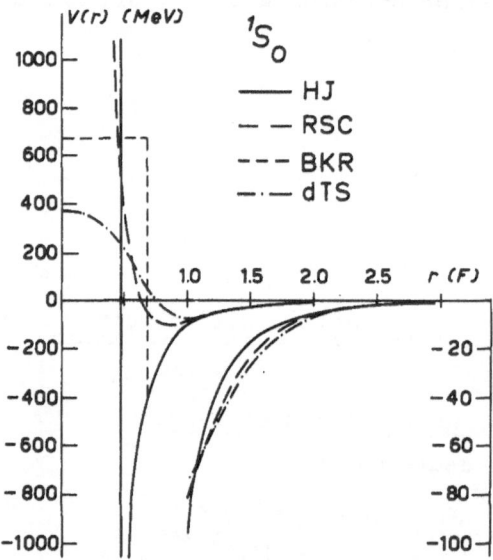

Fig. 2.1 - Radial shape of local two-nucleon interactions in the $^1S_0$ wave. HJ-Hamada-Johnstone; RSC-Reid Soft Core; BKR-Bressel -Kerman and Rouben; dTS-de Tourreil and Sprung (Ref. 2).

and the probabilities of "S and D waves" are

$$P_S = \int u^2 \, dr \qquad\qquad P_D = \int \omega^2 \, dr \qquad\qquad (2.4)$$

The properties of deuteron, calculated with some representative two-body realistic interactions yielding not equal, but comparable values for NN scattering data, are listed in Table 2.1.

In Figs. 2.1 and 2.2 the radial shape of the potential in the $^1S_0$ wave and the corresponding deuteron wave functions are shown. As can be seen, the little known features of the two-body interaction concern the short range behaviour and the D wave probability. Various attempts have been made to learn something about these two quantities by probing the two-body system with high energy particles.

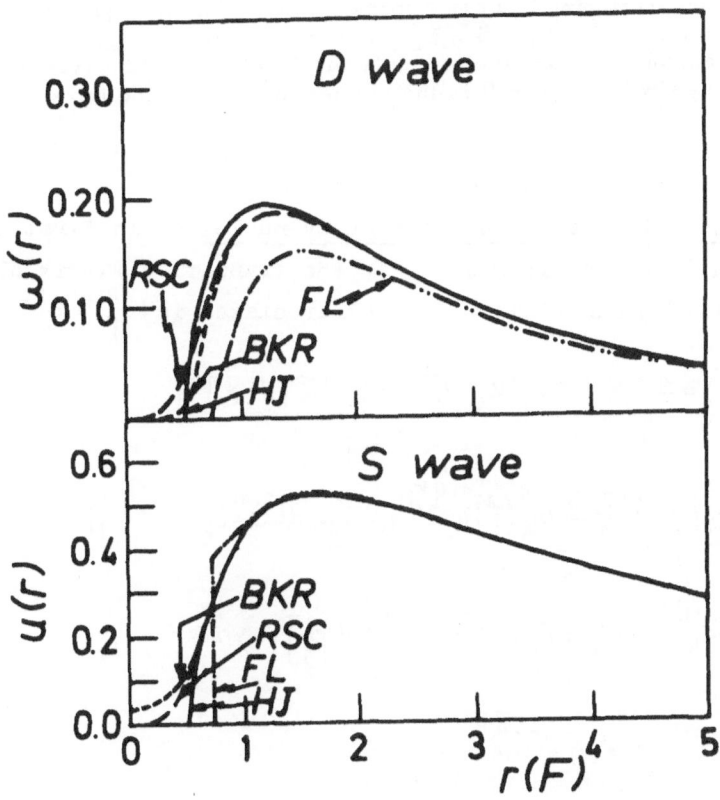

Fig. 2. 2 - Deuteron wave functions for various local interactions. FL-Feshbach and Lomon interaction (Ref. 7). For the other notations see Fig. 2. 1.

2.2. Elastic electron scattering by deuteron.

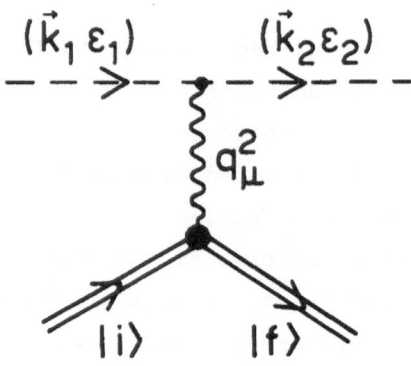

Fig. 2. 3 - Electron scattering in Born Approximation. $|i\rangle$ and $|f\rangle$ are initial and final nuclear states and $(\vec{k}_1, \mathcal{E}_1)$, $(\vec{k}_2, \mathcal{E}_2)$ the momenta and energy of the incident and scattered electrons, respectively; q_μ is the four momentum transfer $q_\mu^2 =$ $= q^2 - \omega^2 = 4\,\mathcal{E}_1\,\mathcal{E}_2\,\sin^2(\theta/2)$ where θ is the scattering angle in the lab system and $\vec{q} = \vec{k}_1 - \vec{k}_2$, $\omega = \mathcal{E}_1 - \mathcal{E}_2$.

Elements of electron scattering by nuclei[3]. In first Born approximation (depicted in Fig. 2.3), the transition matrix element between initial $|i\rangle$ and final $|f\rangle$ nuclear states is $(q_\mu^2 \equiv q^2)$

$$T_{fi} \equiv \langle f|H|i\rangle = \frac{4\pi e^2}{q^2}\left[\langle u_f|u_i\rangle Q - \langle u_f|\vec{\alpha}|u_i\rangle \vec{J}\,\right] \tag{2.5}$$

$$Q = \left(1 - \frac{q^2}{8m^2}\right)\langle f|\sum_j \left[\frac{1+\tau_3(j)}{2}\,G_E^p(q^2)e^{-i\vec{q}\vec{r}_j} + \frac{1-\tau_3(j)}{2}\,G_E^n(q^2)e^{-i\vec{q}\vec{r}_j}\right]|i\rangle \tag{2.6}$$

$$\vec{J} = \langle f|\sum_j \frac{1}{2m}\,G_E^{(j)}(q^2)\hat{e}(j)\left[\hat{\vec{P}}_j\,e^{-i\vec{q}\vec{r}_j} + e^{-i\vec{q}\vec{r}_j}\,\hat{\vec{P}}_j\right]|i\rangle +$$
$$\tag{2.7}$$
$$+ \langle f|\sum_j \frac{i}{2m}\,G_M^{(j)}(q^2)\hat{e}(j)e^{-i\vec{q}\vec{r}_j}(\vec{\sigma}_j \times \vec{q})|i\rangle$$

where

$$\hat{e}(j) = \begin{cases} \dfrac{1+\tau_3(j)}{2} & j - proton \\[4mm] \dfrac{1-\tau_3(j)}{2} & j - neutron \end{cases} \tag{2.8}$$

u_i and u_f are electron spinors, $\vec{\alpha}$ the Dirac operator $\hat{\vec{P}}$ and $\vec{\sigma}$ the Pauli momentum and spin operators. G_E and G_M are the Sachs charge and magnetic form factors of the nucleon, normalized in the following way

$$G_E^p(0) = 1 \qquad\qquad G_M^p(0) = M_p = 2.79278 \tag{2.9a}$$

$$G_E^n(0) = 0 \qquad\qquad G_M^n(0) = M_n = -1.91315 \tag{2.9b}$$

and related to the Pauli form factors F_1 and F_2 by

$$G_E = F_1 - \varkappa\frac{q^2}{4m^2}F_2 \qquad\qquad G_M = F_1 + \varkappa F_2 \tag{2.10}$$

Q (eqn(2.6)) includes the usual Coulomb interaction and the Darwin-Foldy term, whereas $\vec{J}$ (eqn (2.7)) includes the convection and spin current terms, respectively. The nuclear states $|i\rangle$ and $|f\rangle$ must be traslationally invariant , i.e. (cfr. eqn (1.2))

$$|\rangle \equiv \Psi_{JM}(\vec{r}_1 \cdots \vec{r}_A) = e^{i\vec{K}\vec{R}} \Psi_{JM}(\vec{\xi}_1 \cdots \vec{\xi}_{A-1}) \tag{2.11}$$

where the coordinates $\vec{r}_i$ refer to the laboratory system and the set of intrinsic A – 1 coordinates $\vec{\xi}_i$ may be chosen in the Jacobi form

$$\vec{\rho}_1 = \vec{r}_1 - \vec{r}_2 \equiv \vec{r} \qquad \vec{\rho}_2 = \frac{1}{2}(\vec{r}_1 + \vec{r}_2) - \vec{r}_3 \equiv \vec{\rho} \cdots \vec{\rho}_i = \frac{1}{i}\sum_{k=1}^{i}\vec{r}_k - \vec{r}_{i+1}$$

$$\vec{\rho}_{A-1} = \frac{1}{A-1}\sum_{k=1}^{A-1}\vec{r}_k - \vec{r}_A \tag{2.12a}$$

The set of A coordinates $\vec{\xi}_i$'s

$$\vec{\xi}_i = \vec{r}_i - \vec{R} \tag{2.12b}$$

may also be used, in which case the transition matrix element must contain a delta function $\delta\left(\sum_1^A \vec{\xi}_i\right)$ to ensure that only A-1 $\vec{\xi}$'s are linearly independent.

From eqn (2.5) the cross section for elastic scattering by a nucleus with Z protons can be obtained

$$\frac{d\sigma}{d\Omega} = \left(\frac{d\sigma}{d\Omega}\right)_{Mott} \frac{4\pi}{Z^2} \left\{ \sum_{\lambda=0}^{\infty} \frac{|\langle J \| M_\lambda^{Coul}(q) \| J \rangle|^2}{2J+1} + \right.$$

$$\left. + \frac{1}{2}\left(1 + 2tg^2\frac{\theta}{2}\right) \sum_{\lambda=1}^{\infty} \left[\frac{|\langle J \| T_\lambda^{El}(q) \| J \rangle|^2}{2J+1} + \frac{|\langle J \| T_\lambda^{Mag}(q) \| J \rangle|^2}{2J+1} \right] \right\}$$

(2.13)

where the Mott cross section from a point-like nucleus is

$$\frac{d\sigma}{d\Omega} = \frac{e^4}{4\mathcal{E}_1^2} \frac{\cos^2\frac{\theta}{2}}{\sin^4\frac{\theta}{2}} \left\{ 1 + \frac{2\mathcal{E}_1}{m_A} \sin^2\frac{\theta}{2} \right\}^{-1}$$

(2.14)

and M_λ^{Coul}, T_λ^{El} and T_λ^{Mag} are the Fourier transform of the charge and current operators and J the spin of the target nucleus. Due to angular momentum, parity and time reversal conservation, no electric multipoles appear and only even charge and odd magnetic multipoles contribute in elastic scattering.

Electron-Deuteron Scattering. In the case of deuteron (J = 1) only monopole ($\lambda = 0$) and quadrupole ($\lambda = 2$) charge scattering and magnetic dipole ($\lambda = 1$) scattering take place and the elastic cross section can be rewritten in the form

$$\frac{d\sigma}{d\Omega} = \left(\frac{d\sigma}{d\Omega}\right)_{Mott} \left\{ A(q^2) + B(q^2) tg^2\frac{\theta}{2} \right\}$$

(2.15)

where

$$A(q^2) = G_{ES}^2(q^2)\left\{ F_0(q^2)^2 + F_2(q^2)^2 \right\} + \frac{2}{3}\eta(1+\eta) F_M(q^2)^2$$

(2.16)

$$B(q^2) = \frac{4}{3}\eta(1+\eta)^2 F_M(q^2)$$

(2.17)

with $\eta = q^2/4m_d^2$ (m_d is the mass of deuteron). G_{ES} is the isoscalar electric form factor of the nucleon

$$G_{ES} = G_E^P + G_E^m \qquad (2.18)$$

and the charge monopole and quadrupole form factors are

$$F_0 = \int \{ u(r)^2 + \omega(r)^2 \} j_0(q\,r/2)\,dr \qquad (2.19)$$

$$F_2 = 2 \int \omega(r) \{ u(r) - \frac{1}{\sqrt{8}} \omega(r) \} j_2(q\,r/2)\,dr \qquad (2.20)$$

normalized according to $F_0(0)=1$ and $F_2(0)=q^2 Q/3\sqrt{2}$ where Q is the quadrupole moment of deuteron. F_M is the dipole magnetic form factor whose contribution, as shown in Fig. 2.4, is very small and will therefore be hereafter neglected.

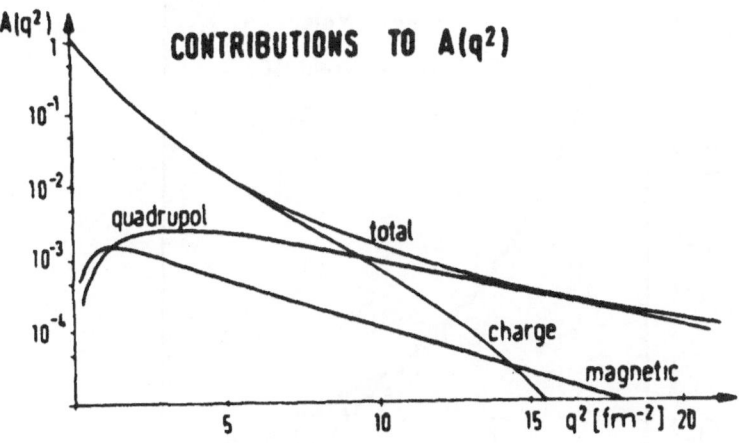

Fig. 2.4 - Contributions to the Deuteron form factor $A(q^2)$ from monopole (F_0), quadrupole (F_Q) and magnetic (F_M) scattering (see eqn (2.16)).

The elastic cross section is thus determined by the total charge form factor

$$A(q^2) = G_{ES}(q^2)^2 \; S(q^2)^2 \qquad (2.21)$$

where

$$S(q^2)^2 = F_o(q^2)^2 + F_2(q^2)^2 \qquad (2.22)$$

is called the body form factor of deuteron. Eqn (2.19)and (2.20) show that elastic electron scattering (in Born Approximation) probes the charge (and quadrupole) distributions of single nucleons measured from the Center of Mass.

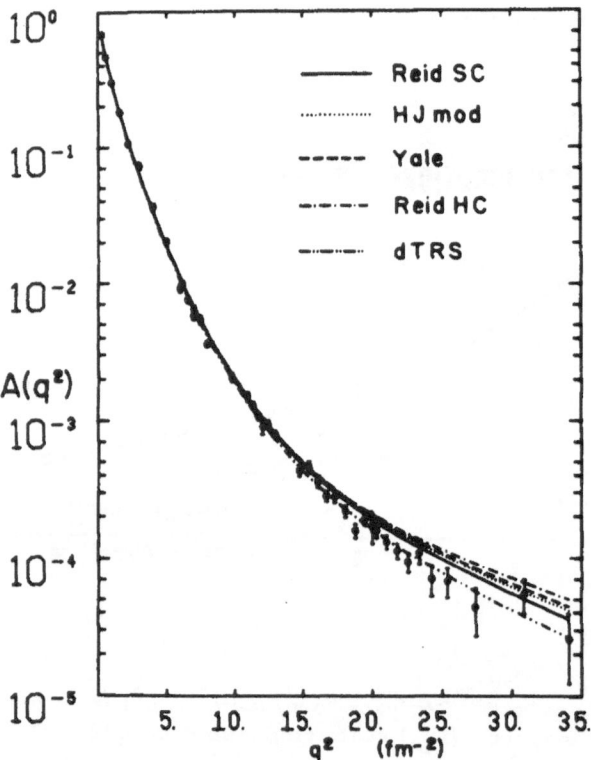

Fig. 2. 5 - Deuteron form factor calculated with various two-body interactions (from ref. 5).

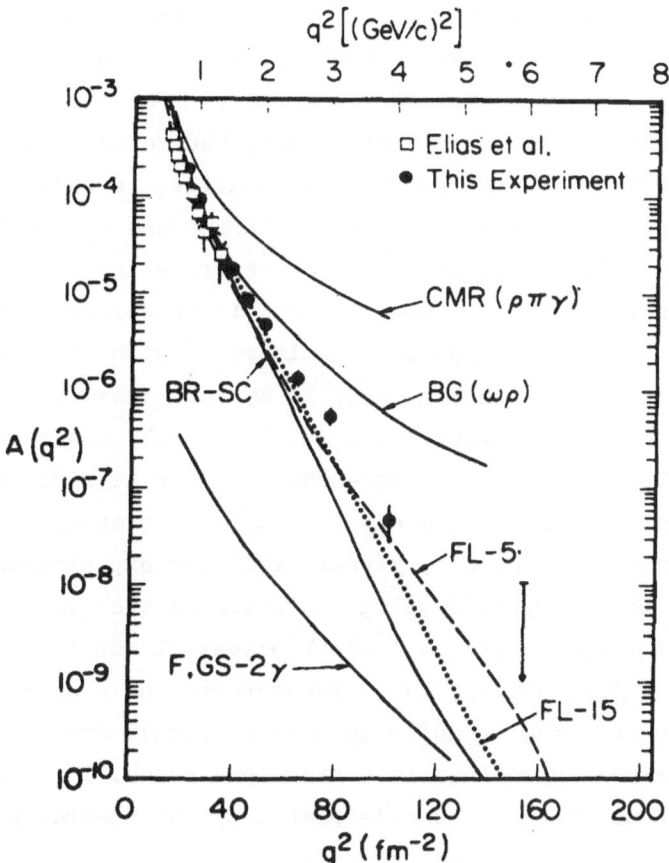

Fig. 2. 6 - Deuteron form factor at high momentum transfer (Ref. 6). The curves labelled BR-SC and FL correspond to the Reid Soft Core and Feshbach-Lomon interactions, respectively, and the curves labelled CMR and BC include the effects of some exchange current contributions (Ref. 13) (from ref. 6).

The position of the diffraction minimum of the monopole form factor F_0 is very sensitive to the short range part of the potential, whereas the quadrupole form factor at low momentum transfer depends upon the D state probability p_D. Unfortunately, the total form factor $A(q^2)$, which is the experimentally measured quantity, is governed at high momentum transfer by the quadrupole scattering and at low momentum transfer by the monopole scattering, so that separate information on the short range behaviour and D state probability is difficult to obtain. The deuteron charge form factor calculated [5] with various realistic interactions is shown in Fig. 2.5; it appears that different core behaviour and D state probabilities result in only small deviations at high momentum transfer. Measurements at very high momentum transfer have recently been carried out at SLAC[6]. A comparison with theoretical calculations (Fig. 2.6) reveals that the experimental data can satisfactorily be interpreted using conventional two-body forces having enough short range repulsion (the FL potential has a hard core radius of $\sim 0.7F$). Unfortunately, the interpretation of electron scattering at very high momentum transfer is very difficult due to the lack of reliable theoretical approaches to various phenomena. In this context, we would like to discuss the role played by the neutron form factor and the effects of exchange currents.

The neutron form factor. The proton charge form factor is determined from elastic electron-proton scattering and the experimental data can satisfactorily be reproduced by the so-called dipole form

$$G_E^p(q^2) = (1 + 0.056\, q^2)^{-2} \qquad \text{q in } F^{-1} \quad (2.23)$$

The neutron form factor G_E^n is instead obtained from elastic electron-deuteron scattering, therefore its form depends ultimately upon various assumptions on the deuteron wave function and on the effects of relativistic and exchange current contributions. As a result, the following models of G_E^n may be compatible with the experimental data [4]

$$-\mu\tau\, G_E^P$$

$$G_E^m(q^2) \quad = \quad
\begin{cases}
-\mu\tau\, G_E^P/(1+4\tau) \\
\\
-\mu\tau\, G_E^P/(1+\alpha\tau)
\end{cases}
\qquad
\begin{array}{l}
\tau = q^2/4m^2 \quad (2.24) \\
\mu = -1.91315
\end{array}$$

The form factor calculated with different G_E^n (and G_E^P corresponding to the dipole form) is shown in Fig. 2.7. It can be seen that the poor knowledge of the neutron form factor makes the interpretation of the high momentum data very ambiguous.

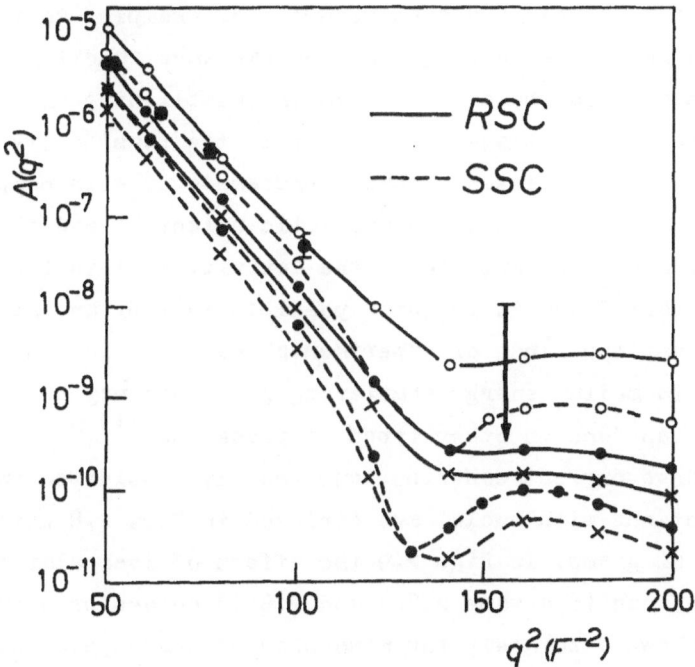

Fig. 2. 7 - Deuteron form factor calculated with two different interactions (SSC denotes dTS interaction) and using various models for the neutron form factor G_E^n. Crosses correspond to $G_E^n = 0$ open dots to $G_E^n = -\mu\tau G_E^P$ and black dots to $G_E^n = -\mu\tau G_E^P/(1+4\tau)$ (See eqns (2. 23) and (2. 24))

Effects of exchange currents. Assuming that nuclear force originate from meson exchange, then it is not permitted to consider the
interaction of photons (real or virtual) only with nucleons, but the
interaction with exchanged mesons should also be considered; in other
words, the external electromagnetic field couples not only with single
nucleon currents, but also with exchanged meson currents. A correct
solution of the "exchange current problem" is outside the range of
present day nuclear physics, since it would imply the solution of a
fully relativistic many-body equation. The problem is therefore tack-
led by means of approximate procedures consisting in the evaluation
of some exchange current contributions, which are believed to be the
dominant ones over a very large number of existing processes. The
treatment of exchange currents within this framework may suffer from
a certain lack of internal consistency, due, for example, to the coexistence of non relativistic (e.g. the nuclear wave function) with
relativistic aspects; to the difficulties in establishing the real
importance of the contributions considered; to the possibility that
these cancel with the neglected exchange processes or with competitive
effects such as various kinds of relativistic effects[8] and/or baryonic
resonance admixtures[9]. Nevertheless, the activity in this field is
developing vary fast[10] and it is quite possible that by an intelligent
comparison with various kinds of experimental data the role of exchange currents in medium energy scattering processes may be clarified,
as has already been done in other types of processes [11].

The exchange current contributions usually considered in electromagnetic interactions with nuclei are depicted in Fig. 2.8 using time
ordered Feynman diagrams. In Fig. 2.9 the effect of isoscalar exchange
current contributions (diagrams 2.8b) and 2.8c)) on deuteron isoscalar
form factor is shown separately for monopole and quadrupole scattering;
in Fig. 2.10 the total form factor is presented and compared with the
experimental data[12]. It can be seen that the exchange currents modify
the form factor at high momentum transfer, but the effect seems to be
of the same order of magnitude as the uncertainties on the neutron form
factor; moreover, at such high momentum transfer neglecting relativistic
effects is not justified. It seems therefore that the effect of exchange

currents on the charge form factor of deuteron is not large enough
to be independent of the assumptions made on other effects, whose
theoretical description is not yet clear (experimental data can in
principle be explained by a proper choice of the neutron form factor

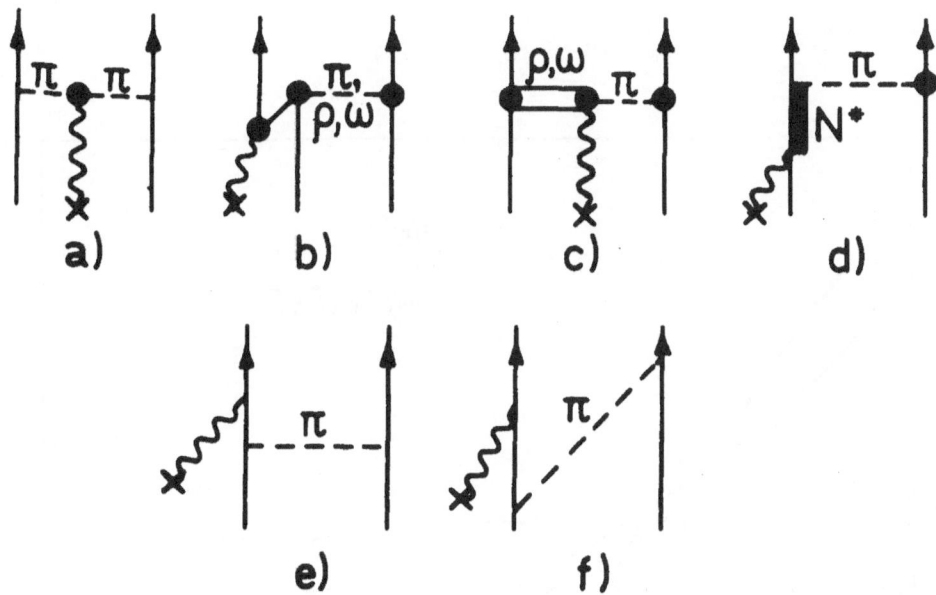

Fig. 2.8 - Some exchange current contributions considered in elastic
electron scattering by nuclei. In this figure upward directed lines denote
nucleons, wavy lines photons and the black dots the momentum distribution
(described by a form factor) of the corresponding vertex. The π exchange
diagram a) and the N* diagram d) have isovector character whereas the
nucleon-antinucleon diagram b) and the π-ρ (ω) exchange diagram c)
have both isovector and isoscalar parts. Thus diagrams a) and d) do not
contribute to the charge form factor of deuteron and other isoscalar nuclei,
whereas all of diagrams a), b), c) and d) have to be considered for nuclei
such as ^{3}He and ^{3}H. Diagram e), which represents the impulse approximation, is implicity included in the many-body wave function obtained from
realistic interactions having OPE tail. Diagram f), called "recoil diagram",
is not included in the many body wave function since it describes the interaction of the photon with a nucleon which is recoiling after emitting a meson
which is not yet absorbed by another nucleon. When the recoil diagram is
considered, the many body wave function has to be properly normalized
(Ref. 13). The inclusion of the recoil diagram has recently been questioned
(Ref. 12).

without invoking exchange currents). In conclusion, we believe that the experimental data on e-d charge elastic scattering neither call for, nor rule out exchange current contributions; only some previous models of exchange currents[13], based on a wrong coupling constant for the $\rho\pi\gamma$ vertex (Fig. 2.8c)) and predicting therefore a very large effect (cf. Fig. 2.6) were definitely excluded by the experimental data.

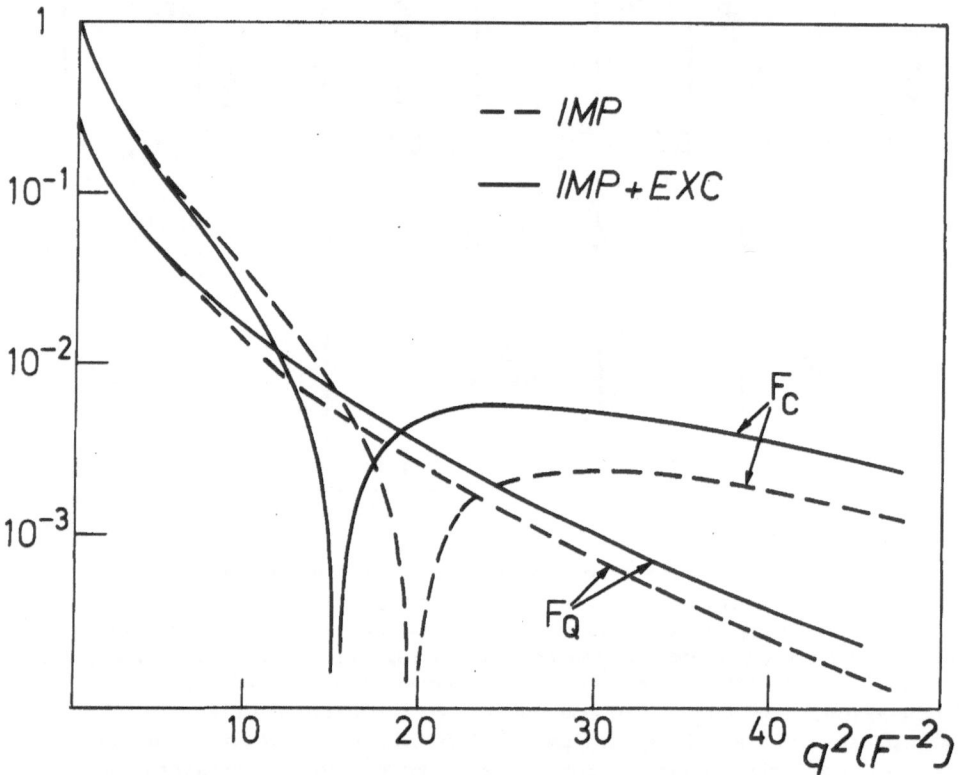

Fig. 2.9 - Monopole $F_C = G_{ES} F_0$ and quadrupole $F_Q = G_{ES}\, 3\sqrt{2}\, F_2/q^2$ form factor of Deuteron calculated without (IMP) and with (IMP+EXC) the effect of exchange current (diagrams 2.8b and 2.8c) using the RSC interaction (Adapted from ref. 12).

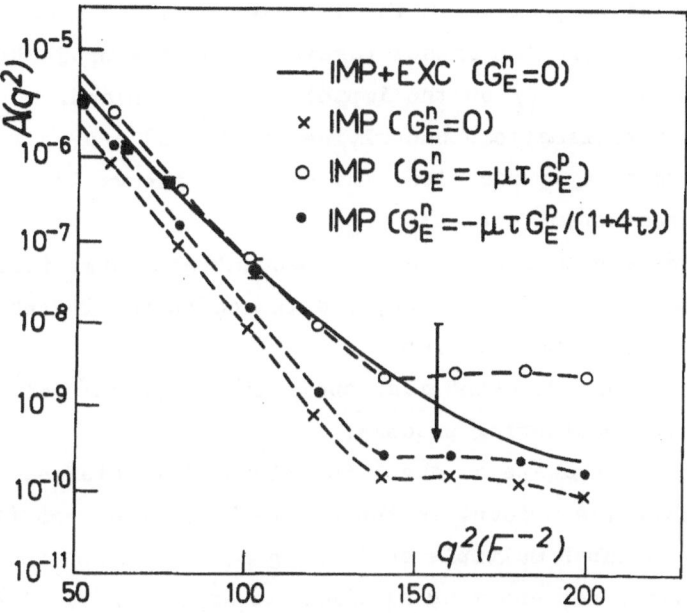

Fig. 2. 10 - Effect of the neutron form factor and the exchange currents on the Deuteron form factor using the RSC interaction (the curve labelled IMP+EXC is from ref. 12 and the experimental data are from ref. 6).

2.3. Elastic scattering of protons by deuteron.

Elements of Glauber 's theory. The theoretical analysis of medium energy hadron scattering by nuclei, is usually based on Glauber's theory[14] which leads to the following scattering amplitude

$$F_{fi}(q) = \frac{ik}{2\pi} \int e^{i\vec{q}\vec{b}} d^2b \left\{ \psi_f^*(\vec{\beta}_1 \cdots \vec{\beta}_{A-1}) \left[1 - \prod_{j=1}^{A}(1 - \Gamma(\vec{b} - \vec{\zeta}_j)) \right] \psi_i(\vec{\beta}_1 \cdots \vec{\beta}_{A-1}) d\tau \right\} \quad (2.25)$$

where Γ is the profile function, i.e. the Fourier transform of the elementary nucleon-nucleon scattering amplitude

$$\Gamma(\vec{b}) = \frac{1}{2\pi ik} \int e^{i\vec{q}\vec{b}} f(\vec{q}) d\vec{q} \quad (2.26)$$

k is momentum of the projectile in the lab system, q the momentum transfer and b and s_j are the impact parameter and the projection of the intrinsic coordinate ξ_j on the impact parameter plane.

The basic approximations underlying eqn (2.25) are:

i) the eikonal approximation, which limits the validity of the theory to small scattering angles;

ii) nucleons are considered as independent scatters, i.e. the potential associated to each nucleon, and the corresponding phase shifts they produce, do not overlap;

iii) nucleons are "frozen" i.e. their motion (Fermi motion) does not affect the scattering process;

iv) nucleons propagate on shell in intermediate states.

By expanding the product in the square bracket of eqn (2.25) $\prod(1-\Gamma)=1-\sum\Gamma+\cdots$, the Glauber multiple scattering series is obtained, the first term of which, representing single scattering contributions only, is equivalent to the Born Approximation of electron scattering.

Elastic proton-deuteron scattering. Placing eqn (2.26) in eqn (2.25) one obtains

$$F_{ii}(E\,\vec{q}) = F^{(1)}(E\,\vec{q}) + F^{(2)}(E\,\vec{q}) \tag{2.27}$$

where the single ($F^{(1)}$) and double ($F^{(2)}$) scattering terms are[14]

$$F^{(1)} = \int e^{i q r/2} \, \psi_d(r) \left\{ f_{pp}(E\,q^2) + f_{pn}(E\,q^2) \right\} \tag{2.28a}$$

$$F^{(2)} \propto f_{pp}(E\,q/2) \, f_{pn}(E\,q/2) \tag{2.28b}$$

where ψ_d is the deuteron wave function. As is well known (see for example Fig. 4 of Ref. 14), the low momentum part of the scattering amplitude is dominated by single scattering, whereas at high momentum transfer, due to the rapid fall off of the body form factor, the double scattering term becomes the most important. In Fig. 2.11 Glauber's theory is compared with the experimental data: there is no agreement at large momentum transfer, and particularly at backward angles where the theory is not reliable, being based on the assumption

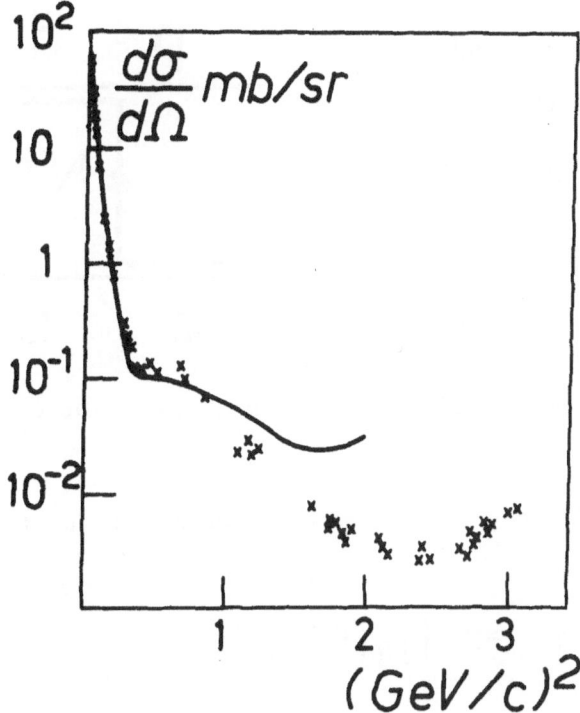

Fig. 2. 11 - Proton-Deuteron elastic scattering cross section calculated within Glauber's theory (from ref. 17. The experimental points are from ref. 15).

that the scattering angle be small. For the interpretation of the backward scattering, two approaches have been proposed: in the first, depicted in Fig. 2.12, a neutron pick-up (or exchange) by the incoming particle has been considered[16]: the effect produced was however not sufficient to explain the experimental data, so that the exchange of N* was added, and the N* content of deuteron estimated. Recently, an interpretation of the backward scattering has been obtained without invoking N* exchange, but simply improving the multiple scattering theory and adding neutron exchange[17]. The improvements with respect to Galuber's theory were:

i) the eikonal approximation (i.e. the limitation to small scattering angles) is removed and the Fermi motion considered. As a result of these improvements the energy at which the elementary

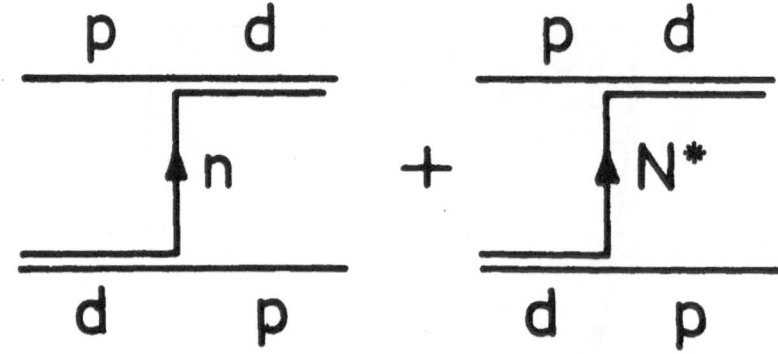

<u>Fig. 2.12</u> - Neutron and N* exchange in elastic proton-Deuteron scattering.

scattering amplitude has to be evaluated becomes a function of the momentum transfer

$$f_{pN}(E\,q^2) \Longrightarrow f_{pN}(E'(q), q^2) \tag{2.29a}$$

with

$$E'(q^2) = \frac{q^2}{4m^2} + \left(E - \frac{q^2}{4m_d^2}\right)\left(\frac{1 + q^2/4m^2}{1 + q^2/4m_d^2}\right) \tag{2.29b}$$

ii) off-shell propagation and the effect of the overlapping potentials are considered.

Using the scattering amplitude in the form (2.29) and recent information on the nucleon-nucleon scattering[18] has the important consequence that at high momentum transfer single scattering is much larger than double-scattering, which can therefore be neglected and the elastic cross section simply becomes

$$\frac{d\sigma}{d\Omega} \approx S(q^2)\left\{ f_{pp}(E'(q^2), q^2) + f_{pn}(E'(q^2), q^2) \right\} \tag{2.30}$$

where S^2 is the deuteron body form factor (eqn (2.21)).

Backward scattering at various energies calculated using the Hamada-Johnstone potential to generate the body-form factor and inclu-

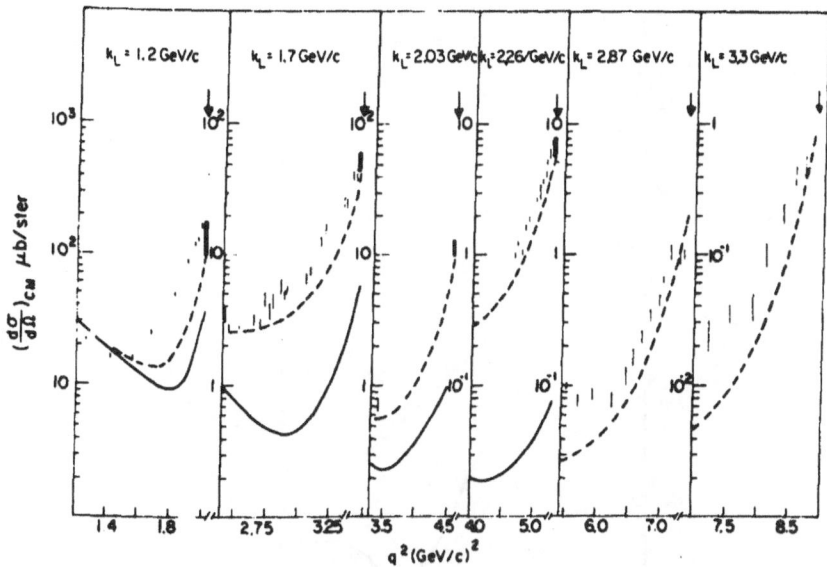

Fig. 2.13 - Backward p-d scattering calculated in ref. 17. Continuous lines correspond to a deuteron body form factor obtained from the HJ interaction (continuous line in Fig. 2.14) and dashed lines to a phenomenological body form factor chosen to fit the experimental data (dashed line in Fig. 2.14).

ding neutron pick-up is represented by the continuous lines in Fig. 2.13. Since the agreement with the experimental data is not very good, the authors of Ref. 17 introduced a phenomenological body form factor (shown in Fig. 2.14) in order to reproduce the experimental data without invoking N* exchange. The results are presented in Fig. 2.13 by the dashed lines. For a given choice of the neutron form factor, the chosen deuteron form factor does not contradict electron scattering data (Fig. 2.15); these, however, can also be interpreted in terms of body form factors obtained from conventional two-body interactions, if other choices for G_E^n are made. Moreover, it should be noted that the phonomenological $S(q)$ which reproduces p-d scattering data, does not seem to correspond to any of the nucleon-nucleon interactions usually adopted. Thus we think that before drawing final conclusions on the deuteron form factor, further improvements of the hadron-nucleus scattering theory and the effect of N* exchange should be considered, in order to establish whether the backward p-d experimental data can

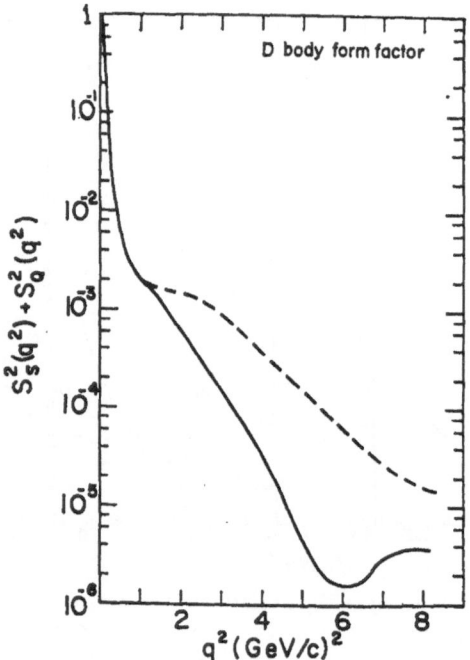

Fig. 2.14 - The body factor of deuteron (eqn (2.21)) corresponding to the HJ potential and to the phenomenological form chosen to reproduce the backward proton-deuteron experimental data (see Fig. 2.13) (from ref. 17).

be interpreted in terms of conventional body form factors of deuteron.

To summarize, it appears from the analysis of elastic electron and proton scattering by deuteron that a non relativistic approach in terms of conventional two-body forces seems to account for the experimental data up to relatively high momentum transfer, but, at the same time, it does not provide any new information on the unknown features of the two-body force; at very high momentum transfer, the Gordian knot manifests itself in all its intricacy: the effects from the neutron form factor, the exchange currents and relativistic corrections may become important, but since one of these processes can be studied only by making assumptions on the others, theoretical speculations become very ambiguous.

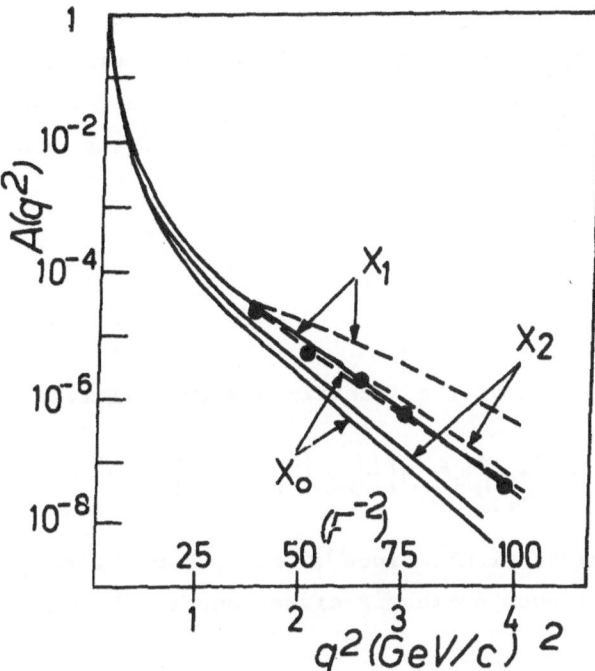

Fig. 2. 15 - Deuteron form factor (eqn (2. 21)) with legends as in Fig. 2. 14 calculated with different choices X_i of the neutron form factor G_E^n and the dipole choice for G_E^p. X_0 corresponds to $G_E^n = 0$, X_1 and X_2 to the first and third of eqn (2. 24) (

2.4. Deuteron disintegration

Elements of nuclear disintegration by medium energy particles. Those processes will be considered in which a projectile (electron, photon or hadron), interacting with a bound A – nucleon system, leads to a final state characterized asyntotically by a free nucleon and a recoiling residual (A–1) nucleon system (which may contain one or more particles in the continuum). The momentum and energy transfer are

$$\vec{q} = \vec{k}_N + \vec{k}_R \tag{2.31a}$$

$$\omega = \vec{k}_N^2 / 2m + (\vec{k}_N - \vec{q})^2 / (2 m_{A-1}) + (E_{A-1} - E_A) \tag{2.31b}$$

$$\approx q^2 / 2Am + \vec{k}_N'^2 / 2\mu + (E_{A-1} - E_A) \tag{2.31c}$$

$$E_{A-1} - E_A = m_{A-1} + m + E_{A-1}^* - m_A \tag{2.31d}$$

where m_A is the rest mass of the target, m_{A-1} the mass of the final $(A-1)$ -nucleon system, E^*_{A-1} its energy and the second term in the rhs of eqn (2.30b)(written in the lab system) its recoil energy; eqn (2.30c) has been obtained by introducing the momentum $\vec{k}'_N$ of the knocked-out nucleon relative to the residual nucleus

$$\vec{k}'_N = \vec{k}_N - \frac{1}{A}\vec{q} \qquad (2.31e)$$

and its reduced mass $M = (A-1)m/A$.

The calculation of the transition matrix element

$$T_{fi} = \langle \Psi^f_{A,\vec{k}'_N}(\vec{\beta}_1 \cdots \vec{\beta}_{A-1}) | H_{int} | \Psi^i_A(\vec{\beta}_1 \cdots \vec{\beta}_{A-1}) \rangle \qquad (2.32)$$

is in general very difficult since Ψ^i and Ψ^f should be respectively, the bound and continuum state eigenfunctions of the same A – body Hamiltonian

$$H|\Psi^i_A\rangle = \left[\sum \frac{p^2}{2m} + \sum v\right]|\Psi^i_A\rangle = E_0|\Psi^i_A\rangle \qquad E_0 < 0 \qquad (2.33a)$$

$$H|\Psi^f_{A\vec{k}'_N}\rangle = \left[\sum \frac{p^2}{2m} + \sum v\right]|\Psi^f_{A\vec{k}'_N}\rangle = E^f_{\vec{k}'_N}|\Psi^f_{A\vec{k}'_N}\rangle \qquad E^f_{\vec{k}'_N} = \frac{\vec{k}'^2_N}{2\mu} + E_{A-1} \qquad (2.33b)$$

Eqn (2.33a) can be solved exactly for $A \leqslant 3$, eqn (2.33b), besides the trivial case of $A = 2$, can be solved for $A = 3$ only for some special two-body interactions (separable interactions), whereas for $A > 3$ none of the two equations can be solved. However reasonable approximations can be adopted when the conditions for the _quasi elastic_ scattering[19]

$$\omega \gg B \qquad\qquad q \ll 1/a \qquad\qquad (2.34)$$

are fulfilled (B is the average nucleon binding energy in the target and a its average size). In this case, the emitted nucleon has a high kinetic energy and its motion can be treated in the distorted wave approximation, or even in the plane wave epproximation

$$\Psi^f_{A\vec{k}'_N} = \mathcal{A} \Psi^f_{A-1}(\vec{\beta}_1 \cdots \vec{\beta}_{A-2}) exp(i\vec{k}'_N \vec{\beta}_{A-1}) \qquad (2.35)$$

where $\hat{\mathcal{A}}$ is the antisymmetrisation operator and Ψ_{A-1} the solution of the Schroedinger equation for (A-1) nucleons, obtained with the same two-body force used to generate the A- body ground state. A quasi-elastic experiment may provide very useful information when the ejected nucleon is detected in coincidence with the scattered particle (electron, proton, etc.).

Using the Plane Wave Born Approximation (PWBA) and considering for ease of presentation only Coulomb interactions, the cross section for a quasi-elastic coincidence process induced by electrons reads

$$\frac{d\sigma_f}{d\varepsilon_p \, d\Omega_p \, d\varepsilon_2 \, d\Omega_2} = K \frac{d\sigma}{d\Omega} \frac{1}{2J_i+1} \sum_{M_i \alpha} \left| \langle \Psi_{A-1}^{\vec{k}_N}(\vec{p}_1 \cdots \vec{p}_{A-1}) | \sum_{j=1}^{A} \hat{e}_j \, e^{i \vec{q} \vec{r}_j} | \Psi_A^i (\vec{p}_1 \cdots \vec{p}_{A-1}) \rangle \right|^2$$

$$\cdot \delta(E - (E_{A-1} - E_A)) \tag{2.36}$$

where K is a kinematical factor α, stands for the magnetic numbers of all particles in the final state and the removal energy E, which is known in a coincidence experiment, is

$$E = \omega - \varepsilon_p - \varepsilon_R \tag{2.37}$$

with the help of eqn (2.35), considering that $\exp(i\vec{q}\vec{r}_j) = \exp[i\vec{q}(\vec{r}_A - \vec{R})] = \exp[i(A-1)\vec{q}\,\vec{p}_{A-1}/A]$ one has ($\vec{k} \equiv \vec{q} - \vec{k}_p$)

$$\frac{d\sigma_f}{d\varepsilon_p \, d\Omega_p \, d\varepsilon_2 \, d\Omega_2} = K \left(\frac{d\sigma}{d\Omega}\right)_{Mott} P(kE) \tag{2.38}$$

$$P(kE) = \frac{1}{2J+1} \sum_{M_i \alpha} |\rho_{fi}(\vec{k})|^2 \, \delta(E - (E_{A-1} - E_A)) \tag{2.39a}$$

$$\rho_{fi}(\vec{k}) = (2\pi)^{-3/2} \int d\vec{p}_{A-1} \, e^{i \vec{k} \vec{p}_{A-1}} \, \phi_{fi}(\vec{p}_{A-1}) \tag{2.39b}$$

$$\phi_{fi}(\vec{p}_{A-1}) = A^{1/2} \langle \Psi_{A-1}^f (\vec{p}_1 \cdots \vec{p}_{A-2}) | \Psi_A^i (\vec{p}_1 \cdots \vec{p}_{A-1}) \rangle \tag{2.39c}$$

$P(kE)$ is called the Spectral Function: it represents the probability that if a nucleon with momentum k were removed instantaneously from the target, the final state would remain in the excited state E. When the final-state interaction between the nucleon and the residual nucleus is taken into account, the simple picture we have just outlined is no longer valid, although a satisfactory description of the process can still be achieved in terms of the Distorted Wave Born Approximation.

Quasi elastic processes induced by hadrons can theoretically be described by the Plane Wave Impulse Approximation (PWIA), in which case the above equations for the cross section remain essentially the same with the Mott cross section replaced by the free hadron-nucleon cross section. An improved description of these processes can be obtained by using distorted waves for protons.

Deuteron disintegration. For deuteron, the overlap integral (eqn (2.39c)) is nothing but the two-body wave function and eqn (2.38) becomes

$$\frac{d\sigma}{d\Omega \, d\Omega_p \, d\varepsilon_2 \, d\Omega_2} = K \left(\frac{d\sigma}{d\Omega}\right)_{ep} \left\{ \rho_0(k) + \rho_2(k) \right\} \delta(E - 2.2245) \qquad (2.40)$$

where the S and D wave momentum distributions ρ_0 and ρ_2 are

$$\rho_0(k) = (2\pi^2)^{-1} \left| \int u(r) \, j_0(kr) \, r \, dr \right|^2 \qquad\qquad (2.41a)$$

$$\rho_2(k) = (2\pi^2)^{-1} \left| \int \omega(r) \, j_2(kr) \, r \, dr \right|^2 \qquad\qquad (2.41b)$$

The sensitivity of the momentum distribution upon the two-body potential is shown in Fig. 2.16. An appreciable difference, due almost entirely to the D wave, can only be seen at high momenta k, where the experiment is difficult to perform (the neutron-proton final state interaction will certainly change the high momentum part, nevertheless the qualitative features of the process remain unchanged). The existing, very limited experimental data on the d(e,e'p)n reaction, are shown in Fig. 2.17; as expected, they agree with a PWBA calculation based on a conventional model of the two-body force. The same agreement up to $k \sim 1 \text{ F}^{-1}$, appears in the analogous reaction d(p,2p) calculated within the framework of the PWIA (cf. Fig. 2.18). The disagreement at higher momenta is most likely due to the inadequacies of the impulse approximation and to the importance of multiple scattering and more complicated processes.

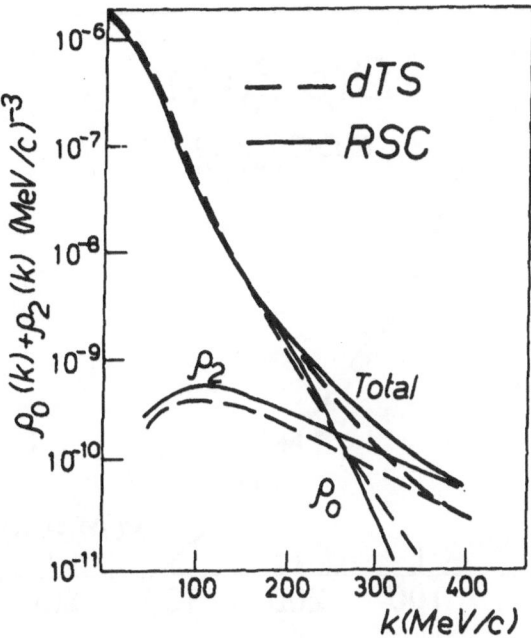

Fig. 2. 16 - Nucleon momentum distributions in Deuteron (eqns (2. 41)) calculated with two different nucleon-nucleon interactions.

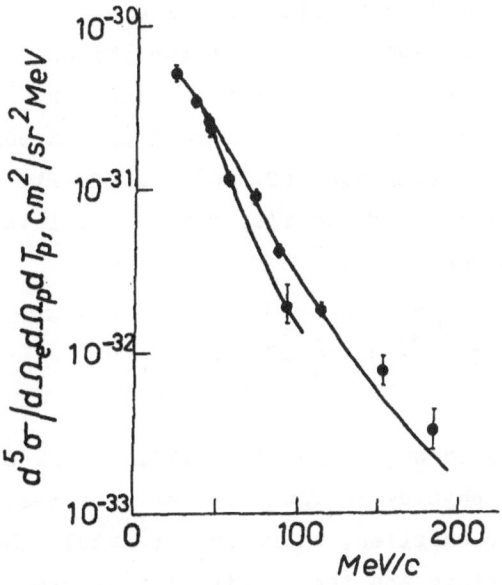

Fig. 2. 17 - Momentum distributions in Deuteron from the reaction d(e, e'p)n. Experimental points from Refs. 20 and 21 a. Theoretical curves from Ref. 21 b.

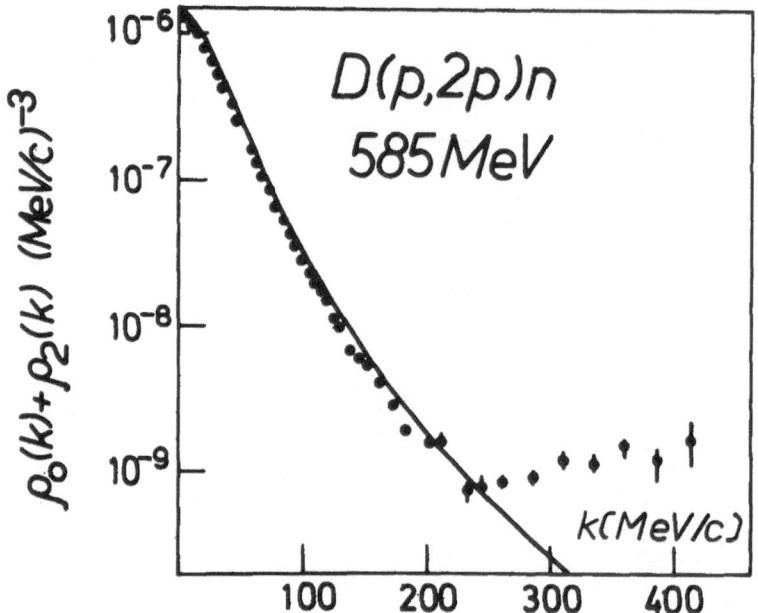

Fig. 2.18 - Momentum distribution in Deuteron from the reaction d(p, 2p)n.
Experimental data from Ref. 22. The theoretical curve has been obtained
using the Plane Wave Impulse Approximation and the RSC interaction.

As is well known, various experimental data exist on the photodisintegration of deuteron at low energies of the photon $E_\gamma \lesssim 100$MeV[23];
at such energies it is obviously incorrect to neglect the neutron-proton
final state interaction. Using realistic interactions, a complete calculation has been performed solving eqns (2.32a) and (2.32b) and including
all multipoles in the expansion of the electromagnetic interaction
Hamiltonian appearing in eqn (2.32)

$$H_\gamma = \frac{ie\hbar}{mc}\sqrt{\frac{2\pi\hbar c}{E_\gamma}}\sum_j e^{i\vec{k}\vec{\xi}_j}\,\varepsilon_\alpha\,\vec{v}_j\,\frac{1+\tau_3(j)}{2} \qquad (2.42)$$

Typical results are shown in the Fig. 2.19. A marked agreement
with conventional nuclear physics is found at low energies. Some disagreement exists at higher energies, where unfortunately the experimental data are very uncertain and the possibility of removing these
discrepancies by considering baryonic resonances in deuteron is
discussed in Ref. 9.

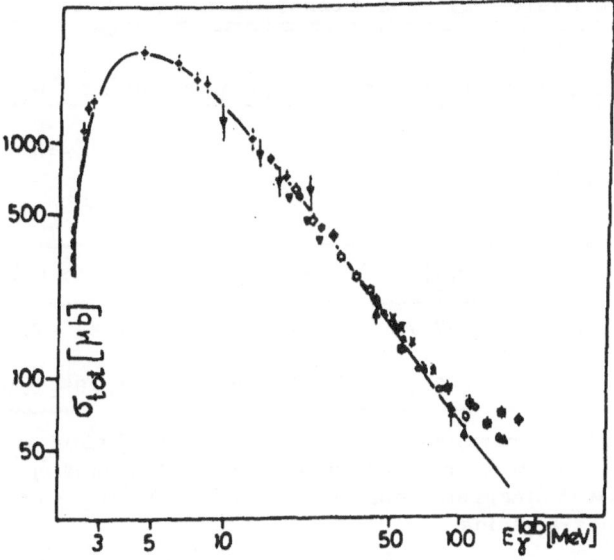

Fig. 2.19 - Total cross section of the reaction d (γ , p)n. The theoretical curve is from Ref. 24.

In conclusion, it appears that much theoretical and experimental work remains to be done in the field of quasi elastic scattering of electrons and protons by the deuteron. As for the photodisintegration process, more precise experiments at higher photon energy would probably be very useful in clarifying the role of baryonic resonance admixtures.

3. The three-body system and its electromagnetic interactions.

3.1. Realistic interactions and static properties of ^{3}He and ^{3}H

	J	E_3(MeV)	$\mu(\mu_N)$	$\langle R \rangle$ (F)
^{3}He	$1/2^+$	-7.72	-2.1276	1.87 ± 0.05
^{3}H	$1/2^+$	-8.48	2.9789	1.70 ± 0.05

Table 3.1 - Experimental values of some bound state properties of the three-nucleon system. E_3 is the binding energy, μ the magnetic moment and $\langle R \rangle = \langle r^2 \rangle^{1/2}$ the root mean square radius.

The interest in the three-body system, whose basic properties are listed in Table 3.1, is twofold: first, it might provide the information that the two-body system was unable to provide (e.g. the core behaviour and the D wave probability of the nucleon-nucleon interaction); second, the importance of typical "many-body" effects (e.g. three-body forces, off-shell behaviour, etc.) might be studied. However, before three-body experimental data and theoretical calculations are used as a source of physical information, the question whether the Schroedinger equation for A = 3 can be solved exactly, should first of all be asked. The answer seems to be a positive one, for different methods which have been developed to solve the three-body problem converge nowadays to very similar results[25,26]. The coordinate system which most computational methods use, is shown in Fig. 3.1, and the corresponding wave function is usually represented as an expansion in terms of the relative motion of a nucleon pair and the motion of the third particle with respect to the pair

$$\Psi_{1/2 M}(\vec{r}\,\vec{\rho}) = \sum_k C_k \,|(L\ell)\mathcal{L}\,(S\tfrac{1}{2})\mathcal{S};\tfrac{1}{2}M\rangle \qquad k \equiv \{L\ell\mathcal{L}S\mathcal{S}\} \qquad (3.1)$$

the most important components being those with $\mathcal{L}=0,1,2$

$$\Psi_{\frac{1}{2}M}(r\,\varsigma) = a_{S}\,\phi_{S}^{\frac{1}{2}} + a_{S'}\phi_{S'}^{\frac{1}{2}} + a_{P}\phi_{P}^{\frac{1}{2}} + a_{\mathcal{D}}\phi_{\mathcal{D}}^{\frac{1}{2}} \qquad (3.2)$$

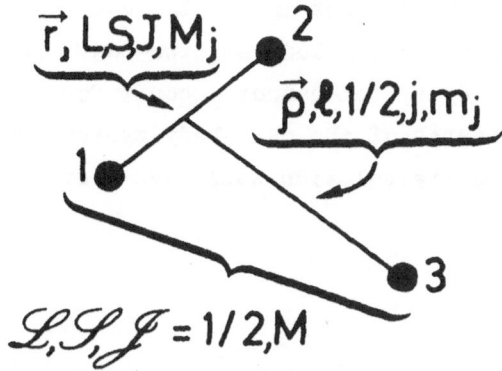

Fig. 3.1   Coordinate system and angular momenta in the three-body systems.

where the first $\mathcal{L}=0$ component describes the totally symmetric state, and the second one the mixed symmetry state. The probability of a given wave is $P_{\mathcal{L}}=|a_{\mathcal{L}}|^{2}$, and according to theoretical estimates[25]

$$P_{S}\sim 85-95\% \qquad P_{S'}\sim 0.5-2\% \qquad P_{D}\sim 0-0.2\% \qquad P_{\mathcal{D}}\sim 4-10\% \qquad (3.3)$$

The results of some relevant calculations on the three-body system are presented in Table 3.2. It is gratifying to observe that very different computational methods lead to equivalent results; therefore, it appears that the three-body problem can be solved, and the three-body system becomes a serious candidate for the study of various physical phenomena without any ambiguity arising from approximate mathematical treatments. The most striking aspect of the results presented in Table 3.2 concerns the underbinding of about ~ 0.5 MeV/A predicted by the Reid interaction, which, by definition, yields the correct binding energy of the two-body system. Many other types of realistic nucleon-nucleon potentials have been investigated, but none of them seems to be able to reproduce the binding energy of the three-body system, which, as shown in Fig. 3.2, is particularly sensitive to the unknown D wave

probability and to the short range repulsion in deuteron. Very soft
interactions yield binding energies in closer agreement with the experimental data but, as we shall see, such potentials do not seems
to be in agreement with medium-energy scattering processes.

Binding energy calculations in the three-body systems have
thus led to a very important conclusions: two-body realistic forces
deduced from the two-body system cannot account for such basic properties, as the binding energy of the many-body system. The origin of this
puzzle is not known and its solution will have a strong impact on
nuclear physics.

	$-E_3$	$\langle R \rangle\,^3H$	$\langle R \rangle\,^3He$	P_s	$P_{s'}$	P_p	P_D
Faddeev [27a]	6.98		2.25	90.2	1.7		8.1
Faddeev [27b]	7.	1.65	1.9	89.2	1.8		9.
Hyper Harmonics [27c]	6.64	1.77	1.9	90.64	0.4	0.06	8.9
Harmonic Oscillator [27d]	7.3	1.85	2.07	89.8	1.4		8.8
Harmonic Oscillator [27e]	7.3	1.85	1.92	89.9		0.1	1.0
Variational [27f]	7.75	1.78		89.5	1.		8.8
Experiment	8.48	1.70± 0.05	1.87± 0.05				

Table 3.2 - Bound state properties of the three-nucleon system calculated
using the Reid Soft Core interaction and various computational methods. E3
and $\langle R \rangle$ are the binding energy and the root mean square radius, respectively,
and p_L the probability of the L wave. Energies in MeV, radii in F.

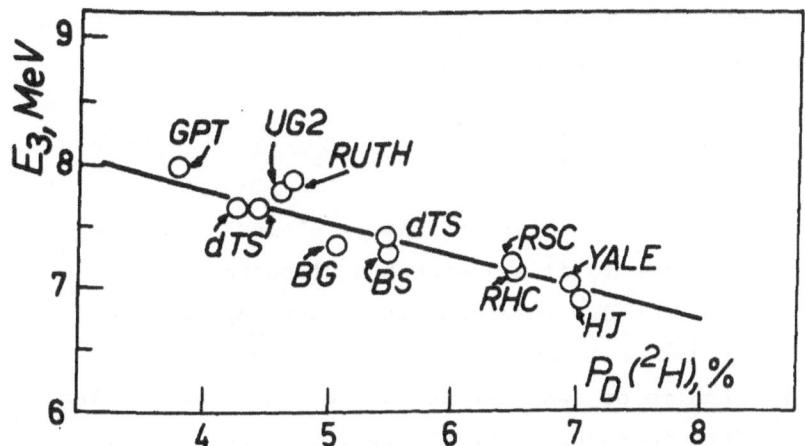

<u>Fig. 3. 2</u> - Binding energy of the three-body system corresponding to various two-nucleon interactions yielding p_D probability for the D wave of Deuteron (Adapted from Ref. 27e).

3.2. <u>Elastic electron scattering by the three-body systems.</u>

Since $J = 1/2$, we have to consider only monopole charge scattering CO and magnetic dipole scattering M1, which experimentally can be measured separately. The charge form factor of ^{3}He and ^{3}H are shown in Fig. 3.3. The diffraction pattern is now very clearly visible unlike the deuteron case, where it is superimposed by the quadrupole form factor. The position of the diffraction dip and the height of the bump after it, are two important experimental features which can be related to the short range behaviour of the force. The cross section for the monopole charge scattering is

$$\frac{d\sigma}{d\Omega} = \left(\frac{d\sigma}{d\Omega}\right)_{Mott} F_C(q^2)^2 \tag{3.4}$$

$$F_C(^3He) = \frac{1}{2}\left\{(2G_E^p + G_E^n)S_1 - (G_E^p - G_E^n)S_2\right\} = \frac{1}{2}\left\{2G_E^p(S_1 - \frac{1}{2}S_2) + G_E^n(S_1 + S_2)\right\} \tag{3.5}$$

$$F_C(^3H) = (G_E^p + 2G_E^n)S_1 + (G_E^p - G_E^n)S_2 = G_E^p(S_1 + S_2) + 2G_E^n(S_1 - \frac{1}{2}S_2) \tag{3.6}$$

where the body form factor $S_1 = F_{SS} + F_{S'S'} + F_{DD} + \ldots$ contains the diagonal form factors of the various waves

$$F_{LL} \propto \int e^{i\vec{q}\cdot\vec{r}_3} |\Phi_L(\vec{\xi}_1,\vec{\xi}_2,\vec{\xi}_3)|^2 \delta(\sum_i^3 \vec{\xi}_i) d\vec{\xi}_1 d\vec{\xi}_2 d\vec{\xi}_3 = \int e^{i\frac{2}{3}\vec{q}\cdot\vec{\rho}} |\Phi_L(\vec{r},\vec{\rho})|^2 d\vec{r} d\vec{\rho} \qquad (3.7)$$

whereas S_2 contains non diagonal matrix elements coupling different waves with the same total spin-isosping function (there is no inter-ference between S and D waves due to the orthogonality of the spin dublet ($S = \mathcal{I} = 1/2$) and quartet ($|\vec{S}| = |\vec{\mathcal{I}} + \vec{L}| = |1/2 + \vec{2}| = 3/2$) states).

The results of two representative theoretical calculations of [3]He and ^{3}H form factors, using the Reid soft core interaction and two different computational approaches, are shown in Fig. 3.4.

As far as these results are concerned, it can be noted that:

i) the Reid soft core interaction can explain low momentum transfer data but neither the position of the minimum nor the high momentum part are reproduced correctly;

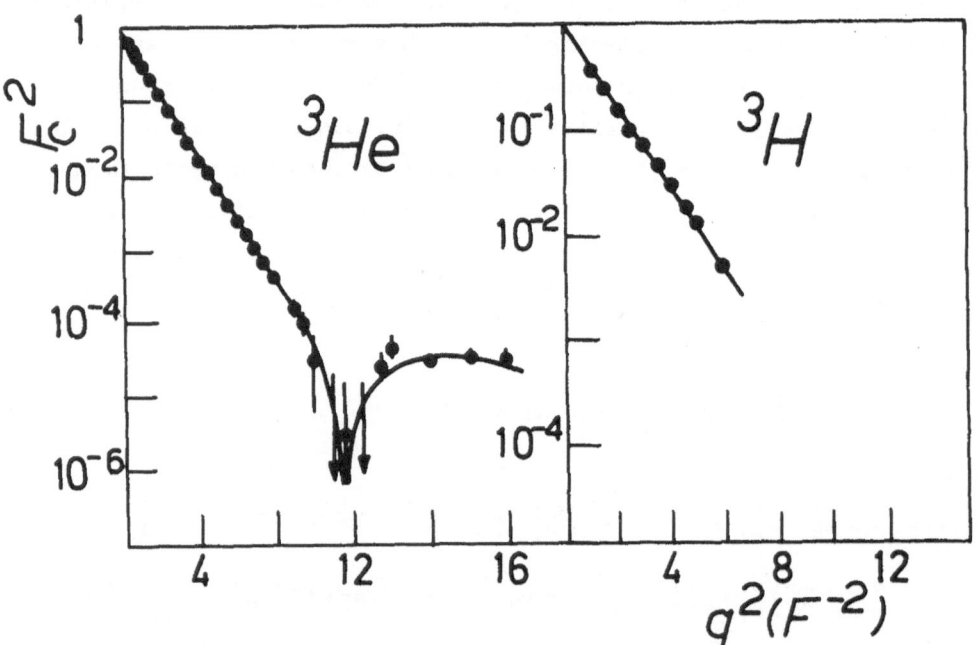

Fig. 3.3 - Experimental charge form factor of [3]He (Ref. 28 a) and [3]H (Ref. 28 b). The lines have been drawn to guide the eye.

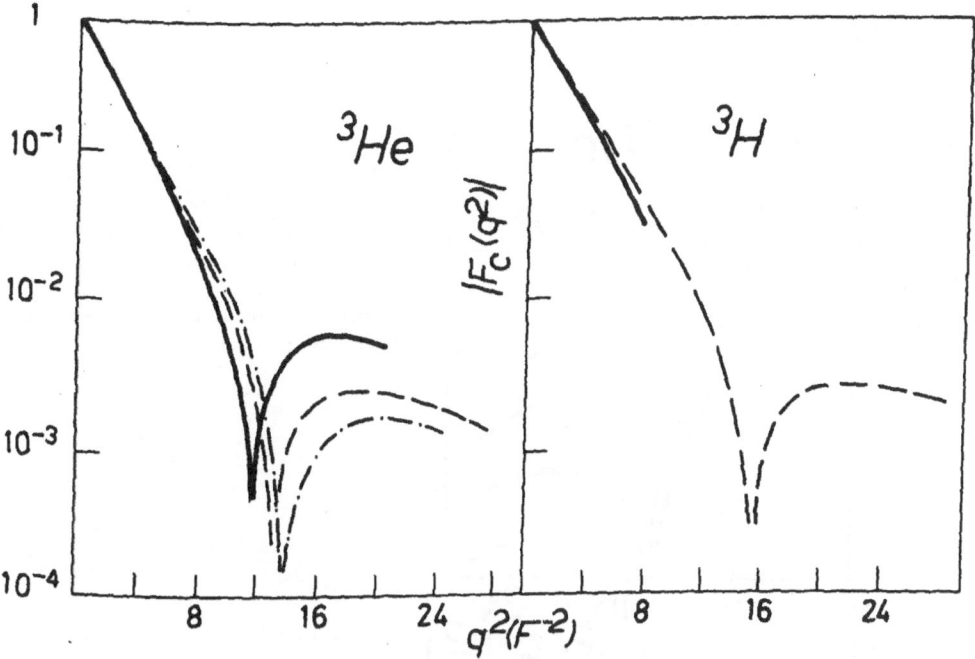

Fig. 3. 4 - The charge form factor of the three-body systems calculated
in Ref. 27 e) using a harmonic oscillator expansion for the three-body
wave function (dashed line) and in Ref. 27 a) using Fadeev's technique (dot-
dashed line). The two-body interaction is the Reid Soft Core and the continuous line represents the experimental data (Fig. 3. 3).

ii) the two computational methods yield slightly different results
at high momentum transfer. Such differences might be due to various
reasons, including convergence problems, wich are particularly important
at such high momentum transfer, and effects due to the different number
of waves included in the two calculations ($L+\ell \leqslant 4$ in one calculation[27a]
whereas practically no restrictions are present in the other[27e]).

The differences between the experimental properties of the three-
body system and the calculated values using the Reid Soft Core interaction are summarized as follows

$$\frac{\Delta E}{E} \equiv \frac{E_{calc} - E_{exp}}{A} \sim -0.5 \ MeV \qquad \Delta q_o^2 \equiv q_{o_{calc}}^2 - q_{o_{exp}}^2 \sim 1.8 \ F^{-2}$$

$$R = \frac{|F_{exp}(q_{Mx}^2)|^2}{|F_{calc}(q_{Mx}^2)|^2} \sim 3 \tag{3.8}$$

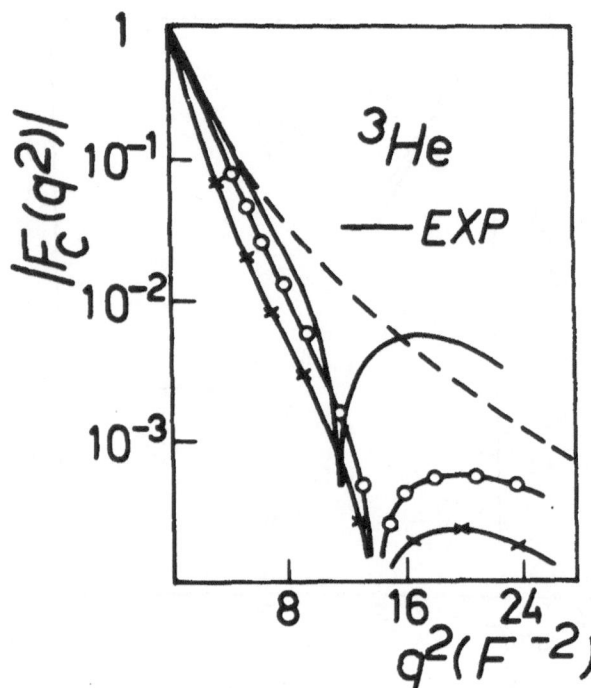

Fig. 3.5 - The ³He charge form factor calculated (Ref. 25) with the Tabakin interaction (dashed line) and two other models of soft separable interactions (for references to these interactions see Ref. 25).

Fig. 3.6 - The ³He charge form factor calculated with a boundary condition model for the two-nucleon interaction using several values of the core radius r_c (from Ref. 29).

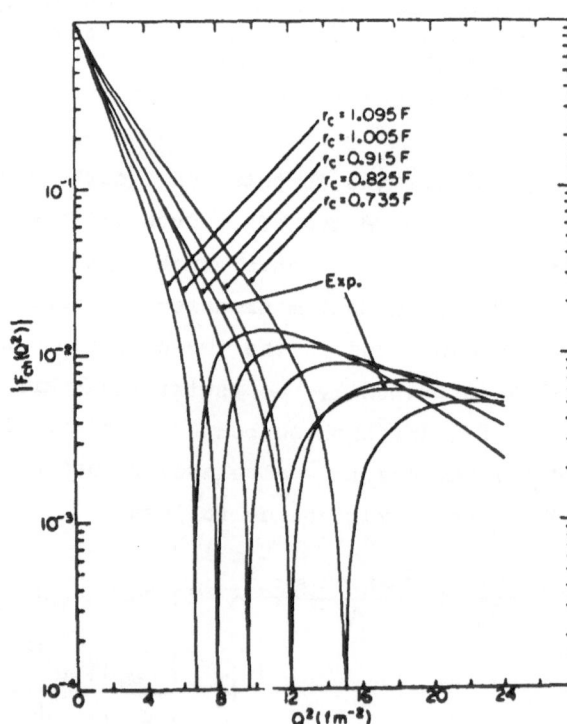

In Fig. 3.5 the charge form factor of ^{3}He, calculated with soft separable interactions, is presented in order to show that such interactions predict a totally wrong high momentum content of the ^{3}He wave function; this should be ascribed to the absence of a substantial short range repulsion which, as qualitatively shown in Fig. 3.6, has marked effects on the high momentum components. The difference between experimental data and theoretical calculations performed with different two-body forces, are summarized in Table 3.3 and the effects of the neutron form factor are shown in Fig. 3.7; these effects are not very important in ^{3}He but become relevant in ^{3}H at high momentum transfer.

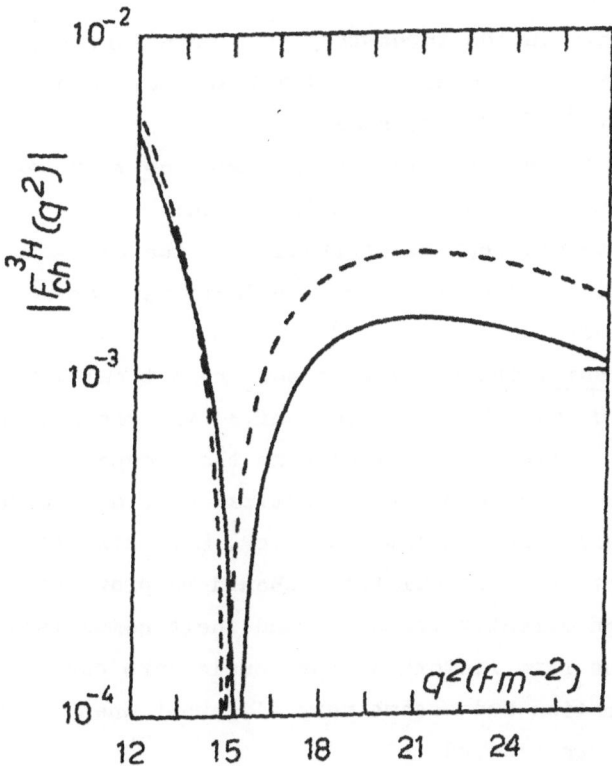

Fig. 3.7 - The ^{3}H charge form factor using two extreme values for the neutron form factor compatible with electron-Deuteron scattering data (From Ref. 30).

It clearly appears that the energy and form factor of the three-body system cannot be explained using non relativistic wave functions resulting from conventional models of the two-body potentials, and including only nucleon degrees of freedom in the electron-nucleus interaction Hamiltonian. Moreover, any improvement of the energy (obtained for example by using a softer interaction) has negative consequences for the form factor (particularly in the high momentum part).

Various effects may be considered to improve this situation:

i) off-shell effects. Existing calculations [31] seem to indicate that reasonable off-shell extrapolations of conventional nuclear forces (e.g. Reid soft core) may improve the energy but, again, the form factor gets worse;

ii) N* components in the three-body wave function: published calculations [32] show that the effects of N(1236) component on the charge form factor of ^{3}He are very small;

iii) three-body forces. Due to our ignorance on this topic, no serious attempts have been made in this direction;

iv) explicit mesonic degrees of freedom in the electron-nucleus interaction Hamiltonian. Exchange current effects were shown to be important at large momentum transfer [33].

Exchange currents. The results of some recent calculations [33] are shown in Fig. 3.8, and they seem to confirm the trend found in deuteron calculations, i.e. large effects on the monopole form factors; moreover, for ^{3}He the effects bring theoretical curves in closer agreement with the experimental data. However, more stringent evidence on the correctness of these calculations should be provided by measurements on ^{3}H at high momentum transfer, and their comparison with ^{3}He data; in fact the form factors of both nuclei are equivalent in the impulse approximation and become very different when exchange currents effects are considered.

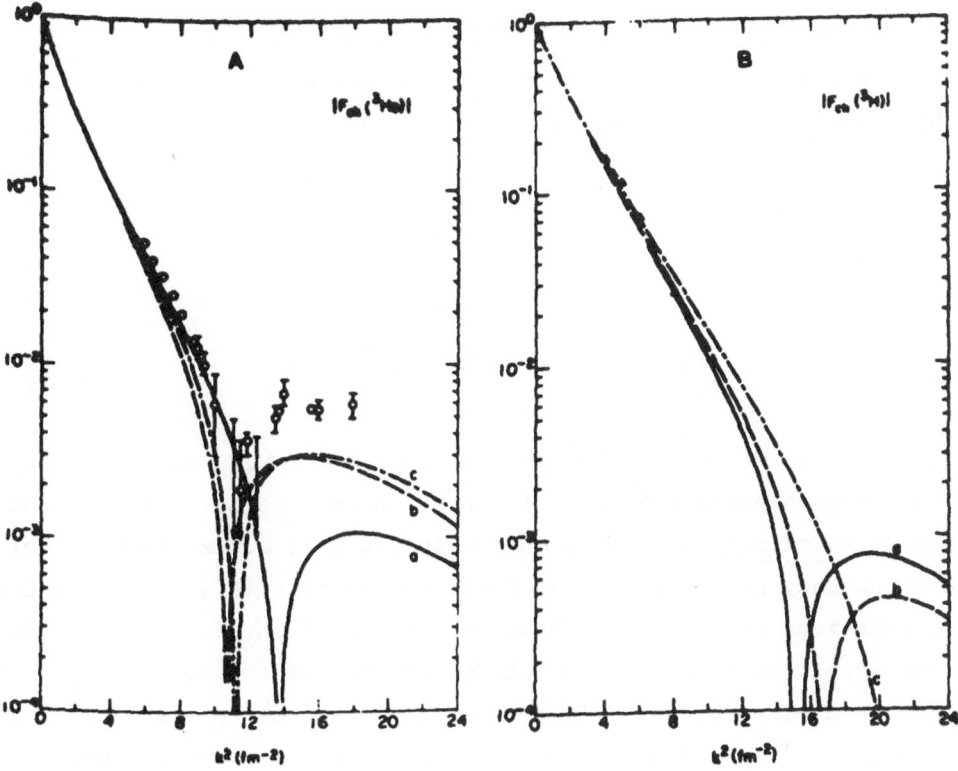

<u>Fig. 3.8</u> - Effects of exchange current contributions on the three-body form factors (Ref. 33). The continuous line in both figures is the impulse approximation result whereas the contribution from diagrams 2.8a), b), c) and d) are represented by the dashed line in Fig. A and by dot-dashed line in Fig. B

3.3. Photodisintegration of the three-body systems.

Experimental data on the total and differential cross sections at 90° concerning the two-body

$$\gamma + {}^{3}He\, ({}^{3}H) \longrightarrow p\,(n) + d \qquad (3.9a)$$

and three-body

$$\gamma + {}^{3}He\, ({}^{3}H) \longrightarrow p\,(n) + n + p \qquad (3.9b)$$

disintegration are available, both at low (E ≤ 60MeV) and high (E ≥ 100MeV) photon energies[23]. In the past there have been many calculations of these reactions but, due to the lack of actual solution of the three-body problem, most of them were restricted to phenomenological wave functions and to approximate treatments of the final state. Few of these calculations[34,35] are based on the exact wave functions obtained from eqns (2.33b) solved by means of Faddeev's techniques. Unfortunately, due to computational difficulties, simple (and unrealistic) separable interactions (Tabakin's) acting only in ^{1}S and ^{3}S waves, were used in these calculations, and no realistic estimates of the importance of the higher angular momentum waves and the short range repulsion were made. Only recently has the solution of the three-body problem with the realistic Reid Interaction been used to calculate the 90° differential cross section at low and high energies [36] describing the final state in the Plane Wave Approximation (eqn (2.35)); although this may be valid only at high photon energies, the results of these calculations at low energies are also very useful, in that, for the first time, the contributions from the deuteron D wave and the ^{3}He $\mathcal{D}$ wave are treated in a realistic way.

Low energies. In this energy region both processes (3.9a) and (3.9b) proceed mainly by an electric dipole transition (in fact, the total cross section up to 170 MeV is in agreement with the dipole sum rule; the experimental angular distributions show contributions from electric quadrupole transitions, but these (being proportional to cos θ) do not contribute to the 90° angular distribution, which is the process for which extensive experimental data have been obtained). If only the zero angular momentum wave of the trinucleon and the deuteron is considered in eqn (2.32), then expanding the electromagnetic operator and retaining only the electric dipole term

$$\mathcal{D} = i\, \hat{\mathcal{E}}_\alpha \sum_j \frac{1 + \tau_3(j)}{2}\, \vec{\xi}_j \qquad (3.10)$$

the final state which can be reached is a ^2P wave, with the spectator pair in relative $L = 0$ motion and the ejected nucleon in $\ell = 1$ motion relative to the pair. Using for ease of illustrantion the Plane Wave Approximation, the 90° differential cross section can be obtained in the form

$$\frac{d\sigma}{d\Omega}(90°) \propto m\, k'_N E_\gamma \left| \int_0^\infty j_1(k'_N \rho)\, u(r)\, \phi_S(r\rho)\, r\, dr\, \rho^3 d\rho \right|^2 \qquad (3.11)$$

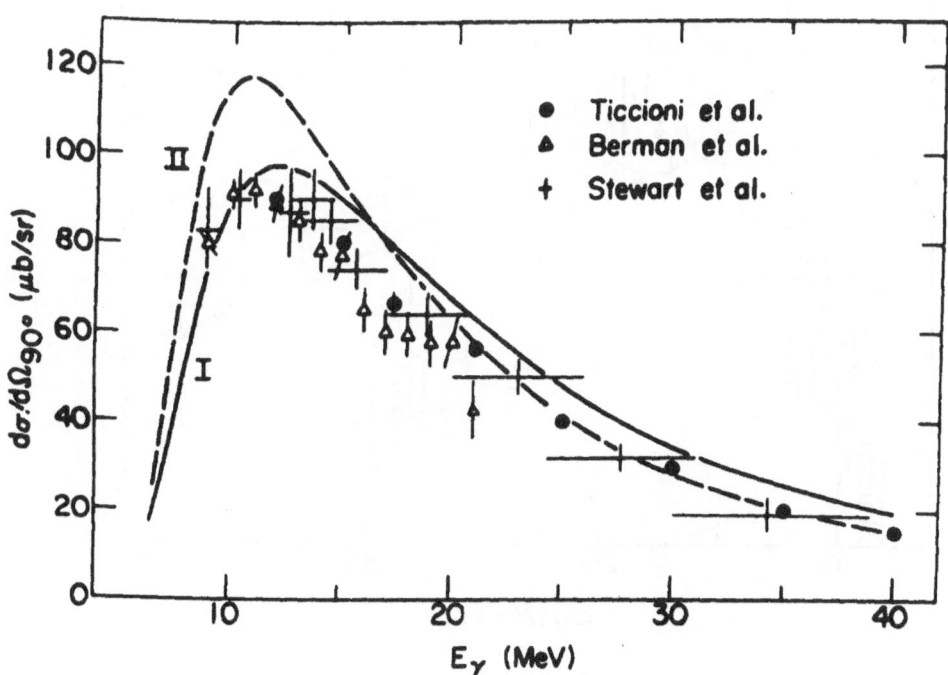

Fig. 3.9 - 90° differential cross section of the reattion ^{3}He (γ, p) d. The theoretical curves were obtained in Ref. 34 (dashed line) and Ref. 35 (continuous line) by means of Faddeev's technique using completely symmetric S state wave functions and Tabakin's interaction. The experimental data are from Ref. 37 a.

where $k'_N = \left[4m(E_\gamma - B_3 - E_\gamma^2/6m)/3\right]^{1/2}$ (see eqn 2.31)) with $B_3 = E_{A-1} - E_A$.

In Ref. 35 the two- and three-body disintegration of ^{3}He have been calculated by a correct consistent treatment of the initial and final states solving eqns (2.32) by means of the Faddeev techniques and Tabakin's separable interaction acting in zero angular momentum waves only (the triplet state parameters were adjusted in order to reproduce the trinucleon binding energy and radius). Only the symmetric S state of deuteron and the S state of ^{3}He were considered. The results are compared with the experimental data in Fig. 3.9 and 3.10, together with the results of similar calculations[34] where the final state is the same as in Ref. 35 but analytical wave functions are used for the spectator pair in ^{3}He. Another consistent calculation of initial and final (T=3/2)

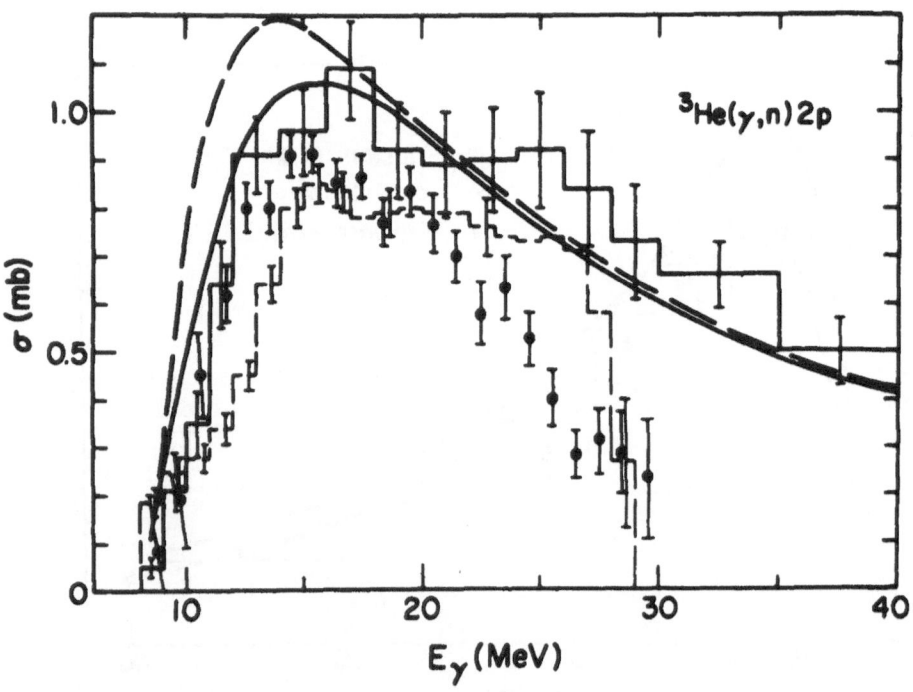

Fig. 3.10 - Total cross section of the three-body photodisintegration of ^{3}He. The theoretical curves were obtained in Ref. 34 (dashed line) and Ref. 35 (continuous line) by means of Faddeev's technique using completely symmetric S state wave function and Tabakin's interaction. The experimental data are from Ref. 37b.

states has been performed in Ref. 38 within the framework of the Hyper-spherical Harmonics approach, using a two-body interaction with a soft-core repulsion. The effect of the final state interaction, which appears to be particularly relevant in the three-body break-up, is illustred in Fig. 3.11.

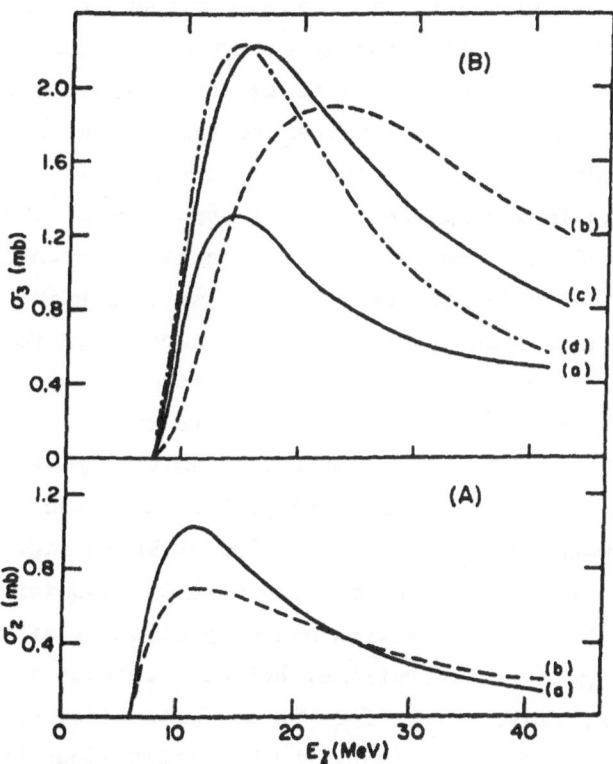

Fig. 3.11 - Dependence of the photodisintegration cross section (Ref. 34) of the three-body system on the structure of the final state.(A) shows the two body cross sections corresponding to the exact Faddeev type solution with separable interaction (continous line) and the plane wave approximation (dashed line), (B) shows the three-body cross section corresponding to the exact solution (lower continuous line), the plane wave approximation (dashed line) the exact solution with ^{3}S interaction equal to the ^{1}S interaction (upper continuous line) and the first rescattering approximation (dot-dashed line).

An overall agreement with the experimental data of theoretical calculations based on separable interactions or simple local interactions is generally observed. Due to the unrealistic character of these interactions (which, moreover, predict a wrong electron scattering form factor (see Fig. 3.5)) and to the approximate ground state description of ^{3}He and d (D waves neglected), one wonders whether the agreement with the experimental data is simply accidental or whether, as commonly believed, the low energy photodisintegration is not sensitive to various nuclear structure details and to the two-body forces, being determined mainly by the overall features (binding energy and radius) of ^{3}He and deuteron in the two-body channel, and to the final state interaction in the three-body channel. A clear answer to these questions, can only be obtained by a comparison with a calculation performed with a realistic interaction and including the complete structure of the two and three-body systems. Such a calculation has appeared recently[36]. The wave function of ^{3}He, obtained from the Reid soft core interaction by means of Faddeev's theory and predicting the charge form factor shown in Fig. 3.4, has been used to calculate the photodisintegration processes at low and high energies. The final state in this calculation is handled in the Plane Wave approximation so that only at high energies a meaningful comparison with the experimental data is possible. The results for the low-energy process, which are shown in Fig. 3.12, clearly demonstrate the importance of high angular momentum states in deuteron and ^{3}He: the D-$\mathcal{D}$ transition not only strongly affects the magnitude of the angular distributions, but also changes its shape, making the peak very flat; this effect is not present in calculations with phenomenological wave functions. The D-$\mathcal{D}$ transitions have not been considered previously; only estimates of electric dipole transition between the deuteron S wave and the ^{3}He $\mathcal{S}$ wave have been made, using for the latter phenomenological wave functions[39] (the D-$\mathcal{S}$ transition is zero since the ^{3}He $\mathcal{S}$ wave contains the spectator pair in L=0 state, which is orthogonal to the L = 2 state in deuteron). The results of Ref. 36 indicate that these transitions can be neglected. The calculated cross section in Fig. 3.12 is lower than the experimental data, which is probably due to the approximate treatment of the final state interaction (see Fig. 3.11A)

    <u>High energies</u>. The high energy angular distributions are shown in Fig. 3.13. The importance of high angular momentum waves is again very clearly demonstrated.

    Since the Plane Wave Approximation is a reasonable approximation at such energies, then a marked discrepancy with the experimental data remains to be explained.

    In conclusion, it appears from realistic calculations of the ^{3}He two-body photodisintegration, that the low energy experimental data, which are mostly sensitive to the overall characteristics of ^{3}He and deuteron, can be interpreted in terms of ground state wave functions

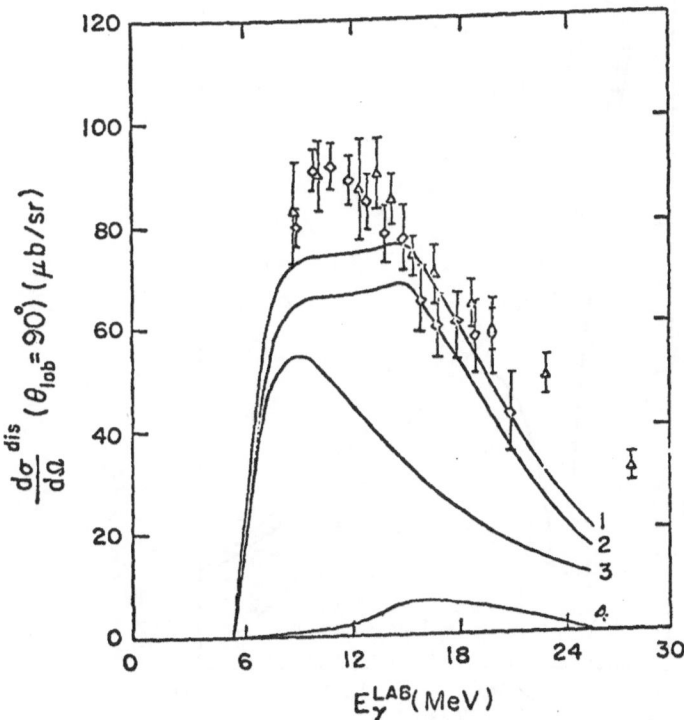

<u>Fig. 3. 12</u> - The 90° differential cross section of the reaction ^{3}He (γ,p) d calculated (Ref. 36) using the three-body wave function corresponding to the Reid Soft Core interaction (Ref. 27a) and the plane wave approximation. Curves 1 and 2 are calculated with the complete wave function, but with and without the spin transitions, respectively. In curve 3 only the S-$\mathcal{S}$ transitions were retained and in curve 4 only the D-$\mathcal{D}$ transitions are present.

obtained from realistic conventional two-body interactions, provided
the deuteron D-wave and ${}^3\text{He}\,\mathcal{D}$ wave are considered. On the contrary, the
high energy process, which is mostly sensitive, besides the D and $\mathcal{D}$
waves, to the high momentum part of the S ${}^3\text{He}$ wave, is in sharp disagreement with theoretical predictions based on conventional two-body
interactions.

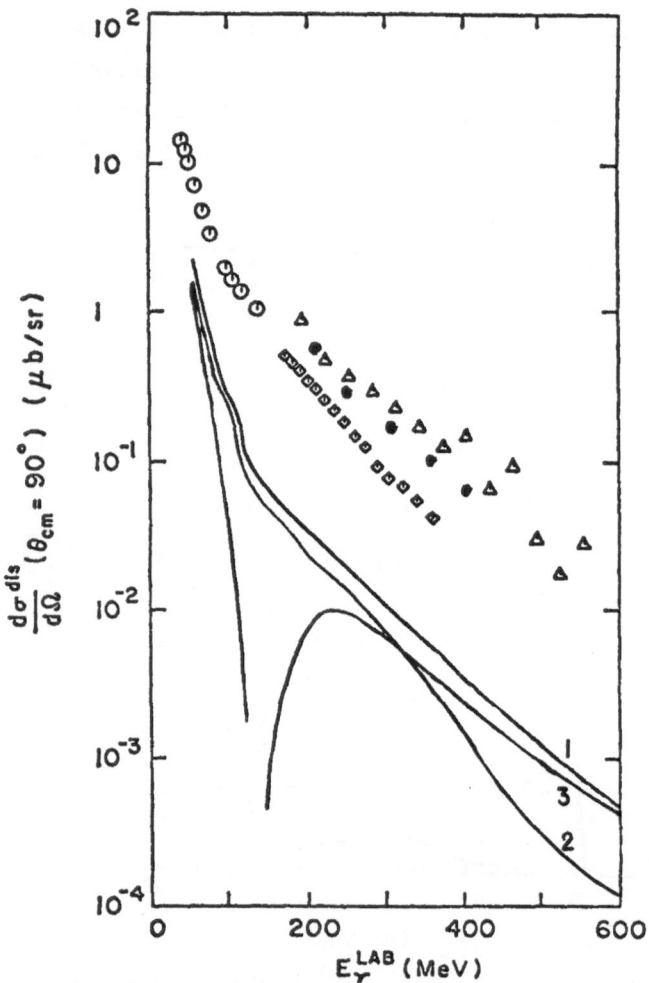

<u>Fig. 3.13</u> - The high energy 90° differential cross section of the reaction
${}^3\text{He}\,(\gamma,\ \text{p})\,\text{d}$ calculated in Ref. 36 using the Reid Soft Core interaction and
the Plane Wave Approximation. Curve 1 is the complete calculation. In
curve 2 and 3, respectively the deuteron D state and the ${}^3\text{He}\,\mathcal{D}$ state were
excluded. Experimental points from Ref. 44. O D'Fallon et al. ; △ Picozza
et al. ; ◇ Argan et al. ; ● Heusch et al.

3.4. Electrodisintegration of the three-body systems.

The electrodisintegration of ^{3}He can proceed via the following processes

$$^3He + e \rightarrow p + d + e \qquad (3.12a)$$

$$p + (pn) + e \qquad (3.12b)$$

$$p + (pp) + e \qquad (3.12c)$$

and the electrodisintegration of ^{3}H via

$$^3H + e \rightarrow n + d + e \qquad (3.13a)$$

$$n + (pn) + e \qquad (3.13b)$$

$$p + (nn) + e \qquad (3.13c)$$

where the two interacting nucleons in the (pn),(pp) and (nn) pairs of the three-body disintegration may be in singlet or triplet spin states depending on their angular momentum. The basic equations describing (in Plane Wave Approximation) the coincidence quasi elastic scattering, were given in Section 2 (eqns (2.31)-(2.39)). For a three-nucleon target, the overlap integral (eqn (2.39c)) appearing in the Spectral Function (eqn (2.39a)) is

$$\phi_2(\vec{p}) = \sqrt{3} \int \psi_2(\vec{r})^* \, \psi_3(\vec{r}\,\vec{p}) \, d\vec{r} \qquad (3.14)$$

where ψ_2 is the deuteron wave function (eqn (2.1)) or the wave function of the two-nucleon scattering state, and ψ_3 the wave function of the three-body ground state (eqn (3.1)). Energy conservation for two-body (eqn (3.12a)) and three-body (3.12b) disintegration of ^{3}He reads (see eqn (2.31)

$$E = \omega - \varepsilon_p - \varepsilon_R = 5.5 \qquad \text{two-body} \qquad (3.15)$$

$$E = \omega - \varepsilon_p - \varepsilon_R = 7.7 + \frac{t^2}{m} \qquad \text{three-body} \qquad (3.16)$$

where t is the relative momentum of the (pn) pair (conservation of energy for the other processes is trivial). In a coincidence experiment, the removal energy E is a measurable quantity, so that the cross section, which theoretically is described by eqn (2.38), can be studied as a function of E in which case it will show some structure corresponding to the two and three-body channels. For a given value (or interval) of E, the cross section can be measured as a function of k (by varying for example the proton emission angle) so that the overlap integral (eqn (3.14)) (or its Fourier transform, the Spectral Function) can be studied separately for the two and three-body processes. Coincidence experiments are exclusive experiments, whereas electrodisintegration with only electron detection is an inclusive experiment and in this case the cross section measured as a function of the energy transfer $\omega = \mathcal{E}_P + \mathcal{E}_R - E$ is determined by the coherent sum of all necessary processes (3.12) (or (3.13)).

The theoretical work in the field of electrodisintegration is similar to that of the photodisintegration, in the sense that only very recently was a calculation with realistic wave functions obtained from the Reid interaction made[41], all previous calculations being

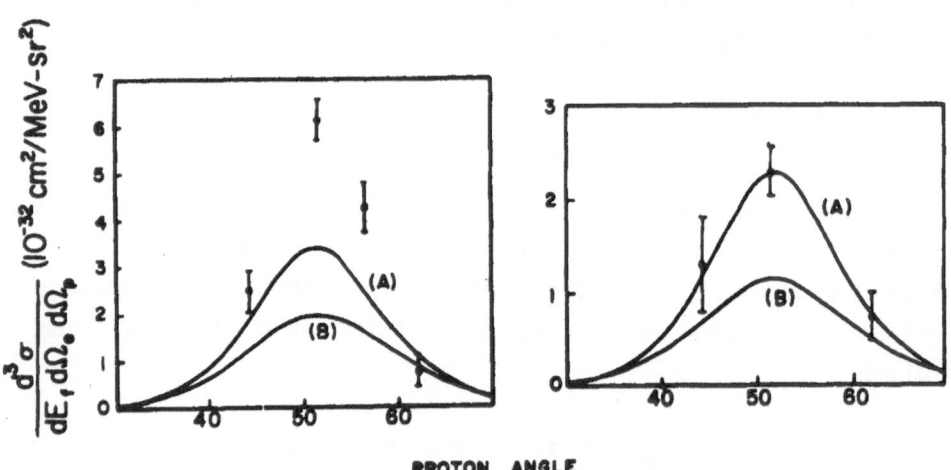

Fig. 3. 14 - Left: the coincidence cross section for the reaction ^{3}He (e, e'p)d. Right: the coincidence cross section for the three-body electrodisintegration of ^{3}H. The experimental points are from Ref. 45a) and the theoretical curves were obtained in Ref. 43 using the Plane Wave Impulse Approximation and Tabakin (A) and Sitenko-Karchenko (B) interactions.

based on phenomenological wave functions or on wave functions obtained
from separable interactions, e.g. Tabakin's [40,43]

The experimental data are very few. Fig. 3.14 shows the angular
distributions of a coincidence experiment and Fig. 3.15 the quasi-elastic
peak. In both figures the experimental data are compared with theoretical
predictions based on Tabakin's separable interaction in the framework
of the Plane Wave Approximation. On account of the inaccuracy of the
experimental data, any comparison with theoretical predictions is of
little significance; moreover, as we have seen in the case of the photodisintegration, calculations which are based on separable interactions
and which neglect the high angular momentum states in the ground state
wave functions, are not realistic. Very recently, accurate experimental
data[46] and realistic calculations[41] of the quasi-elastic electrodi-

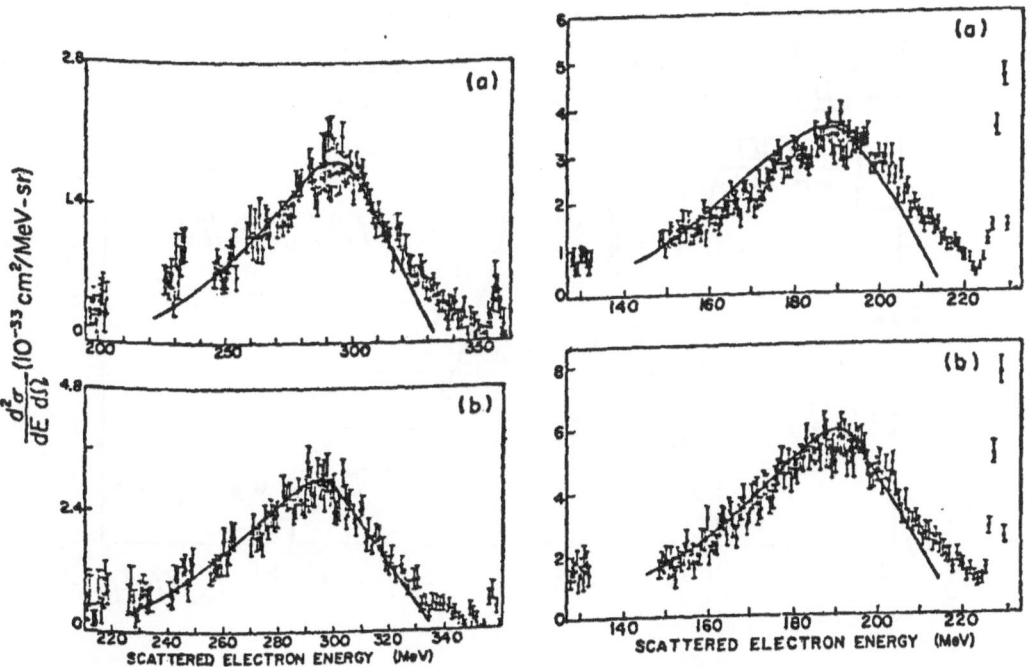

Fig. 3.15 - Cross section for quasi-elastic-electron scattering by ^{3}H(a)
and ^{3}He (b) (Ref. 45 b). Theoretical curves were obtained (Ref. 43) within
the framework of the Plane Wave Impulse Approximation using Tabakin's
interaction. Left: $\mathcal{E}_1$ = 398.4 MeV, θ = 75°; Right: $\mathcal{E}_1$ = 248.8 MeV,
θ = 90°.

sintegration of [3]He appeared. The results of the theoretical calculations, which are based on the wave function corresponding to the Reid
Soft Core potential and used in the description of the elastic electron
scattering and the two-body photodisintegration previously described
are compared in Fig. 3.16 with the experimental data, which are now
sufficiently precise to rule out the predictions by the separable interaction, shown in the same figure.

In spite of the satisfactory results obtained with the realistic
interaction, there is still a strong disagreement with the experimental
data at high energy transfer; such a disagreement is not believed to
be removed by the contribution from meson production, which starts to
be important at higher ω . The tail of the spectrum is very sensitive
to high angular momentum components in the three-body ground state; in

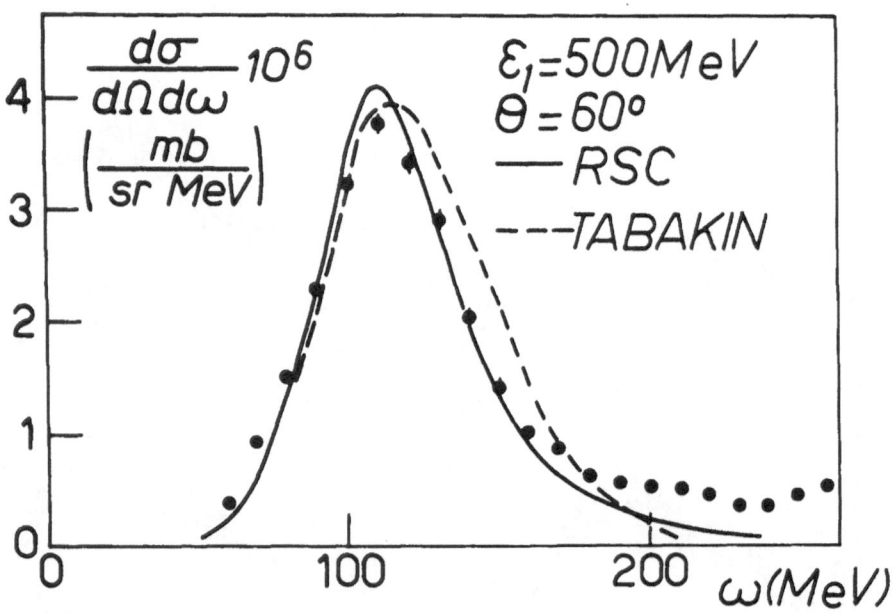

Fig. 3.16 - Quasi elastic electron scattering by [3]He (Ref. 46). The continuous
line has been obtained(Ref. 41) using a three-body wave function corresponding to the Reid Soft Core interaction (Ref. 27a) whereas the dashed line
correspond to the Tabakin interaction (Ref. 43a).

fact, high energy transfer leads to final states characterized by high momenta k and by high relative energy of the spectator pair; on the other hand side, high angular momentum components in the "relative" motion $\vec{p}$ of $\psi_3(\vec{p}\,\vec{\rho})$, strongly overlap with the high energy continuum state appearing in the overlap integral (3.14), whereas the high angular momentum components of the "single particle" motion $\vec{\rho}$ increase the Fourier transform of the overlap integral (i.e. the Spectral Function) at large values of k. For these reasons, the disagreement at high ω might be due to the limited number of angular momentum states considered in the calculation ($L+\ell \leqslant 4$). It cannot be excluded, however, that this disagreement might be due to exotic nuclear physics effects, therefore if we consider the disagreement found in the elastic electron scattering and the two-body photodisintegration, it can be concluded that the electromagnetic and hadronic interaction with the three-body system at intermediate energies cannot be explained by conventional nuclear physics.

4. The four-body systems and heavier nuclei.

4.1. Problems for A⟩3 nuclei. The Center-of-Mass motion.

For A ⟩ 3 the non relativistic Schroedinger equation (1.1) can not be solved exactly, so that the interpretation of the experimental data in terms of many-body wave functions may be obscured by the approximations introduced in order to solve the many-body equation.

The validity of the commonly adopted approximations, i.e. the Brueckner-Hartree-Fock (BHF) approximation, has not been proved, since it can ultimately be established only by a comparison with methods of different nature (e.g. variational). Moreover, these approximate methods present a defect which might be particularly disturbing in the context of the interpretation of medium energy scattering processes:

the lack of traslation invariance in the wave function that they produce. In fact, this wave function is not of the correct form (1.2) or, more specifically, it is not defined, as in the case of A = 3, in terms of intrinsic coordinates (Jacobi or the set $\vec{\xi}_i$; eqns (2.12a) and (2.12b)) but rather in terms of "external" or "Shell model" A independent coordinates. The intrinsic transition matrix element cannot be calculated; only various approximate prescriptions can be adopted to remove the Center-of-Mass (CM) coordinate, and the extent to which they lead to different results is a measure of the importance of spurious Center-of-Mass components. This topic has been thoroughly discussed in many papers[47], here we would like only to recall that spuriosity effects may be very large at high momentum transfer, not only in light nuclei such as ^{4}He, but also in nuclei such as ^{16}O. An example[48] is given in Fig. 4.1, which presents the charge form factor of ^{16}O calculated in terms of Breuckner-Hartree-Fock wave functions[49] and the Reid force. Two different prescriptions have been used to remove the CM coordinate and it can be seen that at high momentum transfer they lead to different results; this means that in this momentum transfer region theoretical calculations are affected by spuriosity effects, in spite of the fact that the overlap of the BHF wave functions with pure harmonic oscillator orbitals (for which no spuriosity effects are present) is larger than 0.9999. Therefore, the frequently adopted procedure, consisting in the use of the harmonic oscillator correction with BHF wave functions, is highly questionable at high momentum transfer. The results presented in Fig. 4.1 clearly show that if the BHF approximation predicts a form factor similar to that of the harmonic oscillator (no second diffraction minimum, i.e. the curve corresponding to b_0 = 1.981F in Fig. 4.1), then the harmonic oscillator correction can be used; if, on the contrary, the BHF form factor is different from that of the harmonic oscillator (which can be the case even if the overlap is very high, see the second diffraction dip corresponding to b_0=1.415F) then the harmonic oscillator correction or any other type of approximate correction cannot be used since spuriosity effects are very important. Fig. 4.1 also shows that the popular procedure of expanding many-body

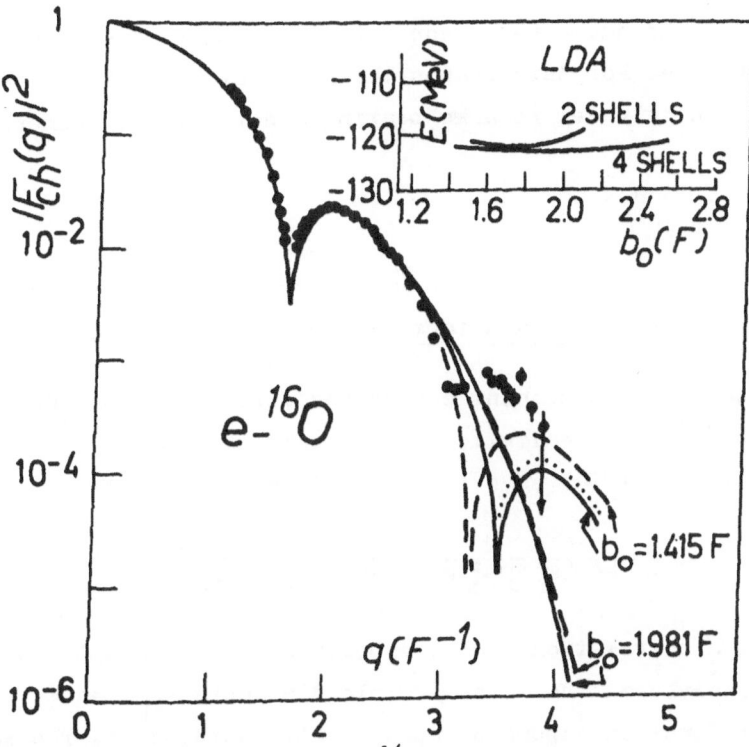

Fig. 4.1 - The charge form factor of ^{16}O calculated (Ref.48) with Brueckner-Hartree-Fock wave functions corresponding to the Reid Soft Core interaction and obtained within the framework of the Local Density Approximation (LDA) (Ref. 49). The self consistent single particle wave functions are expanded in harmonic oscillator basis. The dash, dot and full lines correspond to two approximate prescriptions for the removal of the Center-of-Mass coordinate and the dot line to the harmonic oscillator correction (in the case of $b_0 = 1.981$ F it practically coincides with the full line). The inset shows the dependence of the binding energy on the shell mixing versus the harmonic oscillator parameter b_0 - Experimental data are from Ref. 54.

wave functions in terms of a limited set of harmonic oscillator states which may be a reasonable approximation as far as the energy is concerned (see the inset of Fig. 4.1), is of dubious validity for the calculation of the form factor at high momentum transfer. Apart from the validity of the BHF scheme, a series of approximations which have to be introduced within this scheme in order to make computation feasible, may, generally speaking, mainly influence the calculation of high energy and momentum transfer processes, thus obscuring the information that these processes can provide on "fundamental" problems, such as the

character of the nucleon-nucleon force and the relevance of exotic
nuclear physics. For this reason the A>3 systems will not be discussed
in detail and only few remarks bearing a sufficiently general validity
will be made.

4.2. The binding energy and form factor of ^{4}He and ^{16}O

Recently, the binding energy and form factor of ^{4}He have been
calculated [50] with realistic interactions by applying the variational
principle to the trial wave function of improved Jastrow type

$$\Psi(\vec{r}_1 \vec{r}_2 \vec{r}_3 \vec{r}_4) = F \, \Phi(\vec{\rho}_1 \vec{\rho}_2 \vec{\rho}_3) \qquad (4.1)$$

where F is a correlation factor embodying central and tensor correlations
depending only upon the interparticle distances $\vec{r}_{ij}$ and Φ a harmonic
oscillator function depending upon Jacobi coordinates. The wave function
(4.1) is therefore a traslational invariant and the calculations are
free from Center-of-Mass motion problems. The calculated binding energy
and radius of ^{4}He and ^{3}He, corresponding to the Hamada Johnstone interaction, are listed in Table 4.1; the results for ^{3}He are of the
same quality as those obtained with more sophisticated variational
calculations. The calculted form factor for ^{4}He is shown in Fig. 4.2.

The disagreement between calculated and experimental quantities
of ^{3}He and ^{4}He is very similar, as can be seen from Table 4.1. No calculations exist to date on photon and hadron interaction with ^{4}He
using realistic wave functions, and it is not the purpose of our lectures to discuss the description of these processes in terms of phenomenological wave functions or densities[51,17]. Calculations with
realistic wave functions in this field are highly desirable, and
our experience with the three-body system, coupled with the similarity
of the disagreement between calculated and experimental binding energy
and form factor of ^{3}He and ^{4}He, indicate the need for careful treatment
of nuclear structure effects (D-waves, etc.) in order to correctly
describe photon and hadron interaction with the four-body system.

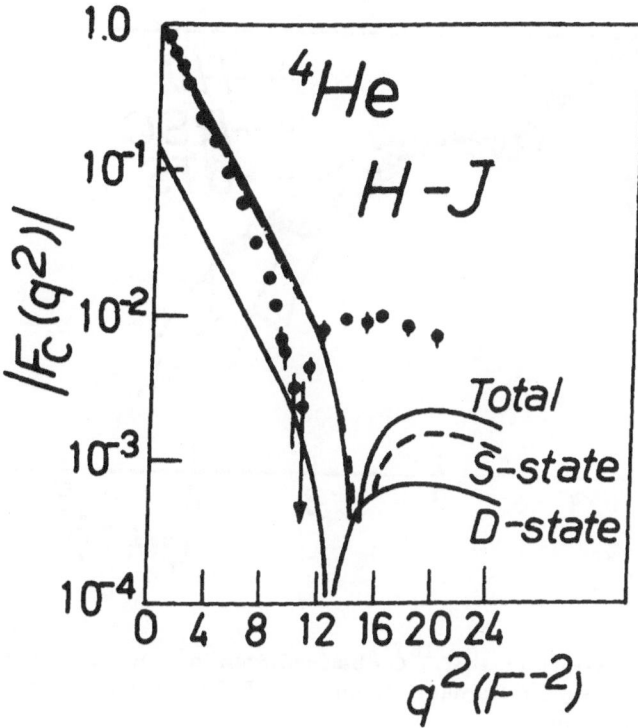

Fig. 4. 2 - Charge form factor of ^{4}He calculated using the Hamada-Jhonstone interaction (Ref. 50). Experimental data from Ref. 55.

	E (MeV)	$\langle r^2 \rangle^{1/2}$ (F)	P_S (%)	$P_{S'}$ (%)	P_{∞} (%)	q_0^2	R
			HELIUM - 3				
Akaishi 50)	-6.00	1.95	90.2	1.21	8.59	14.5	~5.4
Delves 27f)	-6.5±0.2	1.9	89.17	1.8	9	12.8±0.3	~6
Experiment	-8.48	1.87± 0.05				11.6	
			HELIUM - 4				
Akaishi 50)	-20.6	1.65	87.1	0.1	12.8	15	~5
Experiment	-28.2	1.63± 0.04				10	

Table 4. 1 - Ground state properties of ^{3}He and ^{4}He calculated by Akaishi et al. using the wave function (4. 1) and the Hamada-Jhonstone interaction. The results for ^{3}He are compared with the variational calculation by Hennel and Delves.

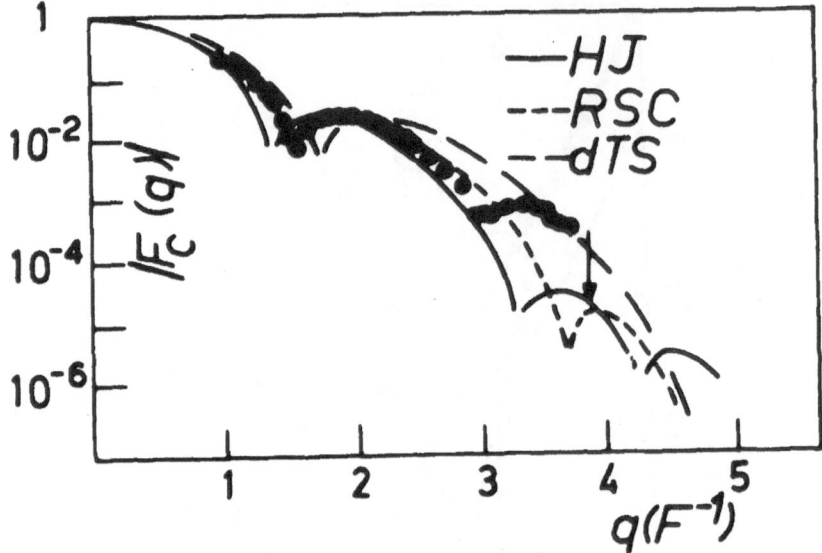

Fig. 4. 3 - Charge form factor of ^{16}O obtained from a Brueckner type calculation (Ref. 52) using the Hamada-Jhonstone, Reid Soft Core and de Tourreil-Sprung interactions.

The form factor of ^{16}O calculated within the framework of the BHF approximation using several two-body forces, is presented in Fig. 4.3[52]. Bearing in mind the approximations previously discussed (Center-of-Mass, harmonic oscillator expansion, etc.) as well as the recent criticism to Brueckner theory[56], we think that the results of these calculations and, particularly, the type of disagreement with the experimental data, should be considered with some caution. Fig. 4.3 is mainly presented in order to show that the monopole form factor of complex nuclei at high momentum transfer may be a powerful tool for obtaining information on the short range part of the two-nucleon interaction.

Elastic electron scattering at high momentum transfer, as well as other experiments measuring single particle momentum distributions (in particular (γ, p) reactions[57]) have been regarded as possible tools "to study Short-Range Correlations in nuclei". This statement has been the origin of some controversy[58]. In fact these experiments depend upon the one-body density distribution

$$\rho^{(1)}(\vec{\xi}_1) = \int \rho^{(2)}(\vec{\xi}_1, \vec{\xi}_2) \, d\vec{\xi}_2 \qquad (4.2)$$

where $\rho^{(2)}$ is the two-body density distribution

$$\rho^{(2)}(\vec{\xi}_1, \vec{\xi}_2) = \int |\Psi(\vec{\xi}_1, \vec{\xi}_2 \cdots \vec{\xi}_A)|^2 \, \delta(\sum_i^A \vec{\xi}_i) \, d\vec{\xi}_3 \cdots d\vec{\xi}_A \qquad (4.3)$$

and therefore cannot determine the two-body correlation function

$$\Delta^{(2)}(\vec{\xi}_1, \vec{\xi}_2) = \rho^{(2)}(\vec{\xi}_1, \vec{\xi}_2) - \rho^{(1)}(\vec{\xi}_1) \rho^{(1)}(\vec{\xi}_2) \qquad (4.4)$$

The main interest in having an experimental determination of $\Delta^{(2)}$ is ultimately to compare its short range behaviour with the theoretical behaviour predicted by two-nucleon interaction models, and the "study of Short Range Correlations" means ultimately the experimental study of the short range behaviour of the two-nucleon force; in this respect, valuable information can also be obtained by the analysis of single particle high momentum components (see Figs. 3.5 and 4.3 and Fig. 1 of Ref. 51), provided that such analysis is carried out within the framework of a correct many-body approach; in fact in a model approach, the type of experimental data we are concerned with can in principle always be interpreted by changing the "shell model" well which generates $\rho^{(1)}(\vec{\xi})$, although the requirement for it to be in agreement with other experimental data and to be traslationally invariant (the latter requirement being very important at high momentum transfer), strongly reduces the freedom in the choice of $\rho^{(1)}$. Unfortunately, the possibility of studying the two-body force by the analysis of high momentum components, might be inhibited by the importance of other effects not

strongly related to the short range part of the interaction but nevertheless producing high momentum components; in this respect the effects
of exchange currents should be considered with particular attention.

Corrections to the impulse approximation due to exchange currents
have been considered for ^{4}He and ^{16}O in Ref. 52 (see also Ref. 53 for
^{4}He), using BHF wave functions obtained from various two-body forces
(see Fig. 4.3). The effect of exchange contribution, as shown in Fig.
4.4, is very large at high momentum transfer, it is much larger than
the impulse approximation and is almost independent of the two-body
force, since the largest contribution comes from the π exchange (π
-pair current diagram shown in Fig. 2.8b), i.e. from the long range
exchange which is nearly model independent. These results are very
interesting but the calculation should be improved by the inclusion
of relativistic effects and by considering more reliable wave functions
and, furthermore, more evidence (rather than a shift toward the exper-

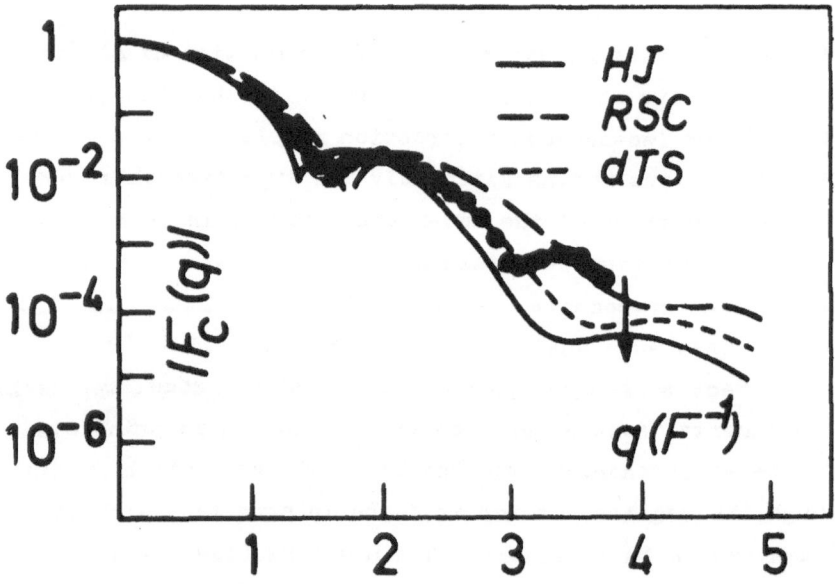

Fig. 4.4 - The same as in Fig. 4.3 with the effect of exchange currents
(diagrams 2.8 b and c) included (Ref. 52).

imental data) should be provided on the correctness of the proposed exchange current models. In this respect, as already noted, the knowledge of the high momentum part of the ^{3}H charge form factor, and its comparison with the corresponding ^{3}He form factor, would be very useful.

5. Summary and conclusions.

A summary of the most important aspects of the problems touched upon in this review can be made by listing the following points:

1) a series of important static and dynamic properties of the two-body system can satisfactorily be interpreted in terms of conventional nuclear physics and two-body forces;

2) conventional nuclear physics and two-body forces cannot on the other hand explain either the static or the dynamic properties of the A>2 systems. For the three-body systems, the discrepancy between conventional nuclear physics and experimental data is quantitatively known, whereas for heavier nuclei the exact extent of such a discrepancy is not known, due to the approximations which are usually introduced in order to solve the many-body non relativistic Schroedinger equation. Calculations with realistic wave functions for the three-body system indicate the importance of the ground-state high angular momentum components in various scattering processes and the ambiguities which may be present in calculations with phenomenological S-state wave functions;

3) attempts have recently been made to improve the agreement between experimental data and theoretical calculations by considering non conventional nuclear physics effects. We have presented some results concerning the effects of exchange currents models in electromagnetic interactions at intermediate energies, pointing out that on the

basis of present calculations and experimental data, a conclusive answer concerning such effects cannot be given, and that further calculations and particularly experimental data are necessary before definite
conclusions in this field can be drawn. At this point it should be mentioned that part of the discrepancy between experimental data on medium energy scattering by nuclei and theoretical calculations, might
be due to some inadequacy of current models of the two-nucleon interaction from which the nuclear wave function is obtained. This consideration is suggested by recent results[56] on nuclear matter calculations which seem to indicate that the Reid Soft Core interaction over-
binds nuclear matter instead of underbinding it, as suggested by current Brueckner-type calculations.

In closing, we should like to indicate some of the directions along
which, in our opinion, the field we are concerned with should develop
in the near future. Regarding experimental developments, we think
that, as discussed in Ch. 3, measurements of the ^{3}H form factor at
high momentum transfer and its comparison with the ^{3}He form factor
would be of particular importance in establishing the correctness of
the exchange current models recently proposed, it would also be useful
to extend measurements to higher momentum transfer for doubly closed
shell nuclei, in order to see whether the form factors flattens, as
predicted by exchange current effects (cfr. Figs. 4.4 and 4.3). Other
experimental developments should be made in the field of coincidence
experiments on deuteron and ^{3}He and ^{3}H, in order to provide new information on the extent to which conventional nuclear physics breaks
down.

As far as theoretical developments are concerned, these involve
fundamental and very general aspects in nuclear physics, which range
from the problem of a correct solution of the many body equation for
$A > 3$, so as to obtain realistic wave functions to be used in scattering
processes, to the problem of the development of consistent approaches
to mesonic, relativistic and many-body forces effects in electromagnetic and hadronic interactions with nuclei. This is a long term pro-
gramme, the practical results of which are difficult to predict; what
has to be done at present is to extend the application of realistic

wave functions of the three-body system to the calculation of various electromagnetic and hadronic processes, thus limiting the use of phenomenological wave functions, which are of little use in the understanding of fundamental problems such as the validity or the breaking down of conventional nuclear physics, the character of the nucleon-nucleon interaction and the role played by mesonic degrees of freedom and relativistic effects.

Acknowledgements. I am grateful to Profs.L. Lovitch and S.Rosati for supplying a computing program for the deuteron wave functions, to Dr. O. Benhar for performing some of the calculations shown in Figs. 2.7, 2.8 and 2.16 and to Dr. G. Salmé for careful reading the manuscript of these lectures.

References

Ch. 1 – Introduction. The Gordian knot of Nuclear Physics.

1) J.S. Levinger, Springer Tracts in Modern Physics 71 (1974) 88

Ch. 2 – The two-body system and its interaction with electrons, photons and hadrons.

2) F. Tabakin, Ann. Phys. (N.Y.) 30 (1964) 51
   R.V. Reid, Jr., Ann. Phys. (N.Y.) 50 (1968) 411
   C.N. Bressel, A.K. Kerman and B. Rouben, Nucl. Phys. A124 (1969) 624
   R. de Tourreil and D.W.L. Sprung, Nucl. Phys. A201 (1973) 193
   T. Hamada and I.D. Johnstone, Nucl. Phys. 34 (1962) 382
3) T.W. Donnelly and J.D. Walecka, Ann. Rev. Nucl. Sci. 25 (1975) 329
4) S. Galster et al., Nucl. Phys. B32 (1971) 221
5) D.W.L. Sprung and K. Srinivasa Rao, Phys. Lett. 53B (1975) 397
6) R.G. Arnold et al., Phys. Rev. Lett. 35,(1975) 776
7) E.L. Lomon and H. Feshbach, Ann. Phys. (N.Y.) 48 (1968) 94
8) F. Gross, Phys. Rev. 142 (1966) 1025, 152 (1966), 1517
   J.L. Friar, Ann. Phys. (N.Y.) 81 (1972) 332
9) H. Arenhövel, This School and Proceedings Symposium on Interaction Studies in Nuclei, Mainz 1975 Eds. H. Jochim and B. Ziegler, North Holland 1975
10) D.O. Riska, Proceedings Mainz Symposium on Interaction Studies in Nuclei, North Holland 1975
    E. Hadjimichael, Proceedings Mainz Symposium on Interaction Studies in Nuclei, North Holland 1975
11) D.O. Riska and G.E. Brown, Phys. Lett. 38B (1972) 183
    J. Hockert et al., Nucl. Phys. A217 (1973) 14
12) M. Gari and H. Hyuga, Preprint RUB TP II/132, Ruhr Universität Bochum, November 1975
13) M. Chemtob and M. Rho, Nucl. Phys. A163 (1971) 1
    M. Chemtob, E.J. Moniz and M. Rho, Phys. Rev. C10 (1974) 344
14) R. Glauber, in High Energy Physics and Nuclear Structure Ed. S. Devons,Plenum Press 1970

15) E.T. Boschitz et al., Phys. Rev. $\underline{C6}$ (1972) 457

16) A.K. Kerman and L.S. Kisslinger, Phys. Rev. $\underline{180}$ (1969) 1483

    J.S. Sharma, U.S. Bashin and A.N. Mitra, Nucl. Phys. $\underline{B35}$ (1971) 466

    G. Barry, Phys. Rev. $\underline{D7}$ (1973) 1441

17) S.A. Gurvitz, Y. Alexander and A.S. Rinat, Ann. Phys. (N.Y.) $\underline{98}$

    (1976) 346

18) NN and ND Interactions, Particle Data Group, UCRL -20000 NN

19) J. Jacob and Th. A. Maris, Revs Mod. Phys. $\underline{45}$ (1973) 6

20) P. Bounin, Ann. Phys (Paris) $\underline{10}$ (1965) 475

21a)Yu P. Antufiev et al., JEPT Letters $\underline{19}$ (1974) 657

21b)V.F. Ksenzov, JEPT Letters $\underline{22}$ (1975) 80

22) T.R. Witten et al., Nucl. Phys. $\underline{A254}$ (1975) 269

23) See M.S. Weiss, Proceedings Mainz Symposium on Interaction Studies

    in Nuclei, North Holland 1975

24) F. Partovi, Ann. Phys. (N.Y.) $\underline{27}$ (1964) 79

Ch. 3 - The three-body system and its electromagnetic interactions.

25) P. Nunberg, E. Pace and D. Prosperi, in The Nuclear Many Body

    Problem vol. 1 Eds. F. Calogero and C. Ciofi degli Atti, Editrice

    Compositori,Bologna 1973

26) Y.E. Kim and A. Tubis, Ann. Rev. Nucl. Sci. $\underline{24}$ (1974) 69

27a)Y.E. Kim et al., Phys. Lett. $\underline{49B}$ (1974) 205

27b)A. Laverne and Gignoux, Phys. Rev. Lett. $\underline{29}$ (1972) 436: Nucl.

    Phys. $\underline{A203}$ (1973) 597

27c)V.F. Demin et al., Phys. Lett. $\underline{44B}$ (1973) 227; $\underline{47B}$ (1973) 394;

    $\underline{49B}$ (1974) 217

27d)Strayer and P. Sauer, Nucl. Phys. $\underline{A231}$ (1974) 1

27e)P. Nunberg, E. Pace and D. Prosperi Nucl. Phys. in press

27f)M.A. Hennel and L.M. Delves, Nucl. Phys. $\underline{A246}$ (1975) 490

28a)J.S. McCarthy et al. Phys. Rev.Lett. $\underline{25}$ (1970) 884

28b)Collard H., et al. Phys. Rev. $\underline{138}$ (1965) B57

29) Y.E. Kim and A. Tubis, Phys. Lett. $\underline{38B}$ (1972) 354

30) R.A. Brandenburg and P.V. Sauer, Phys. Rev. $\underline{C12}$ (1965) 1101

31) E.P. Harper, Y.E. Kim and A. Tubis, Phys. Rev. C6 (1972) 1601

32) A.J. Kallio, P. Toropainen , A.M. Green and T. Kouki, Nucl. Phys. A231 (1974) 77

33) W.M. Kloet and J.A. Tjon, Phys. Lett 61B (1976) 356

34) I.M. Barbour and A.C. Phillips, Phys. Rev. C1 (1970) 165

35) B.F. Gibson and D.R. Lehman, Phys. Rev. C11 (1975) 29; C13 (1976) 477

36) B.A. Craver, Ph.D Thesis, Purdue University (1976)
    B.A. Craver, Y.E. Kim and A. Tubis, to be published

37a)G. Ticcioni et al., Phys. Lett. 46B (1973) 369
    L. Berman, L.J. Koester, and J.H. Smith, Phys. Rev. 133 (1964) B117
    J.R. Stewart, R.C. Morrison and J.S. O'Connell, Phys. Rev. 138 (1965) B372

37b)H.M. Gorstenberg and J.S. O'Connell, Phys. Rev. 144 (1966) 834
    B.L. Berman, S.C. Fultz and P.F. Yergin, Phys. Rev. C10 (1971) 2221
    A.N. Gorbunov, in Photonuclear Processes, Nauka, Moscow 1974

38) M. Fabre de la Ripelle and J.S. Levinger,Nuovo Cimento 25A (1975) 555; Lettere Nuovo Cimento 16 (1976) 413

39) B.F. Gibson, Nucl. Phys. B2 (1967) 501

40) Y.K. Tartakowskii, Sov. Journ. Nucl. Phys. 18 (1973) 409 and references therein cited

41) A.E.L. Dieperink et al., Phys. Lett. 63B (1976) 261

42) R.M. Haybron, Phys. Rev., 130 (1963) 2080

43) D.R. Lehman, Phys. Rev. C3 (1971) 1827

44) N.M. O'Fallon et al., Phys. Rev. C5 (1972) 1926
    P. Picozza et al., Nucl. Phys. A157 (1970) 190
    P.E. Argan et al., Nucl. Phys. A237 (1975) 447
    C.A. Heuschet al., University of California, Santa Cruz, preprint (1973)

45a)A. Johansson, Phys. Rev. 136 (1964) B1030

45b)E.B. Hughes et al. Phys. Rev. 151 (1966) 841

46) J.S. McCarthy et al. Phys. Rev. C13 (1976) 712

Ch. 4 - The four-body system and havier nuclei.

47) F. Palumbo, in The Nuclear Many Body Problem vol. 2 Eds. F. Calo-
    gero and Ciofi degli Atti, Editrice Compositori Bologna 1973
    C. Ciofi degli Atti, Proceedings Symposium on Correlation in Nuclei,
    Balatonfured, Ed. J. Nemeth, Budapest 1974
48) Ciofi degli Atti and R.Guardiola, Phys. Lett. 51B (1974) 26
49) N.E. Reid, M.K. Banerjee and G.J. Stephenson, Phys. Rev. C5
    (1972) 41
50) T. Katayama et al., Progr. Theor. Phys. (supplement) 56 (1974)
    54
51) See C. Ciofi degli Atti, in High Energy Collisions Involving
    Nuclei, Eds. C. Bellini, L. Bertocchi and P.G. Rancoita, Editrice
    Compositori Bologna 1975
52) M. Gari, H. Hyuga and J.G. Zabolitzky, Preprint RUB TP 11/145
    Ruhr Universität February 1976
53) J. Borysowicz and D.D. Riska, Nucl. Phys. A254 (1975) 301
54) I. Sick and J.S. McCarthy, Nucl. Phys. A150 (1970) 631
55) R.F. Frosch et al., Phys. Rev. 160 (1967) 874
56) S.D. Bäckman et al. Phys. Lett. 41B (1972) 247
    O. Benhar et al., Phys. Lett. 60B (1976) 129
    V.R. Pandharipande and R.B. Wiringa, Nucl. Phys. A266 (1976) 269
57) J.L. Matthews et al. Nucl. Phys. A267 (1976) 51
58) See C. Ciofi degli Atti, in The Nuclear Many Body Problem vol. 2
    Eds. F. Calogero and C. Ciofi degli Atti, Edictrice Compositori
    Bologna 1973

Baryon Resonances in Nuclei

Hartmuth Arenhövel

Institut für Kernphysik

Universität Mainz

D-6500-Mainz

Abstract: The field of baryon resonances in nuclei is reviewed. Theoretical developments and experimental evidence as well are discussed. Special emphasis is laid on electromagnetic processes for the two nucleon system. Some aspects of real isobars in nuclei are touched upon.

1. Introduction

In this series of lectures I will give a survey on the concept of isobar
configurations (IC) in nuclei and its role in nuclear and intermediate
energy physics. In speaking of baryon resonances or isobars in nuclei
it is often useful to distinguish between virtual isobars which are far
off mass shell - usual several hundreds of MeV - and real isobars which
are essentially on mass shell except for small interaction effects,
though there are energy regions where this distinction becomes meaning-
less. I will mainly discuss virtual isobars in nuclei, and only at the
end I will make a few remarks on some aspects of real isobars in nuclei.
In the past few years there has been a rapid development of the idea of
considering baryon resonances or isobars (N^*'s) as nuclear constituents
[1-5]. In particular, present medium energy facilities with their possibilities of studying nuclear reactions at high energy and large momentum transfers have given considerably impact on this field. Indeed, in
such kinematic regions one investigates the short range structure of
nuclei and, hopefully, one can measure effects arising from the mutual polarization of nucleons in close collisions, i.e., effects from excitations of internal nucleon degrees of freedom.

The idea of explicit introduction of isobars as nuclear constituents is
to extend the conventional nuclear theory in considering also the internal nucleon degrees of freedom by taking into account the whole spectrum of internally excited nucleon resonances. It is an attempt to shed
some light on the structure of nuclei in the region of small internu-
cleon distances.

In the classical picture the nucleus is viewed as an ensemble of particles, whose internal degrees of freedom can be neglected. They inter-

act weakly through two-body forces and thus nonrelativistic kinematics can be used. Within this framework one essentially faces two problems: (i) what is the nucleon-nucleon force, (ii) how can one solve the many-body problem. This frame was quite successful and many detailed features of nuclei in the low energy region have been understood.

However, it was clear from the beginning that this picture has a limited validity only.First of all, our present understanding of the nuclear forces as being mediated by exchange of mesons leads to additional degrees of freedom in nuclei, the mesonic degrees of freedom, which show up in the form of so-called meson exchange currents. But it is still possible to stick to the old picture by eliminating the meson degrees of freedom and introducing instead effective operators. This procedure leads to the well-known OBE and TBE potentials for the NN-force [6]. In addition, the exchanged mesons will contribute to various nuclear properties and reactions, e.g., to electromagnetic and weak processes in the form of these meson exchange currents (MEC) [7] as illustrated in fig.1.

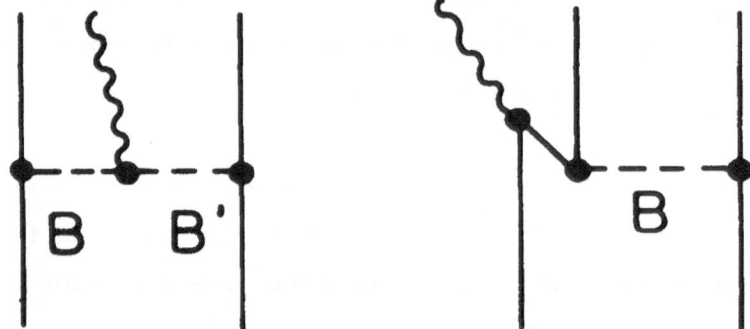

Fig.1. Meson exchange current contributions

Secondly, as one knows from, e.g., π-nucleon scattering, the nucleons itself are not rigid but have quite a complicated internal dynamical structure in the form of a rich spectrum of excited nucleon states, the so-

called baryon resonances or isobars (see Fig.2.). In view of this internal structure one may expect that during close collision two nucleons

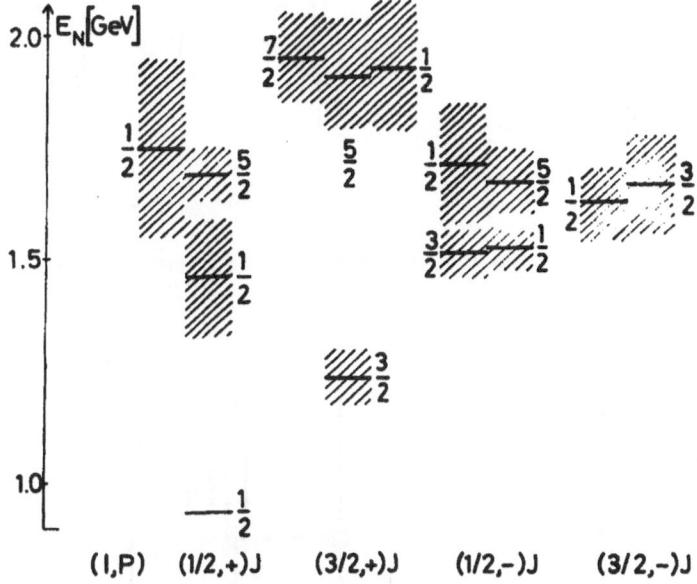

Fig.2. Spectrum of nucleon isobars. I is isospin, P parity and J spin
of the isobar. The shaded region indicates the width.

may change their internal structure, i.e., they may deform or polarize
each other. In other words we may expect the virtual excitation of a
baryon resonance for a short time during a collision. As a consequence
of this a certain though rather small fraction of all nucleons in a
nucleus will be internally excited, i.e., present as isobars.

Again, one may eliminate these internal nucleon degrees of freedom
and introduce instead additional effective operators. For example, in
the case of the two-body NN-force this would give a new contribution
to the TBE potential with intermediate N^{*}'s (see fig.3.) and would
also lead to a three-body force (fig.4.). Furthermore one obtains
effective operator contributions for electromagnetic and weak pro-

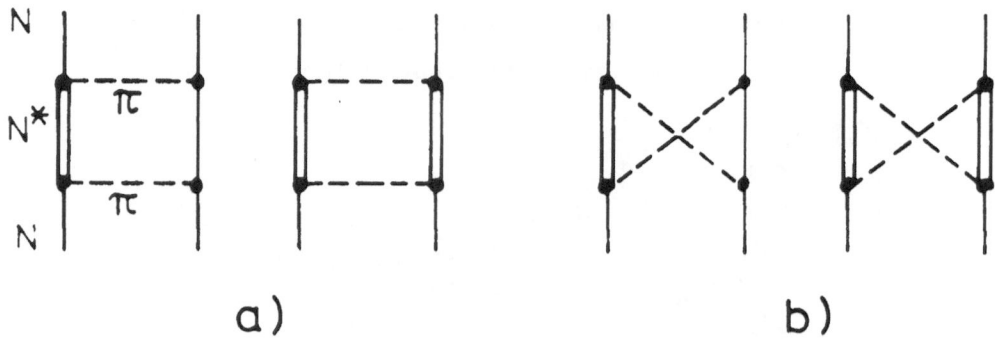

Fig.3. TBE contributions to the two-body force from intermediate iso-

bars.

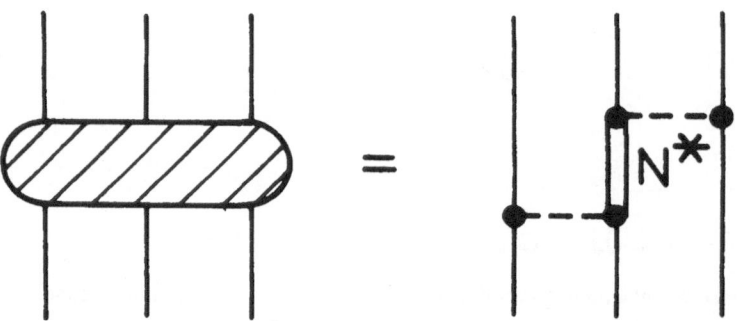

Fig.4. Three-body force contribution from intermediate isobars.

cesses [7] (see fig.5.) and other reactions.

An alternative method to take into account internal nucleon degrees of
freedom is to admit these additional degrees of freedom explicitly in
the nuclear Hamiltonian. Since there does not yet exist a reliable model of internal nucleon dynamics one might use a semi-phenomenological
approach by introducing explicitly the experimentally known isobars as
nuclear constituents. This essentially constitutes the model of isobar
configurations in nuclei which might be regarded as a model to describe the off-mass-shell behaviour of bound nucleons inside nuclei.

As a consequence, the conventional nuclear wave function is supplemented by configurations (so-called isobar configurations (IC)) involving

one or several nucleons in an excited baryon resonance state. The admixture probabilities of these configurations will be small, typically a fraction of a percent, due to the low nuclear density and the rather high isobar excitation energy (mass differences). The characteristic feature of these IC is their rather short range structure since due to the high excitation energy an isobar can not live very long, thus can travel only over a short distance. Therefore, these IC will essentially modify only the high momentum components of the two-particle density.

In this model, those diagrams with intermediate isobars and nucleon lines only should not contribute to effective operators since they are already described by the IC of the nuclear wave function, e.g., the box diagrams a) of fig.3. should not be included in the TBE potential while the crossed diagrams b) have to stay since mesons will still be eliminated and their effect included in effective operators. At this point one might suspect some double counting if both IC and MEC are considered simultaneously, since mesonic degrees of freedom are expected to belong to internal nucleon degrees of freedom.

In fact, suppose π-meson theory would be sufficient to describe an isobar as a superposition of a nucleon and several π-meson states (pictorially a nucleon with a vibrating meson cloud)

$$|N^*\rangle = a_1 |N\pi\rangle + a_2 |N\pi\pi\rangle + \cdots \qquad (1)$$

and one could solve the A-nucleon system including meson degrees of freedom. Then nucleon polarization would be included in the nuclear meson cloud and additional introduction of isobars would lead to double counting. However, at present we are far from having such a com-

prehensive strong interaction theory and one usually takes into account
lowest order diagrams only. In that case there seems little danger of
serious double counting. But if one goes to higher order contributions
one has to be very careful. Fig.5. shows two examples of diagrams which
are dangerous with respect to double counting.

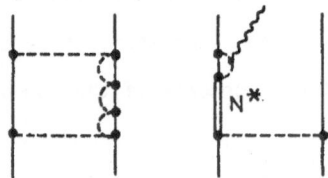

Fig.5. Diagrams which should not be included to avoid double counting.

Thus we shall now adopt the following model. The nucleon isobars will
be considered as stable particles having fixed masses and will be ad-
mitted as nuclear constituents on an equal footing with protons and
neutrons. The two-body NN interaction is extended to include transitions to inelastic channels to describe an isobar excitation in a
collision. Nonrelativistic kinematics will be used. In section 2 an
outline of the general nonrelativistic theory will be given and a
discussion of the transition potentials in section3. The two-body system will be considered in section 4. Both, the bound state problem
(deuteron) as well as nucleon-nucleon scattering will be discussed.
Section 5 is devoted to elastic electron scattering from deuterium
and to electro- and photo-disintegration of the deuteron. Possible experimental evidence of IC from spectator isobar production and other
reactions will be discussed in section 6. Finally, in section 7 we
will touch upon a few interesting aspects of real N^*'s in nuclei.

2. General Nonrelativistic Theory

In the nonrelativistic treatment of isobar configurations in nuclei one starts from the following Hamiltonian of an A-body system

$$H(1,...,A) = \sum_{k=1}^{A} (T(k) + H_{in}(k) + \tfrac{1}{2} \sum_{\ell \neq k} V(k,\ell)) = H_o + V \qquad (2)$$

which differs from the conventional nuclear physics Hamiltonian by the fact, that the kinetic energy T and the potential V depend on intrinsic degrees of freedom and that an additional intrinsic Hamiltonian H_{in} occurs. The intrinsic degrees of freedom need not be specified since it suffices to specify the matrix elements of these operators between intrinsic nucleon states, which are taken from their on-shell-values within this nonrelativistic model, i.e.

$$\langle \nu' | H_{in} | \nu \rangle = (M_\nu - M) \delta_{\nu\nu'}$$

$$\langle \nu' | T | \nu \rangle = \frac{p^2}{2M_\nu} \delta_{\nu\nu'} \qquad (3)$$

$$\langle \nu'\mu' | V | \nu\mu \rangle = V_{\nu'\mu',\nu\mu}$$

where $|\nu\rangle$ labels the different intrinsic states of mass M_ν, which shall be ordered with respect to increasing mass.

The transition interaction $V_{\nu'\mu'\nu\mu}$ $((\nu'\mu') \neq (\nu\mu))$ allows now two nucleons to change their internal structure during a collision. The specific form of the transition potential $V_{\nu'\mu'\ \nu\mu}$ will be discussed later. Because of this possibility of changing the internal structure the nuclear wave function contains in addition to the conventional part components of different configurations where one or several nucleons are internally excited, i.e.,

$$\Phi(1,...,A) = \sum_n \Phi_n(1,..,A) \tag{4}$$

with

$$\sum_n \langle \Phi_n | \Phi_n \rangle = 1 \tag{5}$$

as normalization condition. Here n labels the different intrinsic configuration of A nucleons, i.e.,

$$n = \{n(1), n(2), \ldots\} . \tag{6}$$

The meaning here is that the configuration n contains n(1) nucleons in
the ground state, i.e., neutrons or protons, n(2) nucleons in the first
internally excited state, i.e., $\Delta(1236)$, etc. For the components ϕ_n of
the total wave function one can separate the intrinsic coordinates

$$\Phi_n(1,...,A) = \mathcal{A}_n \psi_n(1,..,A) \phi_n(1,..,A) . \tag{7}$$

Here $\phi_n(1,...A)$ contains the intrinsic degrees of freedom except the
spin-isospin degrees of freedom which will be contained in ψ_n for convenience. ϕ_n is just a product of the different intrinsic nucleon wave functions of the configuration n with the convention that nucleons
1 to n(1) are in the intrinsic state $|1\rangle$, nucleons n(1) + 1 to
n(1) + n(2) in the intrinsic state $|2\rangle$ etc. Thus ϕ_n is symmetric
under permutations of nucleons in the same intrinsic state. ϕ_n is an
orthogonal and complete set with respect to the internal degrees of
freedom, i.e.,

$$\langle \phi_{n'} | \phi_n \rangle = \delta_{n'n}$$
$$\sum_n | \phi_n \rangle \langle \phi_n | = 1 . \tag{8}$$

The orbital and spin-isospin degrees of freedom are contained in ψ_n which shall be antisymmetric under permutations of two nucleons in the same intrinsic state.

Because of the exchange interactions it is useful not to put the distinction between different intrinsic nucleon states at the coordinates but in the wave function. Therefore, we have written the Hamiltonian in a symmetric manner with respect to the numbering of the particles. In this case we have to use a totally antisymmetric wave function corresponding to a further generalization of the general Pauli principle of nuclear physics where neutrons and protons are antisymmetrized. This is just a matter of convenience and does not introduce any unphysical constraints. It corresponds to the assumption that field operators of different intrinsic states anticommute. Since ψ_n is already antisymmetric with respect to nucleons in the same intrinsic state, one needs only to further antisymmetrize nucleons in different intrinsic nucleon states. This is achieved by the operator $\mathcal{A}_n$ which obviously depends on the configuration n.

Inserting ϕ into the Schrödinger-equation and projecting out a specific configuration n one obtains a set of coupled equations

$$\langle \phi_n \mid H - E \mid \mathcal{A}_n \phi_n \rangle \psi_n = - \sum_{n' \ne n} \langle \phi_n \mid V \mathcal{A}_{n'} \phi_{n'} \rangle \psi_{n'} \tag{9}$$

Here $\mathcal{A}_n$, $\mathcal{A}_{n'}$ act also on ψ_n and $\psi_{n'}$, respectively. Except for the two-nucleon system it seems hopeless to try to solve this system of coupled equations. Already the conventional A-body system is too complicated to allow for a complete solution. Since the isobar configurations, i.e., those configurations where one or several nucleons

are in an intrinsically excited state, are expected to be small because
of the large exciattion energy (mass differences), the impulse approximation or first order perturbation (fig.6.) should work quite well.

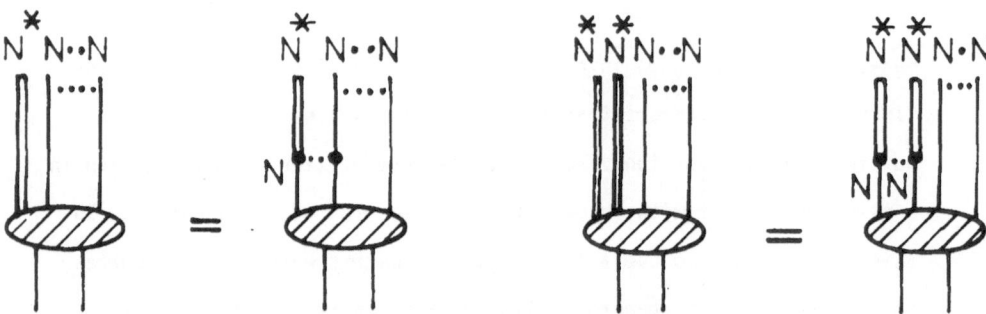

Fig.6. Impulse approximation for calculating IC with a) one N^* and

b) two N^*'S.

That means we keep on the right-hand-side only the term with n' = 1.
This gives for n ≠ 1

$$\psi_n = -\left[\langle\phi_n|H-E|\mathcal{A}_n\phi_n\rangle\right]^{-1}\langle\phi_n|V|\phi_1\rangle \qquad (10)$$

since $\mathcal{A}_1 = 1$.

In second order one obtains also a correction to ψ_1, which can be absorbed in an additional nucleon-nucleon potential corresponding to the
box diagrams a) of fig.3., which is called a dispersive contribution.
Since all nucleon-nucleon interactions are fitted to the experimental
scattering phase shift data this dispersive contribution is already
included in V_{11}. It has been shown that in this case the impulse approximation corresponds almost to the second order treatment up to a
third order term [8]. However, we would like to point out that in a
full coupled channel calculation this dispersive part has to be removed from the NN interaction if it is fit to experiment. Otherwise,

this attractive dispersive part would be counted twice. The restriction to the impulse approximation means that we consider only those isobar configurations where only one or two nucleons are in an intrinsically excited state while all other nucleons stay in the intrinsic ground state (fig.6.).

We shall now derive more detailed equations for isobar configurations n_1 (with one N^*) and n_2 (with two N^*'s) of an A-body nucleus in the impulse approximation. For the conventional part of the wave function we will use a shell model description excluding the 2- and 3-body systems.

For the configuration n_1 with one N^* we use the following expansion

$$\Phi_{n_1}(1,..,A) = \mathcal{A}_{n_1} \sum_\alpha \psi_\alpha^{n_1}(1)\, \psi_{1\alpha}(2,..,A)\, \phi_{n_1}(1,..,A) \tag{11}$$

with

$$\phi_{n_1}(1,..,A) = |N^*(1), N(2),..,N(A)\rangle$$

$$\mathcal{A}_{n_1} = \frac{1}{\sqrt{A}}\left(1 - \sum_{R=2}^{A} P_{1R}\right) \tag{12}$$

where $\{\psi_{1\alpha}(2,....,A)\}$ with energies $\{\epsilon_\alpha\}$ is a complete set of the normal (A-1)-body system, i.e., without isobar configurations, and $\psi_\alpha^{n_1}(1)$ is a single particle wave function of the N^* in the configuration n_1. With this expansion one obtains in the impulse approximation for the different terms

$$\psi_\alpha^{n_1}(1) = -\sqrt{A}\left[p_1^2/2M_{N^*} + M_{N^*} - M - (E - \epsilon_\alpha)\right]^{-1} \tag{13}$$

$$\times (A-1)\int d(2,..,A)\, \psi_{1\alpha}^+(2,..,A)\, V_{N^*N,NN}(1,2)\, \psi_1(1,..,A)$$

which can be evaluated if $\psi_{1\alpha}(2,\ldots,A)$, $\psi_1(1,\ldots,A)$ and $V_{N^*N, NN}$ are known. If one adopts a pure shell model neglecting residual interaction and uses the same single particle states for the (A-1)- and the A-body system then the states $\psi_{1\alpha}$ which contribute are either one-hole-states ($\alpha(i)$, $i < k_f$) or two-hole-one-particle states ($\alpha(ij, k)$ $i,j < k_f < k$) if $\psi(1,\ldots,A)$ is the ground state of a closed shell nucleus. For these two cases one obtains

$$\psi_{\alpha(i)}^{n_1}(1) = -\int d(1'2') \, G_{\alpha(i)}^{n_1}(1,1',E) \sum_{j<k_f} \varphi_j^+(2') V_{N^*N,NN}(1',2') \widetilde{\varphi_j(2')\varphi_i(1')} \quad (14)$$

for $i < k_f$

$$\psi_{\alpha(ij,k)}^{n_1}(1) = -\int d(1'2') \, G_{\alpha(ij,k)}^{n_1}(1,1',E) \, \varphi_k^+(2') V_{N^*N,NN}(1',2') \widetilde{\varphi_j(2')\varphi_i(1')} \quad (15)$$

for $i,j < k_f < k$,

where φ_i denotes the shell model single particle states and

$$G_\alpha^{n_1} = \left[p_i^2/2M_{N^*} + H_{N^*} - M - (E - \epsilon_\alpha) \right]^{-1}. \quad (16)$$

For the configurations n_2 containing two isobars one may start from a similar expansion

$$\Phi_{n_2}(1,2,\ldots,A) = \mathcal{A}_{n_2} \sum_\alpha \psi_\alpha^{n_2}(1,2) \, \psi_{2\alpha}(3,\ldots,A) \phi_{n_2}(1,2,\ldots,A) \quad (17)$$

With

$$\phi_{n_2}(1,\ldots,A) = |N^*(1), N^*(2), N(3),\ldots, N(A)\rangle \quad (18)$$

$$\mathcal{A}_{n_2} = \frac{1}{\sqrt{A(A-1)}} \, (1 - P_{12})\left(1 - \sum_{k=3}^{A} P_{1k}\right)\left(1 - \sum_{k'=3}^{A} P_{2k'}\right). \quad (19)$$

Here the set $\left\{\psi_{2\alpha}(3,\ldots,A)\right\}$ is a complete set of Eigenfunctions of the (A-2)-body normal system with energies $\left\{\epsilon_\alpha\right\}$ and $\psi_\alpha^{n_2}(1,2)$ is a two-body wave function of the two isobars. One obtains

$$\langle N_1^* N_2^* | (H_o - E + \epsilon_\alpha)(1 - P_{12}) | N_1^* N_2^* \rangle \psi_\alpha^{n_2}(1,2) =$$

$$\sqrt{A(A-1)} \int d(3\ldots A)\, \psi_{2\alpha}^+(3,..,A)\, V_{N_1^* N_2^*, NN}(1,2)\, \psi_1(1,..,A) \tag{20}$$

from which $\psi_\alpha^{n_2}$ can be determined if $\psi_{2\alpha}(3,\ldots,A)$ and $\psi_1(1,\ldots,A)$ are given. This expression simplifies considerably if one uses again a simple shell model for the normal parts of the wave functions involved. Then $\psi_{2\alpha}(3,\ldots,A)$ is a two-hole state with respect to $\psi_1(1,\ldots,A)$ and one obtains

$$\langle N_1^* N_2^* | (H_o - E + \epsilon_\alpha)(1 - P_{12}) | N_1^* N_2^* \rangle \psi_\alpha^{n_2}(1,2) = - V_{N_1^* N_2^*, NN}(1,2)\, \widetilde{\varphi_i(1)\varphi_k(2)} \tag{21}$$

In practical calculations the evaluation of eqs.(14) and (15) is still quite involved, in particular the tensor force, and an ansatz according eq.(18) for a configuration of type n_1 is easier to evaluate. Details and applications to specific nuclei are discussed by G. Horlacher [9]. From his work we show in fig.7. the (NΔ)-, (NN(1470)), and ($\Delta\Delta$)-components of ^{4}He.

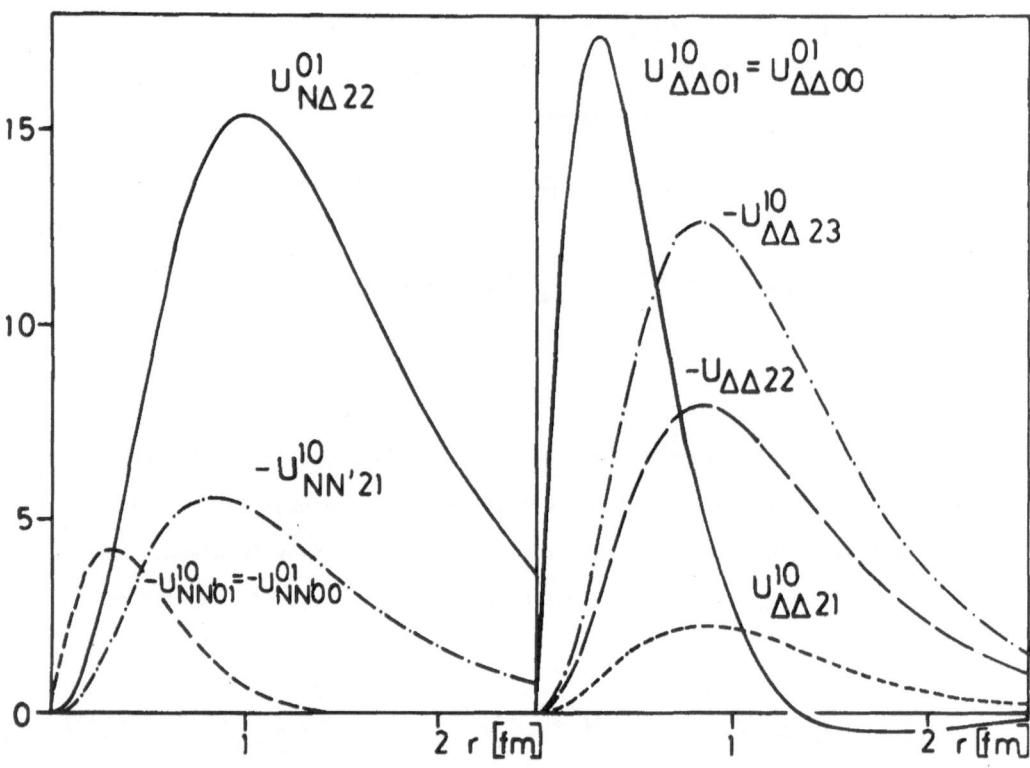

Fig.7. Relative wave functions of isobar configurations in ^{4}He from ref.[9] with total probabilities $P_{N\Delta}$ = 3.86%, $P_{NN'(1470)}$ = 0.48%, and $P_{\Delta\Delta}$ = 4.89%.

3. Transition Potentials

In complete analogy to the OBE potentials of the NN force one can construct transition potentials for the reaction $N + N \rightarrow N_1^* + N_2^*$ by considering boson exchange graphs as illustrated in fig.8. for π-exchange.

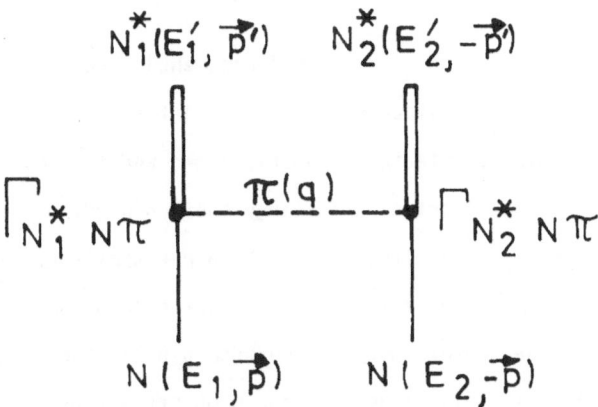

Fig.8. One-pion-exchange diagram for isobar production in NN collisions.

The transition amplitude for the π-exchange will have the following
form

$$T_{N_1^* N_2^* NN} = - \vec{\tau}_{N_1^* N} \cdot \vec{\tau}_{N_2^* N} \frac{\Gamma_{N_1^* N\pi}(p_1', p_1) \, \Gamma_{N_2^* N\pi}(p_2', p_2)}{q^2 - m_\pi^2 + i\varepsilon} \tag{22}$$

where $q = p_1' - p_1$ and $\vec{\tau}_i$ are appropriately chosen isospin transition
operators. The main problem arises in the evaluation of the vertex
function $\Gamma_{N^* N_\pi}(p', p)$. We shall discuss the $\Delta(1236)N\pi$ vertex in some
detail.

Assuming a Rarita-Schwinger description for the Δ one obtains

$$\Gamma_{\Delta N \tau}(p_\Delta', p_N) = -\vec{\sigma}_{\Delta N} \cdot (\vec{p}_\Delta'(\frac{E_N}{M_\Delta} - \frac{\vec{p}_\Delta' \cdot \vec{p}_N}{M_\Delta(E_\Delta+M_\Delta)}) - \vec{p}_N)$$

$$\times (1 - \frac{\vec{p}_\Delta' \vec{p}_N + i\vec{\sigma} \times (\vec{p}_\Delta' + \vec{p}_N)}{(E_\Delta+M_\Delta)(E_N+M)}) \qquad (23)$$

$$\times F_{\Delta N \pi}(p_\Delta'^2, p_N^2, q^2)$$

where the form factor $F_{\Delta N \pi}$ contains the off-mass-shell behaviour of the vertex, which should be normalized to unity if all three particles are on-shell. However, this form factor is poorly known and thus every extrapolation of the vertex function off-energy-shell depends on the specific form chosen for $F_{\Delta N \pi}$ and, therefore, is rather arbitrary. The other problem is at which four momenta of the four external particles one should evaluate the vertex functions because usually the particles are off-mass-shell. This influences also the pion propagator.

Up to now one has used the simplest approximation, i.e., the static limit: $E_\Delta = E_N = M_N$. The choice $E_\Delta = M_N$ instead of $E_\Delta = M_\Delta$ is neccessary to avoid an inconsistency, because the Feynman-diagram of fig.8. contains a total energy conservation $E_1 + E_2 = E'_1 + E'_2$. Otherwise, an unphysical pole would occur. Furthermore, terms of the order P/M are neglected compared to unity. Then the vertex function is proportional to $\vec{\sigma}_{\Delta N}((M_N/M_\Delta)\vec{p}_\Delta - \vec{p}_N)$. The mass ratio is necessary for Galilei invariance. However, usually it is neglected. In this static limit one obtains

$$T_{\Delta\Delta NN}(\vec{q}) = -\vec{\tau}_{\Delta N}(1) \cdot \vec{\tau}_{\Delta N}(2) \frac{f_{\Delta N \pi}^2}{m_\pi^2} \frac{\vec{\sigma}_{\Delta N}(1) \cdot \vec{q} \; \vec{\sigma}_{\Delta N}(2) \cdot \vec{q}}{\vec{q}^2 + m_\pi^2} F^2(\vec{q}^2) \qquad (24)$$

where one uses for $F^2(\vec{q}^2)$ a monopole or dipole form. Taking the Fourier transform one obtains the usual OPE transition potential

$$V_{\Delta\Delta NN} = \vec{\tau}_{\Delta N}(1) \cdot \vec{\tau}_{\Delta N}(2) (\vec{\sigma}_{\Delta N}(1) \cdot \vec{\sigma}_{\Delta N}(2) V_C^{\Delta\Delta NN} + S_{12} V_T^{\Delta\Delta NN}) \qquad (25)$$

where

$$V_c^{\Delta\Delta NN} = \frac{1}{r}\frac{d^2}{dr^2}(rV_0) \tag{26}$$

$$V_T^{\Delta\Delta NN} = r\frac{d}{dr}\left(\frac{1}{r}\frac{dV_0}{dr}\right) \tag{27}$$

$$V_0^{\Delta\Delta NN} = \frac{f_{\Delta N\pi}^2}{12\pi m_\pi^2}\left(e^{-m_\pi r} - \left(1 + \frac{\Lambda^2 - m_\pi^2}{2\Lambda}r\right)e^{-\Lambda r}\right)/r \tag{28}$$

S_{12} is the spin tensor operator and $\vec{\sigma}_{\Delta N}$ and $\vec{\tau}_{\Delta N}$ are defined by their reduced matrix elements

$$\langle \tfrac{3}{2} \| \sigma_{\Delta N} \| \tfrac{1}{2}\rangle = \langle \tfrac{3}{2} \| \tau_{\Delta N} \| \tfrac{1}{2}\rangle = 2 \ .$$

In the explicit expression of eq.(28) a dipole form factor of range Λ has been used. Similar expressions with appropriate spin and isospin dependence hold for transition potentials to other isobar channels. The coupling constants can be determined from the decay width $N^* \to N + \pi$ if available.

The resulting OPE transition potential contains a rather strong tensor force and is quite sensitively dependent on the cut-off-mass. However, it has been observed by Haapakoski [10] that inclusion of ρ-meson exchange leads to a weaker and less cut-off dependent tensor force because the ρ-meson tensor force has the opposite sign. But the central force becomes stronger and there the cut-off dependence is not changed. The potentials are shown in fig.9.

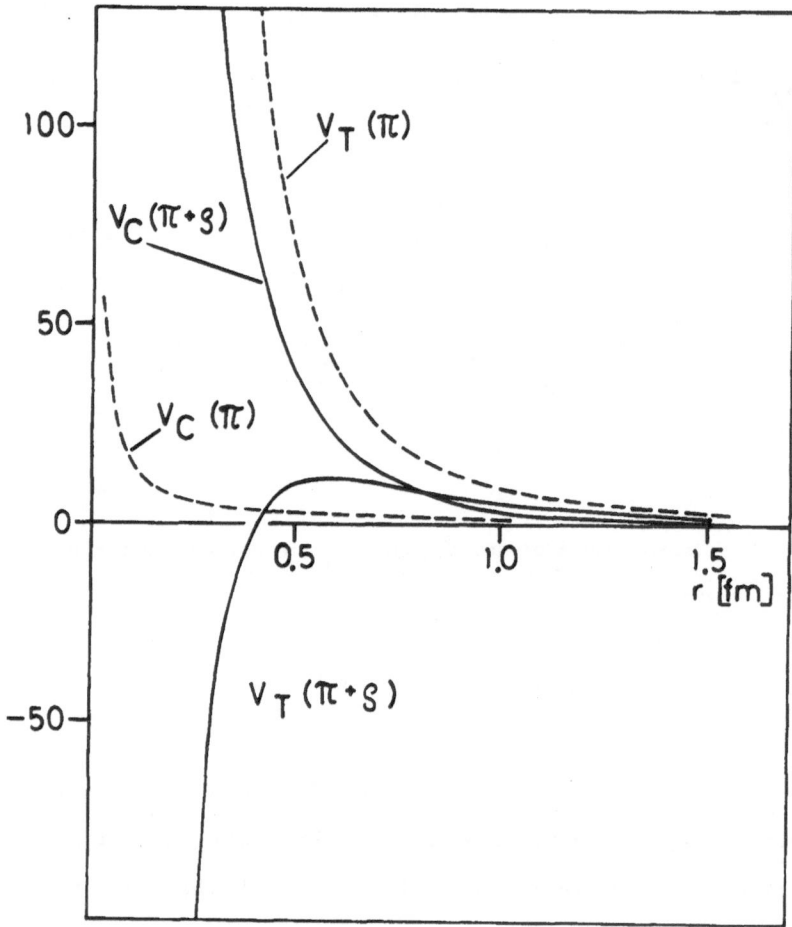

Fig.9. Transition potentials for $N + N \rightarrow \Delta + \Delta$ with π-exchange and $\pi + \rho$-exchange.

In this static limit of the Feynman diagram the range of the transition potential is given by the mass of the exchanged meson. However, one could also use perturbation theory with time ordered diagrams (see fig.10.) which do not implicitly contain total energy conservation. In this case the propagator of the intermediate state is given by $[\omega_q(\omega_q + E_\Delta - E_N)]^{-1}$ where $\omega_q = (\vec{q}^2 + m_\pi^2)^{1/2}$ which gives in the static limit ($E_\Delta = M_\Delta$, $E_N = M_N$) $[\omega_q(\omega_q + M_\Delta - M_N)]^{-1}$. For $M_\Delta = M_N$ the previous result is obtained. However, in this case one does not need to make this approximation. Using this form of the meson propagator one obtains

$$\overline{V}_o^{\Delta\Delta NN} = \frac{f_{\Delta N\pi}^2}{3 m_\pi^2} \frac{1}{(2\pi)^3} \int d^3q \frac{q^2 e^{i\vec{q}\cdot\vec{r}}}{\omega_q(\omega_q + \Delta M)} \quad , \Delta M = M_\Delta - M$$

$$= \frac{f_{\Delta N\pi}^2}{3 m_\pi^2} \frac{1}{2\pi^2} \int_0^\infty dq\, q^2 \frac{j_0(qr)}{\omega_q(\omega_q + \Delta M)} \quad . \tag{30}$$

Rotation of the integration path in the complex plane and change of integration variable yields

$$\overline{V}_o^{\Delta\Delta NN} = \frac{f_{\Delta N\pi}^2}{12\pi m_\pi^2} \frac{2\Delta M}{\pi r} \int_0^\infty du \frac{e^{-\sqrt{u^2+m_\pi^2}\, r}}{u^2 + (\Delta M)^2} \quad . \tag{31}$$

This is a continuous superposition of Yukawa potentials with decreasing range starting at the pion mass and with a decreasing weight function, and is easy to evaluate numerically. The central and tensor forces $(\overline{V}_C, \overline{V}_T)$ are obtained analogous to eqs.(26) and (27) using eq.(31). Fig.11. shows a comparison of these two types of potentials. While the central force $\overline{V}_C$ is somewhat stronger than V_C at small distances the tensor force $\overline{V}_T$ is considerably weaker in the important region. For the transition $N + N \rightarrow N + \Delta$ the potential would contain two terms, one equivalent to eq.(28) corresponding to an intermediate $(NN\pi)$-state and one equivalent to eq.(31) for the intermediate $(N\Delta\pi)$-state, since in this case the two diagrams analogous to fig.10. are different with respect to the intermediate state. The same potential has also been derived by Durso et al. [11] by considering the static limit of the box diagram of fig.3a. They also give approximate analytic expressions for V_o. However, it is by no means clear which transition potential is the more appropriate approximation because of the unresolved problem of off-mass-shell extrapolation.

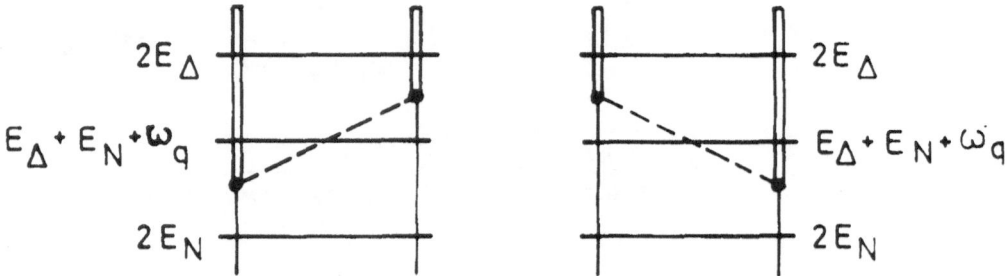

Fig.10. Time ordered diagram for π exchange in the reaction $NN \rightarrow \Delta\Delta$.

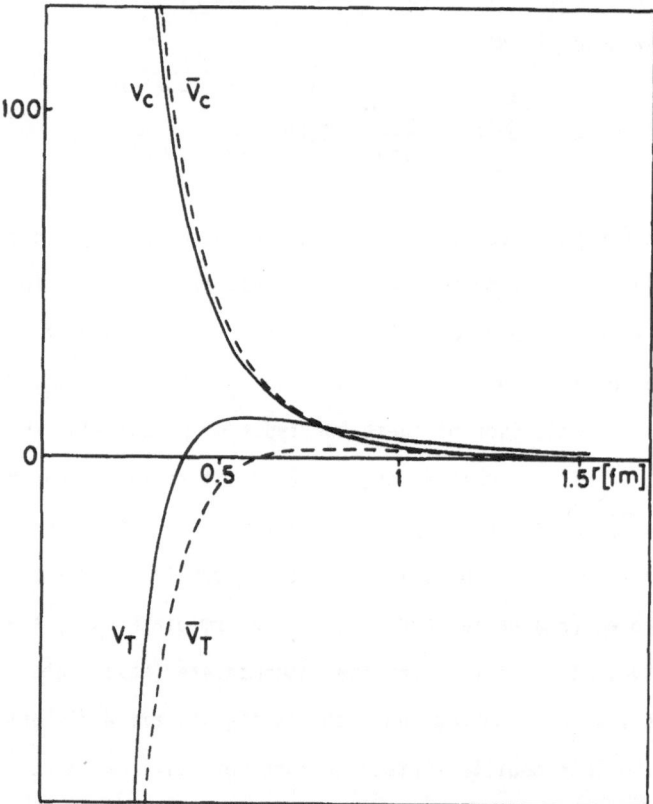

Fig.11. Central and tensor forces for $NN \rightarrow \Delta\Delta$ with $\pi + \rho$-exchange in the static limit of the Feynman diagram and of the time ordered diagrams.

4. The Two-Nucleon System

The simplest nuclear system for studying IC is the two-nucleon system, in particular the deuteron. Because of its rather simple structure the influence of different transition and diagonal potentials with respect to exchanged bosons, coupling constants and cut-off procedures can easily be studied. Furthermore, it can also serve as a good testing ground for the validity of the impulse approximation because in this case a coupled channel calculation is also possible if the number of IC included is limited.

A few isobar configurations of the two-nucleon system are listed in table 1. The energetically lowest configuration (NΔ(1236)) is not allowed in the deuteron which has isospin T = 0.

Tab.1. Isobar configurations up to 600 MeV total mass difference ΔM in the two-nucleon system with possible total spin S, isospin T and intrinsic parity π_i.

	ΔM (MeV)	S	T	π_i
NΔ (1236)	300	1,2	1,2	+
NN (1470)	500	0,1	0,1	+
NN (1520)	585	1,2	0,1	-
NN (1535)	560	0,1	0,1	-
ΔΔ (1236)	600	0,1,2,3	0,1,2,3	+

4.1. The Deuteron

Many of the existing nonrelativistic calculations [8, 12-17] have concentrated on the bound state, i.e., the deuteron problem including NN (1470), $\Delta(1236)$ $\Delta(1236)$, NN (1520), NN(1688) configurations. In one case all NN[*] (t = 1/2) configurations up to NN(2220) have been studied. These various calculations differ with respect to isobar configurations included, transition and diagonal potentials (exchanged bosons, coupling constants, cut-off procedures), and method of computation. The different computational methods are:

(i) Perturbation treatment with a specific analytic form of the isobar wave function [12]. For example

$$\psi_{N_1 N_2}(\vec{r}) = \sum_{LS} R_{N_1 N_2 LS}(r) \langle \hat{r} \mid N_1 N_2 ; (LS) JT \rangle \tag{32}$$

with

$$R_{N_1 N_2 LS}(r) = a_{N_1 N_2 LS} \bar{R}_{N_1 N_2 LS}(r) \tag{33}$$

where

$$\bar{R}_{N_1 N_2 LS}(r) = \begin{cases} A \, j_L(k_0 r) & r < R_0 \\ B \, e^{-k^* r} & r > R_0 \end{cases} \tag{34}$$

with appropriately chosen parameters A, B, k_0 to match the wave function at R_0 and normalize R to unity and some assumption on k^* which determines the asymptotic fall-off. Then one has with $\Delta_{12} = M_1 + M_2 - 2M$

$$a_{N_1 N_2 LS} = -\frac{1}{\Delta_{12}} \sum_{L'S'} \langle N_1 N_2, (LS) J, T \mid V \mid NN; (L'S') J, T \rangle . \tag{35}$$

This method depends sensitively on the chosen form of the isobar wave function and thus in general gives crude estimates only.

(ii) Impulse approximation according to eq.(10) where the diagonal isobar potential is neglected [8, 13, 15]. Explicitly one then has for the radial wave function

$$R_{N_1 N_2 LS}(r) = - \sum_{L'S'} \int dr' r'^2 G_{N_1 N_2 L}(r,r') V_{N_1 N_2 NN}^{LS L'S'}(r) R_{NNL'S'}(r')$$ (36)

with

$$V_{N_1 N_2 NN}^{LS L'S'}(r) = \langle N_1 N_2; (LS)JT | V(\vec{r}) | NN; (L'S')J,T \rangle$$

and the radial propagator

$$G_{N_1 N_2 L}(r,r') = \frac{4}{\pi} k_{12}^2 a_{12} k_L(k_{12} r_>) i_L(k_{12} r_<),$$ (37)

$$k_{12}^2 = 2 \mu_{12} \Delta_{12},$$

$$a_{12} = \mu_{12} / 2 \Delta_{12},$$

$$\mu_{12} = \frac{M_1 M_2}{M_1 + M_2}.$$

The total wave function then has to be renormalized to unity. In this approximation one obtains rather reliable results, if the diagonal potential is of the order of the normal N-N-force. It is even exact if the diagonal potential vanishes and if the effective NN-potential includes the dispersive contributions of intermediate isobars.

(iii) Coupled channel calculation [12, 14, 16, 17]. Here one solves the set of coupled equations given in (9) for a limited number of isobar configuration channels. In some cases a single channel approximation is used, where only one IC-channel is coupled to the normal com-

ponent at a time. In addition to the transition potentials one needs also
the diagonal interaction within a given isobar channel which is even less
well known than the transition potentials.

In the coupled channel calculation a problem arises if one uses for the
N-N channel a potential which is fitted to N-N scattering, because then
it includes as an effective potential already contributions from intermediate N^*'s and one would obtain too much attraction. This is illustrated
in the case of the double-Δ component. The coupled equations read

$$(H_{NN} - E)|NN\rangle = - V_{NN,\Delta\Delta} |\Delta\Delta\rangle$$

(38)

$$(H_{\Delta\Delta} - E)|\Delta\Delta\rangle = - V_{\Delta\Delta,NN} |NN\rangle$$

Elimination of $|\Delta\Delta\rangle$ in the equation for $|NN\rangle$

$$|\Delta\Delta\rangle = - (H_{\Delta\Delta} - E)^{-1} V_{\Delta\Delta,NN} |NN\rangle$$

(39)

gives

$$(H_{NN} + V_{NN}^{DISP} - E)|NN\rangle = 0$$

(40)

where the dispersive potential from intermediate two Δ's (see fig.3a)
is

$$V_{NN}^{DISP} = - V_{NN,\Delta\Delta} (H_{\Delta\Delta} - E)^{-1} V_{\Delta\Delta,NN} .$$

(41)

Therefore, one has to modify in a coupled channel calculation the normal nucleon-nucleon potential, (i.e., weaken the intermediate range attraction) in order to account for the additional attraction from the explicit dispersion contribution to the potential with intermediate N^*'s. This has been done first by Haapakoski & Saarela [14] in an ad hoc manner by changing the intermediate range attraction of the central Reid soft-core potential until the deuteron binding energy is fit to the experimental value. The tensor force has not been changed. If one considers the deuteron only this is a reasonable procedure. However, in a more refined treatment one also has to fit the experimental scattering data as will be discussed in the following section.

A survey of the various results for the deuteron is given in table 2 for the ($\Delta\Delta$)-component and in table 3 for other configurations. Fig.12. shows the radial wave functions of the normal and the ($\Delta\Delta$)-configurations in momentum space for the impulse approximation and a coupled channel calculation using a modified Reid soft core potential for the N-N-interaction and only π-exchange for the transition potential. In this case both methods give almost identical results and cannot be distinguished in the figure. The essential feature of the IC is the enhancement of the high momentum components around 2.5 fm^{-1}. The sometimes appreciable differences of the various theoretical predictions for the most important $\Delta\Delta$-component arise mostly from different ($\Delta N\pi$)-coupling constants, either taken from the quark model or from the decay width of the Δ, and from different cut-off procedures for the short range part of the transition potential on which the results depend sensitively and to a lesser extend from different numerical methods. Also the normal N-N interaction has some influence. A comparative study of various potentials and methods is given in ref.16.

Tab.2. Theoretical predictions of the (ΔΔ)-component of the deuteron. IA means impulse approximation, CC coupled channel, SCC single isobar coupled channel.

NN-Pot.	trans.pot.	cut-off	method	^{3}S(%)	^{7}D(%)	total (%)	ref.
HJ	π	no	IA	0.07	1.50	1.79	28
HJ	π	$\Lambda = 5\,fm^{-1}$	IA	0.04	1.76	0.92	28
RSC	π	$\Lambda = 7.6\,fm^{-1}$	SCC	0.06	0.16	0.25	12
RSC	π+ρ	$r_c = 0.3\,fm$	CC			0.91	14
RSC	π	$\Lambda = 5\,fm^{-1}$	CC	0.04	0.86	0.97	16
RSC	π+ρ	$\Lambda = 5\,fm^{-1}$	CC	0.09	0.46	0.60	16
RSC	π+ρ +diagonal	$\Lambda = 5\,fm^{-1}$	CC	0.06	0.34	0.45	16

Tab.3. Various other IC of the deuteron

configuration	method	probability (%)	ref.
NN(1470)	IA	0.38	28
NN(1470)	SCC	0.17	17
NN(1520)	IA	1.50	13
NN(1520)	SCC	0.26	17
NN(1688)	SCC	0.43	17
NN(1700)	SCC	0.18	17

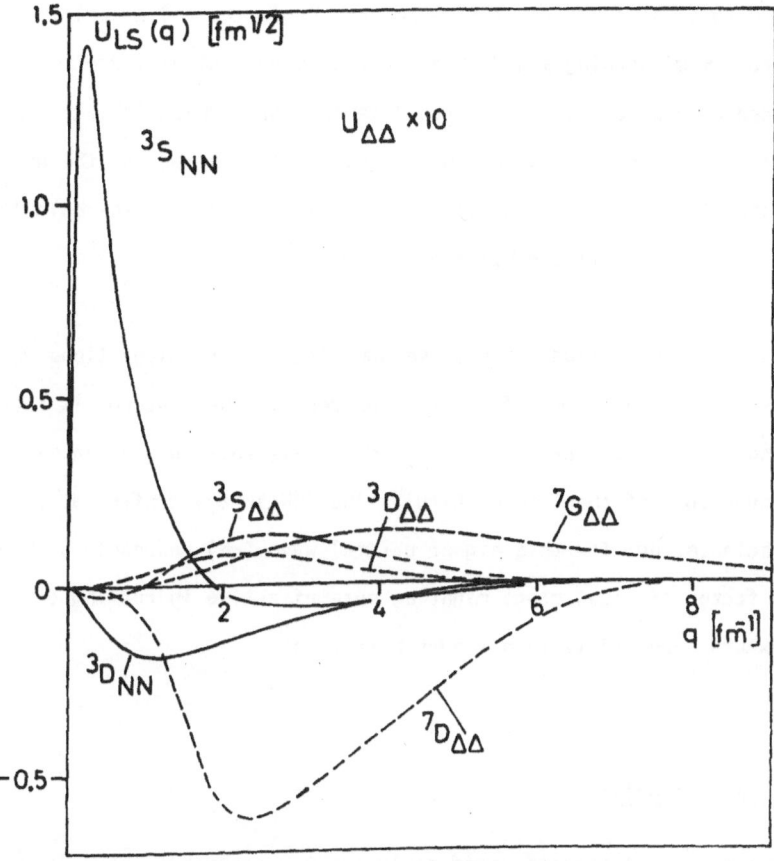

Fig.12. Normal and (ΔΔ)-components of the deuteron in momentum space.

As mentioned before inclusion of ρ-exchange considerably weakens the tensor force and thus cuts down the dominant $^7D_{\Delta\Delta}$-component. However, the $^3S_{\Delta\Delta}$-component is increased. Also the cut-off dependence is somewhat reduced. Inclusion of a diagonal ΔΔ-interaction in a coupled channel calculation leads to a further reduction of the ΔΔ-probability because its central force acts repulsive. This is also true for other configurations. It is satisfying, however, that the impulse approximation gives quite reliable results compared to a full coupled channel calculation provided the diagonal ΔΔ-interaction is not too strong. If the latter condition is not fulfilled, e.g., because of an unreasonably large coupling constant, then also the single channel calculation can be grossly misleading since then the repulsive central force is counterbalanced by the very strong diagonal tensor force which in fact can lead to a very strong binding of the ΔΔ-system. This effect of the diagonal tensor force has unfortunately been neglected also in the calculations of Jena and Kisslinger [12] and of Rost [17].

In general one may conclude from these nonrelativistic calculations that a total ΔΔ-probability of 0.3 to 1 percent in the deuteron is realistic and that energetically higher IC have less admixture probabilities, though some of them, in particular the (NN(1688))-configuration, are appreciable. But for this higher partial wave (πN)-resonance a vertex form factor of usual range might be more effective in cutting down the admixture probability than a pure hard core.

4.2. N-N Scattering

Intermediate virtual isobars, preferably the 3-3 resonance, have long been considered in constructions of effective nucleon-nucleon potentials as a contribution to the two-pion exchange (TPE) amplitude (see fig.3.), mainly in dispersion theoretical models [18, 19]. The interest has also been revived in the more recent developmants of field theoretic models of NN-potentials [6], the one boson exchange potentials (OBEP). A serious drawback of these OBEP's was the necessity to introduce unphysical scalar mesons (σ, σ') in order to account for the intermediate range attraction, particularly necessary to reproduce the $^1S_0$-phase shifts.

An important step to overcome this difficulty has been done by Sugawara and von Hippel [20], who observed that the intermediate range attraction can largely be accounted for by the TPE box contribution with intermediate isobars (fig.3a). The crossed diagram (fig.3b) should still be included in the effective NN-channel potential. They constructed a crude NN-potential from π- and vector-meson (ρ, ω) exchange plus the TPE of intermediate $N\Delta$ and $\Delta\Delta$ channels. Since they were interested in the NN potential only, the inelastic channels were formally eliminated. Using closure and a peaking approximation they arrived at an energy dependent contribution to the NN potential. The energy dependence accounts for the inherent nonlocality due to the propagation of the intermediate isobars. Despite the crudeness of this model the resulting potential was rather similar to the Reid potential [21].

Unfortunately, they used a wrong range in the transition potentials which resulted in an overestimate of the intermediate range contribution from the virtual isobar channels. An approach similar to Sugawara and von Hippel has been studied by Wagner and Winiger [22, 23] using the strong-coupling fixed-source meson field theory with a hard core r_c = 0.5-.55 fm. They included also higher isobars and did an explicit coupled channel calculation with a limited number of inelastic channels. Thus they obtained for the first time explicit wave functions of IC in the continuum with their typical short range structure [22]. Wagner achieved a very good fit to the S-wave phase shifts and except for the P-waves a fair agreement with the higher partial waves [23].

More recently Jena and Kisslinger [12] have done an exploratory study of the effects of the inelastic $N\Delta$ and $\Delta\Delta$ channels in a limited coupled channel approach. In this case also a diagonal OPE interaction within the isobar channels of considerable strength has been included. Again the similarity of the effects to σ-exchange has been noted. At this stage no phase shift fits were intended.

A quantitative description of NN-scattering in the $^1S_0$-state has been obtained by Green and Haapakoski [24] with quite good fits to the $^1S_0$-phase shifts. Haapakoski [10] included the lowest NΔ-channel and solved the system of coupled equations for the $^1S_0$(NN) and $^5D_0$(NΔ) waves.

$$\left(-\frac{d^2}{dr^2} - k^2 + M V_1\right) u_{00}^{NN} = -M V_2 u_{23}^{N\Delta}$$

$$\left(-\frac{d^2}{dr^2} + \frac{6}{r^2} - k^2 + M(M_\Delta - M) + V_3\right) u_{23}^{N\Delta} = -M V_2 u_{00}^{N\Delta} \tag{42}$$

where

$$V_1 = -10.5 \frac{e^{-0.7r}}{0.7r} - 43 \frac{e^{-2.75r}}{2.75r} + 7000 \frac{e^{-3.9r}}{3.9r} - B \frac{e^{-2.1r}}{2.1r} \tag{43}$$

includes π, η and ω contributions and the last term simulates other TPE contributions not included in the model. Parameter B is fitted to the $^1S_0$-phase shift (B = 810 MeV). In the NN→NΔ transition potential

$$V_2 = 36.3 \left(Z(0.7r) Y_0(0.7r) - 220 Z_0(3.85r) Y_0(3.85r) \right) \tag{44}$$

ρ-exchange is included. Finally, for the diagonal NΔ interaction $V_3 = V_1$ was assumed, a choice which did not seem to be crucial. All potentials have a hard core of 0.4 fm.

Green and Haapakoski [24] used a similar model with slightly different parameters and with a soft cut-off factor $(1 - \exp(-B r^2))$ instead of a sharp cut-off. In eliminating the (NΔ)-channels and using a closure approximation they constructed also an equivalent effective NN-potential which is energy dependent.

$$V_{NN}^{eff} = V_1 - \frac{V_2^2}{\Delta E(k^2)}$$ (45)

where $\Delta E(k^2)$ is fitted to the phase shift. This dispersive contribution $V_2^2/\Delta E(k^2)$ resembles remarkably well the σ-exchange potential.

Similar investigations have been done by Smith and Pandharipande [25] who included higher partial waves and the coupling to the $\Delta\Delta$-channel, by Day and Coester [26], and by Holinde and Machleidt [27]. The latter authors avoid a complicated many channel calculation by neglecting the diagonal interactions of the isobar channels which enables them to construct an exactly equivalent effective NN potential. The isobar wave functions are then obtained in the impulse approximation which in this case is then exact. In particular, they found that in order to obtain a reasonable fit for all partial waves the range Λ of the dipole cut-off form factor in the transition potential had to be chosen less than 3.5 fm^{-1}. The remaining σ-contribution was found smaller and shorter ranged compared to the original OBEP.

5. Electromagnetic Processes at the Deuteron

We shall now discuss the consequences of IC in the two nucleon system, mainly the double-Δ component but also the NΔ component in the continuum, on electromagnetic reactions in order to find out whether there exist some regions of energy and momentum transfers at which these IC show significant contributions and whether these lead to better agreement between experiment and theory.

If one takes into account internal nucleon degrees of freedom by admitting IC into the nuclear wave function one also has to modify the operators to allow for possible transitions between different isobar states. In the case of the electromagnetic interactions one has to introduce diagonal and transition isobar currents to account for the electromagnetic processes shown in fig.13. However, the problem is that these additional currents are much less well known than the nucleon current. The usual

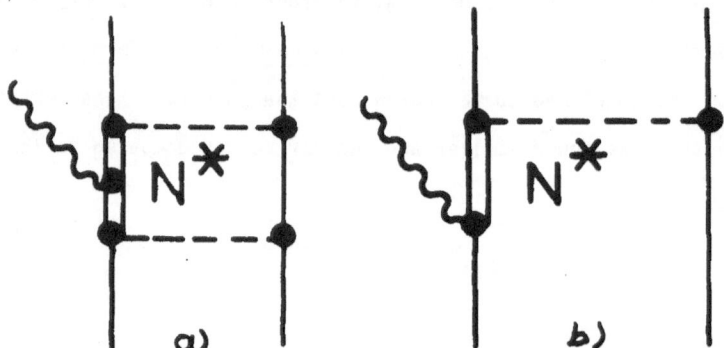

Fig.13. Diagrams for the electromagnetic interaction of an IC involving
a) the form factors of an isobar b) the (NN$\overset{*}{\gamma}$) transition form
factor.

procedure is to derive a general form of the current for the on-mass-shell particles from general principles where the unknown form factors are taken from experiment or some simple model, e.g., quark model, and then this current is used for the off-mass-shell particles, an extrapolation which might not be justified.

As an example, we list the Δ- and the (Δ-N) current which are obtained in this way in the nonrelativistic limit [28].

a) Δ-N current

$$\langle \Delta\, p's'|\; j_o(o)\,|\, N\, ps \rangle = \psi_{s'}^{+}\, [\; \frac{ie}{4HM_\Delta}\; G_{M1}^{\Delta N}\, (\vec{\sigma}_{\Delta N} \times \vec{q})\cdot\vec{P}\,]\chi_s \tag{46}$$

$$\langle \Delta\, p's'|\; \vec{j}(o)\,|\, N\, ps \rangle = \psi_{s'}^{+}\, [\; \frac{ie}{4HM_\Delta}\; G_{M1}^{\Delta N}\; \vec{\sigma}_{\Delta N} \times$$

$$((H+M_\Delta)\vec{q} - (M_\Delta - H)\vec{P})]\chi_s$$

where

$$G_{M1}^{\Delta N} = 1.6\; \frac{2}{3}\sqrt{2}\; G_{M1}^{NN} \tag{47}$$

b) $\Delta\Delta$ current

$$\langle \Delta\, p's'|\; j_o^{}(o)\,|\,\Delta\, ps \rangle = \psi_{s'}^{+}\, e\, G_{E0}^{\Delta\Delta}\; \psi_s \tag{48}$$

$$\langle \Delta\, p's'|\; \vec{j}(o)\,|\,\Delta\, ps \rangle = \psi_{s'}^{+}\, [\frac{e}{2M_\Delta}(G_{E0}^{\Delta\Delta}\, \vec{P} + i\, G_{M1}^{\Delta\Delta}\, \vec{\sigma}_{\Delta\Delta} \times \vec{q})]\psi_s$$

where

$$G_{E0}^{\Delta\Delta} = (\tfrac{1}{2} + t_3)\, G_{Ep}^{NN}$$

$$G_{M1}^{\Delta\Delta} = (\tfrac{1}{2} + t_3)\, G_{Mp}^{NN} \tag{49}$$

$$\langle \tfrac{3}{2}\|\, \sigma_{\Delta\Delta}\, \|\tfrac{3}{2}\rangle = 2\sqrt{15} \tag{50}$$

Here we have already assumed that no quadrupole and octupole form factors $G_{E2}^{\Delta N}$, $G_{E2}^{\Delta\Delta}$, and $G_{M3}^{\Delta\Delta}$ occur. Similar expressions can be obtained for other isobar and transition currents. In the following only the (NΔ), NN (1470) and ($\Delta\Delta$) configurations are considered.

5.1. Elastic Electron-Deuteron Scattering

In the impulse approximation the elastic e-d scattering cross section is described completely by two structure functions $A(q^2)$ and $B(q^2)$.

$$\frac{d\sigma}{d\Omega} = \frac{d\sigma}{d\Omega}\Big|_{Mott} \left(A(q^2) + \tan^2\frac{\Theta}{2} B(q^2) \right). \tag{51}$$

These structure functions depend only on the squared four-momentum q^2 and are related to the deuteron charge G_C^d, magnetic dipole G_M^d, and electric quadrupole G_Q^d form factors by

$$A(q^2) = (G_C^d)^2 + \frac{2}{3} \eta (G_M^d)^2 + \frac{8}{9} \eta^2 (G_Q^d)^2$$

$$B(q^2) = \frac{4}{3} \eta (1+\eta) (G_M^d)^2 \tag{52}$$

$$\eta = q^2/4 M_d^2 .$$

Using nonrelativistic deuteron wave functions the form factors are easily evaluated. Fig.14. shows the results for the Bryan-Scott potential [29]. Here also lowest order isoscalar meson exchange currents (MEC) have been included. While $A(q^2)$ is not changed very much, since it is dominated by the charge and quadrupole form factors, $B(q^2)$ is considerably enhanced.

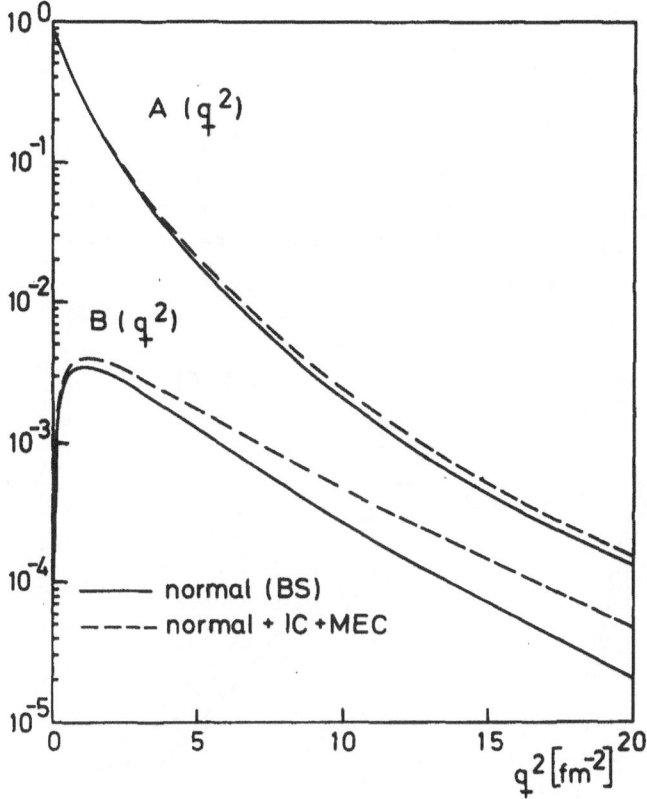

Fig.14. Deuteron structure functions with and without inclusion of IC
and MEC for Bryan-Scott potential (BS).

The region of high momentum transfer up to $q^2 = 150$ fm^{-2} has recently
been explored at SLAC [30]. The experimental points for A(q^2) and some
theoretical results using Reid soft core potential [31] are shown in
fig.15. Also in this region of momentum transfer A(q^2) is not drastically changed by IC and does not lead to serious disagreement with experiment as has been conjectured in ref. [30]. Again B(q^2) is much
more sensitive to interaction effects and, therefore, experimental information would be very desirable. Fig.16. shows the variation of A(q^2)
using different potentials relative to RSC. The differences are con-

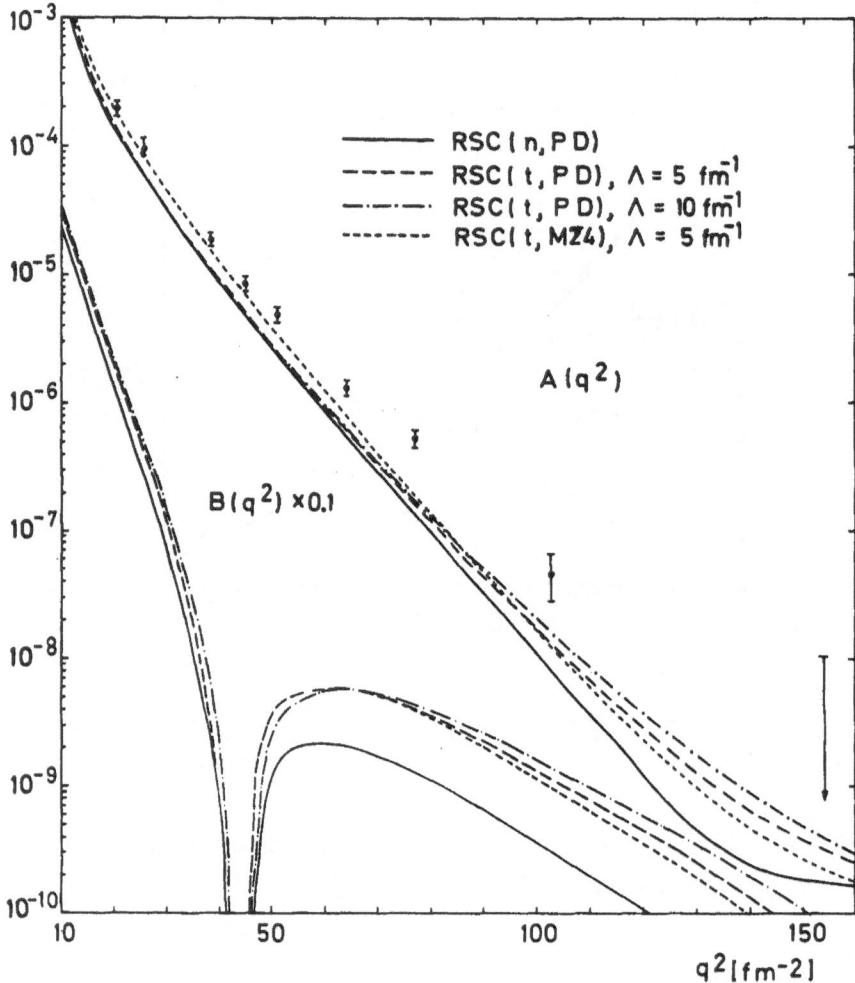

Fig.15. Deuteron structure functions for Reid soft core potential (RSC).
"n" refers to normal deuteron, "t" to inclusion of IC. "PD"
means phenomenological dipole and "MZ4" Mainz four-pole fit [32]
of nucleon electromagnetic form factors.

siderable in this region of high momentum transfer and of the same order than IC and MEC contributions. In addition relativistic corrections
and higher order MEC should be included which have been left out in
these calculations.

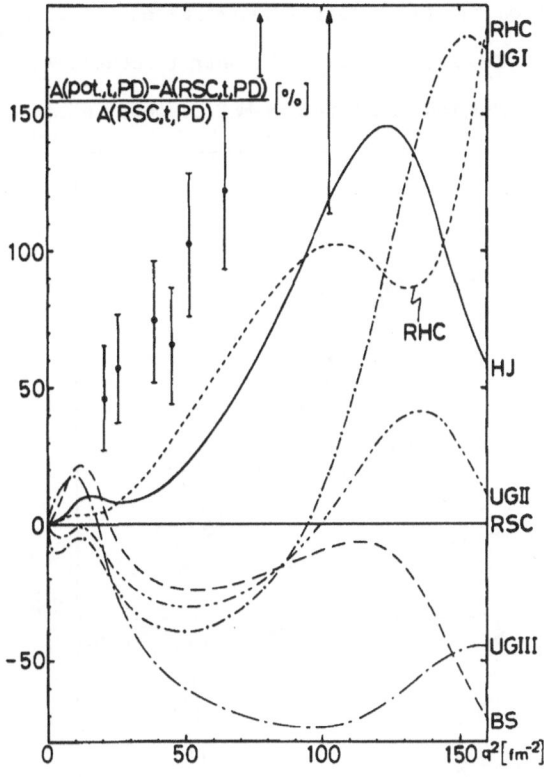

Fig.16. Deviation of $A(q^2)$ from RSC in percent for various potentials:
Reid hard core (RHC), Hamada-Johnston (HJ), Bryan-Scott (BS),
Ueda-Green (UGI, UG II, UG III).

5.2. Deuteron Photodisintegration

More detailed and additional information on effects of IC and MEC are
obtained by studying the photodisintegration since in this process also
isovector transitions do occur, which are particularly important for
the pionic MEC contributions. The most elaborate calculation of $d(\gamma,p)n$
in the framework of conventional nuclear theory has been done by Partovi [33]. This treatment has been extended to include IC and MEC
[34-36].

The total cross section and the relative contributions of IC and MEC are
shown in fig.17. from threshold to 140 MeV. Near threshold where M1 dominates the main contribution comes from π-MEC of about 6% in agreement

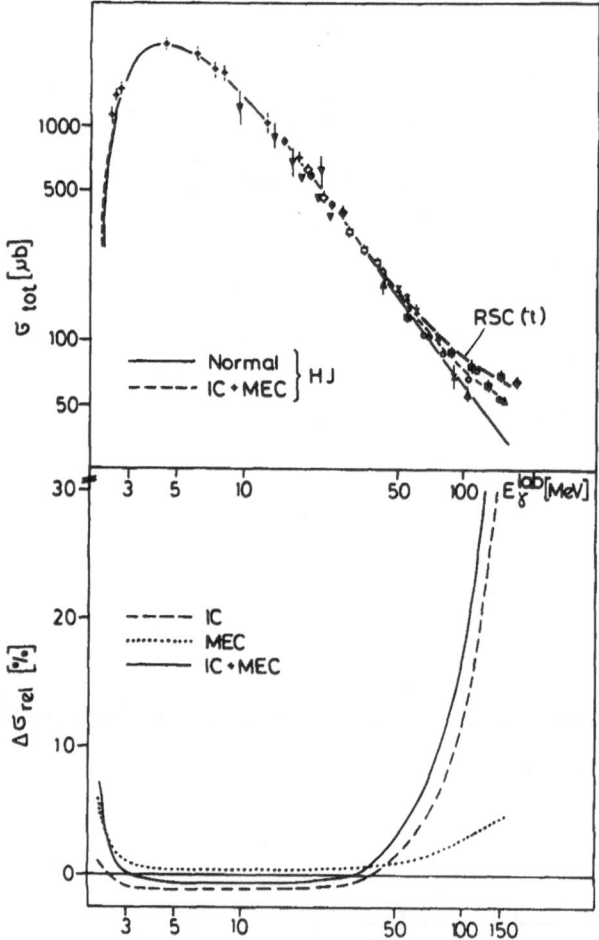

Fig.17. Total cross section for d(γ,p)n with and without interaction
effects for two potentials. Lower part shows relative contributions of IC and MEC for Hamada-Johnston (HJ).

with other calculations [37], while IC contributions are smaller partly
because of wave function renormalization. The largest IC contribution

involves the M1 transition $^3D_1(NN) \rightarrow {}^5D_0(N\Delta)$. Between 4 and 50 MeV where E1 dominates there is only a slight decrease due to renormalization. However, above 60 MeV interaction effects become increasingly important. Here IC dominate because of their higher momentum components. At 140 MeV the total cross section is enhanced by about 30% which seems to be in agreement with experimental data, though they scatter rather widely.

The angular distribution for an outcoming proton at E_γ = 140 MeV is shown in fig.18. One readily sees that inclusion of IC and MEC lead to a much better agreement with experiment. Again more accurate experimental data is desirable. The various potentials give rather similar results though some differences in detail exist.

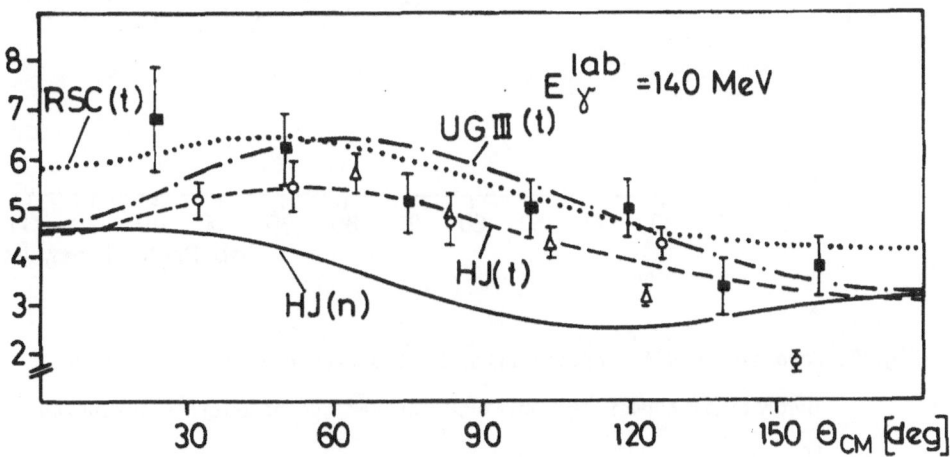

Fig.18. Photoproton angular distribution at E_γ = 140 MeV for various
potentials.

Finally, we would like to mention a recent experiment performed at Mainz by Hughes et al. [38] where the photoproduced protons are measured in forward direction (θ = 0°). Fig.19. shows the experimental results together with theoretical calculations. It seems that interaction effects cannot completely resolve the discrepancy between theory and experiment, though various potentials show different degrees of discrepancy.

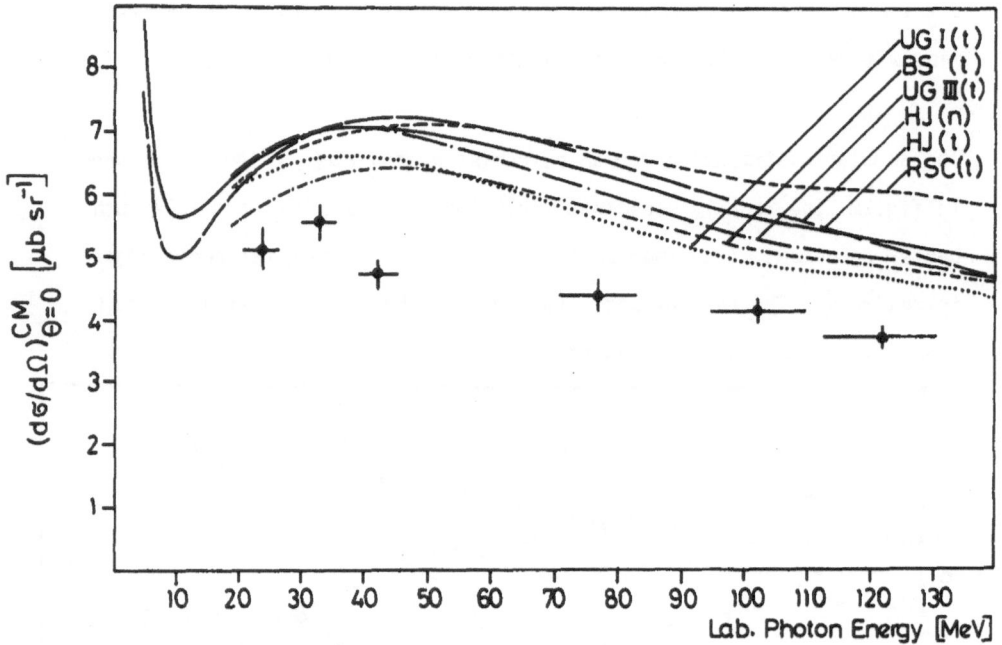

Fig.19. Zero degree differential cross section for d(γ,p)n from ref.38. Theoretical curves are obtained for various potentials including IC and MEC.

5.3. Deuteron Electrodisintegration

Further information is obtained in d(e, e'p)n since energy and momentum transfer can be varied independently. If no nucleon is detected the differential scattering cross section is given in terms of a longitudinal (f_{long}) and a transverse (f_{trans}) form factor. At 180^0 only f_{trans} con-

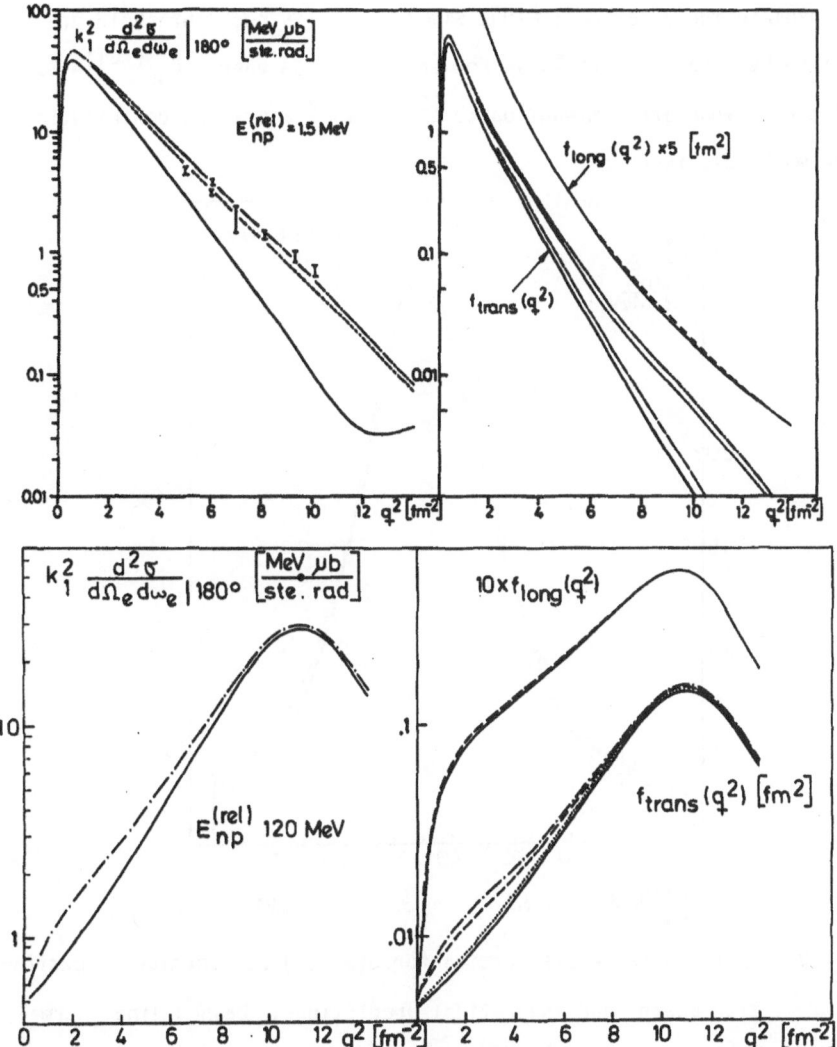

Fig.20. 180°-cross section and inelastic longitudinal and transverse
form factors for two excitation energies using Hamada-Johnston
potential.

tributes. Fig.20. shows the 180°-cross section and the two form factors
near threshold ($E_{np}^{(rel)}$ = 1.5 MeV) and for $E_{np}^{(rel)}$ = 120 MeV [36, 39].
Again one sees near threshold the dominance of π-MEC which are needed

to explain the experiment [40]. Similar results near threshold are obtained by other authors [41]. For an excitation energy $E_{np}^{(rel)} = 120$ MeV IC corrections are dominant up to $q^2 = 6$ fm^{-2} and π-MEC contributions are much less important.

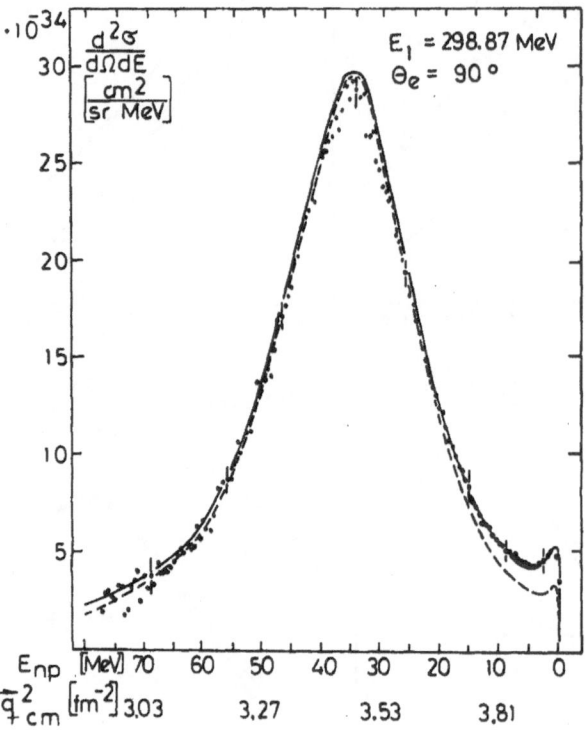

Fig.21. Differential cross section for d(e, e') as function of excitation energy and theoretical predictions. Dashed line: normal, solid line: normal + MEC + IЄ.

Finally, fig. 21 shows experimental results obtained recently by Simon et al. [42] in Mainz. The agreement with the theoretical prediction if interaction effects (IC and MEC) are taken into account is impressive.

6. Spectator Isobar Production

The foregoing discussion has shown that inclusion of IC gives a better theoretical understanding and description of various processes. However, the evidence is rather indirect. A more direct "proof" of the existence of isobars in nuclei might be more convincing. The most direct method for measuring an N* in the deuteron is the so-called spectator isobar production (see fig.22) where a high energetic particle interacts with the other particle and thus shakes loose the N* acting as a spectator and recoiling backward predominantly [4, 43]. At first sight this looks very suggestive and simple. However, there are other reactions leading also to a recoiling isobar without involving the IC in the deuteron. Thus the main problem is to distinguish the spectator signal from other competing background reactions.

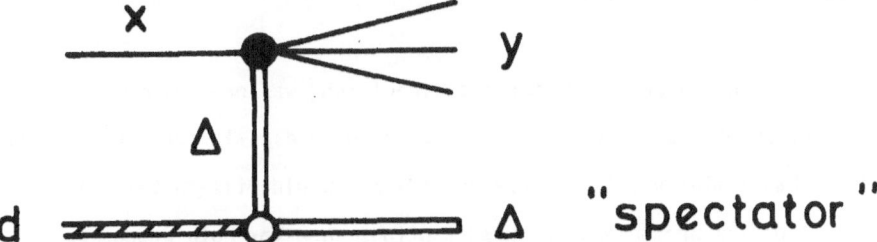

Fig.22. Spectator isobar production for the (ΔΔ) component of the deuteron.

Such an experiment for measuring the double-Δ component of the deuteron has first been proposed by Nath, Weber and Kabir [44], though the experiment itself was done only recently [45]. They discussed the forward proton production from deuterium by incident high enery π^- (fig.23a). This reaction can also be viewed as a double charge exchange which would depend sensitively on the existence of a Δ^{++} in the deuteron. In this reaction the Δ^- would be the spectator, but as spectator signal the forward produced proton is used. Fig.23b shows the most important competing background

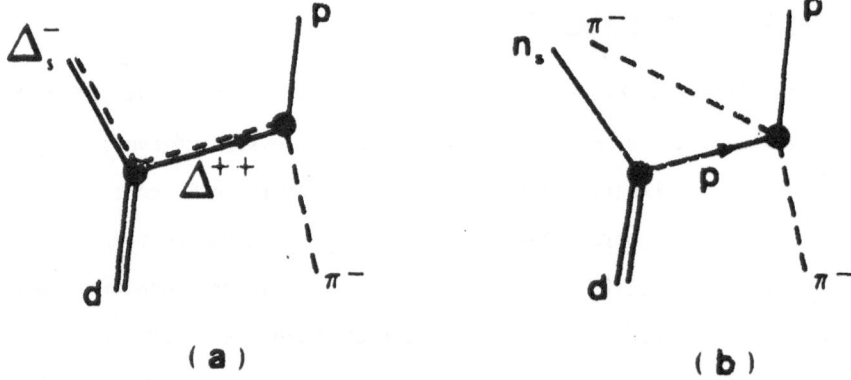

Fig.23. a) Δ^{++} exchange graph for $\pi^- d \rightarrow p\Delta^-$ spectator

b) Nonresonant $\pi^- n_s$ background.

reaction from the quasi-elastic back scattering of the incident π^- from the proton inside the deuteron.

To minimize this background, the experiment [45] was done with the Saclay π-beam at about 1.1 GeV/c where the $\pi^- p$ backward elastic scattering cross section has a minimum. Thus presence of a Δ^{++} would fill in this minimum. In addition the proton angular distributions show for both processes a different behaviour. While protons created from Δ^{++} exchange (fig.23a) have a maximum at 0° the proton from $\pi^- p$ back scattering show for 0° a distinct minimum (see fig.24.)

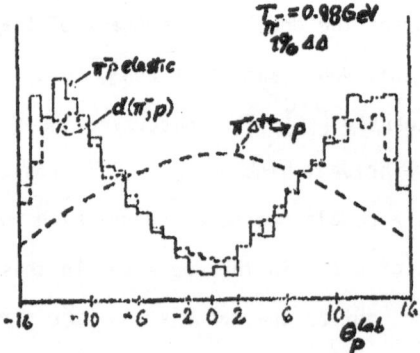

Fig.24. Schematic angular proton distributions for $\pi^- p \rightarrow \pi^-$ (solid), $\pi^- d \rightarrow p\Delta^-$ via Δ^{++} exchange (dashed), and quasi elastic (dotted) from ref.43.

Fig.25. shows the experimental angular distributions indicating the decrease with increasing proton angles as predicted by Δ^{++} exchange. Furthermore, the deuterium cross sections are systematically larger than

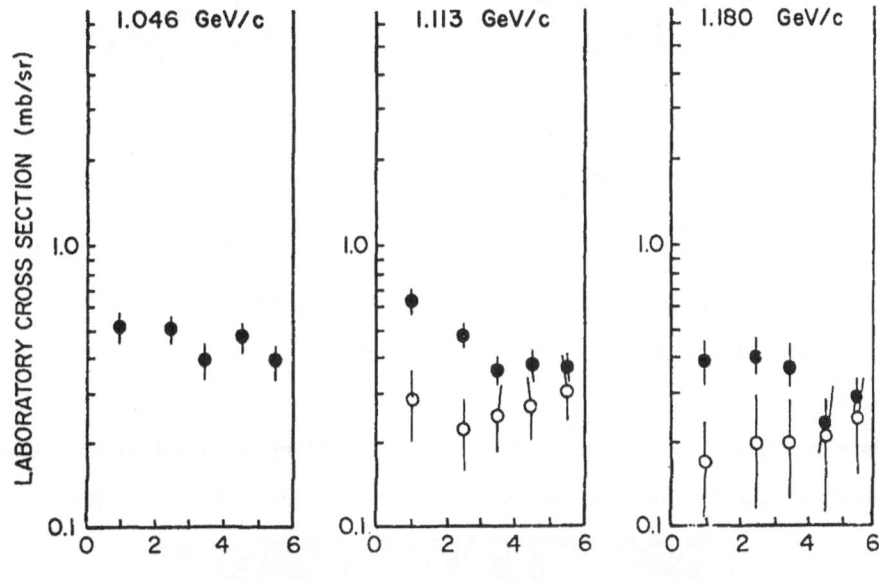

PROTON ANGLE IN LABORATORY (degrees)

Fig.25. Lab. angular distributions of protons for deuterium (full circles) and hydrogen (open circles).

the hydrogen ones. From these data an upper limit of 0.4% for the ($\Delta\Delta$) probability has been extracted.

The first analysis of experimental data in the spirit of spectator isobar production where the spectator particle itself is used as a signal for this process has been done by M. Goldhaber et al. [46]. They looked into deuteron break up ($md \to m\pi^+\pi^- pn$) with high energetic mesons ($\pi^\pm$(15 GeV/c), K^+(12 GeV/c)) to search for events where the meson is scattered "elastically" from the virtual Δ^- in the deuteron leaving behind a spectator Δ^{++} (see fig.26.) recoiling backward.

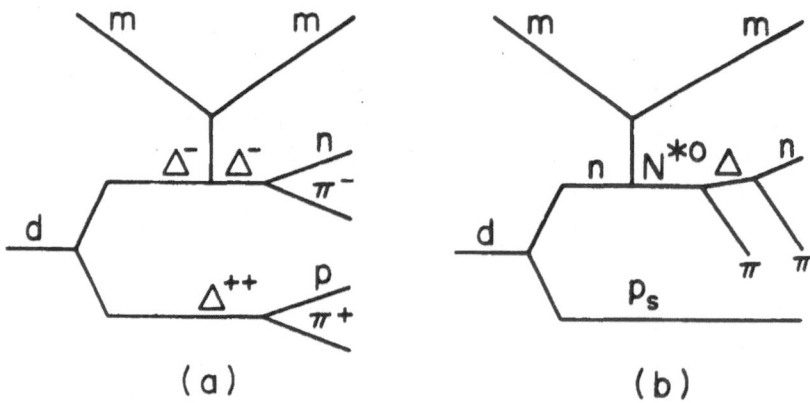

Fig.26. a) Spectator production of Δ^{++} by a meson on the deuteron.

b) Background reaction using a N^* model.

They observed a significant number of Δ^{++} recoiling backward in the laboratory (fig.27). The invariant mass $M(n\pi^-)$ distribution for Δ^{++} (back)

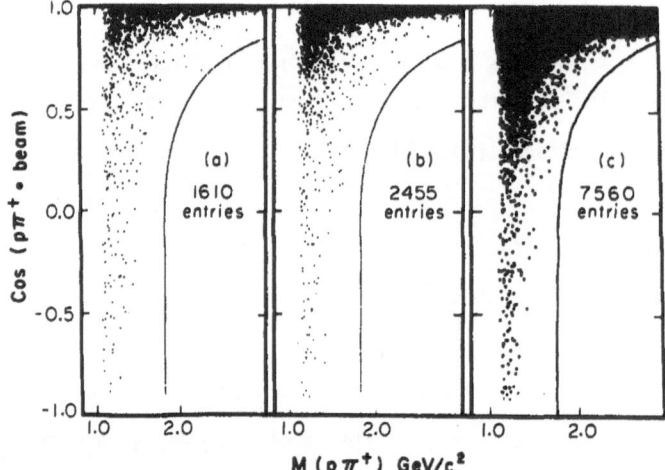

Fig.27. Invariant mass $M(p\pi^+)$ versus cos $(p\pi^+$, beam) in the lab for
md $\rightarrow$ m$\pi^+\pi^-$pn, where m is (a) π^+, (b) π^-, and (c)K$^+$. Curves show
kinematic boundaries (from ref.46).

events (fig.3a) show a distinct peak in the Δ region suggesting that the
$n\pi^-$ combination is a Δ resonance a substantial fraction of the time.

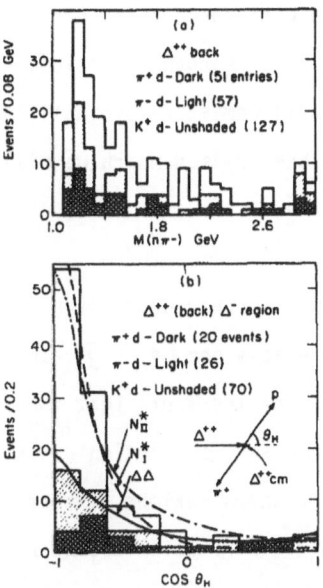

Fig.28. Sum of the data of all three experiments: (a) M(π^-n) for events
with pπ^+ in backward Δ^{++} region, (b) cos θ_H(pπ^+) = $\hat{p} \cdot \hat{v}$ where $\hat{p}$ is
proton direction in the pπ^+ rest frame and $\hat{v}$ is the direction of
pπ^+ in the lab frame for events in the Δ^{++} (back) Δ^- region. The
N* model curves (broken) are normalized to cos θ_H < 0. N*$_I$ uses
pure Hulthén spectrum, N*$_{II}$ uses a modified spectator proton spectrum. The $\Delta\Delta$ model curve (solid) is normalized to cos θ_H > 0 (from
ref.46).

However, not all Δ^{++} (back) and Δ^- events originate from the double Δ component in the deuteron since for the background reaction b) of fig.26.
the spectator proton p_s together with one π^+ can simulate a Δ^{++} (back)
event. Thus one has to analyse the angular distribution of pπ^+ to distinguish genuine Δ^{++} from $p_s\pi^+$ events. This is shown in fig.28b, where
N*$_I$ refers to background production using a Hulthén deuteron wave function, N*$_{II}$ to the same process using a modified proton distribution to
include double scattering corrections. The $\Delta\Delta$ curve gives the prediction
using the double-Δ component which should be symmetric. A slight asymme-

try comes from loss of events with nonstopping protons. The strong asymmetry shows the importance of the background reaction. From this analysis an upper limit of 0.4% for the (ΔΔ) admixture probability in the deuteron is obtained. Some criticism has been raised which we will not discuss here but refer to the literature [4, 43].

This first analysis of spectator isobar production triggered a variety of similar investigations (for reviews see [4, 43]). In most of these investigations one has not made use of the fact, that the spectator proton p_s of the background has the momentum distribution given by the deuteron wave function which has a rather sharp maximum at zero momentum while the Δ spectator has a much broader distribution around 2 fm^{-1} (see fig.12.). Thus eliminating all protons having momentum less than 1 fm^{-1} will reduce the background appreciably without significant loss of Δ events.

This technique was first successfully applied by Benz and Söding [47] in studying the inclusive reaction $\gamma d \rightarrow p\pi^+$ + anything. Their analysis is illustrated in figs.29. and 30. The strong correlation ridge in fig.29.

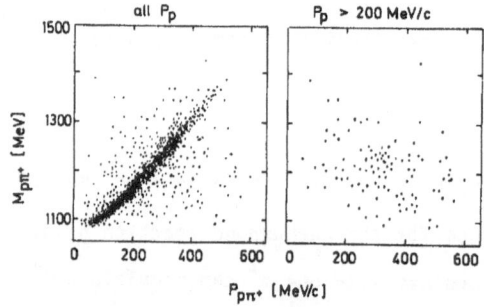

Fig.29. Correlation of momentum with mass of the backward moving $p\pi^+$ systems from $\gamma d \rightarrow p\pi^+$ + anything (from ref.47).

shows the importance of the background reaction since for Δ^{++} such a correlation should not occur and indeed disappears if the slow protons are removed. For a genuine Δ^{++} the decay products $p\pi^+$ have characteristic properties different from the background $p\pi^+$ products. In fact, the effective mass distribution and the decay angular distributions are consistent with the $(\Delta\Delta)$-model. Assuming that all spectator events come

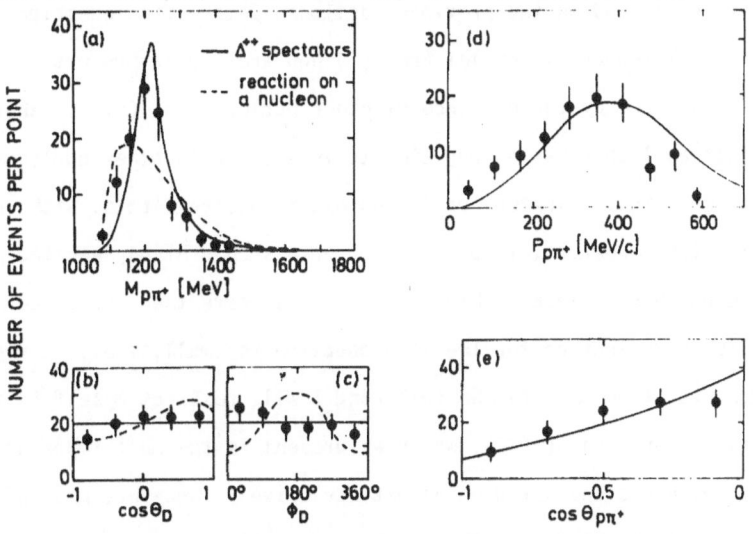

Fig.30.(a) Effective mass distribution, (b) and (c) polar and azimuthal distributions of π^+ in the $p\pi^+$ rest frame with respect to the incident photon in the lab system, (d) lab momentum distribution, and (e) lab production angular distribution with respect to the photon direction for the backward moving $p\pi^+$ system after removal of spectator protons. Solid curves are predictions of $(\Delta\Delta)$-model, dashed curves for background reaction normalized to experiment (from ref.47).

from the double-Δ component and using the quark model prediction $\sigma_{\gamma\Delta} = \sigma_{\gamma p}$ an estimated 3.1% total $(\Delta\Delta)$ probability in the deuteron is obtained.

7. Remarks on Real N*'s in Nuclei

One of the new interesting topics of medium energy physics is the study of real nucleon isobars or baryon resonances inside a nucleus. With real isobars we mean isobars on-mass-shell which have to be distinguished from virtual isobars discussed in the previous sections. Interesting questions associated with real isobars in nuclei are: (i) how are isobar properties and processes influenced by the presence of other nucleons and (ii) is there a chance that an isobar has enough time to interact with other nucleons in order to form a highly excited quasibound exotic nuclear state. With respect to the last question, in order to observe such a highly excited nuclear state one has to look at kinematic regions where the usually dominating impulse approximation for the N* production is small, i.e., at regions of high momentum transfer. In the following I will restrict myself to the lowest mass isobar, the $\Delta(1236)$, which at present is the most important isobar and which has been studied almost exclusively. However, many of the conclusions will be valid for other isobars as well.

The simplest system to study the effect of a surrounding nuclear medium on the properties of a Δ is nuclear matter [48-50]. This question is closely related to the problem of pion propagation in nuclear matter. To study the modification of the Δ one needs a dynamical model. This is usually obtained by coupling a bare $\bar{\Delta}$ to the πN channel, which contributes to the self energy of the real dressed Δ, in particular it determines the width. This is diagrammatically shown in fig.31.

If the Δ is now placed into a medium, e.g., nuclear matter for simplicity, the Δ will wear a different dress than the free Δ because of (i) phase space restrictions on the intermediate (πN) channels because of the Pauli principle (Pauli blocking), and (ii) interaction of πN channels with sur-

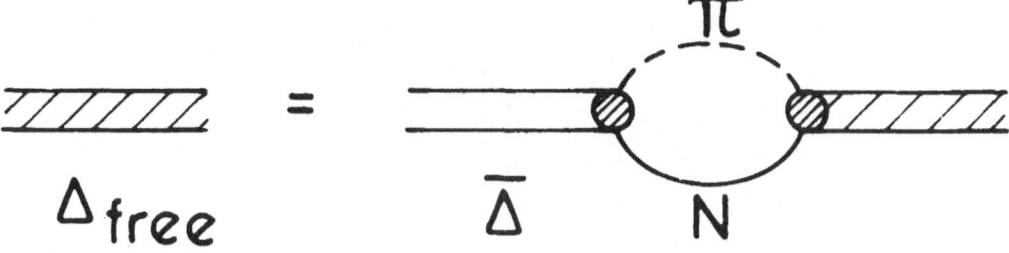

Fig.31. Dynamical model for free dressed Δ propagator.

rounding nucleons (see fig.32.). These two effects lead to a shift of the resonance energy (mass shift) and to a reduction of the width (fig.33).

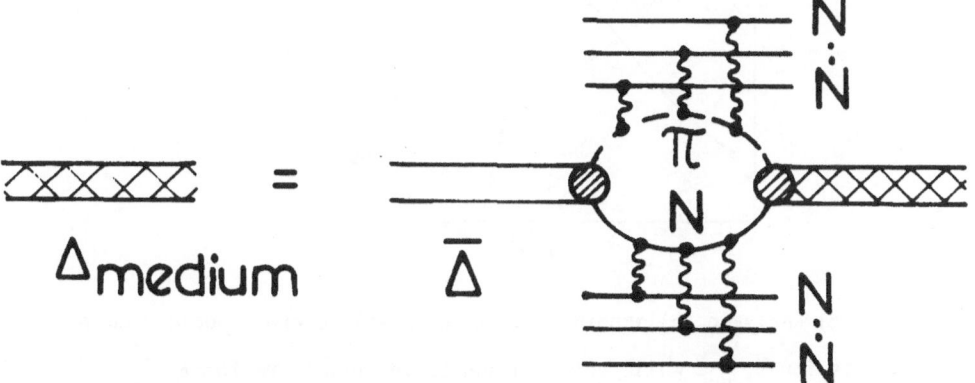

Fig.32. The Δ propagator in a nuclear medium.

However, there will be additional effects arising from the fact that the pion-nuclear matter optical potential is dominated by the Δ resonance and thus is nonlocal. In fact, one has to solve a selfconsistent problem by considering simultaneously the π and Δ propagators. As a result certain modes of mixed π and Δ-hole excitations do occur. The importance of these modes have been stressed by Lenz [48] and I will refer to his work for details of this interesting phenomenon.

I will now turn to Δ's in finite nuclei which have been studied using

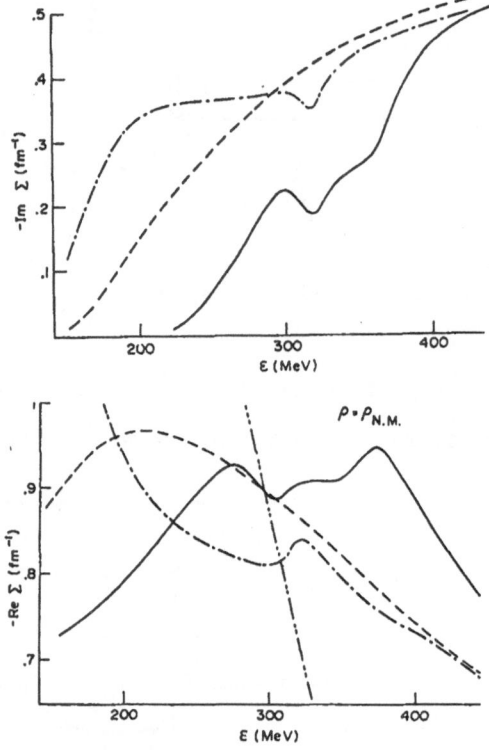

Fig.33. (a) Real and (b) imaginary parts of Δ self energy : dashed curve
is for Δ_{free}, while dot-dash and solid curves are for Δ_{medium}
without and with Pauli blocking, evaluated at nuclear matter den-
sity $\rho_{N.M.}$. Dash-double dot line represents $\overline{\Delta}$ (from ref.49).

similar models in various approximations [51, 52] . While Dillig and
Huber have stressed the possibility of collective Δ-hole excitations
in a schematic model similar to the giant dipole resonance in nuclei, I
have studied in some detail the (NΔ)-system.

These investigations were triggered by an experiment performed at Saclay by P. Argan et al. [53] who studied the reaction γ + ^{4}He →p+π⁻+(ppn)
in the region of the Δ and observed p and π⁻ in coincidence for various

fixed recoil momenta of the residual (ppn)-system.

The experimental results are shown in fig.34. for two recoil momenta of 50 and 200 MeV/c, respectively. While for 50 MeV/c the impulse approximation gives a satisfactory description of the reaction there is an excess above the impulse approximation for a recoil momentum of 200 MeV/c in the form of a rather small "resonance" with a resonance energy slightly downshifted from the free Δ-resonance energy and a width less than 40 MeV. This small width is a particularly striking feature and it has been suggested that this "resonance" could be a resonant state formed by the Δ with other nucleons. Two extreme cases of such a resonant state are dia-

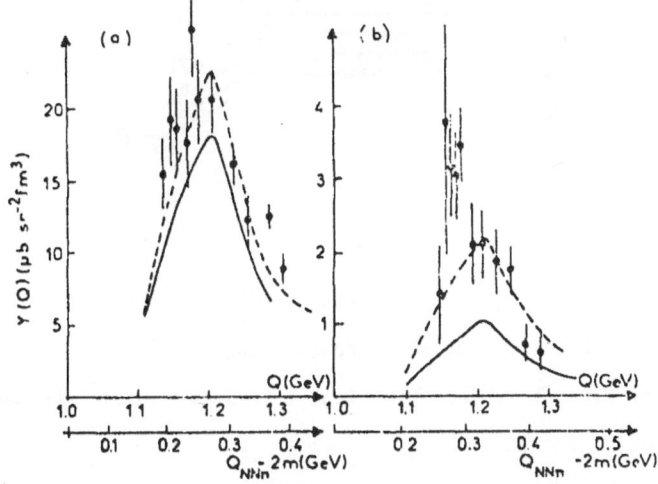

Fig.34. Differential cross section of the reaction ^{4}He(γ, pπ^-) for recoil momenta of 50 MeV/c (a) and 200 MeV/c (b) from ref. 53 . The solid curves are calculated in the impulse approximation and the dashed curves are the same normalized by 1.25 (a) and 2.1 (b), respectively.

grammatically sketched in fig.35. In one case the Δ is produced at a correlated pair forming a (NΔ) state, in the other case the Δ propagates through the whole nucleus.

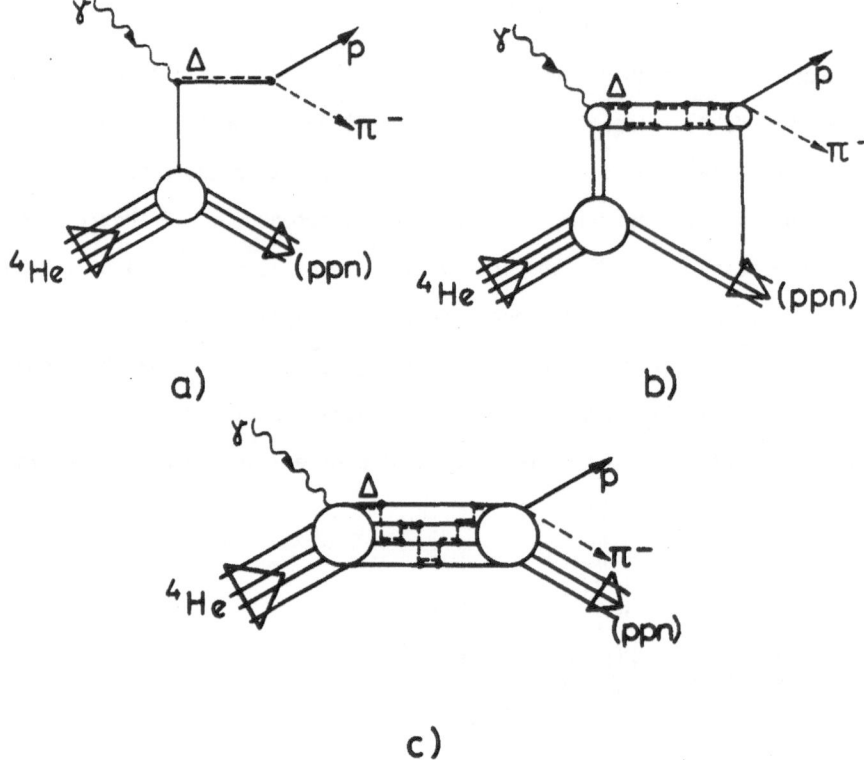

a)

b)

c)

Fig.35. Photoproduction of a Δ-resonance in ^{4}He:

    a) describes the impulse approximation,

    b) corresponds to the formation of an intermediate resonant

       (NΔ)-state and

    c) to an intermediate resonant (3NΔ)-state.

If the interpretation of the experimental results is correct, then I
think, the most striking feature one has to understand in such a model is the small width of the resonance. It means the bound Δ lives
three to four times longer than a free Δ having a width of 120 MeV
corresponding to a lifetime of $\tau_\Delta = 0.6 \times 10^{-23}$s.

A free Δ decays predominantly into $N + \pi$ and the decay width Γ_Δ^{free} is given by

$$\Gamma_\Delta^{free} = \frac{f_{\Delta N\pi}^2}{4\pi m_\pi^2} \frac{(M_\Delta + M)^2 - m_\pi^2}{6 M_\Delta^2} k_o^3 \tag{53}$$

where

$$k_o = [(M_\Delta^2 - M^2 - m_\pi^2)^2 - 4 M^2 m_\pi^2]^{1/2}/2 M_\Delta \tag{54}$$

is the pion momentum in the Δ-rest system, M_Δ, M and m_π are Δ nucleon and pion mass, respectively, and $f_{\Delta N\pi}$ is the ($\Delta N\pi$)-coupling constant.

As in nuclear matter the width Γ_Δ^{bound} of a bound Δ is reduced by two effects restricting the available phase space. First, the Δ is off-mass-shell and one has to replace M_Δ^2 by $S_\Delta = E_\Delta^2 - p_\Delta^2$ in eq.(54) resulting in a smaller momentum $k(p_\Delta)$ for the outcoming pion which goes to zero at approximately $p \approx 2$ fm^{-1} if the Δ is bound by 20 to 30 MeV. Secondly, the Pauli principle forbids some of the final states for the outcoming nucleon due to the presence of the other nucleons. Therefore, the pionic decay probability will be reduced. However, there is another effect, which has the opposite tendency to increase the width. This is the open-ing of new decay modes, essentially the nonpionic decay mode $\Delta + N \rightarrow N + N$. We would like to point out that for a $T = 2$ ($N\Delta$)-pair the nonpionic de-cay is forbidden because of isospin conservation. All these effects are well known in the physics of hypernuclei.

In such models of a Δ-nucleus the Δ is essentially treated as a stable particle interacting with nucleons through a static instantaneously act-ing potential. However, one has to be aware of some shortcomings of

such an approach which might render some of the conclusions to be drawn
only qualitative. First, there is the short lifetime of the Δ which is
at least one order of magnitude shorter than typical nuclear time periods. Thus, it seems very unlikely that the Δ can form a resonant Δ-nu-
cleus before decaying. However, if the pionic decay mode is dominant
the emitted pion might be reabsorbed by another nucleon forming again
a Δ corresponding to the collective Δ-hole mode in nuclear matter and
thus increase its lifetime.

The second point is concerned with the use of an instantaneously acting
potential. Because the interaction time associated with the exchange of
a pion is of the order of 0.5×10^{-23}s and therefore comparable to the
lifetime of a free Δ retardation effects, i.e., nonlocal effects, are
expected to become important. Thus a more realistic description would
have to include explcitly the meson degrees of freedom. For example, in
the (NΔ)-system one would have to couple the (NΔ)-channel with the (NNπ)-
channel completely analogous to what we have mentioned in the case of
nuclear matter [48]. However, finite nuclei are paradoxically a little
bit more complicated than nuclear matter, and, therefore, it seems worth-
wile to start with the simpler approach to obtain an at least qualitative picture of exotic Δ-nuclei.

Following along these lines we have investigated the (NΔ)-system in order to see whether a bound state exists for which the lifetime of the
bound Δ is considerably increased. The (NΔ)-interaction is easily obtained from π- and ρ-exchange (see fig.36.) in the nonrelativistic limit.

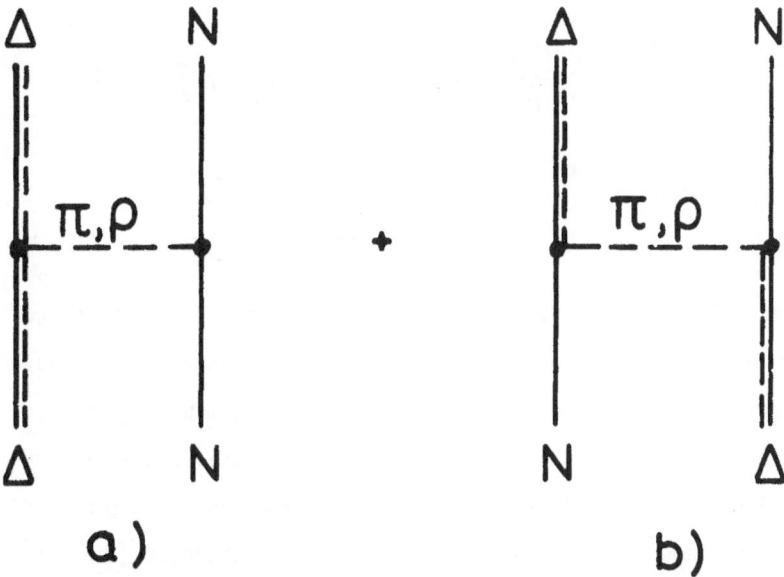

Fig.36. Boson exchange graph for the direct (a) and the exchange (b)
N-Δ-interaction.

From the above discussion we expect the N-Δ-exchange interaction (fig.
36b) to be the most important one. Since the Δ can decay into N + π,
the exchanged pion can be on-mass-shell and there will be a pole in the
physical region in the pion propagator. This pole lies in the complex
plane if one uses a complex Δ-mass $M_\Delta - i\Gamma/2$ in order to schematically include the effect of the finite lifetime. The effect of this pole is that
the (NΔ)-potential becomes complex and will not be of Yukawa-type but
rather long ranged. To investigate the stationary solutions one takes
the real part which for π-exchange is given by

$$Re\ V_{N\Delta}^{ex} = \vec{\tau}_{\Delta N}(1) \cdot \vec{\tau}_{N\Delta}(2) \left(\vec{\sigma}_{\Delta N}(1) \cdot \vec{\sigma}_{N\Delta}(2) V_s^{ex} + S_{12} V_T^{ex} \right)$$

$$+ (1 \leftrightarrow 2)$$

(55)

where

$$V_S^{ex} = \frac{f_{\Delta N\pi}^2}{12\pi m_\pi^2} \frac{1}{r} \frac{d^2}{dr^2}(rf) \, , \quad V_T^{ex} = \frac{f_{\Delta N\pi}^2}{12\pi m_\pi^2} r \frac{d}{dr}\left(\frac{1}{r}\frac{df}{dr}\right), \tag{56}$$

$$f(r) = \cos(mr) e^{-\gamma r}/r \, , \quad m^2 = (M_\Delta - M)^2 - m_\pi^2 - \Gamma^2/4 \, , \tag{57}$$

$$\gamma = (M_\Delta - M)\Gamma/2m \, .$$

This potential has to be supplemented by contribtuions from ρ-exchange and from the direct ($N\Delta$)-interaction (fig.36a). Details of the complete ($N\Delta$)-interaction can be found in ref.[52]. Evaluating the spin and isospin operators one finds that the central force V_S gives attraction for $S = T (= 1$ or $2)$ and repulsion for $S \neq T$. The attraction is strongest for $S = T = 2$ (9 times stronger than for $S = T = 1$). Furthermore, the tensor force V_T which is of crucial importance for binding is also roughly twice as strong for $S = T = 2$ as for $S = T = 1$. Therefore, if binding occurs at all it can be expected most likely for a ($T = 2$, $J^P = 2^+$)-state. The explicit evaluation confirms this conclusion.

The decay width of such a bound ($N\Delta$)-state is easily obtained in the impulse approximation neglecting final state interaction. Since this state has isospin $T = 2$ only the pionic decay mode is allowed: $(N\Delta)_{bound} \rightarrow N + N + \pi$. The two graphs which contribute to the decay amplitude are shown in fig.37. Explicitly one obtains in the nonrelativistic limit

$$\Gamma\left((N\Delta)_{bound} \rightarrow N + N + \pi\right) = \gamma_\Delta \, \Gamma_\Delta^{free} \tag{58}$$

where the reduction factor γ_Δ is given by

$$\gamma_\Delta = \int d^3p \, \chi^2(\vec{p}) \, |\Psi_{N\Delta}(\vec{p})|^2$$

$$- \frac{3}{4\pi} \int d\Omega_\chi \, d^3p \, \Psi_{N\Delta}^+(\vec{p}) \, \vec{\tau}_{\Delta N}(1) \cdot \vec{\tau}_{\Delta N}^+(2) \, (\vec{\sigma}_{\Delta N}(1) \cdot \vec{\chi}') \tag{59}$$

$$\times (\vec{\sigma}_{\Delta N}^+(2) \cdot \vec{\chi}) \, \Psi_{N\Delta}(-\vec{k} - \eta\vec{p})$$

with

$$\vec{x} = \vec{k}/k_0 \ , \ \vec{x}' = (\eta \vec{k} - (1-\eta^2)\vec{p}) \ , \ \eta = M/M_\Delta \ .$$ (60)

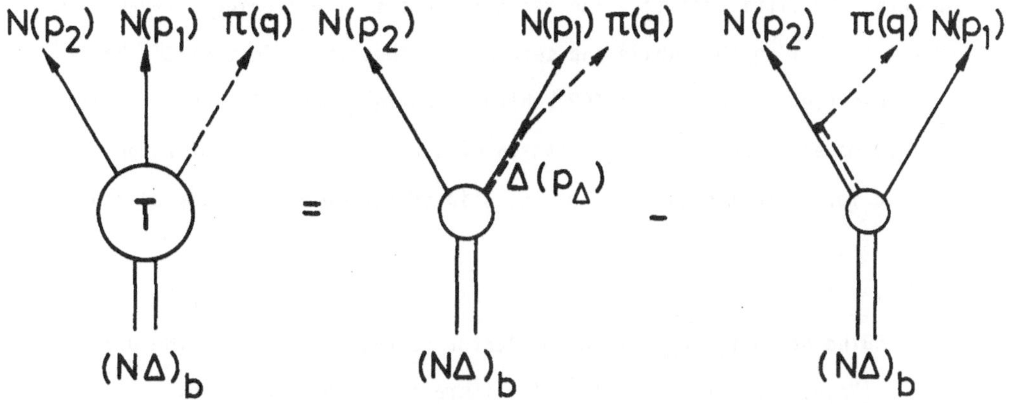

Fig.37. Decay of a bound (NΔ)-state in the impulse approximation.

The first term would be one if $\kappa = 1$. However, since the Δ is off-mass-shell one has $\kappa < 1$ and the integration over dp is restricted to the region $p \lesssim 2$ fm^{-1}. Therefore, this term is less than one and it describes just the effect of the Δ being off-mass-shell. The second term stems from the interference of the two graphs of fig.37. and includes just the effect of the Pauli principle between the two nucleons in the final channel. The interference term is largest for a $S = T = 2(N\Delta)$-state.

Since the (NΔ)-exchange interaction depends on the width of the bound Δ one faces a problem of selfconsistency. However, this does not introduce any additional complications because the exchange interaction is only weakly dependent on the width and thus selfconsistency is reached in two to three steps. This can be seen in table 4 where the numerical results for the $(T = 2, J^P = 2^+)$-state are summarized. The regularization parameter Λ is necessary to suppress the pathological behaviour of the potential near the origin. Fortunately, the results are not so strongly dependent on this parameter Λ. The reduction factor γ_Δ is given as a difference of two numbers. The first describes the reduction due to the Δ being off-mass-shell (first term of eq.(59)) while the second describes the effect of the Pauli principle (second term of eq.(59)) which is quite important. Thus one obtains in this model a bound $T = 2$ ($N\Delta$)-state with a binding energy of roughly 30 MeV and for which the lifetime of the Δ is increased by a factor of four to five.

Tab.4. Binding energy E_b, reduction factor γ_Δ and width $\Gamma_{N\Delta}$ of the $J = T = 2$ ($N\Delta$)-state. For $\Lambda = 6$ fm^{-1} the iteration steps are given.

potential	Γ_{pot} [MeV]	Λ [fm^{-1}]	E_b [MeV]	$\gamma_{N\Delta}$	$\Gamma_{N\Delta}$ [MeV]
$(\pi+\rho)^{ex+dir}$	25.3	11	34.62	0.364-0.152=0.211	25.3
	120	6	19.56	0.498-0.179=0.319	38.3
	38	6	26.62	0.445-0.161=0.283	34.0
	34	6	27.02	0.442-0.160=0.282	33.9

Acknowledgement

It is a pleasure to give much credit to H.J. Weber with whom I am preparing a more extensive and detailed review on this subject matter. The help of W. Fabian and G. Horlacher and interesting discussions with A.M. Green are gratefully acknowledged. Finally, I want to thank Mrs. M. Sturm for her great help in preparing this manuscript which otherwise would not have been finished im time.

References

[1] H.Arenhövel, H.J. Weber, Springer Tracts in Modern Physics, Vol. 65 (1972) 58

[2] L.S. Kisslinger, Proc. Topical Meeting on High Energy Collisions, Trieste (1974)

[3] H. Arenhövel, Proc. Symposion on Interaction Studies in Nuclei, Mainz, ed. by H. Jochim and B. Ziegler, (Amsterdam, 1975), p. 727

[4] H.J. Weber, ibid. p. 749

[5] R. Beurtey, Proc. VI'th Int. Conf. on High Energy Physics and Nuclear Structure, Santa Fe (1975), ed. D. Nagle et al. (AIP, NY) p.653

[6] K. Erkelenz, Phys. Rep. 13C (1974) 191

[7] M. Chemtob, M. Rho, Nucl. Phys. A163 (1971) 1

[8] H. Arenhövel, M. Danos, H.T. Williams, Nucl. Phys. A162 (1971) 12

[9] G. Horlacher, doctoral thesis, Mainz (1976)

[10] P. Haapakoski, Phys. Lett. 48B (1974) 307

[11] J. Durso, M. Saarela, G.E. Brown, A. Jackson, preprint (1976)

[12] S. Jena, L.S. Kisslinger, Ann. Phys. (N.Y.) 85 (1974) 251

[13] H.J. Weber, Phys. Rev. C9 (1974) 1771

[14] P. Haapakoski, M. Saarela, Phys. Lett. 53B (1974) 333

[15] H. Arenhövel, Phys. Lett. 53B (1974) 224

[16] H. Arenhövel, Z. Phys. A275 (1975) 189

[17] E. Rost, Nucl. Phys. A249 (1975) 510

[18] W.N. Cottingham, R. Vinh Mau, Phys. Rev. 130 (1963) 735

[19] M. Chemtob, J.W. Durso, D.O. Riska, Nucl. Phys. B38 (1972) 141

[20] H. Sugawara, F. von Hippel, Phys. Rev. 172 (1968) 1764

[21] R. V. Reid, Ann Phys. (N.Y.) 50 (1968) 411

[22] S. Wagner, P. Winiger, Helv. Phys. Acta 42 (1969) 51

[23] S. Wagner, Phys. Rev. 177 (1969) 2278

[24] A.M. Green, P. Haapakoski, Nucl. Phys. A221 (1974) 429

[25] R.M. Smith, V.R. Pandharipande, Nucl. Phys. A256 (1976) 327

[26] B.D. Day, F. Coester, Phys. Rev. C13 (1976) 1720

[27] K. Holinde, R. Machleidt, preprints, Bonn (1975)

[28] H. Arenhövel, H.G. Miller, Z. Phys. 266 (1974) 13

[29] W. Fabian, H. Arenhövel, H.G. Miller, Z. Phys. 271 (1974) 93

[30] R.G. Arnold et al., Phys. Rev. Lett. 35 (1975) 776

[31] W. Fabian, H. Arenhövel, preprint, Mainz (1976)

[32] F. Borkowski et al., Nucl. Phys. B93 (1975) 461

[33] F. Partovi, Ann. Phys. (N.Y.) 27 (1964) 79

[34] H. Arenhövel, W. Fabian, H.G. Miller, Phys. Lett. 52B (1974) 303

[35] H.G. Miller, Habilitationsschrift, Frankfurt (1974)

[36] W. Fabian, doctoral thesis, Mainz (1975)

[37] D.O. Riska, G.E. Brown, Phys. Lett. 38B (1972) 193

      M. Colocci, B. Mosconi, P. Ricci, Phys. Lett. 45B (1973) 224

      M. Gari, A.H. Huffmann, Phys. Rev. C7 (1973) 994

[38] R.J. Hughes et al., preprint, MPI für Chemie, Kernphysikalische Abteilung,
      Mainz (1976)

[39] W. Fabian, H. Arenhövel, Nucl. Phys. A258 (1976) 461

[40] D. Ganichot et al., Nucl. Phys. A178 (1972) 545

[41] J. Hockert et al., Nucl. Phys. A217 (1973) 14

      J.A. Lock, L.L. Foldy, Ann. Phys. (N.Y.) 93 (1975) 276

[42] G. Simon et al., private communication

[43] H.J. Weber, Proc. VII. Int. Conf. on Few Body Problems in Nuclear and
      Particle Physics, Delhi, 1975, ed. A.N. Mitra (North-Holland 1976);
      Proc. Int. Top. Conf. on Meson Nuclear Physics, Pittsburgh 1976 (to
      be published)

[44] N.R. Nath, H.J. Weber, P.K. Kabir, Phys. Rev. Lett. 26 (1971) 1404

[45] R. Beurtey et al., preprint (1975)

[46] C.P. Horne et al., Phys. Rev. Lett. 33 (1974) 380

[47] P. Benz, P. Söding, Phys. Lett 52B (1974) 367

[48] F. Lenz, Ann. Phys. (N.Y.) 95 (1975) 348

[49] E.J. Moniz, Proc. Effets Mésoniques dans les noyaux, Saclay (1975) p. 123

[50] G.E. Brown, W. Weise, Phys. Rep. C22 (1975) 279

[51] M. Dillig, M.G. Huber, Mesonic Effects in Nuclear Structure, ed. K. Bleuler et al., (Mannheim 1975) p. 80; Proc. Effets mésoniques dans les noyaux, Saclay (1975) p. 111

[52] H. Arenhövel, Nucl. Phys. A247 (1975) 473; Proc. Effets mésoniques dans les noyaux, Saclay (1975) p. 97

[53] P. Argan et al., Phys. Rev. Lett. 29 (1972) 1191

Topics in Applied Physics

Founded by Helmut K. V. Lotsch

This book series is devoted to research achievements of current interest. Each volume deals with a different topic under the editorship of a recognized authority in the field. It covers application-oriented aspects of the topic under consideration, the basic physical principles being summarized in a comprehensive introduction.
The contributors to each volume are internationally known experts. The publication periods are comparable with those of scientific journals to keep pace with the rapidly accumulating results.

Springer-Verlag
Berlin
Heidelberg
New York

Volume 1
Dye Lasers
Editor: F.P. Schäfer
114 figures. XI, 285 pages. 1973
Cloth DM 77,–; US $ 33.90

Volume 2
Laser Spectroscopy
of Atoms and Molecules
Editor: H. Walther
137 figures, 22 tables
XVI, 383 pages. 1976
Cloth DM 97,–; US $ 42.70
ISBN 3-540-07324-8

Volume 3
Numerical and Asymptotic
Techniques in
Electromagnetics
Editor: R. Mittra
112 figures. XI, 260 pages. 1975
Cloth DM 72,–; US $ 31.70
ISBN 3-540-07072-9

Volume 4
Interactions on Metal Surfaces
Editor: R. Gomer
112 figures. XI, 310 pages. 1975
Cloth DM 78,–; US $ 34.40
ISBN 3-540-07094-X

Volume 5
Mössbauer Spectroscopy
Editor: U. Gonser
96 figures. XVIII, 241 pages. 1975
Cloth DM 70,–; US $ 30.80
ISBN 3-540-07120-2

Volume 6
Picture Processing and
Digital Filtering
Editor: T.S. Huang
113 figures. XIII, 289 pages. 1975
Cloth DM 79,80; US $ 35.20
ISBN 3-540-07202-0

Volume 7
Integrated Optics
Editor: T. Tamir
99 figures. XIII, 315 pages. 1975
Cloth DM 79,80; US $ 35.20
ISBN 3-540-07297-7

Volume 8
Light Scattering in Solids
Editor: M. Cardona
111 figures, 3 tables
XIII. 339 pages. 1975
Cloth DM 92,60; US $ 40.80
ISBN 3-540-07354-X

Volume 9
Laser Speckle and
Related Phenomena
Editor: J.C. Dainty
133 figures. XIII, 286 pages. 1975
Cloth DM 94,80; US $ 41.80
ISBN 3-540-07498-8

Volume 10
Transient
Electromagnetic Fields
Editor: L.B. Felsen
111 figures. XIII, 274 pages. 1976
Cloth DM 92.60; US $ 40.80
ISBN 3-540-07553-4

Volume 11
Digital Picture Analysis
Editor: A. Rosenfeld
114 figures. 47 tables.
XIII, 351 pages. 1976
Cloth DM 72,–; US $ 31.70
ISBN 3-540-07579-8

Volume 12
Turbulence
Editor: P. Bradshaw
47 figures, XI, 335 pages. 1976
Cloth DM 97,–; US $ 42.70
ISBN 3-540-07705-7

Volume 13
High-Resolution Laser
Spectroscopy
Editor: K. Shimoda
132 figures. XIII, 378 pages. 1976
Cloth DM 97,–; US $ 42.70
ISBN 3-540-07719-7

Volume 14
Laser Monitoring of the
Atmosphere
Editor: E.D. Hinkley
84 figures. XV, 380 pages. 1976
Cloth DM 97,–; US $ 42.70
ISBN 3-540-07743-X

Volume 15
Radiationless Processes in
Molecules and Condenses
Phases
Editor: F.K. Fong
67 figures XIII, 360 pages. 1976
Cloth DM 97,–; US $ 42.70
ISBN 3-540-07830-4

Volume 16
Nonlinear Infrared Generation
Editor: Y.-R. Shen
134 figures. XI, 279 pages. 1977
Cloth DM 88,–; US $ 38.80
ISBN 3-540-07945-9

Prices are subject to change
without notice

Vol. 431: Séminaire Bourbaki – vol. 1973/74. Exposés 436–452. IV, 347 pages. 1975.

Vol. 433: W. G. Faris, Self-Adjoint Operators. VII, 115 pages. 1975.

Vol. 434: P. Brenner, V. Thomée, and L. B. Wahlbin, Besov Spaces and Applications to Difference Methods for Initial Value Problems. II, 154 pages. 1975.

Vol. 440: R. K. Getoor, Markov Processes: Ray Processes and Right Processes. V, 118 pages. 1975.

Vol. 442: C. H. Wilcox, Scattering Theory for the d'Alembert Equation in Exterior Domains. III, 184 pages. 1975.

Vol. 446: Partial Differential Equations and Related Topics. Proceedings 1974. Edited by J. A. Goldstein. IV, 389 pages. 1975.

Vol. 448: Spectral Theory and Differential Equations. Proceedings 1974. Edited by W. N. Everitt. XII, 321 pages. 1975.

Vol. 449: Hyperfunctions and Theoretical Physics. Proceedings 1973. Edited by F. Pham. IV, 218 pages. 1975.

Vol. 458: P. Walters, Ergodic Theory – Introductory Lectures. VI, 198 pages. 1975.

Vol. 459: Fourier Integral Operators and Partial Differential Equations. Proceedings 1974. Edited by J. Chazarain. VI, 372 pages. 1975.

Vol. 461: Computational Mechanics. Proceedings 1974. Edited by J. T. Oden. VII, 328 pages. 1975.

Vol. 463: H.-H. Kuo, Gaussian Measures in Banach Spaces. VI, 224 pages. 1975.

Vol. 464: C. Rockland, Hypoellipticity and Eigenvalue Asymptotics. III, 171 pages. 1975.

Vol. 468: Dynamical Systems – Warwick 1974. Proceedings 1973/74. Edited by A. Manning. X, 405 pages. 1975.

Vol. 470: R. Bowen, Equilibrium States and the Ergodic Theory of Anosov Diffeomorphisms. III, 108 pages. 1975.

Vol. 474: Séminaire Pierre Lelong (Analyse) Année 1973/74. Edité par P. Lelong. VI, 182 pages. 1975.

Vol. 484: Differential Topology and Geometry. Proceedings 1974. Edited by G. P. Joubert, R. P. Moussu, and R. H. Roussarie. IX, 287 pages. 1975.

Vol. 487: H. M. Reimann und T. Rychener, Funktionen beschränkter mittlerer Oszillation. VI, 141 Seiten. 1975.

Vol. 489: J. Bair and R. Fourneau, Etude Géométrique des Espaces Vectoriels. Une Introduction. VII, 185 pages. 1975.

Vol. 490: The Geometry of Metric and Linear Spaces. Proceedings 1974. Edited by L. M. Kelly. X, 244 pages. 1975.

Vol. 503: Applications of Methods of Functional Analysis to Problems in Mechanics. Proceedings 1975. Edited by P. Germain and B. Nayroles. XIX, 531 pages. 1976.

Vol. 507: M. C. Reed, Abstract Non-Linear Wave Equations. VI, 128 pages. 1976.

Vol. 509: D. E. Blair, Contact Manifolds in Riemannian Geometry. VI, 146 pages. 1976.

Vol. 515: Bäcklund Transformations. Nashville, Tennessee 1974. Proceedings. Edited by R. M. Miura. VIII, 295 pages. 1976.

Vol. 516: M. L. Silverstein, Boundary Theory for Symmetric Markov Processes. XVI, 314 pages. 1976.

Vol. 518: Séminaire de Théorie du Potentiel, Proceedings Paris 1972–1974. Edité par F. Hirsch et G. Mokobodzki. VI, 275 pages. 1976.

Vol. 522: C. O. Bloom and N. D. Kazarinoff, Short Wave Radiation Problems in Inhomogeneous Media: Asymptotic Solutions. V. 104 pages. 1976.

Vol. 523: S. A. Albeverio and R. J. Høegh-Krohn, Mathematical Theory of Feynman Path Integrals. IV, 139 pages. 1976.

Vol. 524: Séminaire Pierre Lelong (Analyse) Année 1974/75. Edité par P. Lelong. V, 222 pages. 1976.

Vol. 525: Structural Stability, the Theory of Catastrophes, and Applications in the Sciences. Proceedings 1975. Edited by P. Hilton. VI, 408 pages. 1976.

Vol. 526: Probability in Banach Spaces. Proceedings 1975. Edited by A. Beck. VI, 290 pages. 1976.

Vol. 527: M. Denker, Ch. Grillenberger, and K. Sigmund, Ergodic Theory on Compact Spaces. IV, 360 pages. 1976.

Vol. 532: Théorie Ergodique. Proceedings 1973/1974. Edité par J.-P. Conze and M. S. Keane. VIII, 227 pages. 1976.

Vol. 538: G. Fischer, Complex Analytic Geometry. VII, 201 pages. 1976.

Vol. 543: Nonlinear Operators and the Calculus of Variations, Bruxelles 1975. Edited by J. P. Gossez, E. J. Lami Dozo, J. Mawhin, and L. Waelbroeck, VII, 237 pages. 1976.

Vol. 552: C. G. Gibson, K. Wirthmüller, A. A. du Plessis and E. J. N. Looijenga. Topological Stability of Smooth Mappings. V, 155 pages. 1976.

Vol. 556: Approximation Theory. Bonn 1976. Proceedings. Edited by R. Schaback and K. Scherer. VII, 466 pages. 1976.

Vol. 559: J.-P. Caubet, Le Mouvement Brownien Relativiste. IX, 212 pages. 1976.

Vol. 561: Function Theoretic Methods for Partial Differential Equations. Darmstadt 1976. Proceedings. Edited by V. E. Meister, N. Weck and W. L. Wendland. XVIII, 520 pages. 1976.

Vol. 564: Ordinary and Partial Differential Equations, Dundee 1976. Proceedings. Edited by W. N. Everitt and B. D. Sleeman. XVIII, 551 pages. 1976.

Vol. 565: Turbulence and Navier Stokes Equations. Proceedings 1975. Edited by R. Temam. IX, 194 pages. 1976.

Vol. 566: Empirical Distributions and Processes. Oberwolfach 1976. Proceedings. Edited by P. Gaenssler and P. Révész. VII, 146 pages. 1976.

Vol. 570: Differential Geometrical Methods in Mathematical Physics, Bonn 1975. Proceedings. Edited by K. Bleuler and A. Reetz. VIII, 576 pages. 1977.

Vol. 572: Sparse Matrix Techniques, Copenhagen 1976. Edited by V. A. Barker. V, 184 pages. 1977.

Lecture Notes in Physics

Vol. 32: Particles, Quantum Fields and Statistical Mechanics. Proceedings 1973. Edited by M. Alexanian and A. Zepeda. V, 132 pages. 1975.

Vol. 33: Classical and Quantum Mechanical Aspects of Heavy Ion Collisions. Proceedings 1974. Edited by H. L. Harney, P. Braun-Munzinger, and C. K. Gelbke. VII, 311 pages. 1975.

Vol. 34: One-Dimensional Conductors GPS Summer School Proceedings, 1974. Edited by H. G. Schuster. VII, 371 pages. 1975.

Vol. 35: Proceedings of the Fourth International Conference on Numerical Methods in Fluid Dynamics, 1974. Edited by R. D. Richtmyer. V, 457 pages. 1975.

Vol. 36: R. Gatignol, Théorie Cinétique des Gaz à Répartition Discrète de Vitesses. II, 219 pages. 1975.

Vol. 37: Trends in Elementary Particle Theory. Proceedings 1974. Edited by H. Rollnik and K. Dietz. V, 472 pages. 1975.

Vol. 38: Dynamical Systems, Theory and Applications. Proceedings 1974. Edited by J. Moser. VI, 624 pages. 1975.

Vol. 39: International Symposium on Mathematical Problems in Theoretical Physics. Proceedings 1975. Edited by H. Araki. XII, 562 pages. 1975.

Vol. 40: Effective Interactions and Operators in Nuclei. Proceedings 1975. Edited by B. R. Barrett. XII, 339 pages. 1975.

Vol. 41: Progress in Numerical Fluid Dynamics. Proceedings 1974. Edited by H. J. Wirz. V, 471 pages. 1975.

Vol. 42: H II Regions and Related Topics. Proceedings 1975. Edited by D. Downes and T. L. Wilson. XII, 488 pages. 1975.

Vol. 43: Laser Spectroscopy. Proceedings 1975. Edited by S. Haroche, J. C. Pebay-Peyroula, T. W. Hänsch, and S. E. Harris. X, 466 pages. 1975.

Vol. 44: R. A. Breuer, Gravitational Perturbation Theory and Synchrotron Radiation. VI, 196 pages. 1975.

Vol. 45: Dynamical Concepts on Scaling Violation and the New Resonances in e^+e^- Annihilation. Edited by B. Humpert. VII, 248 pages. 1976.

Vol. 46: E. J. Flaherty, Hermitian and Kählerian Geometry in Relativity. VIII, 365 pages. 1976.

Vol. 47: Padé Approximants Method and Its Applications to Mechanics. Edited by H. Cabannes. XV, 267 pages. 1976.

Vol. 48: Interplanetary Dust and Zodiacal Light. Proceedings 1975. Edited by H. Elsässer and H. Fechtig. XII, 496 pages. 1976.

Vol. 49: W. G. Harter and C. W. Patterson, A Unitary Calculus for Electronic Orbitals. XII, 144 pages. 1976.

Vol. 50: Group Theoretical Methods in Physics. 4th International Colloquium. Nijmegen 1975. Edited by A. Janner, T. Janssen, and M. Boon. XIII. 629 pages. 1976.

Vol. 51: W. Nörenberg und H. A. Weidenmüller. Introduction to the Theory of Heavy-Ion Collisions. IX, 273 pages. 1976.

Vol. 52: M. Mladjenović, Development of Magnetic β-Ray Spectroscopy. X, 282 pages. 1976.

Vol. 53: D. J. Simms and N. M. J. Woodhouse, Lectures on Geometric Quantization. V, 166 pages. 1976.

Vol. 54: Critical Phenomena. Sitges International School on Statistical Mechanics, June 1976. Edited by J. Brey and R. B. Jones. XI, 383 pages. 1976.

Vol. 55: Nuclear Optical Model Potential. Proceedings 1976. Edited by S. Boffi and G. Passatore. VI, 221 pages. 1976.

Vol. 56: Current Induced Reactions. International Summer Institute, Hamburg 1975. Edited by J. G. Körner, G. Kramer, and D. Schildknecht. V, 553 pages. 1976.

Vol. 57: Physics of Highly Excited States in Solids. Proceedings 1975. Edited by M. Ueta and Y. Nishina. IX, 391 pages. 1976.

Vol. 58: Computing Methods in Applied Sciences. Proceedings 1975. Edited by R. Glowinski and J. L. Lions. VIII, 593 pages. 1976.

Vol. 59: Proceedings of the Fifth International Conference on Numerical Methods in Fluid Dynamics. 1976. Edited by A. I. van de Vooren and P. J. Zandbergen. VII, 459 pages. 1976.

Vol. 60: C. Gruber, A. Hintermann and D. Merlini, Group Analysis of Classical Lattice Systems. XIV, 326 pages. 1977.

Vol. 61: International School on Electro and Photonuclear Reactions I. Edited by C. Schaerf. VIII, 650 pages. 1977.